Matusiewicz, Bulska (Eds.)
Inorganic Trace Analytics

Also of Interest

Handbook of Inorganic Substances 2017
Villars, Cenzual, Gladyshevskii, 2017
ISBN 978-3-11-044408-7, e-ISBN 978-3-11-044540-4

Organic Trace Analysis
Nießner, Schäfer, 2017
ISBN 978-3-11-044114-7, e-ISBN 978-3-11-044115-4

Metal Ions in Life Sciences
The Sigels' Series
Sigel, Astrid / Sigel, Helmut / Sigel, Roland K.O. (Eds.), 2017
ISSN 1559-0836, eISSN: 1868-0402

Handbook of Rare Earth Elements Analytics
Golloch, 2017
ISBN 978-3-11-036523-8, e-ISBN 978-3-11-036508-5

Industrial Inorganic Chemistry
Benvenuto, 2017
ISBN 978-3-11-033032-8, e-ISBN 978-3-11-033033-5

Inorganic Trace Analytics

Trace Element Analysis and Speciation

Edited by
Henryk Matusiewicz and Ewa Bulska

DE GRUYTER

Editors
Prof. Dr. Henryk Matusiewicz
Department of Analytical Chemistry
Poznan University of Technology
Henryk.matusiewicz@put.poznan.pl

Prof. Dr. Ewa Bulska
Faculty of Chemistry
University of Warsaw
ebulska@chem.uw.edu.pl

ISBN 978-3-11-037194-9
e-ISBN (PDF) 978-3-11-036673-0
e-ISBN (EPUB) 978-3-11-039245-6
The articles 1, 2 & 9 in this book have been previously published in the journal *Physical Sciences Reviews* (ISSN2365-659X)

Library of Congress Cataloging-in-Publication Data
A CIP catalog record for this book has been applied for at the Library of Congress.

Bibliographic information published by the Deutsche Nationalbibliothek
The Deutsche Nationalbibliothek lists this publication in the Deutsche Nationalbibliografie; detailed bibliographic data are available on the Internet at http://dnb.dnb.de.

Cover image: Ryan McVay/Photodisc/gettyimages
Typesetting: Integra Software Services Pvt. Ltd.
Printing and binding: CPI books GmbH, Leck
♾ Printed on acid-free paper
Printed in Germany

www.degruyter.com

Preface

The book is comprised of four sections, containing a total of 10 chapters, with two main areas: presampling and sample preparation for different types of matrices for inorganic trace analysis and inorganic and bioinorganic speciation analysis at trace level.

The book discusses a hitherto unrecognized sources of error in analyzing inorganic and biomedical specimens, namely, presampling factors; it accounts for the role of sampling and sample preparation (reviews selected applied techniques for dry and wet ashing of samples) as a vital part of a complete analytical procedure. A systematic approach to understanding contamination, loss, and related measurement problems before starting the analytical work is the only way to minimize the impact of unforeseen errors in the final analytical result. It emphasizes the need for preservation and storage of biological specimens collected for element determination and speciation analysis: cites the major factors affecting preservation such as container material, container pretreatment, storage time, storage temperature, and contamination of the laboratory environment. It gives an extensive review of the most usual sample preparation methods that can be used for trace element determination. While methods for sample preparation in total element determinations are generally well established, the determination of the *chemical forms* of the elements in biological and environmental samples, food, and industry is the main key to future advances in these subject areas and this book gives a review of the state of the art of speciation issues in biological systems, food, industrial materials, and environment sectors. A distinction is recommended between analysis of a specific chemical form of a trace element in several biological, food, environmental, industrial materials and an analysis in which different chemical forms make up the total element content in a compartment. Moreover, the issue of the quality of results is also highlighted toward the accuracy of the results for trace element speciation.

The inevitable delay between writing and publication means that it has not been possible to include all the most recent work in this field. Without doubt, many of the subject areas targeted in this book have already received in-depth treatment by appropriate chapters. Assembling this knowledge into a single source proves advantageous to the user only if it is accomplished concisely and comprehensively.

An acknowledgment must be given for the dedicated work of all the contributing authors that made possible to prepare this book. Finally, we would like to thank the editorial staff of De Gruyter and especially Dr. Katharina Butsch (Acquisitions Editor Chemistry) for the invitation to edit this book and for the dedicated work and contributions to analytical chemistry.

Henryk Matusiewicz
Ewa Bulska

https://doi.org/10.1515/9783110366730-202

Contents

Part II: Matrices – Selected applications

Contributors

Ewa Biaduń
Faculty of Chemistry
University of Warsaw
Warsaw, Poland

Zuzanna Brożek-Mucha
Institute of Forensic Research
Department of Criminalistics
Kraków, Poland

Ewa Bulska
Faculty of Chemistry
University of Warsaw
Warsaw, Poland
E-mail: ebulska@chem.uw.edu.pl

Alma Rosa Corrales Escobosa
University of Guanajuato
Guanajuato, Mexico

Wiktor Dmitruk
Central Forensic Laboratory of the Police
Chemistry Department
Warsaw, Poland
E-mail: wiktor.dmitruk@policja.gov.pl

Armando Gómez Ojeda
University of Guanajuato
Guanajuato, Mexico

Piotr Jamroz
Faculty of Chemistry
Division of Analytical Chemistry
Wroclaw University of Technology
Wrocław, Poland
E-mail: piotr.jamroz@pwr.wroc.pl

Beata Krasnodębska-Ostręga
Faculty of Chemistry
University of Warsaw
Warsaw, Poland
E-mail: bekras@chem.uw.edu.pl

Eliza Kurek
Biological and Chemical Research Centre
University of Warsaw
Warsaw, Poland
E-mail: ekurek@cnbc.uw.edu.pl

Petko Mandjukov
South-West University Neofit Rilski
Department of Chemistry
Blagoevgrad, Bulgaria

Henryk Matusiewicz
Department of Analytical Chemistry
Poznan University of Technology
Poznan, Poland
E-mail: henryk.matusiewicz@
put.poznan.pl

Pawel Pohl
Faculty of Chemistry
Division of Analytical Chemistry
Wroclaw University of Technology
Wrocław, Poland
E-mail: pawel.pohl@pwr.wroc.pl

Anna Ruszczyńska
Faculty of Chemistry
University of Warsaw
Warsaw, Poland
E-mail: aruszcz@chem.uw.edu.pl

Monika Sadowska
Faculty of Chemistry
University of Warsaw
Warsaw, Poland

https://doi.org/10.1515/9783110366730-204

Anna Szymczycha-Madeja
Faculty of Chemistry
Division of Analytical Chemistry
Wroclaw University of Technology
Wrocław, Poland
E-mail: anna.szymczycha-madeja@pwr.wroc.pl

Emilia Vassileva
IAEA Environment Laboratories
Monaco
E-mail: e.vasileva-veleva@iaea.org

Maja Welna
Faculty of Chemistry
Division of Analytical Chemistry
Wroclaw University of Technology
Wrocław, Poland
E-mail: maja.welna@pwr.wroc.pl

Katarzyna Wróbel
University of Guanajuato
Guanajuato, Mexico
E-mail: katarzyn@ugto.mx

Kazimierz Wrobel
University of Guanajuato
Guanajuato, Mexico

Andrzej Wyciślik
Faculty of Materials Science and Metallurgy Silesian University of Technology
Katowice, Poland
E-mail: andrzej.wycislik@polsl.pl

Henryk Matusiewicz and Ewa Bulska

Trace Elements in Environmental, Biological and Industrial Samples

During the past three decades, increasing attention has been focused on enormous changing of the natural environment. The abatement of pollution of the air, of surface, drinking and seawater and of the soil is an important and very complex task, requiring the cooperation of specialists from various disciplines including analytical chemistry. All possible sources of pollution, such as combustion of fossil fuels, municipal waste incineration, industrial processes producing waste sludge and all kinds of waste disposal, need to be investigated, initiating with an elemental analysis of the material involved.

In contrast to the symptoms of acute metal poisoning, the effects of continued exposure to low levels are not always well known and understandable. Since elements considered are mainly taken up by humans through inhalation of air and intake of food and water, it is of vital importance to have analytical techniques that allow the determination of elements in such matrices.

According to the *IUPAC Gold Book*, the term "trace element" is related to any element having an average concentration of less than about 100 parts per million atoms (ppm) or less than 100 µg/g. However, the common understanding of this term depends on the particular discipline, thus besides analytical chemistry, biochemistry is related to those elements which are needed in very minute quantities for healthy growth of living organisms.

The term "trace element" is also commonly used by geologists for those elements that occur in concentrations of less than 0.1 per cent by weight.

The number of analytical tasks performing over the last years in relation to the determinations of trace elements in environmental samples is already enormous and is rapidly increasing. The samples being analysed include water, aerosols and dust, soil, sewage sludge and so on. These analyses are usually rather difficult owing to the complex nature of the samples and to the low concentrations involved.

Ideally the method of analysis applied should be sensitive, specific, accurate and applicable to major, minor and trace constituents. Most analytical techniques only partly meet these objectives, and it is clear that the required performance characteristics of the analytical method depend on the analytical problem at hand. Although for routine applications performed so far, usually only the total amount of an element of interest is determined, its speciation is enjoying an increased interest. Therefore, it is not sufficient to determine the overall metal concentration; it must also be established in which chemical form, for example organometallic, these elements occur.

As mentioned earlier, trace elements are of great importance for the well-being of living organisms, but can also induce several harmful effects. In this context,

https://doi.org/10.1515/9783110366730-001

a distinction is made between essential and non-essential trace elements. Thus, the determination of essential (V, Cr, Mn, Co, Fe, Ni, Cu, Zn, As, Se, Mo, S) and non-essential (Cd, Hg, Pb) trace elements in biological samples is important for our understanding of the physiology, for reaching a diagnosis and for selecting the appropriate treatment of a disease. The desired analytical method and its performance characteristics should be selected to be fit for purpose, depending on the element concerned, on the concentration in the samples of interest, on the variability of this concentration, on the use to be made of the results and on the number of samples to be analysed.

Besides the concern in biological and environmental objects, new developments in the metal, semiconductor and ceramic industries as well as in the nuclear industry require trace and ultra-trace element determinations. The very rapid development in electronics is due largely to semiconductors of higher purity becoming available. The reagents and the water used during the production of semiconductor devices must be of extremely high purity. Also, the metallurgical properties of uranium are influenced to a significant extent by the nature and concentration of contaminants. A lot of analytical efforts were made and are still required during research and development in processes to produce such high purity materials, and the control of the purity of the final product is also an important challenge to the analytical chemist.

It is worth noting that any trace element-oriented research, where there is a need to perform analysis at very low levels of concentration, imposes stringent and exacting restrictions on the analytical community. It is obvious that the analysis must be carried out under strictly controlled conditions in order to obtain meaningful results. Therefore, before enrolling the sample to be subjected to trace element examination, a critical self-examination of the capabilities is required, in order to evaluate whether the laboratory is ready to meet the demands. Then, the entire analytical procedures, including the quality control, must be properly defined, optimized and finally validated. The essential part is also related to the need for constant surveillance of all steps in trace and ultra-trace element analysis, with respect to possible contaminations. Thus, the trace element analytics is simply a disciplined way of life by itself; it is an attitude of mind and calls for a relentless and systematic approach at each stage with concrete, well-documented working procedures.

To conclude, research aimed at understanding the function of trace elements in living organisms as well as supporting the industry, where trace element matters, requires several steps. These include precise quantification even at ultra-trace concentration levels, establishment of any preferential sites for specific elements in the organs of the body or industrial samples and initiation of bioinorganic trace element studies for characterizing chemical forms. It is imperative that the multidisciplinary character of biological/industrial (ultra)trace element research be fully recognized and adopted in all future investigations in this area to obtain meaningful data.

Part I: **Methodology in Trace Element Determination**

Henryk Matusiewicz

1 Sample Preparation for Inorganic Trace Element Analysis

1.1 Introduction

Many modern instrumental techniques require complete sample dissolution prior to an analysis. In other words, these techniques generally involve the introduction of samples as aqueous solutions to the flame, furnace, or plasma. A variety of techniques are employed from ambient pressure wet digestion in a beaker on a hot plate (or hot block) to specialized high-pressure, high-temperature microwave heating. Although the concentrations of the trace elements in the bulk of the sample are still mainly of interest, data on their distribution of the surface of the sample, in microregions, or phase boundaries, have become increasingly important. Further speciation analysis is also required.

In this chapter, the role of sample preparation in trace element analysis, including the sequence of analytical steps, systematic errors, preliminary treatments, wet and dry decomposition, and combustion and fusion decompositions prior to main spectroscopic methods of analysis, will be discussed. Within the general area of inorganic trace analysis, the coverage will be restricted to environmental, bioinorganic, forensic chemistry, and industrial trace element analysis.

In modern trace analysis, the term **sample preparation** cannot be exactly defined and covers a very broad field. It starts with mechanical pretreatment of the sample, for example, cleaning, drying, grinding, sieving, and filtering prior to instrumental methods of analysis, and extends to chemical methods, for example, digestion, decomposition, extraction approaches, separation, and enrichment required for a wet chemical procedure in which the solid, liquid, or gaseous samples are prepared and passed onto the real determination step.

Sample preparation depends on the nature of the sample, the analyte to be determined and their concentrations/amounts, and on the desired determination precision and accuracy. Sample preparation, is, however, inherently expensive and time consuming, and is responsible for the major source of errors in the various stages of an analytical procedure.

We will therefore discuss, in this chapter, mainly the physical and chemical operations that precede the real determination step of the multistage combined procedure.

1.2 Aspects of sampling and sample preservation

The importance of adequate sampling was recognized by leading authorities a considerable time ago, yet it would appear that the level of practice among many

https://doi.org/10.1515/9783110366730-002

investigators left much to be desired. Thus, the warning issued by Thiers in 1957 [1] that "unless the complete history of any sample is known with certainty, the analyst is well advised not to spend his time analyzing it" was largely ignored for several years. No universally accepted definition for this term or other nomenclature in this area exists. Sampling can mean one thing to a statistician and have some other connotation to a technician collecting a sample, an analyst examining a sample, or an administrator determining whether a sample meets the requirements of a law or a contract. Definitions of terms have always been a matter of concern in sampling, and efforts are being made to clarify the situation. For example, the International Union of Pure and Applied Chemistry has prepared a document proposed by Horwitz [2] intended to furnish concepts, terms, and definitions in the field of sampling relevant to analytical chemistry and that are generally applicable regardless of what sampling objective, commodity, location, quality, or form is involved.

Sampling for subsequent trace analysis is without doubt by far the most crucial step in an analytical procedure and is the last step of the preanalytical phase that can possibly affect the accuracy of the analytical results via interference factors. If not properly planned and practically performed by using appropriate sampling tools with the utmost care and expertise, total, systematic, as well as random, errors for sampling can range from a small percentage to several orders of magnitude.

The purpose of sampling is to extract a representative amount of material from a **lot** – **the sampling target**. According to CITAC/EURACHEM [3], the analytical operations can begin by aliquoting a test portion directly from the laboratory sample or from a test sample obtained after one or more pretreatments. A verified sampling plan with well-defined procedures for selection, collection, storage, transport, and preparation of the sample is essential, and the reader should consult expert resources for details [4, 5]. It is clear, in trace analysis, that sampling must always be followed by an appropriate stabilization (preservation) step, with due regard for the nature of the matrix and the analyte and can only be optimized before analysis. It is shown how nonrepresentative sampling processes will always result in an invalid aliquot for measurement uncertainty (MU) characterization [6].

A specific sampling process can either be **representative** (in some cases, it can be replaced and better understood as **appropriate**) or not. If sampling is not representative, we have only undefined, mass-reduced lumps of material without provenance that are not actually worth analyzing. Only representative aliquots reduce the MU of the full sampling and analysis process to its desired minimum; and it is only such MU estimates that are valid. Sampling **correctness** and representatives are essential elements of the sampling process. Representativeness implies both correctness and a sufficient small sampling reproducibility (sampling variance).

The starting point of every measurement process is the primary lot. The lot (also termed the sampling target or decision unit) refers both to the physical, geometrical form and size and the material characteristics of the material being subject to sampling. All lots are characterized by significant material heterogeneity – a concept only fully acknowledged and defined by the theory of sampling (TOS) – where it is crucially subdivided into constitutional heterogeneity and distributional heterogeneity. Heterogeneity is the prime characterization of all naturally occurring materials, including industrial lots. Heterogeneity manifests itself at all scales related to sampling for nearly all lot and material types. The heterogeneity concept is introduced and discussed in complete detail in the pertinent literature [7, 8]. The full pathway from **lot to analytical aliquot** is complex and is subject to many types of uncertainty contributions in addition to analysis. TOS focuses on the conceptual and practical active steps needed to minimize all sampling contributions to MU.

According to the available information regarding the candidate test site and the requested sampling station network resolution, quite a number of sampling approaches have been developed, a few of which are mentioned here, such as random sampling, systematic sampling, representative sampling, and subsampling. All approaches have their merits, and the selection of the most appropriate sampling mode depends on a number of boundary conditions.

1.2.1 Sample

This is a correctly extracted material from the lot, which can only originate from an unbiased, representative sampling process. The term sample should always only be used in this qualified sense of "representative." If there is doubt as to this characteristic, the term "specimen" should be used instead.

1.2.2 Specimen

This is a "sample" that cannot be documented to be the end result of a bona fide representative sampling process. It is not possible to ascertain the representatives status of any isolated small part of a sampling target by itself. It is only the **sampling process** that can be termed representative or not.

1.2.3 Random sampling

Random sampling is the arbitrary collection of samples within defined boundaries of the area of concern and is directed at obtaining an extreme value, such as the best or worst case. The selection of sampling points must be performed in a way that gives all locations within the boundary of the test area the same chance to be selected. The basic assumption for choosing the random sampling approach is that the test area is homogeneous with respect to the pollutants to be monitored. The higher the heterogeneity of the area is, the lesser the random sampling approach shall produce the correct information needed for the complete future cleanout of the area. This means that progressive information with regard to the pollutants of concern is needed prior to sampling, which is usually not the case.

In random sampling of an entire lot of bulk material, the material is divided into a number of real or imaginary segments. Because of its simplicity, sampling at evenly spaced intervals over the bulk is often used instead of random sampling, although results must be closely monitored to prevent errors from periodicity in the material.

1.2.4 Systematic sampling

Each sample collected systematically and analyzed to reflect or test some systematic hypothesis, such as changes in composition with time, temperature, or spatial locations, should be considered representative of a separate, discrete population under the existing conditions. However, the results may still be statistically tested for the significance of any apparent differences.

A carefully designed sampling plan includes the possibility of unanticipated events of phenomena that could prejudice the analyses. For example, measurements at timed intervals are sometimes made with random start or other superimposed random time element. The less known about a given process, the more randomness is merited. Conversely, the more fully a process is understood, the more efficient is a systematic approach to date acquisition.

1.2.5 Representative sampling

Representative sample frequently connotes a single sample of a universe or population expected to exhibit average properties of the population. It is not possible to select such a sample by a random process or to verify whether it is representative. A truly representative sample seems valid only if the sample is defined *a priori* as representing a specific purpose.

Although it may reduce costs, measurement of samples defined as representative yields information not equaling that from valid random samples of the population, except when the population is homogenized before sampling to produce a number of similar subsamples. A properly designated and executed random sampling plan provides sample mean and variation between members, neither of which can be obtained by analysis of one **representative sample**.

1.2.6 Composite sampling

Until here, single-sample approaches have been considered. For a number of good reasons, composite sampling schemes might be preferred, one and not the least of the reasons being identified in analysis cost.

A composite sample may be considered a special type of representative sample. A composite sample represents **physical averaging**, as opposed to arithmetic averaging of analytical results from individual increments. Many sampling procedures assume that only average composition is desired, such as bulk, time-weighted, or flow-proportional averages, and specify collection or preparation of a suitable composite. Elaborate crushing, grinding, mixing, and blending procedures have been developed and standardized for preparing solid composites.

Analysis of individual samples permits determination of the average (at the expense of additional analytical effort), of the distribution of samples within the population (between-sample variability), and of within-sample variability (if replicate analyses are conducted). Composite samples provide limited information, and the consequences should be carefully considered before deciding between this approach and the analysis of individual samples.

1.2.7 Subsampling

Subsample is a correctly mass-reduced part of sample (primary, secondary, etc.). A subsample is a result from a dissociative (disaggregation) process; a composite sample is a result from an integrative process. Subsampling is necessary because the sample received by the analytical laboratory is usually larger than that required for a single measurement. Test portions taken for replicate measurements of different constituents by several techniques must be sufficiently alike so that results are compatible. The effort necessary to reduce particle size, mix, or otherwise process the laboratory sample before withdrawing portions (subsamples) for analysis depends on the homogeneity of the original sample.

It is obvious that a major requirement for trace element analysis is that specimens be taken under noncontaminating conditions and thus avoiding all kind of losses; the specimen and the sample derived from it must exactly reflect the properties of the matrix. The number and types of samples and sampling devices are too diverse to enumerate here, but some general principles and recommendations will be given for collecting samples in which trace metals are to be determined. Ideally, all sampling devices, tools, and containers should be constructed from plastics with a low content of trace metals, such as one of the Teflon, or polyethylene (PE). By removing all metals from the sampler, a major source of contamination is eliminated. Of course, there are many instances in which this is not practical or possible. In these cases, the best alternative is to select a high-purity material that will produce no consequential contamination. In general, for trace analysis, sampling tools made from stainless steel, such as titanium and ceramics for blades, or highly pure materials like nickel for needles should be used. In this case, there is only one element that cannot be analyzed due to high blanks. By now it is obvious that a major requirement for trace analysis is that the sample be taken under noncontaminating conditions. There are those who will object to the cost or dispute the necessity of applying clean room conditions to all sampling operations [9]. The use of noncontaminating implements and samplers, as well as complete portable clean room facilities, has become a well-established procedure among many geochemists and oceanographers working in pristine environment. In addition to contamination by vessels and tools, samples can also be contaminated by reagents, which have to be added as stabilizers, anticoagulants, and preservatives [10], especially because they are often added in excessive amounts.

The second important step of the preanalytical phase is the storage (preservation, stabilization). In trace analysis, sampling must always be followed by an appropriate preservation and/or stabilization step, with due regard for the nature of the matrix and the analyte.

Unfortunately, many materials are not stable once they have been sampled and seldom can samples be analyzed immediately after collection; they need to be stored for a certain period. Samples may deteriorate by trace element adsorption, losses from aqueous samples, particle segregation in heterogeneous powders, dehydration or bacterial growth in biological samples, or by decomposition of the sample matrix and the formation of volatile compounds of the trace element analytes. Stabilization techniques exist for most of these problems, but they must be noncontaminating and the apparatus used for this process must be evaluated for contamination potential.

Liquid samples, such as water (drinking, surface, waste water), beverages, fruit juices, urine samples, etc., should always be acidified with mineral acids to $pH < 2$ (HNO_3, HCl, and $HClO_4$, all of the highest purity grade) immediately after collection for stabilization purposes to avoid losses due to wall adsorption and also inhibit bacterial growth. However, the pH should also be selected in accordance with the requirements of the subsequent analytical steps.

Losses of elements may occur through the formation of insoluble products that are strongly adsorbing to the wall of the container, co-precipitation with the main inorganic and organic constituents of the matrix, the formation of volatile compounds that penetrate the plastic, and also the formation of compounds that are unsuitable for the subsequent analytical procedure. Most of these detrimental effects are diminished or eliminated by the addition of sufficient amounts of preservative. Contamination during storage is due mainly to the material of the containers. A wide variety of sample container materials for bottles, flasks, tubes, and vials can be used, the most common being polyfluorocarbons (polytetrafluorethylene (PTFE), PFA), PE, polypropylene (PP), polyvinyl chloride (PVC), polycarbonate (PC), high-purity quartz, and borosilicate glasses.

Loss of water from aqueous matrices (e.g., tissue, fruit, vegetables, and soil samples) may occur during storage. For this reason, analytical results should always be reported in terms of dry mass to avoid false interpretations. Drying is best conducted immediately after sampling. Water removal can be accomplished by oven drying at elevated temperature, use of desiccating materials, or freeze-drying (lyophilization). Freeze-drying has been shown to be the most satisfactory procedure since it minimizes the loss of highly volatile elements and compounds, and as at low temperatures fewer alterations of the biological material occur and the formation of insoluble substances is also decreased.

Liquid samples requiring preservatives, such as blood and urine, are preferably stored at +4 °C, provided that they will be analyzed within 2–3 weeks. If a longer storage period is necessary, however, the best way to store biological material is to maintain it at temperatures around −20 °C until analysis. In the storage of tissue, it is important to know whether the entire specimen or an aliquot sample representative of the mean value of the specimen will be submitted for analysis. In the first case, no special precaution is needed. After determination of the fresh weight and, if the specimen is dried, the dry weight, the specimen is cooled as soon as possible to −20 °C or below. At this temperature, bacterial and chemical interactions are largely diminished. During freezing and thawing, the structure of the specimen will be altered by the rapture of membranes and outflow of intracellular liquid; however, since the total specimen will be analyzed, the accuracy of the analysis will not be affected. If storage over several years is envisaged, the sample should either be dried at −18 °C or quick-frozen in liquid nitrogen at −196 °C and then stored at or below −70 °C.

Determination of an analyte in serum or plasma requires that the analyte be isolated subsequent to sampling. This is no longer possible after deep-freezing because of the hemolytic nature of blood samples. In case of elements at very low concentration, it is preferable that an investigation be conducted on serum rather than plasma. This reduces the risk of contamination because unlike blood plasma serum is recovered without addition of an anticoagulant.

The risk of contaminations persists during the entire storage period. Thus, if the biological samples cannot be analyzed immediately, they should be kept in an adequately controlled environment or hermetically sealed in a proper material, for example, plastic. So, in general, samples should be stored at low temperatures in cleaned containers made of proper materials [11, 12].

Additional information regarding general principles of sampling design and sample preservation for trace element analysis is discussed by Kratochvil [13] in a book on sample preparation for trace element analysis.

1.3 Error sources during the analytical procedure

Error (systematic error) relating to a trace element analysis may occur at all points from sampling through to the determination step. Systematic error arises whenever the actual nature of the measurement process differs from that assumed. For a measurement to be both accurate and precise, the measured value must be both accurate and precise and must be corrected for all sources of systematic error or bias, and the true value must lie within the stated level of confidence. Systematic errors as a rule become evident at the µg/g concentration range and increase enormously with decreasing absolute amounts or concentrations of the elements to be determined. They can exceed several orders of magnitude, depending on the omnipresence and the distribution of the elements in our environment and in the laboratory. Abnormally high values in the analysis may result from the contamination by airborne dust, reagent blanks, and container material. Of course, as usually a blank value is subtracted, such errors are in principle corrected. A high blank thus primarily affects the reproducibility and the limit of detection of an analysis.

There exists no chance to discern systematic errors by statistic evaluation of the analytical data, especially because the most important condition for a statistical treatment of data, which are supposed to display a normal distribution, very often does not apply. Further, no other simple means for the detection of systematic errors are available. Systematic errors depend strongly on the element to be determined, on the matrix, on the method and procedure used, on the conditions of the laboratory, and on some other parameters.

The most important sources of systematic errors [14] are:

- inadequate sampling, sample handling and storage, in homogeneity of the sample;
- contamination of the sample and/or the sample solution by tools, apparatus, vessels, reagents, and airborne dust during the analytical procedure;
- adsorption and desorption effects at the surface of the vessels and phase boundaries (filters, columns, and precipitates);
- losses of elements (e.g., Hg, As, Se, Cd, and Zn) and/or compounds (e.g., oxides, halides, and hydrides of the elements) due to volatilization;
- unwanted or incomplete chemical reactions (e.g., change of the valence of ions, precipitation, ion exchange, formation of compounds and complexes);
- influences of the matrix on the generation of the analytical signals (incomplete atomization, overlap of peaks); and
- incorrect calibration and data evaluation as a result of incorrect standard materials, unstable standard solutions, or the use of false calibration functions or unallowed extrapolations, respectively.

A survey of systematic errors in decomposition methods [15] is given next:

- Errors resulting from contamination from
  1. the atmosphere (the air, laboratory environment), the sampling;
  2. reagent impurities; and
  3. materials (vessels, tools).
- Errors as a result of losses of elements by
  1. volatilization,
  2. adsorption on the vessel material, and
  3. reactions with the vessel material
- Errors resulting from incomplete sample decomposition or dissolution

This contribution will mainly deal with the most serious sources of systematic errors of multistage procedures: element losses due to volatilization and adsorption as well as the contamination due to the three most important blank sources: tools and vessels, reagents, and laboratory air and dust.

1.3.1 Blank

Masking of components in a sample, severe interferences with detection, uncertain qualitative analysis, and unreliable quantitative measurements are all problems in trace analysis potentially connected through a common factor, **the blank**. Few analytical techniques are free from the influence of these difficulties. Even though the threshold for measuring trace constituents has been lowered significantly with the discovery and development of techniques with high sensitivity, the full potential for measurement by spectrophotofluorometry, polarography, anodic stripping voltammetry, kinetic and spectroscopy methods, and other techniques with sensitivities sufficient for measuring elements at the nano/picogram level is precluded in many practical applications.

In the case of the inability to reproducibly control the blank at levels insignificant in comparison with the constituent being determined or where difficulties associated with quantitatively manipulating and recovering submicrogram quantities of trace elements exist, the limits at which trace elements can be measured by most techniques will be established by these restrictions. In the presence of constantly fluctuating blanks, the accuracy and precision of quantitative trace measurements are also influenced significantly. Thus, first-order improvements in the reliability of measurements at or below the ppm/ppb/ppt region depend greatly upon controlling and reducing the size of the blank and, where possible, eliminating its effects.

Contamination from particulates in air, impurities in reagents, and trace elements from containers is considered to be primarily responsible for the blank. Hazards from less conspicuous sources, including instrumental noise interpreted as a signal from a component of the sample, must be considered as well [16]. Since considerable attention by analytical chemists to these problems is necessary in trace analysis, state-of-the-art techniques for providing pure atmospheres and working conditions, for purifying and storing ultrapure reagents, and for performing routine analytical procedures under ultraclean conditions are subsequently discussed.

1.3.2 Contamination

Contamination in trace analysis is always understood as the increase in the measured amount or concentration of a component, resulting from its introduction at various stages of the analytical procedure, from several independent sources other than the sample and can occur at any point. These are the air, laboratory atmosphere and working areas, reagents, tools, and apparatus associated with sampling and sample preparation, and laboratory ware; these will be discussed in sequence.

The atmosphere of the laboratory is loaded with particulate matter from different sources such as the environment, floor, walls, ceiling (paint), furniture, equipment, clothes, analyst himself, and so on. Accordingly, various inorganic and organic compounds are present and, in principle, any element can be found, depending on the environment, the laboratory itself, and its history. Thus, the composition of the air in the laboratory will often approximate that of the surrounding atmosphere and will fluctuate with prevailing atmospheric conditions. Again the dust particles will contain relatively high concentrations of those elements that show a high abundance in the Earth's crust (e.g., Si, Al, Fe, Ca, Na, K, Mg, P). In addition, all elements of anthropogenic pollution (e.g., Mg, Cu, Cd, Pb, Ni, Co, Zn, Mn) are always present. Dust and particles brought to, and released or created in the laboratory by the activities of personnel, appear as a more abundant and critical source of contamination [17].

When the dust comes in contact with the sample, it is a severe source of contamination. Sometimes, protection may be achieved with a cheap and very simple means, such as closed vessels and apparatus or glove boxes. Ideally, trace analytical work depending on conventional techniques should be carried out in a high-class clean air laboratory. More efficient and convenient are **clean** rooms and clean benches [18], which are flushed with dust-free air. A clean room is an area that is hermetically separated from the outside atmosphere and that is only accessible through an air lock. A filter assembly provides for pure air at an overpressure against the outside atmosphere and for circulation with a turbulent flow. A clean hood also reduces particles and dust in laboratory air. A much cheaper and simpler solution for reducing blank values due to contamination from laboratory air is the use of evaporation chambers [18]. The major development for providing particulate-free air is the so-called high-efficiency particulate air (HEPA) filter [19], which has a definite pore size of 0.3 µm. The dust removal efficiency by the deposition on the filter is about 99.97–99.95 %. The quality of a clean room or clean bench is expressed by the number of particles per cubic foot having a diameter between 0.5 and 5 µm and one distinguishes between the classes 100, 10,000, and 100,000. Typical is "class 100" or "class 10," meaning less than 100 or less than 10 particles per cubic foot of air. The use of a clean bench in a clean room illustrates only one possibility. Another more expensive alternative is to keep the whole clean room as dust-free as possible. This can be achieved with the aid of a laminar air flow that enters the room through a HEPA filter installed over the whole area of one wall or the ceiling [18]. For advanced technologies, however, still much higher-purity demands now exist. For instance, the electronic industry claims a much lower dust content for the production of megabyte chips (class 10 or 1 according to the US Federal Standard). For trace analysis, however, this enormous effort to get such a high air quality is only rarely necessary.

Availability of clean air facilities, however, does not in itself guarantee dust-free conditions. The placement of equipment or containers, too high exhaust velocities, room draughts, or an operator working (the number of persons in the laboratory) near the fume-hood can distort laminar flow, cause turbulences, and dramatically affect the performance of clean air installations. Also, the weather plays an important role; when it is raining, the air generally is essentially purer than when it is dry.

Contamination in trace analysis is comprehensively dealt with in the book by Zief and Mitchell [20] and by Mitchell [21]. It initiated extensive follow-up work, which has made an essential contribution to our understanding of the problem. Advances in contamination control are also the subject of a recent survey by Barnes et al. [22] in a book on microwave-assisted sample preparation for trace element determination edited by de Moraes Flores and Mitchell in his review paper [23]. Supplementary information on systematic errors caused mainly by contamination and losses of elements can be found in comprehensive monographs dealing with trace elemental analysis from Knapp and Schramel [24] and Hoffmann [25].

1.3.3 Reagents

Reagents used in sample preparation are another very common source of contamination. Reagents used that come into contact with the sample or standards, such as water, acids, bases, solvents, salts used as matrix modifiers, buffers, supporting electrolytes, fluxes, oxidants and reductants, chelating agents, and other reagent chemicals, must be of the highest available purity. At least as importantly, they must be stored and handled in such a way as to maintain this level of purity. Some of these needs for pure reagents have been met by commercial suppliers who have focused their attention on this problem and have introduced special lines of ultrapure, super pure, or electronic-grade reagents and metals. Because specific attention has been given to the proper handling, containing, analysis, and storage of reagents, levels of many trace elemental impurities in commercial lots of chemicals have been reduced considerably. The dynamic interplay between preparation, handling, containment, and analysis of ultrapure chemical standards and reagents for the clinical chemical laboratory is described [26]. Even though an overall improvement in the quality of several suppliers' chemicals has been accomplished, for most ultratrace analyses the analyst must still verify that the purity of any purchased reagent is sufficient for the intended analytical application. Reagent impurities capable of interfering with the analytical measurement should be so low that they give a blank value that is less than 10 % of the analyte level. Although reagents sufficiently pure for many applications are available commercially, the analytical laboratory committed to ultratrace determinations of a variety of elements must also be able to produce a wide range of pure reagent chemicals itself. For many analytical problems, the level of a specific contaminant of interest can be adequately controlled only by designing a special laboratory purification method. The basic analytical reagents can now be ultrapurified on the laboratory scale with relative ease by a number of techniques. Accordingly, in extreme trace analysis, we often have to use only those reagents that can easily be purified, such as gases and liquids [27, 28].

Ultrapure water is the most abundantly used analytical reagent, and the preparation of ultrapure water, for example, by distillation below the boiling temperature (sub-boiling, nonboiling) in a quartz [29] or PTFE [29, 30] stills or by membrane processes and its permanent quality control, is of greatest importance. The term sub-boiling distillation was coined and the first publication produced by 1972 [31]. Regarding membrane processes, three membrane processes are widely used in water purification systems: reverse osmosis, ultrafiltration, and microfiltration. A fourth process, ultramicrofiltration, is sometimes identified as lying between ultrafiltration and microfiltration. The processes are differentiated from one another by virtue of the different ranges of pore sizes found in the membranes. The reverse osmosis is a membrane separation technique using a material through which water molecules will pass but solute molecules will not. Unlike ultrafiltration, in which separation of substances is based solely on molecular size and form, reverse osmosis membranes retain ions of a size comparable to that of the water molecule. The next method of water purification is deionization (sometimes called demineralization). Most laboratory deionizers are mixed-bed cartridge units in which cation-exchange and anion-exchange resins act in concert to remove ionic contaminants (mainly inorganic salts) from the feed water. In many laboratories, it may be desirable to preprocess the deionizer feed water using a reverse osmosis unit or mixed bed ion exchanger, thus removing most of the dissolved salts before the final purification step. Normally, the purified water production must be conducted in a laboratory with appropriate air cleanliness and stored in vessels or containers free from contamination. The quality of this reagent, needed in large volumes for dissolving samples and preparing solutions, must be periodically monitored by quantitative analysis [32–36]. Since water of absolutely purity has virtually no conductance, the easiest and fasted method of measuring water purity is by measuring conductance (reciprocal of resistivity) with a meter made for this purpose [37]. Readings are expressed in μs/cm. On the other hand, a resistivity >18.2 MΩ cm is an indication of quality, but is not necessarily a certificate of high-purity water.

High-purity reagents do not always satisfy the standards required by extreme trace element determination since the control of impurities is usually restricted to only a few elements. Liquid decomposition reagents (e.g., nitric, sulfuric, hydrochloric, perchloric acids) and some organic solvents in a highly pure quality can be obtained by the so-called sub-boiling distillation [30]. The distillation made of quartz, PP, or PTFE should be used for the purification of acids (except HF) and water. The purification is based on evaporation of the liquid by infrared heating at the surface to avoid violent boiling [25]. In this way, the formation of liquid aerosols that are transported with the distillate is avoided, which in the conventional distillation technique contaminates the distillate. The residual impurities for sub-boiled liquids are at the pg/mL level, which is sufficient for most of the ultratrace procedures. The yield of such a sub-boiling still amounts to some 100 mL per day. This is sufficient for most purposes in ultratrace analysis purposes due to unavoidable contamination during storage. Therefore, only that volume of acid required for the immediate use should be prepared. Apart from this technique, no other universal (single) purification procedure is capable of removing all metallic or cationic impurities to such a low extent. It should be noted that sub-boiling distillation ensures the separation of impurities of low vapor pressure such as metal ions, but it does not eliminate impurities having high vapor pressure such as organic compounds or some anions [30].

One means that has been used to overcome the formation of liquid aerosols that are transported with the distillate is isothermal (also called isopiestic) distillation [38–40]. In principle, if containers of volatile reagent-grade acids (HCl, HBr, CH_3COOH, HF) and NH_4OH and pure water are placed in a sealed chamber (such as large desiccators), acid vapor will be absorbed in the pure water until an equilibrium vapor pressure is reached. This isopiestic distillation is the room temperature version of the sub-boiling method (which it preceded) and produces a highly pure but medium strength product. The latter problem can be circumvented by saturating pure water or isopiestically prepared reagents with pure gaseous compounds, NH_3, HCl, and HBr, for example. The method for HF purification based on a combination of isopiestic and sub-boiling distillation resulting in increase in production rate has been reported [41]. The purity of the distilled water and purity of the container largely determine the quality of reagent produced.

Reports of quantitative determination of trace elements by electrodeposition and anodic stripping imply general utility of electrochemical methods for ultrapurification of reagent chemicals. Electrodeposition is a very interesting tool for separation of ultratrace concentrations from salt solutions as these are good electrolytes and as electrolysis techniques suffer from contamination only to a very low extent [42]. For example, the constant current electrodeposition of trace elements (Cu, Cd, Co, Fe, Ni, Pb, Zn) onto a graphite tube cathode from highly concentrated NH_4F solutions was studied. The deposition yield of >99 % was achieved in the ng/mL range (and below) so that a solution of high purity was obtained [43, 44].

Electrolysis at the mercury cathode has been recognized for decades as being potentially extremely effective for purifying aqueous solutions of various reagents [27]. In principle, any reagent soluble in water can be purified by electrolysis at the mercury cathode, provided its component ions are not electroactive at the applied cathode potential required for reduction of the impurity ion. Electrochemical inertness at the anode potential is also required. Additionally, the reagent solution must be chemically inert with respect to reaction with mercury (including dissolution of the mercury).

In general, all reagents should be checked for contamination before use by analyzing **reagent** or **sample** blanks containing all the reagents in the same concentrations as would be present in the final sample solution. Any reagent that is shown to be contaminated must be discarded; for this reason, it is often a good idea to purchase these high-purity chemicals in small quantities.

1.3.4 Materials

Materials used in the construction of buildings are known to contribute atmospheric pollutants or produce particle fallout. Calcium in trace quantities can be constantly emitted from materials commonly used in walls and ceilings. Paints containing metallic pigments are chipped, flaked, or abraded from walls and furniture. Stainless steel hoods, sinks, and other furniture, copper faucets and pipes, gas regulators, metal heating and air conditioning units, and other metallic objects usually corrode after prolonged use under normal laboratory conditions. Metallic dusts from these objects contribute significantly to particulates in the atmosphere of the laboratory.

No vessel material is absolutely resistant even to water. The vessel material in contact with the sample during the decomposition or dissolution also frequently causes systematic errors. Elements can be either dissolved from the material or desorbed from or adsorbed on the container surfaces. Therefore, all vessels used to contain samples, standards, or reagents used in the analysis should be checked and shown to be free of contamination before they are used. It is preferable to set aside a set of vessels for trace analysis only. Especially glass, which contains a number of elements as major or minor components and a lot of other elements at a very high trace level, is very impure as compared to quartz, PTFE, PP, and PE [45]. In addition, the losses of elements due to adsorption are very high. Therefore, glass vessels should not be used for element determinations in the extreme trace range.

Vitreous silica, especially artificial quartz, is the most pure material that is commercially available in different purity classes but unfortunately quartz containers (and tools) are also the most expensive ones. They definitely deliver negligible blanks (except for Si), low wall adsorption, temperature resistance up to 1,200 °C, and high resistance to most inorganic acids with the exception of HF and concentrated phosphoric acid.

Fluorinated polymers PTFE (Teflon), PFA (Teflon), and TFM (Hostaflon) are substantially less pure. Nevertheless, they are much more cheaper than quartz and therefore they are preferred in most of the routine laboratories, especially used as the digestion vessels, but exhibits very variable quality depending on its origin. The maximum digestion temperature for such vessels is about 250 °C. The advantage of these materials is their resistance against nearly all acids, and they can therefore be used for sample digestion with HF. Contamination from the vessel material is not a problem. In general, PTFE is not as good as the other two materials because of its porous structure, which arises from the sintering process used for the vessel production. TFM is a chemically modified PTFE (PTFE modified by Hoesch) and does not suffer from the porous structure of PTFE.

FEP is an excellent material for storage containers because of its dense and nonpolar surface. Losses of polar ions via adsorption effects can mostly be neglected. It is also used for liners of digestion bombs, but must not be heated above 200 °C.

The polymer FEP is purer than conventional PTFE and is thus recommended for procedures aiming at the determination of trace elements.

PE and PP are stable up to 135 °C and are successfully used for storage containers, beakers, and flasks, but they are not as good as FEP, although they are less expensive.

Glassy carbon is used for high-temperature digestion vessels. It is also resistant against most acids; however, it is not resistant to oxidation. For example, oxidizing reagents such as nitric acid significantly attack the surface at elevated temperatures. Another disadvantage is a comparatively high risk of contamination; glassy carbon is not as clean as quartz glass or fluorinated polymers.

Especially during sampling, one has to avoid the sample to come into contact with other materials causing severe contamination. Therefore, for example, rubber is not a suitable material because of its relatively high contents of Sb, As, Zn, Cr, Co, and Sc. Nylon contains Co and PVC Zn, Fe, Sb, and Cu at the higher trace level. In addition, contamination in a trace element laboratory can come from a laboratory gloves (vinyl, latex), contaminated samples by Zn and Fe [46], or, for example, from micropipette tips [47].

With regard to the vessels and tools, their bulk material is not the only source of contamination. The other contamination source is impurities at the surface. In addition to the selection of a suitable material (container, vessel, tool), an appropriate preconditioning of the material surface is necessary to minimize adsorption or desorption. To minimize or eliminate such impurities, proper cleaning procedures are necessary. The conventional cleaning technique for laboratory glassware consists of its rinsing and leaching with high-purity acids (HNO_3, HCl) and pure water. In addition, leaching can be supported by applying ultrasonic treatment. Decontaminating the surface of containers by extracting with chelating agents is recommended as well. However, leaching is very expensive and time-consuming and it requires large volumes of pure, high-purity or even ultrapure acids, which will become a waste problem because of environmental contamination. Another disadvantage is the fact that the cleaned vessels remain in contact with the acids now enriched with impurities, by which they are again contaminated. Therefore, in many cases these procedures are not effective enough so as to guarantee for residual blanks down to the lower pg/mL region.

A very effective and much less time-consuming cleaning of quartz vessels can be achieved by steaming with acid vapor described by Knapp [48]. This method can also be applied to vessels and tools made of borosilicate glass, PTFE, TFM, PFA, and glassy carbon. For purification, the vessels are continuously exposed to the hot vapor of the purifying liquid. By steaming the vessel with nitric acid or hydrochloric acid, the surface is cleaned from adsorbed matter and conditioned in such a way that adsorption phenomena are greatly reduced.

Especially during sampling and sample preparation, there is an inevitable risk of contamination by tools for cutting, drilling, milling, sieving, crushing, grinding, and pulverizing. Metal contaminations of biological tissues and fluids with Cr, Ni, Co, Fe,

Mn, Cu, and others due to use of scalpel blades and syringe needles are observed. Therefore, the use of forceps, knives, spatulas, and needles made of plastic, titanium or quartz is recommended. Tools (scalpels and knives) made of corundum single crystals (Al_2O_3) grown artificially were found to be excellent for clean preparation of biological samples [49]. Biological tissues in contact with the surface of a common surgical instruments (scalpels and knives) become contaminated with virtually all trace elements, particularly with Cr and Co, present in the instrument material [50].

The operator himself also represents a very serious source of contamination. The number of particles emitted per minute by a person amounts up to millions. They are released by the skin, hair, clothes, jewelers, cosmetics, disinfectants, talc, etc.

1.3.5 Contamination by sample handling

In spite of the fact that we have available very effective techniques for the cleaning of the vessels and equipment, for the purification of the reagents and for maintaining a clean laboratory and working place, further contamination, for example,, arising from storage of the sample solution [51], from sample handling and from the analytical procedure cannot be totally excluded. At the ppb and still much more at the ppt concentration level, even a very simple working step such as pipetting, shaking, evaporation, filtering, etc., already can increase the blanks considerably [14].

Alarming is not the real value of the blank of one single step but the fact that blanks occur even during very simple operations and accumulate during the whole analytical procedure to amounts in the ng/g range. In extreme trace analysis, the scattering of the blanks can exceed several orders of magnitude.

In principle, the various sources of systematic errors described above are present in all steps of an analytical procedure such as sampling, transport, storage, sample pretreatment, decomposition and dissolution, and separation and preconcentration. Accordingly, the general statement can be made that with decreasing absolute amounts of the elements to be determined, systematic errors increase dramatically and that they are the main problem in extreme trace analysis. Unfortunately, this fact will often not be realized in the daily work of routine laboratories and may then be the reason of wrong results with dramatic consequences with respect to economy, safety, and health.

1.3.6 Losses

Losses of elements are caused by volatilization, chemical reactions, or by reactions with the material of vessels, containers, and tools, and, finally, by adsorption or desorption processes.

Losses of elements by volatilization mainly occur at high temperatures. However, for very volatile elements, these interferences can already be remarkably high at room temperature [52]. The analytes may be lost in the elemental form, oxides, and predominantly as halides or even as hydrides. The extent of the loss depends on the type of sample, and the variable temperature and time. Especially, mercury is well known to be extremely volatile. It can be lost during sampling, storage, and sample preparation, when aqueous solutions are stored in open vessels or in vessels made of organic polymers. Mercury losses can occur within a few hours and, in addition, elemental Hg quickly penetrates through sample containers made of plastic such as PE or PP. Therefore, samples in which Hg is to be determined should not be stored or transported in plastic containers so as to avoid Hg losses by volatilization or prevent contamination by the Hg present in the environment (ambient atmosphere).

The number of elements and compounds that can be lost as a result of volatilization increases with temperature. This must be considered when evaporating solutions or when performing decomposition procedures [53]. Volatile chlorides of Hg^{2+}, As^{3+}, Sb^{3+}, Sn^{4+}, Ge^{4+}, Se^{4+}, Pb^{4+}, and Te^{4-} may be lost during acid solution evaporation or during organic materials combustion.

During the dissolution of metals and alloys with nonoxidizing acids, the hydrides of elements such as S, P, As, Sb, Bi, Se, or Te may escape. Furthermore, hydrides can also be lost from alloys during the sampling step. Also, when drilling or cutting metal samples such as Al or Fe, the well-known smell of H_2S or PH_3 and other volatile hydrides often indicates the loss of these elements.

In general, volatilization can be prevented by application of closed systems (evaporation, closed vessel digestion, etc.). When closed systems are not suitable, volatilization can be reduced or prevented by reducing the temperature (storage, freeze-drying, low-temperature ashing, etc.). One also should avoid all chemical reactions by which volatile compounds can be formed (e.g., the formation of Cr_2OCl_2).

Many problems in elemental analysis are associated with the level of concentration to be determined; the concentration of the trace elements of very diluted solutions may change very quickly as a result of adsorption and desorption effects, which lead to losses or contamination. By these processes, ions or compounds of trace element are bound onto the surface of the container and may be released later on when the composition of the solution changes. The adsorption losses of elements become appreciable at concentrations <10^{-6} mol/L and are of the order of 10^{-9}–10^{-12} mol/cm^2 [14].

In handling aqueous solutions containing low concentrations of ions, the researcher must consider various factors to minimize or prevent container adsorption of ions. The kind of ion and container, the period between collection and analysis, light effects, and dissolved salts are some of the aspects that need to be evaluated. Other factors, such as the sampling technique and contamination effects, must be considered in the analysis. In working with low ion concentrations, only metal-free containers may be used, and scrupulous care must be undertaken in all collection and analytical steps to avoid contamination. For field studies, PE would be preferable to glass because of its handling characteristics.

1.4 Sample treatment after the sampling process

Physical sample preparation is a very important aspect of the chemical analysis process. Methodology depends on the sample type and the reason for the trace analysis. Depending on the type of sample, most of these operations involve predominantly physical methods such as cleaning, drying, grinding, sieving, cooling or freezing, and mechanical agitation and homogenization [20].

Cleaning is most recommended for some kinds of materials, particularly parts of plants, such as roots, leaves, and fruits [53].

When the trace elements determination is carried out on aqueous matrices (e.g., tissues, fruit, vegetables, and soil samples), attention must be paid to changes in sample mass during the storage period. For this reason, analytical results should always be reported in terms of dry mass (weight) to avoid false interpretations. Drying is best conducted immediately after sampling. It is a widely used method that minimizes physical and chemical changes of samples; however, this process irreversibly alters the biological matrix because water and all volatile constituents are eliminated. Such problems are easily solved by oven drying, microwave drying, and freeze-drying (lyophilization).

Oven drying of biological material is performed at temperatures between 60 and 65 °C. The stream of hot clean air was frequently used as a drying procedure in the preparation of biological materials. It is very important to control the temperature because the biological matrix may decompose, depending upon the nature of the sample. Losses of elements during drying may occur owing to the formation of insoluble substances (e.g., aluminum oxides) and the evaporation of volatile elements or compounds of elements. Thus, a large portion of inorganic- and organic-bound mercury is lost by drying. Similar effects can be observed for arsenic, antimony, selenium, and others. On the other hand, drying at lower temperatures will reduce volatilization losses but also expose the sample to the ambient environment for longer periods of time, which serves to increase the risk of the contamination. The drying operation until constant weight is a common requirement for solid samples that contain varying amounts and unknown types of water; drying may be performed at 105 °C.

Microwave-assisted drying, a very fast drying procedure, exhibits the same problems as laboratory oven drying. An exact control of the microwave energy is necessary to prevent overheating of the sample and losses of some elements [54]. Proper equipment and optimized programs for the microwave oven were recommended to prevent losses of elements by overheating of the samples and contamination by the ambient atmosphere. Volatile elements should not be determined in samples dried in a microwave oven or in a drying oven.

Freeze-drying has been shown to be the most satisfactory procedure for trace element analysis of biological materials since it minimizes the loss of highly volatile elements and compounds. It is also known as lyophilization or vacuum drying. Freeze-drying is to be preferred over oven drying as at low temperatures fewer alterations of the biological material occur and the formation of insoluble substances is also decreased. Losses due to evaporation are also reduced, but not excluded. Drying at a temperature as low as 120 °C can result in the loss of up to 10 % of most elements, and losses for mercury, lead, and selenium may be considerably higher (e.g., 20–65 %) with certain matrices [55]. To avoid losses of volatile elements, especially Hg, As, or Se, it is recommended that the sample be cooled during freeze-drying to <10 °C. Without cooling, the temperature of the sample will increase to room (laboratory) temperature after sublimation of the moisture and this may again lead to losses. Contaminations by drying arise mainly from the material of the drying apparatus employed. Thus, specimens for subsequent analyses for Cr, Ni, Co, or Mn should never be dried in a stainless steel oven; in such cases, aluminum is the preferred material.

To avoid systematic errors during the drying process, it is recommended that the original moist sample material be analyzed whenever possible and to correct the analytical result with the factor obtained by the separate determination of the dry mass. In this case, losses of volatile elements or contamination by dust during the drying process are inconsequential.

The homogenization process is another important process in sample preparation for solid sampling analysis, easily leading to contamination and/or losses of elements, and it is easily achieved by grinding process. Homogenization of samples is, in many cases, necessary to provide a representative sample. In solid samples, elements are normally distributed in an inhomogeneous way. Trace element determinations are usually restricted to relatively small samples, which requires that a fairly large sample be comminuted and homogenized prior to removal of an aliquot for analysis. In general, narrow particle size distributions with particles preferably below 10 μm diameter may ensure the desired analytical homogeneity [56, 57]. The choice of grinding technique can vary, depending on the properties of the sample matrix, especially on its hardness, fiber, and fat contents. To avoid possible nickel or chromium contamination from the construction material, all parts of the grinding device coming into contact with the biological materials are made of titanium or PTFE. The cryogenic grinding technique (brittle fracture technique) was introduced by Iyengar and Kasperek [58], and it relies on an increase of hardness of all tissues, the insertion of failures in the crystal structure, and use of very smooth force for the reduction to small pieces [59]. Before starting the grinding process, the whole grinding device (metering trough, mortars and balls, grinding cylinder and rods, etc.) should be cooled with liquid nitrogen for several hours. If the temperature of the components is constant at less than −190 °C, the grinding process is started without any further supply of liquid nitrogen. Therefore, for soft tissues, grinding and milling in vibrating ball mills applying PTFE or PFA containers and PTFE coated balls (made from stainless steel or tungsten, etc.), eventually under cooling with water or liquid nitrogen, are the preferred means. For hard materials, such as bone, teeth, etc., other container materials like Zr, Ti, or W and cooling under liquid nitrogen are necessary. Since then, some cryogenic grinding applications have been described [60–62].

Other mechanical mills such as mortar and pestle, blenda, ball and mortar, and disk mills have been also intensely applied to grind samples. Ball mill was successfully used in the preparation and certification of a pig liver reference material. Fish [63] and bovine liver samples [60] were also homogenized using ball mill. In both cases, good results were obtained for Cd determination in fish samples (slurry-ETAAS) and Cu and Zn determination in bovine liver samples (solid-ETAAS). In the same way, mortar and pestle was successfully used to grind fish samples for Se determination.

Comparing many IAEA and NIST environmental and biological materials ground using difference technique, including cryogenic and other mechanical mills, Fajgelj and Zeissler [64] concluded that jet air milled biological reference materials presented the lowest particle size distribution, making them suitable for small sample analysis.

However, when a representative aliquot is to be taken from the specimen, a homogenization step prior to storage is essential. The structure and composition of the biological material will be changed by both drying and freezing. Homogenization is mostly carried out in mixers with rotating knives. These instruments are one of the main sources of contamination. Often, the composition of the different materials from which the mixer is made is unknown. The most contaminating item is the knife because there will always be some abrasion. The use of mixers with stainless steel blades or knives should be avoided in trace element analysis. Therefore, the knives and/or blades must be made from materials that are of no analytical interest, for example, titanium or tantalum or special ceramic materials. High-speed mixers give rise to considerable cavitation by which water and volatile organic substances, as well as volatile elements (e.g., Hg), can be lost. Cooling during mixing is advised. The efficiency of mixing with respect to homogenization and contamination must be controlled for each new kind of specimen by analyzing samples of different size. For the homogenization of biological material with a low water content, distilled water must be added.

Comminution of the sample presents a significant opportunity for contamination. Contamination of the sample from abrasion of the comminution equipment is fundamentally unavoidable, so efforts must be made to select the best possible equipment for each particular analytical task. Equipment is preferred in which the sample comes into contact only with surfaces fabricated from such high-purity plastics as PTFE since this permits the sample to be used without restriction for the determination of a large number of elements. Friability can be increased by deep-freezing or drying the sample prior to comminution.

The sieving of fresh or previously milled samples is one of the methods most used for the evaluation and classification of the particle size distribution; by knowing the particle size distribution, it is possible to infer the homogeneity of the sample or the viability of the grinding method chosen. In general, microanalytical techniques yield better accuracy for homogeneous particle size distribution. For trace metal analysis, metal sieves can contaminate the samples. Wear-resistant plastic sieves and sieve holders are recommended for all sieving operations of samples for trace metal analysis.

Some analytical techniques require the separation of suspended particles as a primary treatment. The oldest method of water purification is filtration. This method may be divided into adsorption filtration, such as that provided by a carbon filter, and mechanical filtration using a screen or membrane. The carbon filter is effective for removing certain organics, odors and tastes, and chlorine. Mechanical filtration does nothing more than remove undissolved particles. In the water analysis, for example, filtration through a membrane filter with a porosity of 0.20 or 0.45 μm is recommended. The fraction that passes through the filter is designated **soluble**. The use of a fiberglass prefilter with porosity of 1–5 μm is recommended in samples with high levels of suspended solids. The residue remaining on the filter can be leached or decomposed, and subsequently analyzed.

1.5 Decomposition as a sample preparation method for elemental analysis: an analytical perspective

Many modern instrumental techniques require to convert solid (or solid containing) samples to solutions prior to an analysis. The terms decomposition, destruction, digestion, acid digestion, dissolution, ashing, wet ashing, oxidative acid digestion, and mineralization all refer to this process. In this contribution, the general expression will be decomposition, which is specified to be dry or wet ashing. Mineralization refers to those procedures that result in inorganic chemical forms of the analyte only. A variety of techniques are employed from ambient pressure wet digestion in a beaker on a hot plate to specialized high-pressure microwave heating. Traditionally, decomposition of the sample in elemental analysis requires it to be mineralized in order to remove the organic content. Sample decomposition for total element determination therefore appears to be the recommended procedure on every occasion.

In general, it is required from any decomposition procedure to alter the original chemical environment of the sample into a digest, that is, a solution in which the analyte is distributed homogeneously. More specific conditions set to a decomposition technique are

- The decomposition must be complete. Inorganic materials have to be converted completely into soluble compounds, and organic materials have to be totally mineralized.
- Removal of residual matrix components that interfere in the detection; residues should be quantitatively soluble in a small volume of high-purity acid.
- The decomposition procedure has to be as simple as possible and should not require complicated apparatus.
- The decomposition must be adapted in an optimal manner to the whole analytical procedure; possibility of adjustment of the oxidation state of the analyte and, consequently, compatibility with postdecomposition chemistry.

- Preference should be given to procedures where decomposition and separation are achieved in one step.
- In order to minimize the systematic errors in the decomposition procedure (contamination, loss of elements, incomplete decomposition), clean vessels made of an inert material, and the smallest amounts of high-purity reagents should be used, and dust should be excluded. Reaction chambers should be as small as possible. Precautions should be taken so as to minimize losses of analytes (elements) due to adsorption on the vessel material (reactions with the vessel material) and volatilization.
- Execution is not hazardous or dangerous for laboratory personnel.
- The yield from the decomposition step should be checked by using radioactive tracers.

This section gives an overview of decomposition methods and recent developments and applications of the decomposition of different materials. Other sample preparation methods, such as chemical extraction and leaching, solubilization with bases, enzymatic decomposition, thermal decomposition, and anodic oxidation, are beyond the scope of this contribution and will not be discussed here.

There are numerous publications giving useful information on the decomposition of any conceivable combination of matrix and analyte. Some comprehensive books and review articles contain material pertinent to either organic [65–68] or inorganic [69–74] matrices; others, to both [75–85].

Within the scope of this section, a comprehensive discussion on decomposition techniques is not feasible. For more comprehensive information, the following reviews and books are available: books by Šulcek and Povondra [72], Bock [75], and Krakovská and Kuss [81] are dedicated solely to decomposition methods. Other books deal exclusively with a single technique: microwave-assisted sample preparation [85, 86], which has also been reviewed elsewhere [88–98]. Recommended guidelines for sample preparation (methods of digestion) of different matrices are also available from the *Encyclopedia of Analytical Chemistry* [99]. Although it is very difficult to refer to every paper published in this area, the enlisted bibliography of this chapter gives a comprehensive coverage of advance of the topic made to date, its potential application, novel developments, and progress in decomposition techniques.

1.5.1 Sample decomposition techniques

Table 1.1 gives an overview of the different decomposition methods for organic and inorganic sample material. The intent is not to present the procedural details for the various sample matrices, but rather to highlight those methods that are unique to each technique and sample.

Table 1.1: Scheme of decomposition methods.

Decomposition technique	Required reagents	Application
Wet chemical decomposition		
In open systems		
• Acid digestion (thermally convective wet decomposition)	• HNO_3, HCl, HF, H_2SO_4, $HClO_4$	• Inorganic/organic
• Microwave-assisted wet decomposition	• HNO_3, HCl, HF, H_2SO_4, $HClO_4$, H_2O_2	• Inorganic/organic
• Ultraviolet decomposition (photolysis)	• H_2O_2, $K_2S_2O_8$, HNO_3, O_3	• Waters, slurries
• Ultrasound-assisted acid decomposition	• H_2O_2, HNO_3	• Inorganic
In closed systems		
• With conventional heating (thermally convective pressure digestion)	• HNO_3, HCl, HF, H_2O_2	• Inorganic/organic
• With microwave heating	• HNO_3, HCl, HF, H_2O_2	• Inorganic/organic
In flow systems		
• With conventional heating	• HNO_3, H_2SO_4, H_2O_2, HCl	• Inorganic/organic
• With microwave heating	• HNO_3, H_2SO_4, H_2O_2, HCl	• Inorganic/organic
• Ultraviolet decomposition	• H_2O_2, $K_2S_2O_8$, HNO_3	• Waters, slurries?
Vapor-phase acid digestion		
• With conventional heating	• HNO_3, HCl, HF, H_2O_2	• Inorganic/organic
• With microwave heating	• HNO_3, HCl, HF, H_2O_2	• Inorganic/organic
Combustion		
In open systems		
• Dry ashing		• Inorganic/organic
• Low-temperature ashing (combustion in a stream of oxygen)		• Organic
• Cool plasma ashing (Wickbold combustion)		• Organic
In closed systems		
• Oxygen flask combustion (Schöniger flask)		• Organic
• Oxygen bomb combustion		• Organic
• Combustion in a dynamic system (Trace-O-Mat)		• Organic
Fusion decomposition	Fluxes	• Inorganic

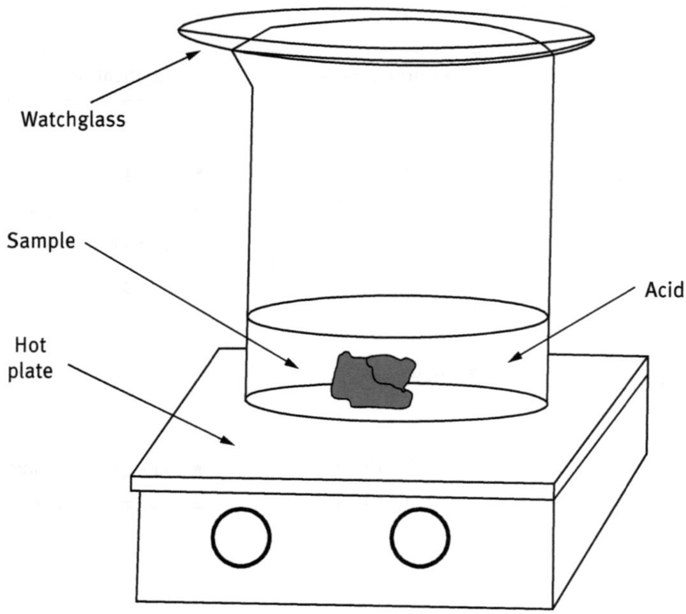

Figure 1.1: Acid decomposition in open system.

One of the oldest and simplest methods, and still most frequently used techniques, is wet decomposition in open systems (Figure 1.1). Wet decomposition can also be used in connection with closed systems.

1.5.1.1 Wet chemical decomposition

Sample wet decomposition is a method of converting the components of a matrix into simple chemical forms. This decomposition is produced by supplying energy, such as heat; by using a chemical reagent, such as an acid; or by a combination of the two methods. Where a reagent is used, its nature will depend on that of the matrix. The amount of reagent used is dictated by the sample size, which, in turn, depends on the sensitivity of the method of determination. However, the process of putting a material into solution is often the most critical step of the analytical process because there are many sources of potential errors, that is, partial decomposition of the analytes present, or some type of contamination from the vessels of chemical products used. It is beyond the scope of this contribution to discuss all possible systematic errors; therefore, further details on how to avoid systematic errors during sample decomposition that cannot be referred to in detail here are discussed in Section 1.5.

The majority of wet decomposition methods involves the use of some combination of oxidizing acids (HNO_3, hot concentrated. $HClO_4$, hot concentrated H_2SO_4) and nonoxidizing acids (HCl, HF, H_3PO_4, dilute H_2SO_4, dilute $HClO_4$) and hydrogen peroxide. All of these acids are corrosive in nature, especially when hot and concentrated,

Table 1.2: Physical properties of common mineral acids and oxidizing agents used for wet decomposition.

Compound	Formula	Molecular weight	Concentration		Density (kg/L)	Boiling point (°C)	Comments
			w/w (%)	Molarity			
Nitric acid (V)	HNO_3	63.01	68	16	1.42	122	68 % HNO_3, azeotrope
Hydrochloric acid	HCl	36.46	36	12	1.19	110	20.4 % HCl, azeotrope
Hydrofluoric acid	HF	20.01	48	29	1.16	112	38.3 % HF, azeotrope
Perchloric acid (VII)	$HClO_4$	100.46	70	12	1.67	203	72.4 % $HClO_4$, azeotrope
Sulfuric acid (VI)	H_2SO_4	98.08	98	18	1.84	338	98.3 % H_2SO_4
Phosphoric acid	H_3PO_4	98.00	85	15	1.71	213	Decomposes to HPO_3
Hydrogen peroxide	H_2O_2	34.01	30	10	1.12	106	

and should be handled with caution to avert injury and accidents. Concentrated acids with the requisite high degree of purity are available commercially, but they can be purified further by sub-boiling distillation [31].

Wet digestion has the advantage of being effective on both inorganic and organic materials. It often destroys or removes the sample matrix, thus helping to reduce or eliminate some types of interference. The physical properties of the common mineral acids used in sample preparation are summarized in Table 1.2.

Most wet decomposition procedures are conducted under conditions that they, in terms of temperature or reagents used, must be considered as extreme. Thus, the material of which the flasks, crucibles, etc., are made must be chosen carefully according to the particular procedure to be employed. The material from which the digestion device is fabricated is also a frequent source of elevated blanks. Elements can be either dissolved from the material or they can be desorbed from the surface. Very important in this respect is the nature of the material. The suitability of materials may be estimated according to the following criteria: heat resistance and conductance, mechanical strength, resistance to acids and alkalis, surface properties, reactivity, and contamination, whereby the specific characteristics of the organic and inorganic material must be also given special consideration. Table 1.3 shows the preferred materials for decomposition vessels. The container material in contact with the sample during the decomposition also frequently causes systematic errors. Elements can be either dissolved from the material or desorbed from or adsorbed on to the container surfaces. This amount will depend on the material, contact time, and temperature. Table 1.4 summarizes the inorganic impurities likely to be encountered with various

Table 1.3: Preferred materials for wet decomposition vessels.

Material	Chemical name	Working temperature (°C)	Heat deflection temperature (°C)	Water absorption (%)	Comments
Borosilicate glass	SiO_2^a, $B_2O_3^b$	<800[c]			Ordinary laboratory glass is not suitable for use in wet decomposition procedures
Quartz	SiO_2^d	<1,200			For all procedures involving wet decomposition of organic material, quarto is the most suitable material for vessels
Glassy carbon	Graphite	<500			Glassy carbon is used in the form of crucibles and dishes for alkaline melts and as receptacles for wet decomposition procedures
PE	Polyethylene	<60			
PP	Polypropylene	<130	107	<0.02	
PTFE	Polytetrafluorethylene	<250	150	<0.03	PTFE is generally used only for decomposition vessels in pressure decomposition systems
PFA	Perfluoralkoxy	<240	166	<0.03	
FEP	Tetrafluorperethylene	<200	158	<0.01	
TFM	Tetrafluormetoxil			<0.01	

[a] SiO_2 content between 81 and 96 %.
[b] B_2O_3 content between 3 and 5 %.
[c] Softens at a temperature of 800 °C.
[d] SiO_2 99.8 %.

Table 1.4: Inorganic impurities in selected vessel materials (data in ng/g).

Element	Borosilicate glass	Quartz	Polyethylene	PTFE Teflon[a]	Glassy carbon
Al	Main	100–50,000	100–3,000		6,000
As	500–22,000	0.1–80			50
B	Main	10–100			100
Ca	10^6	100–3,000	200–2,000		80,000
Cd	1,000	0.4–10			10
Co	100	1	0.5	2	2
Cr	3,000	3–5	20–300	30	80
Cu	1,000	10–70		20	200
Fe	$2 \cdot 10^5$	200–800	1,000–6,000	10–30	2,000
Hg		1		10[b]	1
Mg	$6 \cdot 10^8$	10	100–2,000		100
Mn	6,000	10			100
Na	Main	10–1,000	200–10,000	25,000	350
Ni	2,000				500
Pb	3,000–50,000		200		400
S	Main	Main			85,000
Sb	8,000	1–2		0.4	10
Ti	3,000	100–800			12,000
Zn	3,000	50–100	100	10	300

[a] Teflon is a registered trademark of DuPont.
[b] Strongly dependent on storage conditions.

vessel materials. The borosilicate glass, which contains several major, minor, and trace elements in relatively high concentrations, is usually not suitable for element determinations in the extreme trace range. Quartz can be considered a pure material found in the market and is available in varying degrees of purity. For most sample preparation steps in trace (metal) analysis, high-purity quartz is the preferred container (and tool) material. Alternatively, high-purity synthetic polymers can be used in many decomposition applications: PE, PP (PP), PTFE, and polymers (PFA, FEP, TFM). The apparatus and containers that are used for the wet decomposition procedures must be scrupulously cleaned and tested for any possible contamination. Usually, it is sufficient to boil the flasks in concentrated nitric acid, followed by rinsing several times with ultrapure water before use. In cases where this procedure is not adequate, one of the most powerful cleaning procedures is steaming the vessels with nitric or hydrochloric acid with assembly in a microwave-heated sealed Teflon vessel [48]. This procedure is particularly recommended for quartz, borosilicate glass, and PTFE vessels.

To generalize this section, nitric acid is an almost universal decomposition reagent and the most widely used primary oxidant for the decomposition of organic matter because it does not interfere with most determinations and it is available commercially in sufficient purity. Hydrogen peroxide and hydrochloric acid can be usefully employed in conjunction with nitric acid as a means of improving the quality of a decomposition. Hydrochloric acid and sulfuric acid may interfere with the determination of stable compounds. Mixtures with hydrochloric acid are generally used for samples containing principally inorganic matrices, and combinations with hydrofluoric acid are used to decompose silicate insoluble in the other acids. Safety considerations are particularly important when using perchloric acid.

a) Wet decomposition in open systems

Open vessel acid digestions, one of the oldest techniques, are undoubtedly the most common method of sample decomposition or dissolution of organic and inorganic sample materials used in chemical laboratories. This inexpensive technique is of inestimable value for routine analysis because it can be easily automated; all the relevant parameters (time, temperature, introduction of decomposition reagents) lend themselves to straightforward control.

The main advantage of wet decomposition over dry ashing is its speed. However, systems of this type are limited by a low maximum decomposition temperature, which cannot exceed the ambient pressure boiling point of the corresponding acid or acid mixture. For instance, the oxidizing power of nitric acid with respect to many matrices is insufficient at such low temperatures (boiling point 122 °C). One possible remedy is the addition of sulfuric acid, which significantly increases the temperature of a decomposition solution. Whether or not this expedient is practical depends on the matrix and the determination method. High-fat and high-protein samples are generally not subject to complete decomposition at atmospheric pressure. Other disadvantages relate to the risk of contamination through laboratory air, the necessarily rather large amounts of required reagents, and the danger of losses of trace elements. Losses can be kept low by using an excess of acid combined with a reflux condenser and by optimization of temperature and duration. Nevertheless, systems operated at atmospheric pressure are preferred from the standpoint of workplace safety.

Thermally convective wet decomposition. The conventional approach to wet decomposition entails a system equipped with heated conventional source (Bunsen burner, heating plate, sand bath, etc.) operating either at a fixed temperature or in response to a temperature program. Acid decompositions are often accomplished in any vessel, usually in glass or PTFE (beaker, conical flask, etc.) with or without a refluxing condenser. However, when a sample is decomposed in open wet digestion, refluxing is compulsory. The necessary apparatus has been described by Bethge [100]. Open block decomposition systems have been popular in sample analysis over the past decades, but have consistently suffered from the major drawback of their sensitivity against corrosion and subsequent risk of contamination (Figure 1.1). Therefore, block digestion systems (hotplate techniques) have not been considered state-of-the-art technology in trace and ultratrace sample preparation. Graphite block digestion systems are becoming more frequently considered. These systems overcome the deficiencies of the traditional systems, made from stainless steel or aluminum, because the block is manufactured from graphite and typically coated with a fluoro-polymer to prevent the possibility of metallic contamination from the surface of the system during the handling of the samples. Graphite block systems present an alternative to the current mainstream technology of open- and closed-vessel digestion systems as they allow large numbers of samples to be digested simultaneously, thus overcoming one of the major weaknesses of closed-vessel systems. Commonly employed decomposition agents include nitric acid, sulfuric acid, hydrofluoric acid, perchloric acid, and hydrogen peroxide, as well as various combinations of these. Most applications of wet decomposition involve aqueous or organic matrices, such as surface waters, waste water, biological and clinical samples, food samples, as well as soil, sediment and sewage sludge, coal, high-purity materials, and various technical materials. More recently, open systems have progressed; usual decomposition ramps consist of several vessels equipped with reflux condensers to limit possible volatilization losses of some analytes and to avoid the evaporation of the reactive mixture. Such assembling is entirely satisfactory for ensuring concurrent digestions of a large series of samples. Modern commercially available Hach Digesdahl Digestion Apparatus (Hach Comp., USA) is designed to digest organic and mineral samples for subsequent analysis.

Microwave-assisted wet decomposition. The most innovative source of energy for wet decomposition procedures is microwaves. Because the heating takes place inside the decomposition mixture, microwave decomposition is more efficient than with conventional means of heating. Using microwaves, both the speed and the efficiency of decomposition for some types of samples considered difficult to solubilize are often improved. Additionally, automation becomes possible with some instrumentation.

Since Abu-Samra et al. [101] reported on the application of microwave techniques to wet decomposition of biological samples, there has been a rapid development in microwave-assisted decomposition for elemental analysis. Recent reviews [79–87] detail the application of microwave-assisted decomposition to a wide variety of sample types, such as geological, biological, clinical, botanical, food, environmental, sludge, coal and ash, metallic, and synthetic materials and mixed samples, and present specific experimental conditions as a function of the matrix to be digested. The earliest attempts at microwave-assisted digestion were performed using home appliance microwave ovens. This was necessary because commercial devices were not available at the time. The use of domestic microwave ovens in laboratory experiments should be discouraged because of safety and performance. Microwave-assisted decomposition in open systems at atmospheric pressure is generally applicable only with simple matrices or for strictly defined objectives, and the results are reproducible only if the specified decomposition parameters are strictly observed. The performance of the focused-microwave-assisted systems and a wealth of applications have been reviewed by White and Mermet [88, 89] and very recently by Nóbrega et al. [102]. Focused-microwave-assisted sample preparation is a suitable strategy for dealing with high masses of organic samples (up to 10 g). Losses may be encountered with mercury and possibly also with organometallic compounds. Addition of sulfuric acid is essential in order to achieve a sufficiently high decomposition temperature using atmospheric pressure equipment, where the boiling point of the acid establishes the maximum decomposition temperature, although it is important to remember that the presence of sulfate interferes with many procedures for metal determination. Although nonpressurized microwave systems are limited by a low maximum digestion temperature, which cannot exceed the ambient pressure boiling point of the acid (or the acid mixture), they provide the best option with regard to the safety of personnel because no overpressure can occur. Moreover, nonpressurized microwave-assisted digestion is suitable for online decompositions in continuous-flow systems.

Very recently, Matusiewicz [103] presented an overview of the different microwave-based systems used for solid and liquid sample pretreatment. They comprise relevant publications relating to current research, the unique instrumental approach, and the various commercially available systems including their operating parameters and accessories.

Ultraviolet decomposition. Ultraviolet (UV) decomposition is utilized mainly in conjunction with uncontaminated or slightly contaminated natural water matrices (aqueous solutions), such as sea, surface, fresh, river, lake, ground, estuarine, and costal water. Liquids or slurries of solids are decomposed by UV radiation (light) in the presence of small amounts of hydrogen peroxide, acids (mainly HNO_3), or peroxodisulfate (i.e., beverages, special industrial waste water, water of sewage treatment plants, soil extracts) [104]. Dissolved organic matter and complexes of the analyte elements are decomposed to yield free metal ions. The corresponding decomposition vessel should be placed in the closest possible proximity to the UV lamp (low or high pressure) to ensure a high photon flux. In photolysis, the decomposition mechanism can be characterized by the formation of OH^* and O_2^* radicals from both water and hydrogen peroxide that is initialized by the aid of the UV radiation [104]. These reactive radicals are able to oxidize, to carbon dioxide and water, the organic matter present in simple matrices containing up to about 100 mg L^{-1} of carbon. Complete elimination of the matrix is, of course, possible only with simple matrices or by combining photolysis with other decomposition techniques [105]. The method does not oxidize all organic components possibly present in water; chlorinated phenols, nitrophenols, hexachlorobenzene, and similar compounds are only partly oxidized. Effective cooling of the sample is essential because losses might otherwise be incurred with highly volatile elements. Hydrogen peroxide addition may need to be repeated several times to produce a clear sample solution. Modern UV decomposition systems are commercially available (see Ref. [104], Table 1.1).

Ultrasound-assisted acid decomposition. Although analytical chemists have shown little interest in the use of ultrasound, its potential usually surpasses that of the other conventional auxiliary energies. Thus, ultrasound is of great help in the pretreatment of solid samples as it facilitates and accelerates steps such as dissolution, fusion, and decomposition, among others. An acid decomposition method that uses ultrasonic device (bath or probe) has been developed. The propagation of ultrasonic waves characterized by a minimum frequency of 16 kHz results in rapid fluid movement through compression and rarefaction: an enormous number of microscopic cavities are formed and free radicals are generated, chemical layers are dispersed and the contact between the ingredients of the reaction is accelerated. Usually, ultrasonic effects are much more intense in heterogeneous than in homogeneous chemical systems because emulsification is favored and mass heat transfer in two-phase systems is increased. These effects have been exploited for sample preparation in agriculture, biological, and environmental chemistry [106, 107].

Wet decomposition in closed systems. During the last few decades, methods of wet sample preparation using closed vessels have become widely applied. Closed systems offer the advantage that the operation is essentially isolated from the laboratory atmosphere, thereby minimizing contamination. Decomposition of the sample is essentially ensured by a common wet digestion procedure, which is performed under the synergistic effects of elevated temperature and pressure; decomposition occurs at relatively high temperature due to boiling point elevation. The pressure itself is, in fact, nothing more than an undesirable – but unavoidable – side effect. These techniques are generally much more efficient than conventional wet decomposition in open systems, the loss of volatile elements is avoided, any contribution to blank values may be reduced, and the decomposition of more difficult samples is possible. The principal argument in favor of this form of decomposition is the vast amount of relevant experience acquired in recent decades. Closed-system decomposition is particularly suitable for trace and ultratrace analysis, especially when the supply of sample is limited.

Because the oxidizing power of a decomposition reagent shows a marked dependence on temperature, an arbitrary distinction should be made between low-pressure decomposition and high-pressure decomposition. Low-pressure decomposition (<20 bar) is limited to a temperature of ca. 180 °C, whereas with high-pressure apparatus (>70 bar) the decomposition temperature may exceed 300 °C.

Thermally convective wet pressure decomposition. The decomposition of inorganic and organic substances in sealed tubes was the method first proposed for pressure digestion at the end of 19th century, and some of these applications are still difficult to replace by other digestion methods. The use of sealed glass tubes goes back to Mitscherlich [108] and Carius [109], often referred to as the Carius' technique, first described in 1980. Carius undertook digestion of organic materials with concentrated nitric acid at 250–300 °C. The sample and acid were mixed in a strong (thick)-walled quartz ampule and sealed. The ampule was transferred to a "bomb canister" and heated in what was called a "bomb oven" for several hours, after which it was cooled, opened, and the contents analyzed. Carius tube decomposition involves the generation of internal pressure in excess of 100 bar at 240 °C. For safety, any stainless steel sleeve jacket (along with solid CO_2 pellets, to maintain equal pressure across the tube wall when heated) that is large enough to contain the Carius tube will suffice as an external pressure vessel [110].

With the development of the so-called Carius tube, the field of closed-vessel decomposition was born. Decomposition in autoclaves with metal inner reaction vessels was originally proposed in 1894 by Jannasch [111], but was not widely employed because of a number of drawbacks (such as strong corrosion of the platinum vessel).

Extensive use of pressure decomposition in analytical procedures began in 1960 as a result of the considerable technological progress in the manufacture of organic polymers. Convectively heated pressure vessel systems have proved to be the most valuable systems for guaranteeing complete, or almost complete, digestion of solid samples because they provide elevated digestion temperatures (about 200–230 °C) [112]. Most sample vessels for use in thermally convective pressure digestion are constructed from PTFE [113–115], PFA [116], or PVDF [117], although special quartz vessels with PTFE holders [118] or glassy carbon vessels [119] are available for trace analysis purposes. The sample vessel is mounted in a stainless steel pressure autoclave and then heated, usually in a laboratory drying oven, furnace or heating block, to the desired temperature. Because of the necessity to examine numerous samples, mechanized multisample pressure digestion systems able to process rather large sample numbers of the same matrix type were developed [120]. A cooling circuit can be fitted into the metal casing (jacket) to permit rapid manipulation of the solution formed immediately after removing the "digestion bomb" from the oven or heating block [121]. Dissolution can be also accelerated by mixing the reactants, preferable by using a stirring bar (covered with PTFE) [122]. An alternative design has been proposed by Uchida et al. [123], wherein a small screw cap vial for sample digestion is placed inside the Teflon digestion double vessel. To improve dealing with pressure–temperature evaluation and the carbon balance for some materials, a system with a Teflon-lined membrane pressure meter and a thermocouple was designed [124]. Recently, a digestion vessel for use with a convection oven was proposed [125], which has an unusual design in which the vessel consists of three nested structures: an innermost PTFE container of 30 mL capacity, an intermediate PTFE container of 100 mL capacity, and an outer stainless steel shell.

All thermally initiated digestions have the disadvantage that a considerable amount of time is consumed in preheating and cooling the digestion solutions and sample vessel [126], the limited sample size, and the inability to visually check the progress of the digestion. The contributions of Langmyhr, Bernas, Tölg, and coworkers are worth mentioning with regard to the commercialization of the digestion vessels or "digestion bombs," as they are often called. Today, there are a number of digestion bombs covering the whole market range, including the popular Parr acid digestion bombs (Parr Instrument Company, USA), Uniseal decomposition vessels

(Unseal Decomposition Vessels Ltd, Israel), stainless steel pressure vessels with Teflon inserts (Berghof Laborprodukte GmbH, Germany), the pressure decomposition system CAL 130FEP (Cal Laborgeräte GmbH, Germany), and the pressure digestion system (PRAWOL, Germany).

To avoid the problem of loss of mechanical stability at high temperatures, vessels made of quartz are now being used in a new pressure digestion system [127, 128]. The introduction of a high-pressure ashing (HPA) technique by Knapp [127] has not only reduced the effective digestion time but also opened the way to digestion of extremely resistant materials, such as carbon, carbon fibers, and mineral oils. A perfected system of wet decomposition under high temperature (320 °C) and pressure (130 bar) developed by Knapp is commercially available, the HPA-S High Pressure Asher system (Anton Paar GmbH, Austria) [128, 129].

Very recently, again in respect to complete decomposition of organic waste materials, a potent digestion technique was developed [130] based on a prototype of an HPA device using infrared heating (IR-HP-asher). High-pressure decomposition is conducted in six quartz vessels inside a steel autoclave, with a maximum digestion temperature as high as 300 °C at a pressure of 130 bar. The novelty of this approach lies in the design of an HPA system with IR heating.

As metal autoclaves are expensive, a pressure vessel without an outer metal casing has been designed. The vessel can be sufficiently well sealed by using a screw cap [131]. Volatile components are not lost during heating and the laboratory atmosphere is thus not contaminated by acid vapors. All-Teflon thick-walled PTFE vessels (bombs) have been used in the dissolution of refractory oceanic suspended matter using HCl, HNO_3, and HF [132]. Translucent Nalgene-sealed bottles have been proposed for the "wet pressure digestion" of biological materials (fish, bird, plant tissue) using a combination of $HClO_4$ and HNO_3 [133]. A method utilizing a pressure digestion technique for real sample matrices using linear PE bottles has been proposed [134]. Vessels of PE are transparent, permitting observation of the whole digestion process and reduction of the reaction time to a minimum. A complete decomposition of fatty material with slight overpressure (<4 bar) was possible in a closed system completely made from quartz [135]. A closed PTFE bomb (30 mL capacity, screw-cap vessel machined from molded, stress-relieved Teflon-TFE rod) was designed for the digestion of materials using a conventional heating (drying) oven [136].

Microwave-assisted pressurized wet decomposition. Closed-vessel microwave-assisted decomposition technology has been acknowledged as one of the best solutions for "clean" chemistry applications and has a unique advantage over other closed-vessel technologies. The vessels used for microwave acid digestion are either low-pressure or high-pressure bombs. The current generation of microwavable-closed vessels consists of a two-piece design, liners and caps composed of high-purity Teflon or PFA with casings (outer jacket) made of polyetherimide and polyetherethereketone or other strong microwave transparent composite material. Their practical working temperature is 260 °C (softening point of Teflon), and their pressure limit is 60–100 bar. Closed-vessel decomposition is ideal for those samples that are being dissolved in HNO_3 and/or HCl.

Microwaves only heat the liquid phase, while vapors do not absorb microwave energy. The temperature of the vapor phase is therefore lower than the temperature of the liquid phase and vapor condensation on cool vessel walls. As a result, the actual vapor pressure is lower than the predicted vapor pressure. This sort of sustained dynamic, thermal nonequilibrium is a key advantage of microwave technology as very high temperatures (and, in turn, short digestion times) can be reached at relatively low pressures.

The inspiration for pressure digestion studies came from a US Bureau of Mines report [137], which described how rapid dissolution of some mineral samples had been achieved using a microwave oven to heat samples and an acid mixture contained in PC bottles. Smith et al. [138] substituted Teflon PFA fluorocarbon resin vessels for PC because of its superior chemical and mechanical properties. Buresch et al. [139] used low-pressure-relief-type containers made of PTFE or quartz. Alvarado et al. [140] exploited modified thick-walled Pyrex glass test tubes fitted with PP screw caps as pressurizable vessels. Kojima et al. [141] modified a Teflon digestion bomb by using a double Teflon vessel with a PP jacket to permit leak-free and safe decomposition of samples. A closed-vessel microwave digestion system was described [142]. *In situ* measurement of elevated temperatures and pressures in closed Teflon PFA vessels during acid decomposition of organic samples was demonstrated; temperature and pressure monitoring permitted controlled decomposition and studies of decomposition mechanisms.

Laboratory-made all Teflon bombs, used for low- or medium-pressure work, are also appropriate for microwave-heated digestion purposes [143], some fitted with pressure-relief holes, valves, or membranes (rupture discs).

Low-volume microwave-assisted decomposition methods have found applications for studies involving small sample sizes where loss of sample in large digestion equipment is inevitable. Small quantities of tissue (5–100 mg dry weight) are decomposed in high-purity nitric acid by use of a modified Parr microwave acid digestion bomb with modified Teflon liner [144]. The use of low-volume (7 mL) Teflon-PFA closed vessels designed for the preparation of small-sized (<100 mg dry mass) biological tissues has been described [145].

In order to prevent excessive pressure rises during closed microwave acid decomposition of fairly large (1 g) samples having high organic content, an open-vessel predecomposition technique under reflux was designed to allow the escape of oxidation products, such as carbon dioxide, without incurring evaporation losses of acid or analytes. Following predecomposition, the vessels were capped and subjected to microwaves to complete the decomposition under pressure [146].

In an attempt to minimize the delay in opening Teflon pressure vessels following microwave acid digestion, and thus significantly reduce sample preparation time, digestions with the pressure vessels immersed in liquid nitrogen and the use of liquid nitrogen as a pre- and postdigestion coolant were applied [147]. In other developments, a special type of Teflon bomb was constructed in which the vapor pressure can be maintained at a moderate level (up to 5 bar) by means of an internal quartz or Teflon cooling spiral. During operation, reflux of the condensed acid and water vapors continuously renews the liquid phase over the sample [148].

Several microwave heating configurations were presented by Pougnet et al. [149, 150] based on 500 or 1,200 W, 2.45 GHz fundamental-mode microwave waveguide cavities, which heat pressure vessels currently used in laboratories for sample decomposition and other applications.

The capsule concept was reviewed in detail by Légère and Salin [151, 152]. The sample is handled in an encapsulated form until it is in the digestion solvent. The operation of the capsule-based microwave-assisted digestion system proceeds in several steps, during which temperature and pressure are monitored.

From the previous discussion, it is clear that microwave acid digestion can be easily adapted for closed-vessel digestions; hence, its application has been limited to digestions in closed Teflon-lined vessels made of nonmetallic microwave-transparent materials operating with a maximum upper safe pressure of around 60–100 bar. In response to these limitations, Matusiewicz [153, 154] developed a focused-microwave-heated bomb that would exceed the operational capabilities of existing microwave digestion systems and permit the construction of an integrated microwave source/bomb combination, capable of being water or fluid cooled *in situ*. Another vessel configuration integrates the microwave chamber around the vessel. These consist of one or several microwave transparent vessels (Teflon, quartz), which can be sealed, enclosed in an acid-resistant stainless steel chamber [155]. The steel chamber acts as both the pressure vessel and microwave chamber. Modern systems can handle acid decompositions at temperatures up to 320 °C and pressures of 130–200 bar.

Very recently, a novel microwave-assisted high-temperature UV digestion for accelerated decomposition of dissolved organic compounds or slurries was developed [156, 157]. The technique is based on a closed, pressurized, microwave decomposition device wherein UV irradiation is generated by immersed electrodeless Cd discharge lamps (228 nm) operated by the microwave field in the oven cavity. The immersion system enables maximum reaction temperatures up to 250–280 °C, resulting in a tremendous increase of mineralization efficiency.

Today, there are a number of microwave-digestion bombs and systems available [103].

1.5.1.2 Flow systems

Discrete vessel systems, whether at elevated or atmospheric pressure, require a large amount of handling. Processes such as assembling, closing, opening, and positioning the vessel in the ordinary oven or microwave field are laborious and time consuming. Continuous flow-through thermal digestion, UV decomposition, and microwave digestion systems were designed to overcome some of the limitations by replacing the vessels with flow-through tubing (coil). Samples are digested by pumping them through a coil containing a digestion matrix while being heated (by thermal, UV, or microwave). The continuous flow of a carrier stream through these systems washes the system, removing the need for tedious vessel clean-up procedures. These systems can handle reactions that produce sudden increases in temperature and pressure, or unstable samples. Many different designs of flow digestion systems have been published, but very few meet the prerequisites for high-performance sample decomposition.

b) Thermal heating

Many of the disadvantages of sample digestion can be overcome by automating sample preparation in an enclosed system through the use of flow technology.

A well-established digestion system was presented by Gluodenis and Tyson [158]. Here, the PTFE tubing is loosely embedded in a resistively heated oven. By using PTFE tubing, the maximum digestion temperature is restricted to ca. 210 °C. The limited mechanical strength of the material merely allows maximum working pressures of up to 35 bar. Therefore, the usual working pressure is about 10–20 bar. The potential of the system was illustrated by the decomposition of cocoa powder slurries in 10 % HNO_3 injected into the manifold and decomposed under stopped-flow, medium-pressure conditions.

In a series of papers [159–161], Berndt described development of a new high-temperature/high-pressure flow system for the continuous decomposition of biological and environmental samples. It was shown [159] that temperature up to 260 °C and pressure up to 300 bar can be reached in a flow system when an electrically heated Teflon lined HPLC tube is used as the digestion capillary. Digested biological samples (blood, liver, leaves) were collected at the outlet of the flow system. In subsequent papers [160, 161], an electrically heated Pt/Ir capillary served as a digestion tube at temperatures of 320–360 °C and pressures of about 300 bar, and withstands concentrated acids. Due to the totally glass-free environment, samples having high silicate content can be digested by the addition of hydrofluoric acid.

c) UV online decomposition

UV decomposition is a clean sample preparation method as it does not require the use of large amounts of oxidants. Furthermore, UV decomposition is effective and can be readily incorporated into flow injection manifolds. The sample flows, in the presence of H_2O_2, H_2SO_4 or HNO_3, through a tube (PTFE, quartz) coiled around a fixed UV lamp(s). A short review of such flow systems has appeared recently [162]. Analyzers of this kind are produced by SKALAR Analytical, Holland, for example.

Fernandes *et al.* [163] developed a manifold based on a two-stage online UV/thermal-induced digestion procedure for oxidation purposes. The UV digestion apparatus consisted of a 4-m-long PTFE tube tightly wounded directly around the UV source (15 W) to form a spiral reactor. The thermal digestion apparatus consisted of a 2-m-long PTFE tube coiled in a helix and submerged in a thermostatic bath at 90 °C.

d) Microwave-assisted pressurized flow-through

Many different designs of microwave-assisted flow digestion systems have been published [89, 96, 162], which open up new possibilities, primarily in fully automated sample preparation for elemental analysis.

The earliest work reported in this field was Burguera *et al.* [164], who applied a flow injection system for online decomposition of samples and determined metals (Cu, Fe, Zn) by flame AAS. The methodology involved the synchronous merging of reagent and sample followed by decomposition of serum, blood, or plasma in a Pyrex coil located inside the microwave oven. This approach permits essentially continuous sample decomposition and drastically reduces sample processing time, and is suitable for those samples that require mild decomposition conditions (especially liquids).

According to the location of the digestion unit in the system, there are two types of manifolds described in the literature to date: before and after the injection unit. In the former arrangement, the sample is introduced into the microwave oven in a continuous flow [165] or a stopped flow mode [166]; after decomposition, the injected sample flows to the microwave oven unit together with the reagent(s) to be decomposed, and is then cooled and degassed prior to its delivery to the detector [167]. In both cases, the measurements can be performed partially or totally offline or online.

Solid samples call for more sophisticated flow systems because they need to be digested in the presence of highly concentrated acids, which rapidly destroy organic matrices. A first attempt aimed at simplifying manipulation of the digest was reported in 1988 [168]. Lyophilized, finely ground and weighed samples of liver and kidney were placed in test tubes together with mineral acids and the contents shaken before exposing them to microwave radiation to avoid violent reaction with abundant foam formation. The tubes were loaded into a covered Pyrex jar inside a domestic microwave oven operated for a specified time at a given power. Carbonell et al. [165] initiated the determination of metallic elements in solid samples using the slurry approach coupled with microwave oven digestion in a flow injection system for F-AAS determination of lead. Various natural samples (artichoke, chocolate, sewage sludge, tomato leaves), real and certified, were slurried in a mixture of HNO_3 and H_2O_2 using magnetic stirring, followed by continuous pumping around an open recirculating system, part of which (120 cm PTFE tubing) was located in a domestic oven.

A microwave-heated, flow-through digestion container (coiled Teflon tubing) was designed for a commercial (Prolabo A300) focused microwave system (instead of microwave oven) and applied to the online preparation of biological samples, including milk, blood, and urine [169].

For an extensive oxidation of organic sample constituents with nitric acid, temperature of more than 200 °C is necessary. The PTFE tubes used, however, cannot withstand the vapor pressure of the decomposition mixture at 200 °C or more. Thus, new alternatives had to be found to overcome this limitation. One way to increase the pressure resistance of the tubes is to wrap them with a plastic tape of high mechanical strength. Results from a digestion system (CEM SpectroPrep system) equipped with such tubes have been published [170]. A CEM SpectroPrep system was used at moderate powers to perform online decomposition of slurried samples of biological tissues (0.5 % m/v) and marine sediment (1 % m/v). The pressure thresholds of this system are near 25 bar. To achieve the desired temperatures of approximately 250 °C, however, it is necessary to be able to increase the pressure in the system up to 35 bar or so. A recently developed device enables the application of such high temperatures (250 °C) by means of a new pressure equilibrium system (with a pressure of 40 bar) [171]. The pressure equilibrium system keeps the pressure inside and outside the digestion tube (PTFE or PFA) equal, even for extremely fast oxidation reactions. The system's ability to handle only up to 1 % m/v slurries and lower slurry concentrations for biological materials restricts the type of sample that can be analyzed, unless the most sensitive elemental detection devices are used, such as ICP-MS. Therefore, Mason et al. [172] modified the SpectroPrep oven and developed a wide bore continuous-flow microwave digestion system for the determination of trace metals (Cd, Cr, Mn, Ni, Pb) following *aqua regia* extraction. The described system demonstrated an ability to cope with real soil samples ground to a larger particle size (250 μm) and slurried without the use of surfactants.

Perhaps, the current fascination for using microwave heating for online digestion has led to the introduction of commercial instruments based on this hybrid technique [103].

The advantages of microwave-enhanced flow systems basically include a significant reduction in sample preparation time, the ability to accomplish reactions that would normally be too dangerous in a closed vessel because of sudden increases in temperature and pressure, and the capability to handle transient or readily decomposed samples or intermediates. However, flow-through systems are a problem area because all samples must be homogeneous and small enough to pass through the tube, and the majority of samples requires some form of processing before they can be put into the tube.

1.5.1.3 Vapor-phase acid decomposition

An alternative approach to acid digestion of the sample matrix that prevents introduction of impurities exploits gas-phase reactions. In the past four decades, several novel approaches to sample digestion procedures have been considered using inorganic acid vapor produced in one vessel to attack and dissolve material in another. A review by Matusiewicz [173] summarized analytical methods based on vapor-phase attack in promoting the dissolution and decomposition of inorganic and organic materials prior to determination of their trace element content. This approach is currently used (in open, semi-closed, and closed systems) whenever applicable because digestion using gas-phase reagents is preferred to the solution.

The combination of hydrofluoric acid and nitric acid vapor as a digestion agent has proven effective in the preparation of samples for spectrographic determination of trace impurities in open system. Zilbershtein et al.[174] used this approach to dissolve silicon and to concentrate impurities on a PTFE sheet. The residue and PTFE sheet were transferred to a graphite electrode that subsequently served as one electrode of the dc arc for spectrographic trace analysis.

With respect to semiclosed systems, a PTFE apparatus generating HF vapor has been specifically designed to minimize contamination during trace-element determination of ultrapure silicon, quartz, and glass [175]. The sample is placed in a PTFE beaker mounted on a perforated PTFE plate that is kept above the level of liquid HF in the chamber. Thomas and Smythe [176] describe a simple all-glass apparatus for vapor-phase oxidation of up to 90 % of plant material with nitric acid. Addition of perchloric acid ensured fast and complete oxidation, and the presence of HNO_3 during the final $HClO_4$ oxidation step eliminated any danger of explosion. Klitenick et al. [177] used the same technique, with a simplified pressurized PTFE digestion vessel, for the determination of zinc in brain tissue.

Some materials may not be fully dissolved by acid digestion at atmospheric pressure. A more vigorous treatment involves bomb digestion in pressure vessels designed to incorporate the techniques of a closed pressure vessel and vapor-phase digestion in a single unit. This has the advantage of being easier to construct than the apparatus described in previous papers [174–177], and it requires considerably smaller volumes of acids. Heating can be accomplished in an ordinary oven (with conductive heating) or using a microwave field.

A predecessor of this concept of closed-vessel vapor-phase sample digestion was introduced by Woolley [178]. He described a low-temperature (up to 110 °C) and high-temperature (up to 250 °C) version of the apparatus. Each device consists of an airtight PTFE vessel containing two concentric chambers: an inner chamber that holds the sample cup and an outer chamber. Both vessels were designed for the digestion of high-purity glass using relatively impure solvent acids: a 50:50 mixture of concentrated HNO_3 and HF. A completely closed PTFE bomb or autoclave [179] has been developed with a temperature gradient for digestion of more difficult compounds, such as siliceous material. Marinescu [180] presented an interesting development in which the conventional single-sample pressure digestion bomb was converted for multisample vapor-phase digestion. A multiplace holder for field sampling was developed to fit directly into the digestion bomb. This technique has been used for organic and inorganic solid, semisolid, and liquid samples. Kojima et al. [181] modified a sealed PTFE bomb in which the dissolution of highly pure silica with HNO_3, HCl, and HF acid vapor was possible using a PTFE vial placed in a PTFE outer vessel. A laboratory-made high-pressure digestion bomb with a PTFE microsampling device was developed by Matusiewicz [182]. This simple and inexpensive apparatus was found to be convenient for treating a small number of samples and can be easily made by modifying available PTFE bombs. It should be noted that PTFE microsampling devices can be used for both vapor-phase digestion and discrete nebulization techniques in atomic spectrometry. Vapor-phase digestion in a closed system (bomb) of high-purity materials for spectrographic determination of trace elements is a convenient and useful technique [183]. The method uses graphite electrode with an enlarged cavity and excludes the use of a collector. A technique [184] has been developed that employs the vapor-phase acid generated in the quartz vessel of a commercial high-pressure, high-temperature digestion apparatus (High Pressure Asher HPA, Anton Paar, Austria). Small biological samples (50–165 mg) were digested in a mini-quartz sample holder (3.1 mL volume). When biological standard reference materials were digested at 230 °C and 122 bar, the residual carbon content (RCC) in the digested samples was less than 1.8 %.

Despite methodologies previously proposed for closed systems with conventional heating being successful, very few attempts to employ microwave power for vapor-phase digestions have been described. An early trial with a low-pressure microwave arrangement was unsatisfactory [185], although recently an interesting variant of the digestion vessel design has been proposed for dissolution and decomposition of samples [186]. The method developed was an extension of the acid vapor-phase thermal pressure decomposition of biological materials reported previously by Matusiewicz [187]. Microwave-assisted vapor-phase acid digestion employing a special PTFE microsampling cup, suitable for 250 mg subsamples of marine biological and sediment reference materials were digested with HNO_3 and HNO_3-HF, respectively, at a maximum pressure of ca. 14 bar [186]. Very recently, several papers [188–192] discussed the further application and evaluation of this pioneering concept of Matusiewicz [187], employing either commercial pressurized microwave digestion systems and quartz sample containers [188], quartz inserts [189, 190], TFM inner vessels [191], or focused microwave ovens operating at atmospheric pressure and PTFE microsampling cups [192].

To summarize this section, use of acid vapor-phase digestion and attack of some organic and inorganic matrices as a sample preparation method is a convenient and useful technique. Closed pressure systems are the technique of choice to avoid losses of elements by volatilization while still maintaining extremely low values for the blank (by application of isopiestic distillation of the reagents and technical grade acids).

1.5.1.4 Efficiency of wet decomposition procedures

Quality control is becoming increasingly more significant in analytical chemistry. However, it is presently applied primarily to measurement techniques and not to sample preparation. For quality control in sample decomposition, it is necessary to measure and record certain parameters exactly to be able to subsequently trace the course of the decomposition process.

In spite of that, complete decomposition of the sample is required to achieve reproducible and accurate elemental results by instrumental analytical methods. This is particularly the case for all voltammetric and polarographic determinations [193–196]. Interferences caused by incomplete decomposed organic compounds also occur, to a certain degree, when using atomic spectrometric methods such as AAS [197, 198], ICP-OES [199, 200], and ICP-MS [201, 202]. As noted earlier, nitric acid is the most frequently utilized sample dissolution medium. Unfortunately, the carbon contained in organic materials is only partly converted to CO_2 by HNO_3 at temperatures of up to 200 °C [14]. In these cases, extending the digestion time and increasing the quantity of nitric acid does not improve the extent of decomposition. In principle, temperature and digestion time ultimately determine the effectiveness of a digestion, with RCC serving as a useful measure of quantitative assessment [123, 203]; in other words, the highest temperatures are required to achieve a decomposition as complete as possible [204, 205]. It should be noted here that the usefulness of the decomposition technique should be judged not from a visual point of view because it often happens that a clear, colorless solution, indistinguishable from water still contains significant amounts of carbon. In closed systems, the pressure depends not only on the temperature but also on the type and quantity of the sample, the size of the vessel, and the nature and quantity of the decomposition reagent. This pressure is not responsible for the determination quality, but nevertheless it should be controlled automatically. Würfels et al. [206–209] described the extremely strong impact from residual organic compounds on elemental determinations by means of inverse voltammetry and demonstrated that a temperature of 300–320 °C is necessary for pressurized sample decomposition with pure nitric acid to obtain a solution containing less than 0.1 % carbon. Otherwise, trace elements cannot be determined with inverse voltammetry. This was confirmed by Wasilewska et al. [210], who showed that for complete oxidation of organic compounds with nitric acid, the decomposition temperature should be raised to 300 °C. The influence of the digestion equipment (either thermal or microwave) is negligible if the digestion time employed is long enough to reach the steady-state temperature. Sample digestion with nitric acid between 220 and 250 °C leads to RCCs in the low-percentage range.

The mode of heating of the digestion vessels is more and more supplanted by microwave technology; therefore, microwave-assisted wet decomposition is a frequently used sample preparation technique for trace element determinations in organic materials. Studies of the RCC as a measure of decomposition efficiency have been undertaken [185, 211–215]. Using gas chromatography, Stoeppler et al. [123] quantified the ashing ability of conventional pressurized decomposition. Differences between the carbon content in the original sample and the carbon converted to CO_2 showed that the investigated biological and environmental samples were not completely ashed with nitric acid. Würfels and Jackwerth [216] determined the residual carbon in samples decomposed under pressure or evaporated with HNO_3. In most cases, microwave decomposition of biological material was incomplete. Subsequently, the undigested compounds were identified [207]. Parallel to Würfels and Jackwerth's studies [216], the residual organic species in nitric acid digests of bovine liver were identified by Pratt et al. [217]. Kingston and Jassie [218] evaluated the decomposition of several biological and botanical samples wet decomposed with HNO_3. Free amino acid concentrations of human urine samples were typically reduced by a factor of 10^5. This reflects the comparative efficiency of protein hydrolysis, and is not necessarily equivalent to the total carbon oxidation efficiency. Nakashima et al. [213] investigated the digestion efficiency of HNO_3–$HClO_4$ mixtures. The total RCC in a number of decomposed marine biological reference material (NRCC TORT-1) solutions was determined and used as a relative measure of the efficiency of various decomposition schemes. Two-stage microwave-assisted decomposition was superior to single-stage decompositions. However, even the two-stage procedures were not complete and 24 % carbon remained. The determination of residual carbon in digests of biological material with simultaneous ICP-OES analysis was described by Hee and Boyle [214] and Krushevska et al. [215]. The oxidation efficiencies of different dry and wet ashing procedures for milk samples were compared by Krushevska et al. [219], who noted that the residual carbon concentrations obtained with medium-pressure microwave-assisted decompositions varied between 5 and 15 %. Oxidizing mixtures of H_2O_2 or H_2SO_4 with HNO_3 applied in a medium pressure (11 bar) microwave system did not yield a decomposition efficiency higher than that for pure nitric acid. However, with the high-pressure/temperature-focused-microwave-heated TFM-Teflon bomb device, organic material is totally oxidized with nitric acid in a single-step procedure [153, 155]. Matusiewicz and Sturgeon [220] critically evaluated online and high-pressure/temperature closed-vessel techniques with regard to efficiency of decomposition. The completeness of destruction of biological materials (standard and certified reference materials) was characterized in terms of their RCC in the solution following digestion. Pressurized decomposition in a TFM-Teflon vessel was the most effective procedure (organic material was totally oxidized with nitric acid in a single-step procedure), whereas urine and sewage plant effluent were incompletely decomposed (between 56 and 72 %) with online microwave-heated decomposition using nitric acid, nitric acid and hydrogen peroxide, and peroxydisulfate oxidation. Very recently, the residual weight of a bottom antireflective coating (BARC) sample was successfully used as an indicator to evaluate the decomposition kinetics [221]. The weight degradation rate was independent of the sample weight under various temperatures, but strongly dependent on the digestion acid volume and the digestion temperature. Mathematical modeling for prediction of decomposition efficiency for the BARC sample was achieved by employing decomposition kinetics as the backbone.

Hydrogen peroxide is a very popular oxidizing reagent as it is converted to water and oxygen during the oxidation of biological material [185, 222–224]. However, an experiments with HNO_3–H_2O_2 mixtures conducted by Matusiewicz et al. [222] showed that all versions of pressurized microwave-assisted digestion with HNO_3 and H_2O_2gave an incomplete decomposition. No significant improvement in the efficiency was achieved with 50 % H_2O_2. The extension of this observation to medium-pressure and high-temperature microwave heating provided verification of this observation [225]. Nitric acid digestion with the addition of H_2O_2 did not enhance decomposition efficiency in this study compared to the use of only HNO_3. Thus, an alternative oxidizing reagent is desirable to completely and safely decompose organic carbon residues. It was found that ozone is very effective in destroying natural organic compounds [226–228] and has the potential to be used as an additional decomposition and/or finishing reagent [229].

A single digestion procedure is often insufficient for the complete decomposition of a complex matrix, leading some authors to recommend a combination of two or more techniques. Two examples will suffice to illustrate the principle [156, 194]. First, pressure digestion followed by UV photolysis. Thus, it has been shown that analysis of olive leaves for heavy metals by voltammetric methods leads to distorted results after "pressure digestion" alone. Reliable data can be obtained only by supplementing the digestion with UV irradiation to ensure adequate decomposition of the matrix [194]. Second, a novel microwave-assisted high-temperature UV digestion procedure was developed for the accelerated decomposition of interfering dissolved organic carbon prior to trace element determination in liquid samples. This new technique significantly improved the performance of the process of UV decomposition (oxidation) and is especially useful for ultratrace analysis due to its extremely low risk of contamination [156, 230].

In order to investigate the completeness of dissolution of inorganic materials, the recovery (or incomplete recovery) and accuracy of major, minor, and trace element determinations are usually applied. If silicates are present, which is usually the major inorganic component of many matrices (i.e., soils, sediments, sludges, ceramics, and other similar samples), the use of HF to achieve complete dissolution is mandatory [231, 232].

1.5.1.5 Comparison of wet decomposition techniques

A careful comparison of several digestion techniques is the only way of assuring accurate results, particularly when little experience is available with respect to the digestion of a specific matrix, or existing reports are contradictory. The analyst must choose the sample preparation technique carefully to ensure that the system is optimal for the analyses at hand. However, there is still no universal sample preparation system. With respect to requirements specific to contamination or losses through volatilization or retention, convection-heated or microwave-assisted wet digestion, quartz-lined high-pressure wet digestion, UV digestion, and vapor-phase acid digestion seem to be the best choice. However, all of these techniques require considerable investment for apparatus. Digestion of samples in an open vessel presents a serious risk of significant analyte loss despite the use of a reflux condenser. As far as economic aspects are concerned (low procurement, short digestion time, high sample throughput), microwave-assisted wet digestion and especially microwave-assisted pressurized online digestion appear to rank high. According to completion of the digestion, complete degradation of many samples is achieved only through high-pressure, high-temperature Teflon- or quartz-lined pressure vessel digestion, or by combination of a closed wet digestion system with UV irradiation.

Table 1.5 summarizes the advantages and disadvantages of the wet digestion techniques discussed in Section "Wet decomposition" with respect to losses of analytes, blank levels, contamination problems, sample size, digestion time, degree of digestion, and economic aspects.

1.5.1.6 Decomposition systems

Presently, the instrumentation market offers many devices to make wet decomposition more efficient and easier to manage by means of possible automation, but this is achieved principally with microwave energy.

Wet decompositions in open vessels are undertaken with or without refluxing. Because it is very critical to adhere very closely to the optimized time and temperature digestion parameters, mechanization of the digestion not only leads to higher sample throughput with less human intervention but also to the avoidance of errors. The simplest form of mechanization can be implemented through a time (programmable timer) and temperature (via an autotransformer) controlled heating block. There are many models of heating blocks on the market. A greater degree of mechanization would also incorporate control of reagent reflux during digestion.

These procedures operate batchwise. Continuous sample handling has some advantages over discontinuous handling; the former generally better matches analytical needs. The automated wet digestion device (VAO, Anton Paar, Austria) is such a continuously operating digestion system, and an ideal instrument for laboratories requiring high throughput of similar samples with which all methods of wet chemical decompositions can be performed [233]. With the help of a microprocessor, all important digestion parameters are controlled. Automation controls the time-temperature/pressure program for sample digestion, so that different sample materials can be processed under optimum conditions. The loading or charging of the high-pressure asher with sample material is achieved manually. A fully automated version of this high-pressure asher is not available. Berghof pressure digestion systems [234] serve for sample preparation of inorganic and organic matrices at high temperature (max. 200–250 °C) and high pressure (max. 100 and 200 bar) in pure, isostatically pressed PTFE or quartz vessels.

As noted already, three basic types of microwave-assisted digestion systems have evolved: atmospheric pressure, elevated pressure (closed vessel), and flow-through, working in the two common modes: multimode cavity and focused-type (waveguide). Reviews of commercially available microwave-assisted digestion systems and vessels (summary of the vessels, ovens, and oven systems) are given in Refs. [103, 235] together with specifications and features for elevated pressure, atmospheric pressure, and flow-through units. The simplicity and efficacy of microwave digestion easily

Table 1.5: Advantages and disadvantages of wet decomposition methods.

Decomposition technique	Possible ways of losses	Source of blank	Sample size (g)		Maximum		Decomposition time	Degree of decomposition	Economic aspects
			Organic	Inorganic	Temperature (°C)	Pressure (bar)			
Open systems									
Conventional heating	Volatilization	Acids, vessels, air	<5	<10	<400		Several hours	Incomplete	Inexpensive, needs supervision
Microwave heating	Volatilization	Acids, vessels, air	<5	<10	<400		<1 h	Incomplete	Inexpensive, needs supervision
UV decomposition	None		Liquid		<90		Several hours	High	Inexpensive, needs supervision
Ultrasound-assisted decomposition	Volatilization	Acids, vessels, air					Several minutes	Incomplete	Inexpensive, needs supervision
Closed systems									
Conventional heating	Retention	Acids (low)	<0.5	<3	<320	<150	Several hours	High	Needs no supervision
Microwave heating	Retention	Acids (low)	<0.5	<3	<300	<200	<1 h	High	Expensive, needs no supervision
Flow systems									
Conventional heating	Incomplete decomposition	Acids (low)	<0.1 (slurry)	<0.1 (slurry)	<320	>300	Several minutes	High	Expensive, needs no supervision
UV online decomposition	Incomplete decomposition	None	Liquid		<90		Several minutes	High	Inexpensive, needs no supervision
Microwave heating	Incomplete decomposition	Acids (low)	<0.1 (slurry)	<0.3 (slurry)	<250	<40	Several minutes	High	Expensive, needs no supervision
Vapor-phase acid decomposition									
Conventional heating	None	None	<0.1	<0.1	<200	<20	<1 h	High	Needs no supervision
Microwave heating	None	None	<0.1	<0.1	<200	<20	<20 min	High	Needs no supervision

lends itself to automation and robotics. Systems have been developed that are capable of weighing samples, adding acids, capping and uncapping vessels, accomplishing microwave-assisted digestion, diluting digestates, transferring vessels, and even cleaning and reusing the vessels. Once such a system is operational, the only thing the analyst has to do is supply and place the representative sample(s) in locations recognized by the system and then initiate the controlling program.

1.5.1.7 Safety of acid decomposition

The reagents, instruments, and operations employed in the digestion of materials are potentially hazardous, even when used as directed. The operator must always be properly protected with a laboratory coat, gloves, and safety glasses or, better still, face protection. Some concentrated fuming acids (HF, HNO_3, HCl) are to be handled only in a well-ventilated hood. Oxidizing acids (HNO_3, $HClO_4$) are more hazardous than nonoxidizing acids (HCl, H_3PO_4, HF), being more prone to explosion, especially in the presence of reducing agents, such as organic matter. Perchloric acid is oxidizing only when it is concentrated and hot; it must never be brought into contact with organic matter unless diluted with nitric acid.

Acid digestion must be conducted in a fume cupboard with efficient scrubbers installed. The evaporation of perchloric acid is to be performed only in an appropriate stainless steel, stoneware, or PP hood, with washing facilities to eliminate any perchlorate deposit.

Great care should be taken when using "pressure digestion" methods. Pressure digestion vessels (bombs) contain the acid fumes and are useful for rapid, one-step digestions without losses. But, again, there are restrictions; in some reactions (especially spontaneous), potentially explosive gases are produced that exceed the safety limits of the vessels. For instance, nitric acid and especially the spontaneous HNO_3 and H_2O_2 decomposition of organic matter in a closed vessel may result in explosion due to unintended pressure build-up within the vessel. These systems produce high-pressure spikes, which can be avoided by decreasing the sample weight or by applying a gradual temperature increase.

Microwave-assisted sample digestion has its own safety requirements. As a result of the direct energy absorption and rapid heating, microwave techniques introduce unique safety considerations that are not encountered in other methods. Differences in conditions between traditional laboratory practices and microwave-implemented methods should be examined before microwave energy is applied to heat reagents or samples. An excellent summary of this aspects is extensively reviewed in Refs. [82, 83, 236].

1.5.1.8 Combustion

a) Combustion in open systems

Dry ashing. The term **dry ashing** is intended to encompass all processes based on gaseous or solid ashing reagents. Such a distinction relative to wet decomposition processes is not absolutely essential, but it does offer certain practical advantages. Strictly speaking, dry ashing refers to the oxidation (combustion) of a substance in air at a temperature of several hundred degrees Celsius, often in a muffle furnace or similar apparatus.

For samples that contain much organic matter, which are being analyzed for non-volatile metals, dry ashing is a relatively simple method of removing the organic matter that can be used for relatively large samples (2–10 g) and requires little of the analyst's time. Classical dry ashing relies on the pyrolysis and combustion of the organic sample in a muffle furnace or laboratory flame, with the oxygen in air at 400–600 °C to remove the organic constituents. The organic is converted into CO_2 and H_2O [237]. The resulting inorganic "ash" residue is generally soluble in dilute acid. Crucibles used for ashing are usually made of silica, quartz, porcelain, platinum, zirconium, or Pyrex glass.

Dry ashing is rarely applied anymore and has largely been replaced by wet decomposition (ashing) because it has several disadvantages, such as losses due to volatilization, very low ashing of some materials, difficult dissolution of ashed materials, and contamination.

Advantages of this method are that no reagents are used and little operator attention is required.

Recently, analytical instruments have been developed to dry ash samples based on thermal and microwave heating: dry mode mineralizer (TESSEK, APION A), microwave ashing furnace (MILESTONE MLS 1200 PYRO), and microwave ashing system (CEM, MAS 7000).

Low-temperature ashing. For the determination of volatile elements, such as Se, As, Sb, Cd, Zn, and Tl, in organic materials, a very gentle treatment is required. For this, a low-temperature (<200 °C) ashing with excited oxygen at a pressure of 1–5 Torr is very suitable [238].

The oxygen plasma can be produced either by a radio-frequency power supply or by microwave energy (up to 300 W at 13.5 MHz); created reactive oxygen species (free oxygen radicals and excited oxygen) reacts effectively with the organic sample surface (up to 2 g), thus forming an organic ash residue. The ash and the elements adsorbed onto the cooling finger are then solubilized by refluxing with acid. An advantages of this method is that the elements are obtained in a comparatively high concentration. This method can be used for sample preparation of all kinds of combustible solids such as wood, paper, coal, food, or polymers.

One commercial system based on this technique is available: the Cool Plasma Asher CPA-4 (Anton Paar)

Cool plasma ashing (Wickbold). The Wickbold combustion technique is very suitable for processing liquid combustible samples, which are hardly decomposed by other techniques such as petroleum products [239]. In the Wickbold combustion system, an oxygen-hydrogen flame is used to sample decomposition at high temperatures

(2,000 °C). Liquid samples are directly introduced to flame while for solid samples a preliminary pyrolysis step in a precombustion unit is necessary. After the combustion, the resulting products are condensed in a quartz surface and absorbed in a suitable solution.

A Wickbold combustion apparatus V5 (Heraeous) is commercially available.

b) Combustion in closed systems

Oxygen flask combustion (Schöniger). Combustion in an oxygen flask, or commonly called Schöniger technique, offers advantages when readily volatilized elements such as halogens, Se, S, P, B, Hg, As, and Sb, are to be determined. The combustion is performed with oxygen in a sealed container and the reaction products are absorbed in a suitable solvent before the reaction vessel is opened.

A simple apparatus for performing such oxidation has been suggested by Schöniger [240]. It consists of a flask (500–1,000 mL capacity) fitted with a ground glass stopper. Attached to the stopper is a platinum gauze basket that holds from 2 to 200 mg of sample. If the sample is a solid, it is wrapped in a piece of low ash-content filter paper. Liquid samples can be weighed into gelatin capsules that are also wrapped in a filter paper. A small volume of an absorbing solution is placed in the flask. During the combustion, the flask is inverted to prevent the escape of the volatile oxidation products. Subsequently, the vessel is opened and the resultant solution containing the analytes is removed and diluted for analysis. The necessary time for the decomposition using the oxygen flask is typically less than 10 min; in addition, materials and equipment are relatively inexpensive. However, the procedure needs a continuous attention of the analyst during the combustion and is commonly applied for processing only one sample each time.

The apparatus of Schöniger system is a commercially available, flask-type combustion apparatus HERAEUS MIKRO K.

Oxygen bomb combustion. Combustion bomb is the classical technique successfully applied for several matrices and analytes [241].

In this technique, the samples, as pellets, are introduced in an ignition cup with two platinum wires that are connected to two electrodes. About 10 mL of absorbing solution is added at the bottom of vessel that is made in stainless steel or covered with platinum. After closing, the system is pressurized with oxygen at 20–30 bar. Then, the ignition is performed by electric current and after the combustion the gaseous are absorbed. After cooling, the system is opened and the absorbing solution is removed. In the combustion bomb, a relatively high sample mass (>0.5 g) can be burnt with high decomposition efficiency and the procedure takes place in less than one hour.

Commercially available apparatus such as oxygen combustion bomb PAAR and pressure decomposition device SIEMENS BIOKLAV can be employed for sample preparation applications.

Combustion in a dynamic system (Trace-O-Mat). The dynamic system (also called Trace-O-Mat) was developed by Knapp et al. [242] in the early 1980s and allows the sample combustion with a minimum contamination, which enables not only combustion in a closed system but also further treatment. The Trace-O-Mat combustion unit VAE-II is commercially available.

The sample is burnt in a stream of pure oxygen in a system made entirely of quartz. An essential feature is the cooling system above the combustion chamber having a volume of only 75 mL, which is filled with liquid nitrogen and condenses all volatile traces together with the combustion products CO_2 and H_2O. After combustion, the nitrogen is evaporated and the residual ash and the condensed volatile elements are dissolved by refluxing in 1–2 mL of high-purity HCl or HNO_3 and are then collected in the reagent vessel placed below. Solid samples are pressed into pellets. For liquid organic samples, a special sample holder is available. Maximum sample amounts of 0.7–0.8 g and a minimum of 1 mL of acid result in solution with high element concentrations.

1.5.1.9 Fusion decomposition

Fusion (especially alkaline fusion) is a powerful technique especially both for organic matrices and those with a high silica and alumina content having relatively high trace element contents. In general, a salt fusion is performed by mixing a sample with salts, melting the mixture with heat, cooling it, and, finally, dissolving the solidified melt. The fusion-flux properties range from acidic to basic, according to the Lewis acid–base definition, in which an acid can accept and a base can donate an electron pair. The flux properties may also be termed as oxidizing or reducing.

While acid attack was a classical means of dissolving silicate samples, the use of lithium metaborate fusion was a new departure, originating in the work of Ingamells [243], who showed that a clear aqueous solution could be easily and quickly prepared from silicates. Fluxes will decompose most substances at the high temperature required for their use (500–1,000 °C) and the high concentration of reagent brought into contact with the sample. The sample in the form of a very fine powder is mixed with perhaps a tenfold excess of the flux in a graphite of platinum (sometimes nickel or zirconium) crucible. The crucible is then placed in a muffle furnace at 500–1,000 °C for a few minutes to several hours to give a "melt." After cooling, the melt is dissolved.

Different fluxes can be used. Basic fluxes employed for the attack of acidic materials include carbonates, hydroxides, peroxides, and borates. As acidic flux, pyrosulfates can be used. If an oxidizing flux is required, sodium peroxide can be used. As an alternative, small quantities of the alkali nitrates or chlorates can be mixed with sodium carbonate. Basic and acidic fluxes are respectively dissolved in an acid or basic medium.

The addition of fluxes increases the risk of raising the blank value, owing to the amount of flux required for a successful fusion. In addition, the final aqueous solution obtained from the fusion will have a high salt concentration, which may cause difficulties in the subsequent steps of the analysis. The high temperatures required for a fusion increase the danger of volatilization losses.

These disadvantages make fusion a less-than-ideal technique for extreme trace element determination. However, for the determination of major, minor, and even some trace elements in such matrices as fly ash, silicates, slags, and dust good results can be obtained.

1.6 Conclusions and future trends

Trace element research imposes a stringent discipline on its practitioners. Provided, of course, that one has modern equipment, mistakes today in analyzing trace elements are generally not so often caused by equipment. They occur, on the contrary, during preparation of the test material, especially when the sources of sampling, contamination, and converting solid or liquid samples to solutions are not recognized and done away with. The analyses must be performed under rigorously controlled conditions to protect the samples from artifacts due to the containers, the reagents, or the ambient air.

If we want to improve our understanding of the role of trace elements in nature, we will have to learn to work more precisely. Along with this, we must not put restraints on our work in microanalysis through contamination in our own laboratories. This requires a change in thinking and a change in the way we have worked up to now. We should take as an example modern surgeons who operate as germ-free as the present level of technology permits in sterile oxygen tents with helmets and breath removal suction – and we should in our own way with the help of modern methods try to carry out our analysis with as few impairing factors as possible.

The chief methods used for the decomposition of organic and inorganic samples have been evaluated. A brief summary of applications of these techniques to various sample matrices is presented in Table 1.6. The variety of approaches currently available for the decomposition of solid and liquid samples allows the most suitable method to be selected for each application, depending on both the matrix and type of analyte, and the subsequent steps to be developed in order to complete the analytical process. In spite of that, it is fair to point out that sample decomposition must not be looked at as an isolated step, but one that needs to be integrated into the entire analytical process.

Attention has been focused on decomposition at elevated temperature and pressure. High-pressure decomposition with its large decomposition temperature range is the most universal decomposition system at present. This is the technique of choice from the vast majority of both inorganic and organic materials. New ways to further increase the efficiency of sample preparation should continue with the development of hyphenated decomposition technique. A novel, microwave-assisted, high-temperature UV digestion for accelerated decomposition of dissolved organic

Table 1.6: Summary of applications of total wet decomposition procedures to the analysis of materials (determination of elements).

Material/matrix/sample	Required acid(s)[a]	Decomposition technique (mode)[b]
Water(s)	H_2O_2, HNO_3	UV irradiation
Environmental samples		
Coal	HNO_3, HCl, HF	Open or closed system
Coal fly ash	*Aqua regia*[c] + HF[d]	Open or closed system
Dust	*Aqua regia* + HF	Open or closed system
Catalysts	*Aqua regia*	Open systems
Waste materials		
Sewage sludge	HNO_3, HCl	Open or closed or flow systems
Waste water	HNO_3	Flow systems
Botanical systems		
Botanicals	$HNO_3 + H_2O_2$ + HF	Open or closed system
Plants	$HNO_3 + H_2O_2$ + HF	Open or closed system
Clinical	HNO_3	Open or closed system
Marine	HNO_3	Open or closed system
Forensic	HNO_3	Open or closed system
Food(s)	HNO_3	Open or closed system
Beverages	HNO_3, H_2O_2	Open or closed or flow systems
Silicates		
Soils	*Aqua regia* + HF	Open and/or closed systems
Sediments	*Aqua regia* + HF	Open and/or closed systems
Glasses	HF	Open systems
Geological samples		
Rocks	*Aqua regia* + HF[e]	Open or closed systems
Ores	*Aqua regia* + HF	Open or closed systems
Minerals	HF + H_2SO_4, HCl	Open systems
Petroleum products		
Fuels	HNO_3+ HCl	Open or closed systems
Oils	HNO_3+ HCl	Open or closed systems
Drugs and pharmaceuticals	HCl, HNO_3	Open systems
Metals		
Ferrous	HNO_3+ (HF lub HNO_3	Open systems
Nonferrous	lub H_2SO_4)	Open systems
Alloys	HCl lub HNO_3lub HF	Open systems
Steels	woda królewska+ HF	Open systems
	HCl + HNO_3, $HClO_4$[f]	
Chemicals	HCl, HNO_3, HF, H_2SO_4	Open or closed systems
Polymers	HCl, HNO_3, HF, H_2SO_4	Open or closed systems
Refractory compounds[g]		
Ceramika	HNO_3, HCl, HF, H_2SO_4, H_2O_2	Open or closed systems
Kompozyty	HNO_3, HCl, HF, H_2SO_4, H_2O_2	Open or closed systems
Nuclear materials	HNO_3 or HCl, H_3PO_4, $HClO_4$	Open or closed systems

[a]Concentrated acids are usually employed; H_2O_2 is 30 %; in most cases alternative decompositions are possible depending on requirements of analyst.
[b]Conventional or microwave.
[c]Unstable.
[d]Use only Teflon vessels, the addition of HF is required to obtain quantitative recoveries for Cr.
[e]Addition of H_3BO_3 to neutralizm of HF by forming tetrafluoro-boric acid.
[f]Danger of explosion.
[g]Certain refractory materials are not decomposed; these must be solubilized by fusion.

compounds or slurries was developed [156]. This new technique is ideal for extreme trace analysis due to the low blank values and low acid concentration. In addition, this digestion method can be used for the determination of nonmetals by ion chromatography. Alternatively, within the limits of the Teflon-lined digestion vessels, improvement in the decomposition efficiency can be achieved by adding optimum concentrations of strong oxidizing agents, such as ozone or oxygen, which appear to be efficient decomposition agents for the treatment of biological material. Again, this has the advantage that the agent does not contribute to the analysis blank. It should be mentioned that vapor-phase acid digestion offers an alternative solution to these problems: reduced concentration of acid in the digestate and the possibility of using a technical grade acid without any deterioration of the analytical blank. Another example where significant improvement in decomposition and dissolution was obtained is the use of a reactor that combines microwave and ultrasound energy [103]. It is expected that these two methods could open a new research field **combined digestion techniques.**

It can be said with certainty that the majority of all digestions will be performed in the future by means of microwave assistance. Progress has been made over the past several years in reducing systematic errors and improving detection limits with microwave digestion, as well as its automation. A noticeable trend toward pressurized closed-vessel systems permitting high-temperature decomposition compatible with trace analysis has occurred. While some researchers advocate high-pressure (>100 bar) digestion at 250–300 °C to destroy interferences in refractory compounds, manufacturers are working to device sample vessels that can withstand these conditions.

There has been a growing trend in recent years toward the development of fully automated online analysis techniques. Microwave-assisted high-pressure flow digestion system with PTFE or PFA tubes for digestion temperatures up to 250 °C opens up new possibilities for fully automated sample preparation [103]. On the other hand, the development of new high-temperature/high-pressure flow digestion systems that incorporate resistively heated capillaries for the continuous digestion of various samples, coupled with atomic spectrometric instruments, has arisen [159–161]. It is predicted that flow systems will become dominant for liquid samples and slurries and extend the analytical capabilities of instrumental methods by combining sample preparation with simultaneous analysis using only micrograms of sample and microliters of reagents. The final goal of these studies should be the adaptation of standard batch digestion methods to online systems combining flow-through digestion directly to analyzers.

It is evident that wet decomposition methods will remain a fertile area for development. New digestion techniques need to be designed that address the limitations for the instrumentation and maximize its potential. Development trends for conventional and microwave instruments will focus on sample throughput, enhanced vessel performance specifications, the use of new materials, further refinement of *in situ* vessel control (direct temperature and pressure, incident and reflected microwave power), and computer-controlled sample digesters with automated capability.

Finally, the development of automated methods for wet decomposition of solid samples without human participation can only be achieved with the use of a robotic station [244]. Nevertheless, a number of auxiliary energies and commercially available modules can facilitate and/or accelerate one of the most time-consuming steps of the analytical process, that is, to obtain the analyte(s) from a solid sample in the form of a solution.

Funding

Financial support by the National Science Centre (NCN) Poland (grant no. UMO-2012/06/A/ST4/00382) is gratefully acknowledged.

References

[1] Thiers RE. Contamination in trace element analysis and its control. In: Glick D, ed. Methods of Biochemical Analysis. New York, Interscience, 1957, 274–83.
[2] Horwitz W. Nomenclature for sampling in analytical chemistry (Recommendations 1990). Pure Appl Chem 1990, 62, 1193–208.
[3] CITAC/EURACHEM, Guide to quality in analytical chemistry. An aid to accreditation. Teddington, 2002.
[4] Soeppler M. ed. Sampling and sample preparation. Practical guide for analytical chemists. Springer, Berlin, 1997.
[5] Anderson R. Sample pretreatment and separation. Chichester, John Wiley & Sons, Chichester, 1991.
[6] Esbensen KH, Wagner C. Theory of sampling (TOS) versus measurement uncertainty (MU) – A call for integration. Trends Anal Chem 2014, 57, 93–106.
[7] Gy PM. Sampling for analytical purposes. Chichester, John Wiley & Sons, 1998.
[8] Petersen L, Dahl CK, Esbensen KH. Representative sampling for reliable data analysis. Theory of sampling. Chemometr Intell Lab Syst 2005, 77, 261–77.
[9] Boutron CF. A clean laboratory for ultralow concentration heavy metal analysis. Fresenius J Anal Chem 1990, 337, 482–91.
[10] Bowen HJM. Trace Elements in Biochemistry. New York, Academic Press, 1966.
[11] Anand VD, White JM, Nino HV. Some aspects of specimen collection and stability in trace element analysis of body fluids. Clin Chem 1975, 21, 595–602.
[12] Cornelis R, Heinzow B, Herber RFM, Christensen JM, Poulsen OM, Sabbioni E, Templeton DM et al. Sample collection guidelines for trace elements in blood and urine. J Trace Elements Med Biol 1996, 10, 103–27.
[13] Kratochvil B. Sampling and sample preservation for trace element analysis. In: Mester Z, Sturgeon R, eds. Sample Preparation for Trace Element Analysis. Amsterdam, Elsevier, 2003, 1–21.
[14] Tölg G, Tschöpel P. Systematic errors in trace analysis. In: Alfassi ZB, ed. Determination of Trace Elements. Weinheim, VCH Verlagsgesellschaft mbH, 1994, 1–38.
[15] Knapp G. Systematic errors in decomposition methods. Trends Anal Chem 1984, 3, 182–91.

[16] Robertson DE. Contamination problems in trace-element analysis and ultrapurification. In: Zief M, Speights R, eds. Ultrapurity. Methods and Techniques. New York, Marcel Dekker, 1972, 207–53.
[17] Rodushkin I, Engström E, Baxter DC. Sources of contamination and remedial strategies in the multi-elemental trace analysis laboratory. Anal Bioanal Chem 2010, 396, 365–77.
[18] Kosta L. Contamination as a limiting parameter in trace analysis. Talanta 1982, 29, 985–92.
[19] Vandecasteele C, Block CB. Modern Methods for Trace Element Determination. Chichester, John Wiley & Sons, 1993.
[20] Zief M, Mitchell JW. Contamination Control in Trace Analysis. John Wiley & Sons, New York, 1976.
[21] Mitchell JW. Ultrapurity in trace analysis. Anal Chem 1973, 45, 492A–9A.
[22] Barnes RM, Santos Júnior D, Krug FJ. Introduction to sample preparation for trace element determination. In: de Moraes Flores ÉM, ed. Microwave-Assisted Sample Preparation for Trace Element Determination. Amsterdam, Elsevier, 2014, 1–58.
[23] Mitchell JW. State-of-the-art contamination control techniques for ultratrace elemental analysis. J Radioanal Chem 1982, 69, 47–105.
[24] Knapp G, Schramel P. Sources of analyte contamination and loss during the analytical process. In: Mester Z, Sturgeon R, eds. Sample preparation for trace element analysis. Amsterdam, Elsevier, 2003, 23–45.
[25] Hoffmann J. Experience with the sources of contamination when preparing samples for the analysis of trace elements. Fresenius Z Anal Chem 1988, 331, 220–2.
[26] Zief M, Michelotti FW. Clinical chemistry: A challenge for high-purity standards and reagents. Clin Chem 1971, 17, 833–40.
[27] Mitchell JW. Purification of analytical reagents. Talanta 1982, 29, 993–1002.
[28] Moody JR, Beary ES. Purified reagents for trace metal analysis. Talanta 1982, 29, 1003–10.
[29] Moody JR, Wissink CE, Beary ES. Design principles for a large high-efficiency sub-boiling still. Anal Chem 1989, 61, 823–7.
[30] van Belle JC, Boelrijk NAIM. Compact Teflon still for sub-boiling distillation of acids. Analyst, 1988, 113, 1145–7.
[31] Kuehner EC, Alvarez R, Paulsen PJ, Murphy TJ. Production and analysis of special high-purity acids purified by sub-boiling distillation. Anal Chem 1972, 44, 2052–56.
[32] Hughes RC, Mürau PC, Gundersen G. Ultra-pure water. Preparation and quality. Anal Chem 1971, 43, 691–6.
[33] Pickford CJ, Rossi G. Analysis of high purity water by flameless atomic-absorption spectroscopy. Part I. The use of an automated sampling system. Analyst 1972, 97, 647–52.
[34] Kim JI, Böhm G, Henkelmann R. Analysis of high-purity water: A direct determination of ultratrace inorganic impurities by monostandard neutron activation analysis. Fresenius Z Anal Chem 1987, 327, 495–8.
[35] Hoelzl R, Fabry L, Kotz L, Pahlke S. Routine analysis of ultra pure water by ICP-MS in the low- and sub-ng/L level. Fresenius J Anal Chem 2000, 366, 64–9.
[36] Yoshida C, Nomura A, Endou M. State-of-the-art on ultra-trace analysis of ultrapure water. Bunseki Kagaku 2010, 59, 349–56.
[37] Otten G. Measuring water purity by specific resistance. Int Lab 1972, 6, 31–5.
[38] Irving H, Cox JJ. The preparation of metal-free acids, alkalis and buffer solutions of high purity. Analyst 1958, 83, 526–8.
[39] Veillon C, Reamer DC. Preparation of high-purity volatile acids and bases by isothermal distillation. Anal Chem 1981, 53, 549–50.
[40] Chaurasia SC, Sahayam AC, Mishra RK. Room-temperature isopiestic distillation of in situ generated arsenious chloride and its application for the determination of trace level impurities in arsenious oxide. Anal Chem 2002, 74, 6102–5.

[41] Chaurasia SC, Sahayam AC, Mishra RK, Venkateswarlu G, Mukherjee T. A combined isopiestic and sub-boiling distillation method for the purification of hydrofluoric acid used for ICP-MS elemental analysis. At Spectrosc 2006, 27, 134–7.

[42] Sioda RE. Concentration limits of electrolytic preconcentration in instrumental analysis Anal Chem 1988, 60, 1177–9.

[43] Hoppstock K, Garten RPH, Tschöpel P, Tölg G. Purification of reagents and separation of element traces by electrodeposition onto a graphite tube cathode. Fresenius J Anal Chem 1992, 343, 778–81.

[44] Hoppstock K, Garten RPH, Tschöpel P, Tölg G. Purification of analytical reagents by constant-current electrodeposition of heavy metals at ultra-trace levels from highly concentrated salt solutions. Anal Chim Acta 1994, 294, 57–68.

[45] Tschöpel P, Tölg G. New Ways for the Analytical Characterization of High-purity Materials. J Trace Microprobe Techn 1982, 1, 1–.

[46] Friel JK, Mercer C, Andrews WL, Simmons BR, Jackson SE, Longerich HP. Laboratory gloves as a source of trace element contamination. Biol Trace Elem Res 1996, 54, 135–42.

[47] Salmela S, Vuori E. Contamination with cadmium from micropipette tips. Talanta 1979, 26, 175–6.

[48] Barnes RM, Quináia SP, Nóbrega JA, Blanco T. A fast microwave-assisted, acid-vapor, steam-cleaning procedure for autosampler cups. Spectrochim Acta Part B 1998, 53, 769–71.

[49] Szymczyk S, Cholewa M. Tools for clean preparation of biological samples (reference materials). Fresenius Z Anal Chem 1987, 326, 744–5.

[50] Gretzinger K, Kotz L, Tschöpel P, Tölg G. Causes and elimination of systematic errors in the determination of iron and cobalt in aqueous solutions in the ng/ml and pg/ml range. Talanta 1982, 29, 1011–8.

[51] Reh P, Gaede J. Difficulties in analyzing traces of chromium in nitric acid 100%. Fresenius J Anal Chem 1992, 343, 715–6.

[52] Bächmann K, Rudolph J. Gas chromatography of inorganic radioactive compounds. J Radioanal Chem 1976, 32, 243–63.

[53] Markert B. Sample preparation (cleaning, drying, homogenization) for trace element analysis in plant matrices. Sci Total Environ 1995, 176, 45–61.

[54] Maichin B, Kettisch P, Knapp G. Investigation of microwave assisted drying of samples and evaporation of aqueous solutions in trace element analysis. Fresenius J Anal Chem 2000, 366, 26–9.

[55] Sansoni B, Iyengar GV. Sampling and storage of biological materials for trace element analysis. Elemental analysis of biological materials: Current problems and techniques with special reference to trace elements (IAEA Technical Report 197), International Atomic Energy Agency, Vienna, Austria, 1980, 57–71.

[56] Kurfürst U. ed. Solid sample analysis – Direct and slurry sampling using GF-AAS and ETV-ICP. Springer Verlag, Berlin, 1998.

[57] Zeisler R. Reference materials for small-sample analysis. Fresenius J Anal Chem 1998, 360, 376–9.

[58] Iyengar GV, Kasperek K. Application of the brittle fracture technique (BFT) to homogenise biological samples and some observations regarding the distribution behaviour of the trace elements at different concentration levels in a biological matrix. J Radioanal Chem 1977, 39, 301–16.

[59] Koglin D, Backhaus F, Schladot JD. Particle size distribution in ground biological samples. Chemosphere 1997, 34, 2041–7.

[60] Nomura CS, Silva CS, Nogueira ARA, Oliveira PV. Bovine liver sample preparation and micro-homogeneity study for Cu and Zn determination by solid sampling electrothermal atomic absorption spectrometry. Spectrochim Acta Part B, 2005, 60, 673–80.

[61] Santos Jr D, Barbosa Jr F, Tomazelli AC, Krug FJ, Nóbrega JA, Arruda MAZ. Determination of Cd and Pb in food slurries by GF-AAS using cryogenic grinding for sample preparation. Anal Bioanal Chem 2002, 373, 183–9.
[62] Santos Jr D, Barbosa Jr F, de Souza SS, Krug J. Cryogenic sample grinding for copper, lead and manganese determination in human teeth by slurry sampling GFAAS. J Anal At Spectrom 2003, 18, 939–45.
[63] Lima EC, Krug FJ, Ferreira AT, Barbosa Jr. F. Tungsten-rhodium permanent chemical modifier for cadmium determination in fish slurries by electrothermal atomic absorption spectrometry. J Anal At Spectrom 1999, 14, 269–74.
[64] Fajgelj A, Zeisler R. Particle size determination of some IAEA and NIST environmental and biological reference materials. Fresenius J Anal Chem 1998, 360, 442–5.
[65] Gorsuch TT. The Destruction of Organic Matter. Pergamon Press, Oxford, 1970.
[66] Sansoni B, Panday VK. Ashing in trace element analysis of biological material. In: Fachetti S, ed. Analytical Techniques for Heavy Metals in Biological Fluids. Elsevier, Amsterdam, 1983, pp. 91–131.
[67] Subramanian KS. Determination of metals in biofluids and tissue: Sample preparation methods for atomic spectroscopic technique. Spectrochim Acta Part B 1996, 51, 291–319.
[68] Iyengar GV, Subramanian KS, Woittiez JRW. Sample decomposition. In: Element Analysis of Biological Samples. Principles and Practice. CRC Press, Boca Raton, 1998, 103–34.
[69] Matusiewicz H. Review of sample preparation methodology for elemental analysis of cellulose type materials. ICP Inf Newsletter 1999, 25, 510–4.
[70] Matusiewicz H. Solubilization: Trends of development in analytical atomic spectrometry for elemental food analysis. In: Caroli S, ed. The Determination of Chemical Elements in Food. Applications for Atomic and Mass Spectrometry. John Wiley & Sons, Chichester, 2007, 19–50.
[71] Šulcek Z, Povondra P, Doležal J. Decomposition Procedures in Inorganic Analysis. CRC Crit Rev Anal Chem 1989, 6, 255–323.
[72] Šulcek Z, Povondra P. Methods of Decomposition in Inorganic Analysis. CRC Press, Boca Raton, 1989.
[73] Povondra P, Šulcek Z. Modern methods of decomposition of inorganic substances. In: Zŷka, P. ed. Instrumentation in Analytical Chemistry. Ellis Horwood, New York, 1991, pp. 188–245.
[74] Chao TT, Sanzolone RF. Decomposition techniques. J Geochem Explor 1992, 44, 65–106.
[75] Bock R, A Handbook of Decomposition Methods in Analytical Chemistry. Translated and revised by Marr I.L, International Textbook Company, Glasgow, 1979
[76] Bajo S. Dissolution of matrices. In: Alfassi ZB, Wai CM, eds. Preconcentration Techniques for Trace Elements. CRC Press, Boca Raton, 1992, pp. 3–29.
[77] Vandecasteele C, Block CB. Modern Methods for Trace Element Determination. John Wiley & Sons, Chichester, 1993, pp. 9–52.
[78] Stoeppler M. ed. Sampling and Sample Preparation. Springer, Berlin, 1997.
[79] Hoenig M, de Kersabiec AM. Sample preparation steps for analysis by atomic spectroscopy methods: Present status. Spectrochim Acta Part B 1996, 51, 1297–307.
[80] Hoenig M. Preparation steps in environmental trace element analysis: Facts and traps. Talanta 2001, 54, 1021–38.
[81] Krakovská E, Kuss HM. Rozklady v Analitickej Chémii. VIENALA, Košice, Slovakia, 2001 (In Slovak)
[82] Matusiewicz H. Wet digestion methods. In: Mester Z, Sturgeon R, eds. Sample Preparation for Trace Element Analysis. Elsevier, Amsterdam, 2003, pp. 193–233.
[83] Arruda MAZ. ed. Trends in Sample Preparation. Nova Science Publisher, Inc, New York, 2007.
[84] de Oliveira E. Sample preparation for atomic spectroscopy: Evolution and future trends. J Braz Chem Soc 2003, 14, 174–82.

[85] Korn MGA. Sample preparation for the determination of metals in food samples using spectroanalytical methods. A review. Appl Spectrosc Rev 2008, 43, 67–92.

[86] Kingston HM, Jassie LB, eds. Introduction to Microwave Sample Preparation. Theory and Practice. ACS, Washington, DC, 1988.

[87] Kingston HM, Haswell SJ, eds. Microwave-Enhanced Chemistry. Fundamentals, Sample Preparation, and Applications. ACS, Washington, DC, 1997.

[88] White, Jr RT. Open reflux vessels for microwave digestion: Botanical, biological, and food samples for elemental analysis. In: Kingston HM, Jassie LB, eds. Introduction to Microwave Sample Preparation. Theory and Practice. ACS, Washington, DC, 1988.

[89] Mermet JM. Focused-microwave-assisted reactions: Atmospheric-pressure acid digestion, on-line pretreatment and acid digestion, volatile species production, and extraction. In: Kingston HM, Haswell SJ, eds. Microwave-Enhanced Chemistry. Fundamentals, Sample Preparation, and Applications. ACS, Washington, DC, 1997.

[90] Matusiewicz H, Sturgeon RE Present status of microwave sample dissolution an decomposition for elemental analysis. Prog Analyt Spectrosc 1989, 12, 21–39.

[91] Kuss HM. Applications of microwave digestion technique for elemental analyses. Fresenius J Anal Chem 1992, 343, 788–93.

[92] Zlotorzynski A. The application of microwave radiation to analytical and environmental chemistry. Crit Rev Anal Chem 1995, 25, 43–76.

[93] Smith FE, Arsenault EA. Microwave-assisted sample preparation in analytical chemistry. Talanta 1996, 43, 1207–68.

[94] Burguera M, Burguera JL. Microwave sample pretreatment in analytical systems. A review. Quim Anal 1996, 15, 112–22.

[95] Mamble KJ, Hill SJ. Microwave digestion procedures for environmental matrices. Analyst 1998, 123, 103R–33R.

[96] Burguera M, Burguera JL. Microwave-assisted sample decomposition in flow analysis. Anal Chim Acta 1998, 366, 63–80.

[97] Chakraborty R, Das AK, Cervera ML, de la Guardia M. Literature study of microwave-assisted digestion using electrothermal atomic absorption spectrometry. Fresenius J Anal Chem 1996, 355, 99–111.

[98] Kingston HM, Walter PJ. The art and science of microwave sample preparations for trace and ultratrace elemental analysis. In: Montaser A, ed. Inductively Coupled Plasma Mass Spectrometry. Wiley-VCH, New York, 1998, pp. 33–81.

[99] Meyers RA. Encyclopedia of Analytical Chemistry. John Wiley & Sons, Chichester, 2000.

[100] Bethge PO. An apparatus for the wet ashing of organic matter. Anal Chim Acta 1954, 10, 317–20.

[101] Abu-Samra A, Morris JS, Koirtyohann SR. Wet ashing of some biological samples in a microwave oven. Anal Chem 1975, 47, 1475–7.

[102] Nóbrega JA, Trevizan LC, Araújo GCL, Nogueira ARA. Focused-microwave-assisted strategies for sample preparation. Spectrochim Acta Part B 2002, 57, 1855–76.

[103] Matusiewicz H. Systems for microwave-assisted wet digestion. In: de Moraes Flores ÉM, ed. Microwave-Assisted Sample Preparation for Trace Element Determination. Elsevier, Amsterdam, 2014, 77–98.

[104] Golimowski J, Golimowska K. UV-photooxidation as pretreatment step in inorganic analysis of environmental samples. Anal Chim Acta 1996, 325, 111–33.

[105] Trapido M, Hirvonen A, Veressinina Y, Hentunen J, Munter R. Ozonation, ozone/UV and UV/H_2O_2 degradation of chlorophenols. Ozone Sci Eng 1997, 19, 75–96.

[106] Priego-Capote F, Luque de Castro MD. Analytical uses of ultrasound I. Sample preparation. Trends Anal Chem 2004, 23, 644–53.

[107] Seidi S, Yamini Y. Analytical sonochemistry; developments, applications, and hyphenations of ultrasound in sample preparation and analytical techniques. Cent Eur J Chem 2012, 10, 938–76.
[108] Mitscherlich A. Beitrage zur analytischen chemie. J Prakt Chem 1860, 81, 108–17.
[109] Carius GL. Über elementar analyse. Ber Dtsch Chem Ges 1870, 3, 697–702.
[110] Long SE, Kelly WR. Determination of mercury in coal by isotope dilution cold-vapor generation inductively coupled plasma mass spectrometry. Anal Chem 2002, 74, 1477–83.
[111] Jannasch P. Über die aufschliessung von silikaten unter druck durch konzentrierte salzsäure. Z Anorg Allgem Chem 1894, 6, 72–3.
[112] Jackwerth E, Gomišček S. Acid pressure decomposition in trace element analysis. Pure Appl Chem 1984, 56, 479–89.
[113] Ito J. A new method of decomposition for refractory minerals and its application to the determination of ferrous iron and alkalies. Bull Chem Soc Jpn 1962, 35, 225–8.
[114] Langmyhr FJ, Sveen S. Decomposability in hydrofluoric acid of the main and some minor and trace minerals of silicate rocks. Anal Chim Acta 1965, 32, 1–7.
[115] Bernas B. A new method for decomposition and comprehensive analysis of silicates by atomic absorption spectrometry. Anal Chem 1968, 40, 1682–6.
[116] Okamoto K, Fuwa K. Low-contamination digestion bomb method using a Teflon double vessel for biological materials. Anal Chem 1984, 56, 1758–60.
[117] Ravey M, Farberman B, Hendel I, Epstein S, Shemer R. A vessel for low-pressure acid dissolution of mineral and inorganic samples. Anal Chem 1995, 67, 2296–8.
[118] Uhrberg R. Acid digestion bomb for biological samples. Anal Chem 1982, 54, 1906–8.
[119] Kotz L, Henze G, Kaiser G, Pahlke S, Veber M, Tölg G. Wet mineralization of organic matrices in glassy carbon vessels in a pressure-bomb system for trace element analysis. Talanta 1979, 26, 681–91.
[120] Stoeppler M, Backhaus F. Pretreatment studies with biological and environmental materials I. Systems for pressurized multisample decomposition. Fresenius Z Anal Chem 1978, 291, 116–20.
[121] Kotz L, Kaiser G, Tschöpel P, Tölg G. Decomposition of biological materials for the determination of extremely low contents of trace elements in limited amounts with nitric acid under pressure in a Teflon tube. Fresenius Z Anal Chem 1972, 260, 207–9.
[122] Tomljanovic M, Grobenski Z. The analysis of iron ores by atomic absorption spectrometry after pressure decomposition with hydrofluoric acid in a PTFE autoclave. At Absorption Newslett 1975, 14, 52–6.
[123] Uchida T, Kojima I, Iida C. Determination of metals in small samples by atomic absorption and emission spectrometry with discrete nebulization. Anal Chim Acta 1980, 116, 205–10.
[124] Stoeppler M, Müller KP, Backhaus F. Pretreatment studies with biological and environmental materials III. Pressure evaluation and carbon balance in pressurized decomposition with nitric acid. Fresenius Z Anal Chem 1979, 297, 107–12.
[125] Takenaka M, Kozuka S, Hayashi M, Endo H. Determination of ultratrace amounts of metallic and chloride ion impurities in organic materials for microelectronics devices after a microwave digestion method. Analyst 1997, 122, 129–32.
[126] Lechler PJ, Desilets MO, Cherne FJ. Effect of heating appliance type on Parr bomb temperature response performance. Analyst 1988, 113, 201–2.
[127] Knapp G. Development of mechanized sample preparation for plasma emission spectrometry. ICP Information Newslett 1984, 10, 91–104.
[128] Knapp G, Grillo A. A high pressure asher for trace analysis. Am Lab 1986, 3, 1–4.
[129] Knapp G. Sample preparation techniques: An important part in trace element analysis for environmental research and monitoring. Intern J Environ Anal Chem 1985, 22, 71–83.

[130] Strenger S, Hirner AV. Digestion of organic components in waste materials by high pressure ashing with infrared heating. Fresenius J Anal Chem 2001, 371, 831–7.
[131] Hale M, Thompson M, Lovell J. Large batch sealed tube decomposition of geochemical samples by means of a layered heating block. Analyst 1985, 110, 225–8.
[132] Eggimann DW, Betzer PR. Decomposition and analysis of refractory oceanic suspended materials. Anal Chem 1976, 48, 886–90.
[133] Adrian WJ. A new wet digestion method for biological material utilizing pressure. At Absorption Newslett 1971, 10, 96.
[134] Kuennen RW, Wolnik KA, Fricke FL, Caruso JA. Pressure dissolution and real sample matrix calibration for multielement analysis of raw agriculture crops by inductively coupled plasma atomic emission spectrometry. Anal Chem 1982, 54, 2146–50.
[135] May K, Stoeppler M. Pretreatment studies with biological and environmental materials IV. Complete wet digestion in partly and completely closed quartz vessels for subsequent trace and ultratrace mercury determination. Fresenius Z Anal Chem 1984, 317, 248–51.
[136] Matusiewicz H. Modification of construction of Teflon digestion bomb. Chem Anal (Warsaw) 1983, 28, 439–52.
[137] Matthes SA, Farrell RF, Mackie AJ. A microwave system for the acid dissolution of metal and mineral samples. Tech Prog Rep US Bur Mines 1983, 120, 1–9.
[138] Smith F, Cousins B, Bozic J, Flora W. The acid dissolution of sulfide mineral sample under pressure in a microwave oven. Anal Chim Acta 1985, 177, 243–5.
[139] Buresch O, Hönle W, Haid U, v. Schnering HG. Pressure decomposition in the microwave oven. Fresenius Z Anal Chem 1987, 328, 82–4.
[140] Alvarado J, León LE, López F, Lima C. Comparison of conventional and microwave wet acid digestion procedures for the determination of iron, nickel and vanadium in coal by electrothermal atomization atomic absorption spectrometry. J Anal At Spectrom 1988, 3, 135–8.
[141] Kojima I, Uchida T, Iida C. Pressurized microwave digestion of biological samples for metal determination. Anal Sci 1988, 4, 211–4.
[142] Levine KE, Batchelor JD, Rhoades CB, Jones BT. Evaluation of a high-pressure, high-temperature microwave digestion system. J Anal At Spectrom 1999, 14, 49–59.
[143] Xu L.Q, Shen WX. Study on the PTFE closed-vessel microwave digestion method in food elemental analysis. Fresenius Z Anal Chem 1988, 332, 45–7.
[144] Nicholson JRP, Savory MG, Savory J, Wills MR. Micro-quantity tissue digestion for metal measurements by use of a microwave acid-digestion bomb. Clin Chem 1989, 35, 488–90.
[145] Baldwin S, Deaker M, Maher W. Low-volume microwave digestion of marine biological tissues for the measurement of trace elements. Analyst 1994, 119, 1701–4.
[146] Reid HJ, Greenfield S, Edmonds TE, Kapdi RM. Reflux pre-digestion in microwave sample preparation. Analyst 1993, 118, 1299–302.
[147] Reid HJ, Greenfield S, Edmonds TE. Liquid nitrogen cooling in microwave digestion. Analyst 1993, 118, 443–5.
[148] Heltai G, Percsich K. Moderated pressure microwave digestion system for preparation of biological samples. Talanta 1994, 41, 1067–72.
[149] Pougnet M, Michelson S, Downing B. Microwave digestion procedure for ICP-AES analysis of biological samples. J Microwave Power Electromagn Energy 1991, 26, 140–5.
[150] Pougnet M, Downing B, Michelson S. Microwave irradiation systems for laboratory pressure vessels. J Microwave Power Electromagn Energy 1993, 28, 18–24.
[151] Légère G, Salin ED. Capsule-based microwave digestion system. Appl Spectrosc 1995, 49, 14A–20A.

[152] Légère G, Salin ED. Design and operation of a capsule-based microwave digestion system. Anal Chem 1998, 70, 5029–36.
[153] Matusiewicz H. Development of a high pressure/temperature focused microwave heated Teflon bomb for sample preparation. Anal Chem 1994, 66, 751–5.
[154] Matusiewicz H. New technology for in situ visualization, monitoring and controlling microwave chemical reaction progress using a focused microwave high pressure-temperature closed-vessel digestion system. Analyst 2009, 134, 1490–7.
[155] Matusiewicz H. Development of a high-pressure asher focused microwave system for sample preparation. Anal Chem 1999, 71, 3145–9.
[156] Florian D, Knapp G. High-temperature, microwave-assisted UV digestion: A promising sample preparation technique for trace element analysis. Anal Chem 2001, 73, 1515–20.
[157] Matusiewicz H, Stanisz E. Characteristics of a novel UV-TiO_2-microwave integrated irradiation device in decomposition processes. Microchem J 2007, 86, 9–16.
[158] Gluodenis TJ, Tyson JF. Flow injection systems for directly coupling on-line digestion with analytical atomic spectrometry. Part 1. Dissolution of cocoa under stopped-flow, high-pressure conditions. J Anal At Spectrom 1992, 7, 301–6.
[159] Gräber C, Berndt H. Development of a new high temperature/high pressure flow systems for the continuous digestion of biological samples. J Anal At Spectrom 1999, 14, 683–91.
[160] Haiber S, Berndt H. A novel high-temperature (360 °C) high-pressure (30 MPa) flow system for online sample digestion applied to ICP spectrometry. Fresenius J Anal Chem 2000, 368, 52–8.
[161] Jacob P, Berndt H. Online element determination in biological and environmental samples by flame AAS coupled with a high-temperature/high-pressure flow digestion system. J Anal At Spectrom 2002, 17, 1615–20.
[162] Burguera JL, Burguera M. Flow injection systems for on-line sample dissolution/decomposition. In: Sanz-Medel A, ed. Flow Analysis with Atomic Spectrometric Detectors. Elsevier, Amsterdam, 1999, 135–49.
[163] Fernandes SMV, Lima JLFC, Rangel AOSS Spectrophotometric flow injection determination of total phosphorus in beer using on-line UV/thermal induced digestion. Fresenius J Anal Chem 2000, 366, 112–5.
[164] Burguera M, Burguera J, Alarcón OM. Flow injection and microwave-oven sample decomposition for determination of copper, zinc and iron in whole blood by atomic absorption spectrometry. Anal Chim Acta 1986, 179, 351–7.
[165] Carbonell V, de la Guardia M, Salvador A, Burguera JL, Burguera M. On-line microwave oven digestion flame atomic absorption analysis of solid samples. Anal Chim Acta 1990, 238, 417–21.
[166] Karanassios V, Li FH, Liu B, Salin ED. Rapid stopped-flow microwave digestion system. J Anal At Spectrom 1991, 6, 457–63.
[167] Haswell SJ, Barclay D. On-line microwave digestion of slurry samples with direct flame atomic absorption spectrometric elemental detection. Analyst 1992, 117, 117–20.
[168] Burguera M, Burguera J, Alarcón OM. Determination of zinc and cadmium in small amounts of biological tissues by microwave-assisted digestion and flow injection atomic absorption spectrometry. Anal Chim Acta 1988, 214, 421–7.
[169] Martines Stewart LJ, Barnes RM. Flow-through, microwave-heated digestion chamber for automated sample preparation prior to inductively coupled plasma spectrochemical analysis. Analyst 1994, 119, 1003–10.
[170] Sturgeon RE, Willie SN, Methven BA, Lam WH, Matusiewicz H. Continuous-flow microwave-assisted digestion of environmental samples using atomic spectrometric detection. J Anal At Spectrom 1995, 10, 981–6.
[171] Pichler U, Haase A, Knapp G, Michaelis M. Microwave-enhanced flow system for high-temperature digestion of resistant organic materials. Anal Chem 1999, 71, 4050–5.

[172] Mason CJ, Coe G, Edwards M, Riby P. Development of a wide bore flow through microwave digestion device for the determination of trace metals in soil. Analyst 2000, 125, 1875–83.

[173] Matusiewicz H. A review of acid vapour-phase sample digestion of inorganic and organic matrices for elemental analysis. Spectroscopy 1991, 6, 38–46.

[174] Zilbershtein KhI, Piriutko MM, Jevtushenko TP, Sacharnova IL, Nikitina ON. Spectrochemical analysis of high purity silicon. Zavod Lab 1959, 25, 1474–6.

[175] Mitchell JW, Nash DL. Teflon apparatus for vapor phase destruction of silicate materials. Anal Chem 1974, 46, 326–8.

[176] Thomas AD, Smythe L. Rapid destruction of plant material with concentrated nitric acid vapour (vapour phase oxidation). Talanta 1973, 20, 469–75.

[177] Klitenick MA, Frederickson CJ, Manton WI. Acid-vapor decomposition for determination of zinc in brain tissue by isotope dilution mass spectrometry. Anal Chem 1983, 55, 921–3.

[178] Woolley JF. PTFE apparatus for vapour-phase decomposition of high-purity materials. Analyst 1975, 100, 896–8.

[179] Karpov YuA, Orlova VA. Analytical autoclaves. Vysokochist Veshchestva 1990, 2, 40–56.

[180] Marinescu DM. Multisample pressure decomposition in the vapor phase. Analusis 1985, 13, 469–70.

[181] Kojima I, Jinno F, Noda Y, Iida C. Vapour-phase acid decomposition of highly pure silicas in a sealed PTFE bomb and determination of impurities by "one-drop" atomic spectrometry. Anal Chim Acta 1991, 245, 35–41.

[182] Matusiewicz H. PTFE microsampling device for acid-vapour-phase decomposition in Teflon digestion bomb. Chem Anal (Warsaw) 1988, 33, 173–5.

[183] Pimenov VG, Pronchatov AN, Maksimov GA, Shishov VN, Shcheplyagin EM, Krasnova SG. Atomic emission spectrochemical analysis of high purity germanium with preconcentration of impurities by vapour phase autoclave sample decomposition in the electrode. Zh Anal Khim 1984, 39, 1636–1640.

[184] Amaarsiriwardena D, Krushevska A, Argentine M, Barnes RM. Vapour-phase acid digestion of micro samples of biological material in a high-temperature, high-pressure asher for inductively coupled plasma atomic emission spectrometry. Analyst 1994, 119, 1017–21.

[185] Matusiewicz H, Sturgeon RE, Berman SS Trace element analysis of biological material following pressure digestion with nitric acid-hydrogen peroxide and microwave heating. J Anal At Spectrom 1989, 4, 323–7.

[186] Matusiewicz H, Sturgeon RE, Berman SS. Vapour-phase acid digestion of inorganic and organic matrices for trace element analysis using a microwave heated bomb. J Anal At Spectrom 1989, 6, 283–7.

[187] Matusiewicz H. Acid vapour-phase pressure decomposition for the determination of elements in biological materials by flame atomic emission spectrometry. J Anal At Spectrom 1989, 4, 265–9.

[188] Amaarsiriwardena D, Krushevska A, Barnes RM. Microwave-assisted vapor-phase nitric acid digestion of small biological samples for inductively coupled plasma spectrometry. Appl Spectrosc 1998, 52, 900–7.

[189] Czégény Zs, Berente B, Óvári M, Garcia Tapia M, Záray Gy. Microwave-assisted vapor-phase acid digestion of cellulose nitrate filters for elemental analysis of airborne dust samples. Microchem J 1998, 59, 100–6.

[190] Eilola K, Perämäki P Microwave heated vapor-phase digestion method for biological sample materials. Fresenius J Anal Chem 2001, 369, 107–12.

[191] Han Y, Kingston HM, Richter RC, Pirola C. Dual-vessel integrated microwave sample decomposition and digest evaporation for trace element analysis of silicon material by ICPMS: Design and application. Anal Chem 2001, 73, 1106–11.

[192] Araújo GCL, Nogueira ARA, Nóbrega JA. Single vessel procedure for acid-vapour partial digestion in a focused microwave: Fe and Co determination in biological samples by ETAAS. Analyst 2000, 125, 1861–4.

[193] Würfels M, Jackwerth E, Stoeppler M. About the problem of disturbances of inverse voltammetric analysis after pressure decomposition of biological samples. Fresenius Z Anal Chem 1987, 329, 459–61.

[194] Hertz J, Pani R. Investigation of the completeness of digestion procedures prior to voltammetric trace metal analysis of olive leaves. Fresenius Z Anal Chem 1987, 328, 487–91.

[195] Schramel P, Hasse S. Destruction of organic materials by pressurized microwave digestion. Fresenius J Anal Chem 1993, 346, 794–9.

[196] Reid HJ, Greenfield S, Edmonds TE. Investigation of decomposition products of microwave digestion of food samples. Analyst 1995, 120, 1543–8.

[197] Nève J, Hanocq M, Molle L, Lefebvre G. Study of some systematic errors during the determination of the total selenium and some of its ionic species in biological materials. Analyst 1982, 107, 934–41.

[198] Welz B, Melcher M, Nève J. Determination of selenium in human body fluids by hydride generation atomic absorption spectrometry: Optimization of sample decomposition. Anal Chim Acta 1984, 165, 131–40.

[199] Knapp G, Maichin B, Baumgartner U. Interferences in ICP-OES by organic residue after microwave-assisted sample digestion. At Spectrosc 1998, 19, 220–2.

[200] Machat J, Otruba V, Kanicky V. Spectral and non-spectral interferences in the determination of selenium by inductively coupled plasma atomic emission spectrometry. J Anal At Spectrom 2002, 17, 1096–102.

[201] Ashley D. Polyatomic interferences due to the presence of inorganic carbon in environmental samples in the determination of chromium at mass 52 by ICP-MS. At Spectrosc 1992, 13, 169–73.

[202] Begerov J, Turfeld M, Duneman L. Determination of physiological palladium, platinum, iridium and gold levels in human blood using double focusing magnetic sector field inductively coupled plasma mass spectrometry. J Anal At Spectrom 1997, 12, 1095–8.

[203] Matusiewicz H, Golik B, Suszka A. Determination of the residual carbon content in biological and environmental samples by microwave-induced-plasma atomic emission spectrometry. Chem Anal (Warsaw) 1999, 44, 559–66.

[204] Carrilho ENVM, Nogueira ARA, Nóbrega JA, de Souza GB, Cruz GM. An attempt to correlate fat and protein content of biological samples with residual carbon after microwave-assisted digestion. Fresenius J Anal Chem 2001, 371, 536–40.

[205] Costa LM, Silva FV, Gouveia ST, Nogueira ARA, Nóbrega JA. Focused microwave-assisted acid digestion of oils: An evaluation of the residual carbon content. Spectrochim Acta Part B 2001, 56, 1981–5.

[206] Würfels M, Jackwerth E, Stoeppler M. Residues from biological materials after pressure decomposition with nitric acid. Part 1. Carbon conversion during sample decomposition. Anal Chim Acta 1989, 226, 1–16.

[207] Würfels M, Jackwerth E, Stoeppler M. Residues from biological materials after pressure decomposition with nitric acid. Part 2. Identification of the reaction products. Anal Chim Acta 1989, 226, 17–30.

[208] Würfels M, Jackwerth E, Stoeppler M. Residues from biological materials after pressure decomposition with nitric acid. Part 3. Influence of reaction products on inverse voltammetric element determination. Anal Chim Acta 1989, 226, 31–41.

[209] Würfels M. Voltammetric determination of metal traces in marine samples after nitric acid decomposition. Marine Chem 1989, 28, 259–64.

[210] Wasilewska M, Goessler W, Zischka M, Maichin B, Knapp G. Efficiency of oxidation in wet digestion procedures and influence from the residual organic carbon content on selected technique for determination of trace elements. J Anal At Spectrom 2002, 17, 1121–5.

[211] Matusiewicz H, Suszka A, Ciszewski A. Efficiency of wet oxidation with pressurized sample digestion for trace analysis of human hair material. Acta Chim Hung 1991, 128, 849–59.

[212] Krushevska A, Barnes RM, Amarasiriwaradena CJ, Foner H, Martines L. Comparison of sample decomposition procedures for the determination of zinc in milk by inductively coupled plasma atomic emission spectrometry. J Anal At Spectrom 1992, 7, 851–8.

[213] Nakashima S, Sturgeon RE, Willie SN, Berman SS. Acid digestion of marine samples for trace element analysis using microwave heating. Analyst 1988, 113, 159–63.

[214] Hee SSQ, Boyle JR. Simultaneous multielement analysis of some environmental and biological samples by inductively coupled plasma atomic emission spectrometry. Anal Chem 1988, 60, 1033–42.

[215] Krushevska A, Barnes RM, Amarasiriwaradena CJ. Decomposition of biological samples for inductively coupled plasma atomic emission spectrometry using an open focused microwave digestion system. Analyst 1993, 118, 1175–81.

[216] Würfels M, Jackwerth E. Investigations on the carbon balance in decomposition of biological materials with nitric acid. Fresenius Z Anal Chem 1987, 322, 354–8.

[217] Pratt KW, Kingston HM, MacCrehan WA, Koch W. Voltammetric and liquid chromatographic identification of organic products of microwave-assisted wet ashing of biological samples. Anal Chem 1988, 60, 2024–7.

[218] Kingston HM, Jassie LB. Microwave energy for acid decomposition at elevated temperatures and pressures using biological and botanical samples. Anal Chem 1986, 58, 2534–41.

[219] Krushevska A, Barnes RM, Amarasiriwaradena CJ, Foner H, Martines L. Determination of the residual carbon content by inductively coupled plasma atomic emission spectrometry after decomposition of biological samples. J Anal At Spectrom 1992, 7, 845–50.

[220] Matusiewicz H, Sturgeon RE. Comparison of the efficiencies of on-line and high-pressure closed vessel approaches to microwave heated sample decomposition. Fresenius J Anal Chem 1994, 349, 428–33.

[221] Ko FH, Chen HL. Study of microwave digestion kinetics and establishment of a model for digestion efficiency prediction. J Anal At Spectrom 2001, 16, 1337–40.

[222] Matusiewicz H, Barnes RM. Tree ring wood analysis after hydrogen peroxide pressure decomposition with inductively coupled plasma atomic emission spectrometry and electrothermal vaporization. Anal Chem 1985, 57, 406–11.

[223] Sah RS, Miller RO. Spontaneous reaction for acid dissolution of biological tissues in closed vessels. Anal Chem 1992, 64, 230–3.

[224] Veschetti E, Maresca D, Cutilli D, Santarsiero A, Ottaviani M. Optimization of H_2O_2 action in sewage-sludge miocrowave digestion using pressure vs. temperature and pressure vs. time graphs. Microchem J 2000, 67, 171–9.

[225] Matusiewicz H. Critical evaluation of the effectiveness of nitric acid oxidizing systems: Pressurized microwave-assisted digestion procedures. Chem Anal (Warsaw) 2001, 46, 897–905.

[226] Clem RG, Hodgson A. Ozone oxidation of organic sequestering agents in water prior to the determination of trace metals by anodic stripping voltammetry. Anal Chem 1978, 50, 102–10.

[227] Filipović-Kovačević, Ž, Sipos L. Voltammetric determination of copper in water samples digested by ozone. Talanta 1998, 45, 843–50.

[228] Sasaki K, Pacey GE. The use of ozone as the primary digestion reagent for the cold vapor mercury procedure. Talanta 1999, 50, 175–81.

[229] Jiang W, Chalk SJ, Kingston HM. Ozone degradation of residual carbon in biological samples using microwave irradiation. Analyst 1997, 122, 211–6.
[230] Matusiewicz H, Stanisz E. Evaluation of high pressure oxygen microwave-assisted wet decomposition for the determination of mercury by CVAAS utilizing UV-induced reduction. Microchem J 2010, 95, 268–73.
[231] Schramel P, Lill G, Seif R. A complete HF digestion method for element and trace element determination in soils, sediments, sludges and other similar samples using a closed system. Fresenius Z Anal Chem 1987, 326, 135–8.
[232] Matusiewicz H. High-pressure microwave dissolution of ceramics prior to trace metal determinations by microwave induced plasma atomic emission spectrometry. Mikrochim Acta 1993, 111, 71–82.
[233] Budna KW, Knapp G. Continuous decomposition of organic materials with H_2O_2-H_2SO_4. Fresenius Z Anal Chem 1979, 294, 122–4.
[234] Stainless steel pressure vessels with Teflon inserts, Berghof Laborprodukte GmbH, Eningen, Germany.
[235] Matusiewicz H. Development of high-pressure closed-vessel systems for microwave-assisted sample digestion in microwave-enhanced chemistry. In: Kingston HM, Haswell SJ, eds. Microwave-Enhanced Chemistry. Fundamentals, Sample Preparation, and Applications. ACS, Washington, DC, 1997, 353–369.
[236] de Moraes Flores ÉM. ed. Microwave-Assisted Sample Preparation for Trace Element Determination. Elsevier, Amsterdam, 2014.
[237] Jorhem L. Dry ashing, sources of error, and performance evaluation in AAS. Mikrochim Acta 1995, 119, 211–8.
[238] Gleit CE, Holland WD. Use of electrically excited oxygen for the low temperature decomposition of organic substances. Anal Chem 1962, 34, 1454–7.
[239] Wickbold R. Neue Schnellmethode zur Halogenbestimmung in Organischen Substanzen. Angew Chem 1952, 64, 133–5.
[240] Schöniger W. Eine Mikroanalytische Schnellbestimmung von Halogen in Organischen Substanzen. Mikrochim Acta 1955, 1, 123–9.
[241] Souza GB, Carrilho ENVM, Oliveira CV, Nogueira ARA, Nóbrega JA. Oxygen bomb combustion of biological samples for inductively coupled plasma optical emission spectrometry. Spectrochim Acta Part B 2002, 57, 2195–201.
[242] Knapp G, Raptis SE, Kaiser G, Tölg G, Schramel P, Schreiber B. A partially mechanized system for the combustion of organic samples in a stream of oxygen with quantitatively recovery of the trace elements. Fresenius Z Anal Chem 1981, 308, 97–103.
[243] Ingamells CO. Absorptiometric methods in rapid silicate analysis. Anal Chem 1966, 38, 1228–34.
[244] Luque de Castr MD, Luque Garcia JL. Automation of sample preparation. In: Mester Z, Sturgeon R, eds. Sample Preparation for Trace Element Analysis. Elsevier B.V., Amsterdam, 2003, 649–80.

Ewa Bulska and Anna Ruszczyńska

2 Analytical Techniques for Trace Element Determination

2.1 Introduction

A lot of elements occur in different matrices at low levels of content, and a lot of these elements were not detectable by analytical methods for a long time. The knowledge about their presence appeared with the development of analytical technology and caused the origin of the term "trace elements." Trace element defined by IUPAC [1] is any element having an average concentration of less than about 100 parts per million atoms or less than 100 mg/kg. In the second half of the 20th century, together with rapid increase of detection capabilities of analytical techniques, a new term of ultratrace elements appeared. Even though the term exists and is commonly used, there is no rigid definition. Ultratrace concerns elements at mass fraction below 1 ppm.

The knowledge of trace and ultratrace elements is very important in various fields of science, industry, and technology. Ultralow concentrations of elements might be as well essential as hazardous doses for organisms; some traces can dramatically change properties of designed devices. Therefore, the need for accurate measurements at low amount of contents is required and very important.

The common use of extremely sensitive instrumentation needs the adequate control of contamination and verification of the accuracy of determination. The gain of analytical sensitivity multiplied contamination as well as other problems. Therefore, correct precautions should be taken to determine trace elements in the parts per billion concentration range and below. Errors during trace and ultratrace elemental analysis can be caused by improper sampling, storage, sample preparation, and, finally, by analysis itself. Therefore, an accuracy of an analytical determination should be always established.

Collecting a representative sample without contaminating is a key to the meaningful analysis and Thiers' words from 1957 "unless the complete history of any given sample is known with certainty, the analyst is well advised not to spend his time analyzing it" [2] is always up to date.

Nowadays, there are a large number of available analytical techniques allowing for trace and ultratrace analysis of elemental composition. For the trace elements that are present in parts per million concentration range, the most widely used technique is probably atomic absorption spectrometry with flame atomization. For ultratrace elements present in concentration of parts per billion and below, the number of suitable techniques drops due to the required analytical sensitivity. The determination of trace elements is commonly held with potentiometry, voltammetry, atomic spectrometry, X-ray, and nuclear methods. Electrochemical methods can measure either free ions

https://doi.org/10.1515/9783110366730-003

in solution (potentiometry) or free ions together with ions bound in labile complexes (voltammetry), and they can also provide analysis of the oxidation state of some of the elements. Atomic spectrometric techniques are very sensitive and can be used to measure the total element content within a sample; however, accuracy of these techniques can be affected by the matrix of the sample. X-ray and nuclear techniques provide very low limit of detections and matrix insensitivity and are used for comparison of results due to their principles fundamentally different from those of the other analytical techniques. Therefore, they are less likely to be prone to the same systematic biases. Benefits and losses of each technique should concern the number of analytes possible to measure with the use of the technique, occurrence of interferences and difficulties, detection limits, throughput of samples, and expenses.

The determination of trace elements and contaminants in complex matrices often requires extensive sample preparation and/or extraction prior to instrumental analysis. A large number of samples that need to have determined the concentration of essential and toxic elements belong to food [3, 4], environmental [5, 6], clinical and biological [7–9] samples. Routinely, the determination of trace metals has been carried out by inductively coupled plasma atomic emission spectrometry (ICPAES), inductively coupled plasma mass spectrometry (ICPMS), electrothermal atomic absorption spectrometry (ETAAS), and flame atomic absorption spectrometry (FAAS). However, matrix of many samples (biological, clinical, environmental, etc.) is complex and consists of high amounts of soluble solid substances and large amounts of inorganic compounds (i.e., salts of Ca, K, Na, Mg, chlorides, phosphates, sulfates). The direct analysis of these kinds of samples gives many difficulties in sample introduction, nonspectral and spectral interferences in measurements by atomic absorption spectroscopic methods, inductively coupled plasma atomic emission or mass spectrometry. Therefore, samples have to be mineralized before analysis to destroy the organic matter or at least diluted to decrease the content of concomitant substances.

2.2 Spectroscopic techniques

2.2.1 Atomic absorption spectrometry

FAAS is one of the most conventional techniques for the determination of trace metal ions because of the relative simplicity and inexpensiveness of equipment. In this technique, a sample is introduced into a flame where it is dissociated into constituent atoms. Electromagnetic radiation in the UV/Vis part of the spectrum is directed through the flame and partially absorbed by the atoms. Methodology for most elements is well known and allows to use the FAAS technique to determine trace elements directly in various samples' materials [10–12]. However, in many cases the available analytical instrumentation does not have enough sensitivity for the analysis of natural samples and suffers from the matrix interferences. Several procedures have been developed for preconcentration and separation of traces of metals required prior to instrumental determination to lower the detection limits, to improve the precision and accuracy of analytical results, and to bring the analyte concentration into the dynamic range of the detector. The preconcentration methods like solvent extraction, ion exchange, adsorption, and coprecipitation were frequently used for trace analysis of lead, cadmium, copper, cobalt, chromium, nickel, tin, gold, palladium, iron, and zinc in different research materials [13–20]. The coprecipitation method is useful for the preconcentration of trace metal ions and is one of the most useful ways for the preconcentration as well as separation of trace elements from the sample matrix. For the determination of Cr(III), Cu(II), Fe(III), Pb (II), Pd(II), and Zn(II) in food samples, the use of Ni(II)-α-benzoin oxime as a coprecipitation agent can be successfully applied without too much prolongation of the procedure [21]. Liquid–liquid extraction transferring analyte from the aqueous sample to a water immiscible solvent is widely used for samples' preparation. Cloud point extraction (CPE), similar to liquid–liquid extraction, transferring analyte from the aqueous sample to a water immiscible solvent, is widely used for samples' preparation and coupled with AAS technique. CPE is based on the property of surfactants to form micelles, which under certain conditions (temperature and concentration) separate into two phases: a surfactant-rich phase of a small volume and a large aqueous phase. Hydrophobic complexes of metallic elements present in such media are trapped in the hydrophobic micellar core and extracted in the surfactant-rich phase, which is directed to AAS detector. The small volume of the surfactant-rich phase obtained after the CPE methodology seems to be ideal for coupling with electrothermal AAS, even though there are applications of CPE coupled with FAAS [16].

Nevertheless, the above-mentioned methods are time consuming and require at least some chemical additives and complex equipment. Miniaturization of liquid extraction methods can be achieved by a drastic reduction of the extractant phase volume by single-drop microextraction, hollow fiber liquid-phase microextration, and dispersive liquid–liquid microextraction (DLLM) allowing for separation and preconcentration of organic and inorganic contaminants in a single step. Due to the need of a sample volume between 2 and 4 mL for FAAS analysis, a microsample injection system (MIS) in case of small volumes obtained after preconcentrations methods might be applied [17, 22]. Online separation and preconcentration based on the adsorption of an analyte on an appropriate material representing the solid phase (SP) and then followed by an elution step directing the solution to the FAAS detector allows the use of wide range of sorbents, chelating agents, and eluents; it is also easily implemented and controlled. The analytes can be retained in their complexed or ionic forms on sorbents or functionalized with specific ligands. The online separation or preconcentration system requires stable material packed in minicolumns placed immediately after the injection valve or its sampling loop. Due to the high enrichment factor, high recovery, low cost, low consumption of organic solvents, and the ability to combine with different detection techniques, an SP extraction (SPE) has been commonly coupled with FAAS. The analytical parameters such as selectivity, affinity, and capacity depend on the sorbent chosen for SPE. Different materials can be used as sorbents for online preconcentration: modified silica gel, modified chitosan resins, chelating resins, magnetic nanoparticles, carbon materials, cellulose, and egg shell membrane.

Due to the high surface to volume ratio, easy derivatization procedures and unique thermal and mechanical stability carbon nanotubes (CNTs) have attracted a lot of attention. CNTs have diameters from fractions to tens of nanometers and lengths not bigger than several micrometers. The surface areas range from 150 to 1,500 m^2/g, which is a basis for serving as good sorbents. To improve selectivity, CNTs can be functionalized with different organic molecules. However, CNTs need to be modified with a specific ligand to improve the performance of sorbents by increasing the sorbent capacity and selectivity [15]. The chelating resins are superior in selectivity to solvent extraction and ion exchange due to their triple function of ion exchange, chelate formation, and physical adsorption. The functional group atoms capable of forming chelate rings usually include oxygen, nitrogen, and sulfur. The properties of selectivity and sorption of these resins can be affected by different factors: chemical activity of the complexing group, the nature and kind of the metal ions, the pH of the solution, ionic strength, or polymetric matrix [23].

Among other carbon materials, carbon dots (CDs) turned out to be a selective and sensitive method to separate and determine Cr in various samples. Due to the unique physical and chemical properties, CDs functionalized can facilitate the adsorption of analytes in consequence of electrostatic interaction, anion exchange, chelate interaction, or physical structure and can be employed in SPE as separation and preconcentration material in online or offline modes. Novel water-soluble CDs capped with branched polyethyleneimine polymer with dispersed particle extraction coupled with slurry sampling technique and followed by FAAS detection were employed on Cr(VI). CDs modified with cationic surfactant promoting small droplets generation during the aspiration and nebulization processes acted as a selective sorbent for separation and preconcentration of Cr(VI) enhancing the sensitivity of its determination [19].

An interesting solution for FAAS technique for elemental analysis is modification of instrumentation and the use of thermospray flame-furnace (TS-FF) AAS improving the efficiency of sample introduction. In this case of TS-FF consisting of a nickel tube, a sample solution is nebulized via a ceramic capillary to a standard burner head of an FAAS instrument, onto which the tube is placed [24]. Comparing to a standard FAAS systems, the TS-FF introduces a complete sample to the atomizer and provides a much longer residence time of the sample in the flame. As a result, the sensitivity of measurements may increase by an order of magnitude [25].

For the determination of elements forming hydrides or volatiles species (As, Bi, Ge, Pb, Se, Sb, Sn, Te, Hg, Cd, Co, Cu, Ag, Au), a chemical vapor generation (CVG) system is applied for various samples. The direct transfer of the volatile compounds to any atomizer eliminates the need of other steps but atomization, thus improving the sensitivity. In contrast to ETAAS, the FAAS technique is compatible with the online generation systems of volatile species. Analytes are easily preconcentrated before atomization by trapping directly on the flame atomizer. Even though there are solutions of coupling CVG for both techniques: FAAS and ETAAS [26].

ETAAS differs from FAAS by the use of much higher atomization temperatures, which reach up to 3,000 K. ETAAS is typically used for determining low concentrations of elements (e.g., Al, Ca, Co, Cr, As, Cd, Cu, Fe, Mn, Ni, Pb, Zn) [27–32] and can be applied without the need of earlier preconcentration of analytes. Measurements commonly apply to only one or two elements [33, 34]. However, relatively small quantities of solid and liquid samples may be analyzed. The technique, similarly to FAAS, has various types of interferences including background absorbance, matrix effect on atomic absorbance values, and differences of the chemical form in elements. There are various tools used to eliminate or to reduce these interferences: preliminary mineralization of a sample, separation of elements to be determined from interfering components, chemical modifiers, Zeeman effect background correction, and devices with separated zone of evaporation and atomization including graphite "filter furnace" (FF) atomizer and a L'vov platform.

Graphite is the most commonly used material in ETAAS. Due to the porosity of its surface, various species in the subsurface agglomerate in tubes are present. As a consequence, several corrosion changes have been observed when tubes were exposed to different compounds. Due to the surface and in-depth corrosion of tube and platform, their total lifetime can be significantly reduced. Usually, chemical modifiers are applied to the graphite furnace by adding to the solution with or after the sample or standards. But chemical modifiers can be also applied as a metal deposit on the graphite tube surface or on the L'vov platform acting as a permanent modifier and making the pyrolysis and atomization steps possible without repeating the treatment of the tube or platform. It was demonstrated that electrodeposited noble metals could serve as permanent modifiers by intercalation and remaining near the surface region of the tube for the total tube's lifetime. This observation developed permanent performance of various modifiers [35]. There are several permanent modifiers recommended – noble metals as well as their mixtures of less volatile metals: palladium, molybdenum, magnesium, molybdenum-iridium, molybdenum-ruthenium, vanadium, iridium, ruthenium, vanadium-iridium, vanadium-ruthenium, tungsten, tungsten-vanadium, and tungsten-magnesium [36]. In case of liquid samples, palladium and tungsten were used successfully [37–39], while the direct determination of antimony in solid samples was ineffective [40]. Problems in the determination of arsenic and antimony result from background spectral interferences and may be resolved by the selection of alternative analytical lines by the use of high-resolution-continuum source (HR-CS) ETAAS or by the application of a combination of Zeeman background correction together with the selection of an appropriate modifier.

In comparison with a platform, graphite FF atomizer provides increase in sensitivity. Filtration of atomized sample through the porous graphite improves performance of the FF atomizer because during this process molecular species are eliminated from the atomization volume. A wider range of sample volumes, which can be introduced into the graphite FF atomizer, provides additional advantage of this atomizer compared with the platform. The use of FF with Pd-Mg chemical modifier in the determination of Pb, As, and Cd during direct ET AAS analysis in various food samples provided a nearly twofold increase in sensitivity in comparison with a conventional heated graphite furnace with a platform. Additionally, it significantly eliminated matrix effects including background absorbance [29].

The main disadvantage of ETAAS and FAAS analysis of solid samples is the sample pretreatment, which is often the most time consuming and problematic (e.g., incomplete dissolution, precipitation of insoluble analyte, loss of elements during heating, contamination) step. The total concentration of analyte can be determined after acid digestion or alkali fusion. Microwave-assisted sample dissolution has been employed commonly for sample dissolution but barring some obvious advantages it still causes some problems (cost, short lifetime of digestion vessels, explosions, losses and contamination, long time of cooling, small sample throughput, corrosion of microwave parts, constant supervision during the digestion), which are exaggerated during trace elements analysis. The ideal method for the analysis of solid samples would eliminate the sample dissolution minimizing the sample preparation and improving the analytical results. Solid sampling can be applied to materials of different physical structures, whereas slurry sampling is dependent on the size and structure of the particles to be analyzed. In both cases, the use of an appropriate modifier mixture and optimized pyrolysis and atomization temperatures eventually avoids background absorption caused by the complex matrix. Slurry or solid sampling with ETAAS method has been extensively used for analysis of biological materials, sediments, and soils slurries in order to simplify sample preparation and to avoid problems associated with sample dissolution procedures [34, 41, 42]. The ETAAS method with solid sampling into a graphite furnace ensures the rapid and reliable determination of metals in soils or precipitates providing strict control of the sampling efficiency, particle size, number of particles present in the injected volume, analyte homogeneity, suspension medium, slurry concentration, stirring and sample depth. Analyses of environmental samples by solid sampling used either two-stage atomization with the use of permanent modifiers or slurry sampling. ETAAS analysis of volatile elements (as arsenic or antimony), which evaporate as oxides in the temperature of graphite furnance higher than 400°C, need the use of chemical modifiers (e.g., noble metals: palladium, nickel and magnesium nitrates, high melting of tungsten or zirconium carbides or mixtures of some of them) stabilizing analyte and facilitating the removal of the matrix by the increase of the temperature of atomization. Even ETAAS with the injection of sample suspensions into a graphite furnace ensures the rapid and reliable determination of metals; it is worth mentioning that it significantly shortens the life of tubes and lowers the sample throughput. The use of ultrasound-assisted extraction procedure can allow to avoid these problems because the sample matrix is not introduced into the atomizer, avoiding the buildup of carbonaceous residues or silicates on the graphite platform [43].

HR-CS AAS is an innovation that improves the performance of AAS. After the introduction of AAS technique in 1955 by Walsh, the use of spectral line instead of CS was necessary due to required spectral resolution, which was not achieved with the available monochromators at that time [44]. Together with efficient background corrector (like least-square background correction), it has been applied to the analysis of many analytes in a great variety of samples [45–47].

HR-CS AAS is a valuable tool due to the visibility of the spectral environment of the analytical line at high resolution. With the use of line source ETAAS with Zeeman effect background correction, a high signal indicating the presence of spectral interference for lead in fertilizer samples is observed. Due to this interference, there is no possibility in the determination of lead at 217.0 nm because of the presence of phosphorus monoxide (PO) under the magnetic field and splitting of the rotational lines of the molecular spectrum. The background absorption without and with magnetic field is not the same, resulting in background correction errors. HR-CS ETAAS is a tool that allows for investigation of these interferences due to the visibility of the spectral environment of the analytical line at high resolution. With the use of HR-CS ETAAS, it is possible to detect the presence of spectral interferences (S and N) and storage of the spectra of diatomic molecules with rotational fine structure that coincide temporally and spectrally with the analyte absorption [46]. To eliminate the fine structured background using least-squares background correction, it is mandatory to identify the molecule that is responsible for the spectral interference. Then, the reference spectrum of the interfering molecule is recorded and subtracted from the sample spectrum and a spectrum of pure analyte is obtained. The fine structured background depends on the chemical composition of each sample. In case of arsenic, the spectral interferences were corrected using a reference spectrum obtained from NaCl and PO, while in the case of selenium NO and PO reference spectra were used to correct the interferences [48]. In both cases, it was possible to obtain accurate results. Sulfur and nitrogen-containing molecules were responsible for the fine-structured background, which was completely corrected in case of lead determination. Using the most sensitive line at 217.001 nm, some unknown spectral interferences were observed. The use of HR-CS ETAAS made it possible to verify that the digestion of samples did not avoid the presence of spectral interferences since digested samples presented fine-structured background similar to the samples prepared as slurry. Comparing HR-CS ETAAS and line source ETAAS with Zeeman effect background correction, the results can be considered similar, indicating that the latter technique was able to correct the herein found spectral interferences to a reasonable extent [46].

2.2.2 Inductively coupled plasma atomic emission spectrometry

ICPAES, also referred to as inductively coupled plasma optical emission spectrometry (ICPOES), provides an excellent scope for the determination of trace elements with high sensitivity. This is due to very high temperatures (up to 8,000 K) of plasma used for atomization of analyte present in a sample. The ICP is created by argon gas, which is ionized in the intense electromagnetic field and flows in a particular rotationally symmetrical pattern toward the magnetic field of the radio-frequency coil. A stable plasma is generated as the result of the collisions between the neutral argon atoms and the charged particles. While the sample is introduced to the plasma, it immediately collides with the electrons and charged ions and is broken down into charged ions. Various molecules break up into their respective atoms, which then lose electrons and recombine repeatedly in the plasma. Excited atoms emit electromagnetic radiation at wavelengths characteristic of a particular element. The intensity of this emission is indicative of the concentration of the element within the sample and detected by a photomultiplier or a semiconductor detector.

Even ICPAES has similar limits of detection to FAAS; it can detect many elements simultaneously and has a much larger dynamic range. On the other hand, an ICPAES suffers from a lot of interferences and is much more expensive than the AAS techniques. Since the condition of ICP is changed by matrix elements from sample solution, signal intensities derived from analytical elements might be influenced. In analysis of high-viscosity solutions, injection volume of sample might be not constant, which may result in not reproducible analytical results or higher limit of detection. Therefore, the investigation of the influence of matrix elements and high viscosity of sample must be indispensable for accurate, sensitive, and reproducible determination by ICPAES. One of the solutions to reduce the influence of some matrix elements might be vaporization of Cl^- and CO_3^{2-} ions, decreasing of the viscosity of a sample, and lowering the pH of samples or decomposing organic compounds present in a sample. This could be obtained by the use of H_2SO_4-fume treatment [49].

Not without significance is the level of limit of detections that makes this technique not suitable for direct analysis of extremely low element levels. Therefore, prior to detection with ICPAES, an effective preconcentration step is required, similarly like in FAAS analysis described earlier [50, 51]. Various factors affecting the preconcentration process such as sample volume, concentration of the eluent, sample and eluent flow rates, as well as the accuracy of the method have to be always optimized and investigated in this case.

2.2.3 Inductively coupled plasma mass spectrometry

ICPMS is widely used in routine multielemental determination at the trace and ultratrace level in liquid samples with different matrix composition. The use of separation and enrichment techniques for analytes improves its limits of detection from the level of sub-μg/L even to sub-pg/L. Due to the excellent sensitivity, low detection limits, possibility of isotopic determination, and small sample volume ICPMS, wide dynamic range is widely used in clinical and biological [52–54], food [55, 56] environmental, geological, industrial analysis [57–59], and in a variety of different tasks [60]. Most of the elemental analysis with the use of ICPMS described in the literature concerns easy available materials [61–63], but a great number belongs also to the limited samples [64].

In the ICPMS technique, the sample is ionized in the same type of argon plasma as in the ICPAES technique. At the first stage of the process, the liquid sample is nebulized with an effective nebulizer transforming it into a fine aerosol, which is then transported with argon to ICP torch. In the plasma, nebulized water matrix and chemical compounds are evaporated, molecules dissociated into atomic constituents, and then ionized into positively single-charged ions. Ions are extracted from the argon plasma into mass analyzers: quadrupole (QMS), double focusing sector field (SFMS), and time of flight (TOFMS). In mass analyzer, ions are separated according to their mass-to-charge ratio or energy-to-charge ratio in double focusing SF instruments. The separated ion beams are detected by photomultiplier or Faraday cups.

Among a large variety of sample introduction systems developed for ICPMS, the most common and most economical is liquid solution nebulization. Therefore, there is a need of previous digestion of solid samples that is a very important stage for the whole analytical procedure. There are a lot of different liquid sample introduction systems developed [65–67]. The most frequently used for mineralized samples is pneumatic nebulizer (concentric, cross-flow, V-groove, sonic spray, or multi-microspray) combined with spray chamber (double pass, single pass, and cyclonic) with a solution uptake rate of 0.5–2 mL/min and very low transport efficiency (1–20 %) [68]. Higher sample introduction efficiency is provided by high-efficiency nebulizers like ultrasonic nebulizers [69, 70] or electrothermal vaporization, which allows for *in situ* sample preparation and preconcentration [71, 72]. Due to the trace element determination in micro- or nanoliters of sample, there are also micronebulizers available with solution uptake rate of 0.1 μL/min, which is a great advantage because of reduction of contamination problems (memory effects, deposition, clogging) or some interferences caused by solvents or possibility of coupling with techniques like electrophoresis requiring low sample consumption.

Besides the very high expenses, the ICPMS technique has a lot of advantages: sensitivity, very low limit of detections, throughput, multielemental analysis, and isotopic information, even though it suffers from atomic and molecular isobaric and multielemental interferences [73, 74]. This can be overcome simply by the choose of noninterfered isotope in case of multi-isotopic elements, by the subtraction of blanks, appropriate sample preparation [75], the use of mathematical correction [76], cold plasma conditions [77], by the use of collision or reaction cell technology [78–81], or by the use of high-resolution mass spectrometers that resolve elements and interferences. In order to overcome some physical interferences, an internal standard, standard addition method, the choice of sample introduction system, or simply dilution of the sample is frequently used [62, 82–84]. Extremely important for the achieved detection limits (LODs) as well as precision (RSD) of measurements is the kind of mass analyzer used in the ICPMS system. Samples with complexed matrix are the source of many interferences. Due to the low resolution commonly used, quadrupole analyzers have a lot of limitations compared to the systems with high-resolution mass analyzers [85–88].

Some of the most repeatable and accurate analytical measurements achieved today are thanks to the conjunction of an ICPMS with an isotopic dilution quantification methodology (Table 2.1).

2.2.3.1 Isotopic dilution inductively coupled plasma mass spectrometry (ID-ICP MS)

The capability of measuring precise and accurate isotope ratios by ICPMS is applied in the determination of element concentrations by the isotope-dilution (ID) technique, which is regarded as a primary method of measurements. ID ICPMS provides the possibility of absolute quantification for elements with two or more isotopes in any sample material. One or two enriched isotope tracers of the element of interest in known concentrations are added to the sample. The determination is performed after precise mixing by measuring changed isotope ratios in the sample spike mixture compared to those in the sample and highly enriched isotope tracer. The precision and accuracy of the trace element concentrations determined by ID ICPMS are in the low percentage range [72, 89–91]. Other advantages of the IDMS technique are no need of the preconcentration or dilution factor once complete isotope equilibration between the sample and the spike has been achieved and no influence of the instrumental sensitivity on the final value for the element concentration [92, 93]. IDMS is well documented for total elemental analysis.

2.2.4 Laser ablation ICPMS

Direct solid sampling is possible with an ICPMS system due to the laser ablation (LA) application. This technique is based on surface ablation of sample material by the use of focus laser beam. First, the sample is placed in a special ablation cell isolating it from the ambient environment. Then, the material is ablated and the formed dry aerosol is introduced to the plasma with the aid of a gas stream allowing for surface analysis or depth analysis of studied materials. Different ablation cells with different internal volumes and geometries mainly depending on the sample sizes influence the overall transport efficiency and signal profile. Volume of ablation cell affects mainly the dispersion of the signal [94].

LA ICPMS is not completely accepted for quantitative analyses mainly due to the fractionation effects and the persistent lack of adequate reference materials for the wide variety of samples [95, 96]. Laser wavelength, pulse duration, power, and spot size influence fractionation during the LA process. The size of the aerosol determinates the particle size distribution of the generated aerosol because of its chemical

Table 2.1: Overview of analytical performance of selected techniques used for trace elements determination.

Sample	Analyte	Technique	LOD [µg L^{-1}]	RSD [%]	Ref
Food, natural waters	Cr(III), Cu(II), Fe(III), Pb(II), Pd(II), Zn(II)	Coprecipitation FAAS	0.1–5.3	<5	20
Tap water, groundwater, industrial effluent	Cr(VI), Pb(II)	DLLM MIS-FAAS	0.037–0.054	<4	22
Fruit (apple, grape, nectarine, green beans, cantaloupe)	Cu(II)	Functionalized CNTs FAAS	0.65	3	15
Environmental (water samples)	Cr (VI)	CD FAAS	0.21	2.8	19
Water, biological (hair, nail), liver	Zn(II)	Flow injection FAAS	2.2	<1.2	23
Fruit juices, seawater	Cu	Chelating resign TS-FFAAS	0.15	2.7–6	25
Wine, beer, milk, kefir, yoghurt, juice, lemonade	Pb, As, Cd	FF-ETAAS	0.1–2.0	3–6	27
Juices, alcoholic beers	Sn(IV)	CPE FAAS	0.33	2.1–6.2	16
Vegetable oil, fat samples	Mn, As, Pb, Cu, Cd, Zn	FF ETAAS	4–0.8	3–8	28
Vinegar	Cd	ETAAS	0.004	1.07–2.33	32
Water	Multielemental	SPE ICPAES	0.01–0.39	1.2–2.2	50
Body fluids	Cr, Ni, Al, V	H_2SO_4 fume ICPAES	1–2	1	49
Body fluids	Al, Be, Cd, Co, Cr, Hg, Mn, Ni, Pb, V	SF ICPMS	0.001–0.05	1–5	61
River water	Pb, As, Cu, Hg	Stripping voltammetry	0.07–0.4	2–12	113
Tap water	As(III)	Boron-doped diamond stripping voltammetry	1.5	4.56	111
Freshwater and seawater	Ir(III)	Catalytic adsorptive stripping voltammetry	0.002	6	117
Atmospheric filters	Multielemental	TXRF	0.0003–0.009	–	123

composition, transport efficiency and decomposition in the ICP. The thermal character of the LA process might lead to the formation of agglomerates and molten spherical particles of different sizes in dependence of the laser wavelength.

Matrix effects may be induced in the ablation process, during aerosol transport to plasma, or in the ionization process in the plasma. Therefore, the calibration of LA ICPMS has to provide the capabilities to compensate these differences among samples and standards in order to obtain quantitative data. There are several calibration strategies for LA ICPMS; among them are internal standard method, external calibration, standard addition, isotope dilution, or matrix matched standards [96–98]. Additionally, the use of isotope dilution mass spectrometry in combination with LA ICPMS allows an accurate, precise, and time-effective quantitative analysis of trace elements in powdered samples using different isotope dilution calibration strategies.

LA ICPMS is becoming one of the most important direct analytical techniques for the rapid and sensitive determination of stable and radioactive isotopes on solid surfaces [99]. LA ICPMS avoids wet decomposition of a sample as well as the risk of contamination during sample preparation. In fact, it needs little or no sample preparation and offers a very good sample throughput and reduced spectral interferences. A significant feature of LA is the high spatial resolution between 10 and 100 μm for nanosecond lasers and below 1 μm for femtosecond lasers, with very low sample uptake of picograms. This microdestructive character of the technique is important in case of unique samples [100–103]. LA ICPMS has been used to produce images of element distribution in various materials, mainly in clinical and biological samples [104]. The challenging task of LA ICPMS for future applications is the analysis of a single cell or particle with interesting applications in life and material science.

2.3 Electrochemical techniques

Potentiometry using ion-selective electrodes (ISEs) and stripping voltammetry using polarographic techniques are two principal methods for trace detection.

Potentiometric sensors are passive electrochemical devices in which changes in the equilibrium voltage are monitored under zero current conditions. The equilibrium potential of the device is ideally described as a Nernstian function of the activity of the target ions to be measured.

ISEs are chemical sensors for the selective determination of ions in complex matrixes and have been applied for many years in clinical chemistry for the measurement of electrolyte concentrations in blood or plasma. Electrodes are available for the most important clinically relevant ions, for example, Na^+, K^+, Cl^-, Ca^{2+}, and H^+. Most of the potentiometric devices had limits of detections in the micromolar range, but numerous new developments of sensor arrays, new reference electrodes, polyion sensing, new ionophores, miniaturization for liquid chromatography, and capillary electrophoresis detectors significantly improved lower detection limits, broadening the range of application like trace metal analysis in environmental and biological samples [105–109].

The LOD for ISEs is defined as the activity at which the extrapolated Nernstian response, for the target ion, intercepts with the extrapolated potentiometric response from an infinitely dilute solution. Possible biases existing in ISE measurements come from the super-Nernstian or sub-Nernstian response of the ISE in the low concentration and from the presence of interfering ions in solution; the nonideal aspects of this technique can be eliminated by special electrode design, use of new ion-selective materials, and robust calibration of the system. The elimination of the effect of interfering ions is challenging by calibration since the composition of the calibration solutions and filling solutions may vary from the sample. In this case, improved electrode and materials design can provide better results.

Stripping voltammetry is a powerful analytical technique for trace metal detection. Its uniqueness comes from remarkably low detection limits, high sensitivity, and the ability to detect trace metal elements in different oxidation states. Another advantage that should be mentioned is portability of the instrumentation and the ability to perform rapid analysis as well as low costs for basic instrumentation and operation [110]. The problem that occurs is due to the common use of mercury film electrode that should be avoided because of its toxic soluble salts. Mercury electrode was substituted by the use of some nonmercurial electrodes, for example, electropolymerized polymer film, boron doped diamond, bismuth, antimony, and silver or gold electrodes [111–113]. Gold electrodes in different forms have been used to detect heavy metals by stripping analysis: solid disk electrodes, films on glassy carbon, graphite or platinum, microelectrodes, modified gold electrodes, gold nanoparticles, and screen-printed electrodes. Many voltammetric studies, including stripping methods, have focused on the modification of the electrode surface or use of a chelating/complexing agent to increase the detection limit. With this aim, different complexing have been used [114].

Stripping voltammetry consists of two steps. An electrolysis/deposition step where the analyte is accumulated at the surface of the working electrode during a certain period either by faradic process (anodic or cathodic) [110, 115] or nonfaradic (adsorptive) [116, 117] process. This preconcentration phase can be as long as required, depending on the concentration of analyte in solution. Low element concentrations will require long accumulation step to buildup enough material on the electrode. The accumulation increases with time, not with concentration of an element in the sample, which allows achieving very low detection limits. This first step of preconcentration is followed by a striping step removing the previously accumulated metal back into the solution by the means of applying an anodic potential sweep. The current generated during the stripping is directly proportional to the concentration of the metal present in the water sample. During the electrolysis, the analyte is reduced at the working electrode together with any other element that can be reduced at this deposition potential. If a low deposition potential is used in acidic conditions, hydrogen is formed at the working electrode surface by the reduction of protons and oxidants are produced at the auxiliary electrode. The hydrogen generation is problematic as it blocks the electrode surface, affects the reproducibility, and increases the noise of the voltammograms. The very few attempts that have been made to measure in such conditions were mostly directed toward the determination of As(V), which is only reducible in acidic conditions. To avoid hydrogen bubbles blocking the electrode, a rotating electrode was developed with the old disk placed on the side to facilitate the removal of hydrogen. The gold disk needs to be polished and cleaned after only 10 measurements. In this regard, the wire electrode has been more successful by using a gold film deposited on a platinum wire electrode. Stripping voltammetry is limited in acidic conditions to relatively high deposition potentials because of the interfering effects of the hydrogen produced at the working electrode. A simple procedure performing reliable, sensitive, and reproducible trace metal analysis and smooth stripping signals made at gold microwire electrode was obtained after applying a simple electrochemical conditioning procedure of using a vibrate electrode instead of stirred or stagnate conditions, which resulted in that hydrogen did not block the electrode in mild acidic conditions (pH <1) [118]. Vibrations at 200 Hz help in removing the hydrogen bubbles.

Determination of more than one metal using stripping analysis is difficult due to the possibility of overlapping peaks, which can occur in the narrow potential range where metals are reduced or oxidized. The multielement detection of As, Cu, Pb, and Hg was performed in the presence of oxygen by differential pulse anodic stripping voltammetry with the use of gold vibrating microwire electrode, which allowed to perform analysis without deoxygenation that simplified the use of stripping and decreased the measurement time [113]. The combination of a vibrator with a gold microwire electrode resulted in a very small stable and reproducible diffusion layer thickness, which improved the sensitivity and reproducibility of the results, allowing making determination with a very short deposition time. The use of HCl 0.1 M with NaCl 0.5 M was found optimum in terms of peak resolution, shape, and sensitivities, and additionally was found similar to the seawater properties, which raised the range of research objects. Possible effects of Al, Cd, Cr, Fe, Mn, Ni, Sb, and Zn in different concentrations from 1.0 µg/L to 1,000 µg/L on 1.0 µg/L of a mixture of As, Cu, Hg, and Pb did not cause any significant interferences. Arsenic was the element that suffered the most from these interferences; however, concentrations of interfering metals tested were very high compared to their expected levels in real samples.

Adsorptive processes can further lower detection limits of stripping voltammetry to the picomolar concentration levels and are very useful for saline matrices like seawater and estuarial water samples as well as for speciation studies. In adsorptive stripping voltammetry, the analyte is adsorbed on the working electrode by means of a nonelectrolytic process prior to the voltammetric scan. This procedure avoids the main difficulties of anodic stripping analysis with metals with a strong tendency to form intermetallic compounds, with low solubility on mercury electrode (e.g., Ni). For trace metals, the method is based on the formation and accumulation of a metal complex on the working mercury electrode and subsequent measurements of the reduction peak of the accumulated complex. The formation of a monomolecular complex layer on electrode increases the sensitivity of voltammetric measurement because the metal is not dissolved in the mercury [114, 116, 117, 119].

Despite many advantages of electrochemical detection, there is a great limitation in sample throughput and its size, which is at least 10 mL. Reducing sample size would ensure the extension to limitedly available samples, concurrently reducing reagent consumption and analysis time. Sample size reduction in voltammetry has been achieved by miniaturization and the development of dedicated electrodes. Microfluidic devices featuring solid-state microelectrodes have been developed significantly reducing sample requirement. Detection capabilities, precision, and accuracy are similar to typical solutions but application might be broadened to a range of limited availability of samples. The electrochemical route allows on-chip integration of the control instrumentation to produce self-contained truly portable microanalytical systems. The introduction of new modified electrodes, the development of novel derivatization schemes, the integration of additional functional elements (preconcentration) on a single microchip platform, and the coupling of new detection schemes and assays enhance the power of electrochemical detectors for microscale analytical systems [120, 121].

2.4 Other techniques used for trace analysis

2.4.1 Techniques uses X-ray

Due to its very low limit of detections and matrix insensitivity, X-ray methods have been widely used for trace analysis in geological materials, steels, cements, archaeological samples, forensic samples, and environmental samples such as airborne particulate matter [122]. A commonly used technique is X-ray fluorescence (XRF) used to analyze for all elements. The sensitivity of XRF depends on the energy of the incident radiation, the geometry of the instrument used, and the efficiency of the detector. The precision of XRF measurements is limited by the statistics of the detected photons. The LODs depend on the sensitivity of the instrument and the background level of the sample matrix. Typical XRF precision is better than 0.1 % and typical LODs are few g/cm^2 for particulate material for a wide range of elements.

In total reflection XRF (TXRF), the incoming radiation is incident on the sample at less than the critical angle and is totally reflected. The low background levels result in improved LODs and even 2 pg may be detected for a variety of elements with a counting time of 1,000 s. Using synchrotron radiation sources (SR), TXRF with polychromatic beam for sample excitation allows for lowering detection limits to even $0.03\ pg/m^3$ and the use of more intense photon fluxes will lower them by several orders of magnitude [123].

For these techniques, samples need to be in a form of thin film and that is a reason for low matrix effects conversely to the above-mentioned techniques. Quantification is performed by the addition of a single element, not present in the sample itself, as internal standard and a calibration curve valid in all matrices. The lack of robustness in these calibration methods leads to large systematic errors that are significant compared to the very low amounts being analyzed.

X-ray emission may be induced by heavy charged particles (particle-induced X-ray emission). Calibration is often by means of thin-film standards or using fundamental physical parameters in conjunction with an experimentally determined efficiency curve. LODs of ng/cm^2 have been claimed for particulate material on ambient air filters for a range of elements.

2.4.2 Activation analysis

Activation analysis refers to the identification and quantitative determination by use of radionuclides produced from a target element. It requires very little or no sample preparation, is nondestructive in nature and largely unaffected by the sample matrix, and allows for multielemental determination.

Neutron activation analysis (NAA) is the most common variant in which neutrons are used to irradiate and to activate the sample. When the measurement is carried out without prior chemical separation, the method is called instrumental NAA (INAA). As a result of a nuclear reaction between the neutron and the isotope of the element of interest, radionuclides with characteristic half-lives may be produced emitting radiation of varying energies that may be measured by a suitable detector and are characteristic of the element form that they were produced. In INAA, the decomposition of the radioactive sample is not necessary; the method is also nondestructive, which in case of unique objects is a valuable feature.

References

[1] IUPAC. Compendium of Chemical Terminology, 2nd ed. (the "Gold Book"). Compiled by A.D. McNaught and Wilkinson A, Blackwell Scientific Publications, Oxford 1997, XML on-line corrected version: http://goldbook.iupac.org (2006–2014) created by Nic M, Jirat J, Kosata B; updates compiled by A. Jenkins. ISBN 0-9678550-9-8. doi:10.1351/goldbook.

[2] Thiers RE. Contamination in trace analysis and its control. In: Glick D, ed. Methods of Biochemical Analysis. Volume V, Interscience Publishers Inc., New York, 1957, 273–337.

[3] Orecchio S, Amorello D, Raso M, Barreca S, Lino C, Di Gaudio F. Determination of trace elements in gluten-free food for celiac people by ICP-MS. Microchem J 2014, 116, 163–72.

[4] Nowak S, Gesell M, Holtkamp M, Scheffer A, Sperling M, Karst U, Buscher W. Low gas flow inductively coupled plasma optical emission spectrometry for the analysis of food samples after microwave digestion. Talanta 2014, 129, 575–8.

[5] Badocco D, Lavagnini I, Mondin A, Tapparo A, Pastore P. Limit of detection in the presence of instrumental and non-instrumental errors: study of the possible sources of error and

application to the analysis of 41 elements at trace levels by ICP-MS technique. Spectrochim Acta Part B 2015, 107, 178–84.
[6] Butler O, Cairns WRL, Cook J, Davidson CM. Atomic spectrometry update. Environmental analysis. J Anal At Spectrom 2013, 28, 177–216.
[7] Taylor A, Branch S, Day M, Patriarca M, White M. Atomic spectrometry update. Clinical and biological materials, foods and beverages. J Anal At Spectrom 2008, 23, 595–646.
[8] El Ati-Hellal M, Hellal F, Hedhili A. Application of Plackett–Burman and Doehlert designs for optimization of selenium analysis in plasma with electrothermal atomic absorption spectrometry. Clin Biochem 2014, 47, 95–100.
[9] Zawisza B, Pytlakowska K, Feist B, Polowniak M, Kita A. Sitko R, Determination of rare earth elements by spectroscopic techniques: a review. J Anal Atomic Spectrom 2011, 26, 2373–90.
[10] Waheed S, Rahman S, Siddique N. Calcium supplements as source of trace elements: adequacy and safety of supplements with vitamin C, vitamin D and phosphate formulations. Appl Radiat Isot 2014, 89, 134–40.
[11] Rahman S, Waheed S. Blood-cooper and zinc levels and consequences of cardiovascular complications: a study by INAA and FAAS. J Radioanal Nucl Chem 2011, 287, 657–64.
[12] Sidique N, Majid A, Chaudhry MM. Tufail M. Determination of heavy metals in air conditioner dust using FAAS and INAA. J Radioanal Nucl Chem 2012, 292, 219–27.
[13] Korn MGA, De Andrade JB, de Jesus DS, Lemons VA, Bandeira MLSF, dos Santos WNL, De Almeida Bezerra M, Amorin FAC, Souza AS, Ferreira SLC. Separation and preconcentration procedures for the determination of lead using spectrometric techniques: A review. Talanta 2006, 69, 16–24.
[14] Pyrzynska K, Kilian K. On-line sorption-based systems for determination of cadmium with atomic spectrometry detectors. Water Res 2007, 41, 2839–51.
[15] Rezvani M, Ebrahimzadeh H, Aliakbari A, Khalilzadeh A, Kasaeian M, Amini M. Novel modified carbon nanotubes as a selective sorbent for preconcentration and determination of trace copper ion in fruit samples. J Sep Sci 2014, 37, 2559–65.
[16] Gurkan R, Altunay N. Determination of total Sn in some canned beverages by FAAS after separation and preconcentration. Food Chem. 2015, 177, 102–10.
[17] Ozdemir C, Sacmci S, Kartal S, Sacmaci M. Determination of gold and palladium in environmental samples by FAAS after dispersive liquid-liquid microextraction pretreatment. J Ind Eng Chem 2014, 20, 4059–65.
[18] Pourjavid MR, Arabieh M, Yousefi SR, Jamali MR, Rezaee M, Hosseini MH, Sehat AA. Study on column SPE with synthesized graphene oxide ad FAAS for determination of trace amount of Co(II) and Ni(II) ions in real samples. Mater Sci Eng C 2015, 47, 114–22.
[19] Liu Y, Hu J, Li Y, Wei HP, Li Zhang X-S, Chen SM, Chen XQ. Synthesis of polyethyleneimine capped carbon dots for preconcentration and slurry sampling analysis of trace chromium in environmental water samples. Talanta 2015, 134, 16–23.
[20] Mendil D, Karatas M, Tuzen M. Separation and preconcentration of Cu(II), Pb(II), Zn(II), Fe(III) and Cr(III) ions with coprecipitation method without carrier element and their determination in food and water samples. Food Chem 2015, 177, 320–24.
[21] Ozdemir C, Sacmaci S, Kartal S. A coprecipitation procedure for the determination of some metals in food and environmental samples by flame atomic absorption spectroscopy. Anal Methods 2013, 5, 3977–83.
[22] Baig JA, Kazi TG, Elci L, Afridi HI, Khan MI, Naseer HM. Ultratrace determination of Cr(VI) and Pb(II) by microsampling system flame atomic spectroscopy in drinking water and treated and untreated industrial effluents. J Anal Methods Chem 2013, ID: 629495, 1–8.
[23] Yilmaz S, Tokaliglu S, Sahan S, Ulgen A, Sahan A, Soykan C. On-line preconcentration/determination of zinc from water, biological and food samples using

synthesized chelating resin and flame atomic absorption spectrometry. J Trace Elem Med Biol 2013, 27, 85–90.
[24] Nascentes CC, Arruda MAZ, Nogueira ARA, Nobrega HA. Direct determination of Cu and Zn in fruit juices and bovine milk by thermospray flame furnace atomic absorption spectrometry. Talanta 2004, 64, 912–17.
[25] Ferreira SLC, Lemos VA, Moreira BC, Spinola Costa AC, Santelli RE. An on-line continuous flow system for copper enrichment and determination by flame atomic absorption spectroscopy. Anal Chim Acta 2000, 403, 259–64.
[26] Dina J. Atomization of volatile compounds for atomic absorption and atomic fluorescence spectrometry. On the way towards the ideal atomizer. Spectrochim Acta Part B 2007, 62, 846–72.
[27] Qadir MA, Ahmed M, Shahzad S. Determination of aluminum by electrothermal atomization atomic absorption spectrometry in serum to characterize hemodialysis toxicity. Anal Lett 2015, 48, 147–53.
[28] Zhuravlev, Zacharia A, Gucer S, Chebotarev A, Arabadji M, Dobrynin A, Direct atomic absorption spectrometry determination of arsenic, cadmium, copper, manganese, lead and zinc in vegetable oil and fat samples with graphite filter furnace atomizer. J Food Compos Anal 2015, 38, 62–8.
[29] Zacharia A, Zhuravlev A, Chebotarev A, Arabadji M. Graphite "Filter Furnance" atomizer with Pd-Mg chemical modifier for direct analysis of foods using electrothermal atomic absorption spectrometry. Food Anal Methods 2015, 8, 668–77.
[30] Parengam M, Judprasong K, Srianujata S, Jittinadana S, Laoharojanaphand S, Busamongko A. Study of nutrients and toxic legumes by instrumental neutron activation analysis and graphite furnace atomic absorption spectrophotometry. J Food Compos Anal 2010, 23, 340–5.
[31] Katskov D. Low resolution continuum source simultaneous multi-element electrothermal atomic absorption spectrometry: steps into practice. Spectrochim Acta Part B 2015, 105, 25–37.
[32] Junior MM, Silva LO, Leao DJ, Ferreira SL. Analytical strategies for determination of cadmium in Brazilian vinegar samples using ET AAS. Food Chem 2014, 160, 209–13.
[33] Suzuki T, Miyada M, Otha K, Kaneco S, Mizuno T. Determination of osmium in waste water by graphite furnace atomic absorption spectrometry. Microchim Acta 1998, 129, 259–63.
[34] Burylin MYu. Specific features of the determination of arsenic in soils by electrothermal atomic absorption spectrometry with the injection of suspensions. J Anal Chem 2015, 70, 39–43.
[35] Tsalev DL, Slaveykova VO, Lampugnani L, Ulivo AD, Georgieva R. Permanent modification in electrothermal atomic absorption spectrometry – advances, anticipation and reality. Spectrochim Acta Part B 2000, 55, 473–90.
[36] Piaścik M, Bulska E. Performance of electrodeposited noble metals as permanent modifiers for the determination of cadmium in the presence of mineral acids by electrothermal atomic absorption spectrometry. Spectrochim Acta Part B 2001, 56, 1615–23.
[37] Volynsky AB. Comparative efficiency of platinum group metal modifiers in electrothermal atomic absorption spectrometry. Spectrochim Acta Part B 2004, 59, 1799–821.
[38] Piascik M, Pyrzyńska K, Bulska E. Determination of selenium in human urine by electrothermal atomic absorption spectrometry. Chem Anal 2005, 50, 235–247.
[39] Wojciechowski M, Piaścik M, Bulska E. Noble metals modifiers for antimony determination by graphite furnace atomic absorption spectrometry in biological samples. J Anal At Spectrom 2001, 16, 99–101.
[40] Cal-Prieto MJ, Carlosena A, Andrade JM, Muniategui S, Lopez-Mahia P, Prada D. Study of chemical modifiers for the direct determination of antimony in soils and sediments by ultrasonic slurry sampling-ETAAS with D2 compensation. At Spectrosc 2000, 21, 93–99.

[41] Barbosa F, Lima EC, Krug FJ, Determination of arsenic in sediment and soil slurries by electrothermal absorption spectrometry using W-Rh permanent modifier. Analyst 2000, 125, 2079–83.

[42] Torok P, Zemberyova M. Utilization of W/Mg(NO3)2 modifiers for the direct determination of As and Sb in soils, sewage sludge and sediments by solid sampling electrothermal atomic absorption spectrometry. Spectrochim Acta Part B 2010, 65, 291–6.

[43] Lima EC, Barbosa F Jr, Krug FJ, Silva MM, Vale MGR. Comparison of ultrasound-assisted extraction, slurry sampling and microwave assisted digestion for cadmium, copper and lead determination in biological and sediment samples by electrothermal atomic absorption spectrometry. J Anal At Spectrom 2000, 15, 995–1000.

[44] Welz, Mores S, Carasek E, Vale M, Orkuss M, Becker-Ross H. High resolution continuum source atomic and molecular absorption spectrometry – A review. Appl Spectrosc Rev 2010, 45, 327–54.

[45] Paz-Rodriguez B, Dominguez-Gonzalez MR, Aboal-Somoza M, Bermejo-Barrera P. Application of high resolution-continuum source flame atomic absorption in tea and tisanes. Food Chem 2015, 170, 492–500.

[46] Borges AR, Becker EM, Francois LL, de Jesus A, Vale MGR, Welz B, Dessuy MB, de Andrade JB. Investigation of spectral interferences in the determination of lead in fertilizers and limestone samples using high-resolution continuum source graphite furnace atomic absorption spectrometry. Spectrochim Acta Part B 2014, 101, 213–19.

[47] da Silva DB, Junior MMS, Silva LOB, Portugal LA, Matos GD, Ferreira SLC. Determination of cadmium in rice by electrothermal atomic absorption spectrometry using aluminum as permanent modifier. Anal Methods 2011, 3, 2495–500.

[48] Becker-Ross H, Florek S, Heitmann U. Observation, identification and correction of structured molecular background by means of continuum source AAS – determination of selenium and arsenic in human urine. J Anal At Spectrom 2000, 15, 137–41.

[49] Ohtsu N, Ashino T, Kimura H, Takada K, Hanawa T. Investigation for analytical procedure for determination of trace metallic ions in simulated body fluids by inductively coupled plasma atomic emission spectrometry. J Mater Sci Mater Med 2007, 18, 429–33.

[50] Nomngongo PN, Ngila JC, Msagati TAM, Moodley B. Preconcentration of trace multi-elements in water sample using Dowex 50W-x8 and Chelex-100 resins prior to their determination using inductively coupled plasma atomic emission spectrometry. Phys Chem Earth 2013, 66, 83–88.

[51] Garcia-Otero N, Barciela-Alonso MC, Dominguez-Gonzalez R, Herbello-Hermello P, Moreda-Pineiro A, Bermejo-Barrera P. Evaluation of offgel electrophoresis, electrothermal atomic absorption spectroscopy and inductively coupled plasma optical emission spectroscopy for trace metal analysis in marine plankton protein. Microchem J 2015, 119, 51–7.

[52] Moens L. Applications of mass spectrometry in the trace element analysis of biological materials. Fresenius J Anal Chem 1997, 359, 309–16.

[53] Siemianowski O, Barabasz A, Kendziorek M, Ruszczynska A, Bulska E, LWilliams E, Antosiewicz DM. HMA4 expression in tobacco reduces Cd accumulation due to the induction of the apoplastic barrier. J Exp Bot 2014, 63, 1125–39.

[54] Barabasz A, Wilkowska A, Ruszczyńska A, Bulska E, Hanikenne M, Czarny M, Kramer U, Antosiewicz DM. Metal response of transgenic tomato plants expressing P1B-ATPase. Physiol Plant 2012, 145, 315–31.

[55] Cubadda F. Inductively coupled plasma-mass spectrometry for the determination of elements and elemental species in food: A review. J AOAC Int 2004, 87(1), 173–204.

[56] Barbaste M, Robinson K, Guilfoyle S, Medina B, Lobinski R. Precise determination of the strontium isotope ratios in wine by inductively coupled plasma sector field multicollector mass spectrometry (ICP-SF-MC-MS). J Anal At Spectrom 2002, 17, 135–7.

[57] Halicz L, Segal I, Yoffe O. Direct REE determination in fresh waters using ultrasonic nebulization ICP-MS. J Anal At Spectrom 1999, 14(10), 1579–81.
[58] Krachler M, Mohl C, Emons H, Shotyk W. Influence of digestion procedures on the determination of rare earth elements in peat and plant samples by USN-ICP-MS. J Anal At Spectrom 2002, 17(8), 844–51.
[59] Simpson LS, Hearn R, Catterick T. The development of a high accuracy method for the analysis of Pd, Pt and Rh in auto catalysts using a multi-collector ICP-MS. J Anal At Spectrom 2004, 19, 1244–51.
[60] Moor Ch, Devos W, Guecheva M, Kobler J. Inductively coupled plasma mass spectrometry: a versatile tool for a variety of different tasks. Fresenius J Anal Chem 2000, 366, 159–164.
[61] Bocca B, Alimonti A, Petrucci F, Violante N, Sancesario G, Forte G, Senofonte O. Quantification of trace elements by sector filed inductively coupled plasma mass spectrometry in urine, serum, blood and cerebrospinal fluid of patients with Parkinson's disease. Spectrochim Acta Part B 2004, 59, 559–66.
[62] Bocca B, Alimonti A, Forte G, Petrucci F, Pirola C, Senofonte O, Violante N. High-throughput microwave-digestion procedures to monitor neurotoxic elements in body fluids by means of inductively coupled plasma mass spectrometry. Anal Bioanal Chem 2003, 337, 65–70.
[63] Yang L, Hang M, Lin S, Chen D, Zheng M. Minor and trace element analysis in breast milk using inductively coupled plasma mass spectrometry (ICP-MS). Microchim Acta 2003, 142, 85–88120.
[64] Czauderna M, Kowalczyk J, Ruszczynska A, Bulska E, Boldizarova K, Niedzwiedzka KM, Leng L. Effect of dietary CLA isomers on selenium, zinc, copper, chromium, magnesium and calcium levels in rat liver. J Anim Feed Sci 2005, 14, 529–32.
[65] Ammann AA, Inductively coupled plasma mass spectrometry (ICPMS): a versatile tool. J Mass Spectrom 2007, 42, 419–427.
[66] Becker JS. Trace and ultratrace analysis in liquids by atomic spectrometry. Trends Anal Chem 2005, 24, 243–54.
[67] Todoli JL, Mermet JM. Elemental analysis of liquid microsamples through inductively coupled plasma spectrochemistry. Trends Anal Chem 2005, 24, 107–116.
[68] Chajduk, Bartosiewicz I, Pyszynska M, Chwastowska J, Polkowska-Motrenko H. Determination of uranium and selected elements in Polish dictyonema shales and sandstones by ICP-MS. J Radioanal Nucl Chem 2013, 295, 1913–19.
[69] Pereira JSF, Pereira LSF, Mello PA, Guimaraes RCL, Guarnieri RA, Fonseca TCO, Flores EMM. Microwave-induced combustion of crude oil for further rare earth elements determination by USN-ICP-MS. Anal Chim Acta 2014, 844, 8–14.
[70] Kunze J, Koelling S, Reich M, Wimmer MA, ICP-MS determination of titanium and zirconium in human serum using an ultrasonic nebulizer with desolvator membrane. At Spectrosc 1998, 19, 164–7.
[71] Okamoto Y. Direct determination of lead in biological samples by electrothermal vaporization inductively coupled plasma mass spectrometry (ETV-ICP-MS) after furnace-fusion in the sample cuvette-tungsten boat furnace. Fresenius J Anal Chem 2000, 367, 300–5.
[72] Turner J, Hill SJ, Evans EH, Firman B, Wolff Briche CSJ. Accurate analysis of selenium in water and serum using ETV-ICP-MS with isotope dilution. J Anal At Spectrom 2000, 15, 743–746.
[73] May TW, Wiedmeyer RH. A table of polyatomic interferences in ICP-MS. At Spectrosc 1998, 19, 150–5.
[74] Wojciechowski M, Krata A, Bulska E. Determination of mercury isotopic profile by inductively couples plasma mass spectrometry: possibilities and limitations. Chem Anal Warsaw 2008, 53, 797–808.
[75] Ebdon L, Fisher AS, Worsfold PJ, Crews H, Baxter M. On-line removal of interferences in the analysis of biological materials by flow injection inductively coupled plasma mass spectrometry. J Anal At Spectrom 1993, 8, 691–5.

[76] Feng X, Wu S, Wharmby A, Wittmeier A. Microwave digestion of plant and grain standard reference materials in nitric and hydrofluoric acids for multi-elemental determination by inductively coupled plasma mass spectrometry. J Anal At Spectrom 1999, 14, 939–46.

[77] Tanner SD. Characterization of ionization and matrix suppression in inductively coupled cold plasma mass spectrometry. J Anal At Spectrom 1995, 10, 905–21.

[78] Moens LJ, Vanhaecke FF, Bandura DR, Baranov VI, Tanner SD. Elimination of isobaric interferences in ICP-MS, using ion-molecule reaction chemistry: Rb/Sr age determination of magmatic rocks, a case study. J Anal At Spectrom 2001, 16, 991–4.

[79] Bandura DR, Baranov VI, Tanner SD. Reaction chemistry and collisional processes in multipole devices for resolving isobaric interferences in ICP-MS. Fesenius J Anal Chem 2001, 370, 454–470.

[80] Baranov VI, Tanner SD. A dynamic reaction cell for inductively coupled plasma mass spectrometry (ICP-DRC-MS). Part 1. The rf-field energy contribution in thermodynamic of ion-molecule reactions. J Anal At Spectrom 1999, 14, 1133–1142.

[81] Hattendorf B, Gunther D. Characteristics and capabilities of an ICP-MS with a dynamic reaction cell for dry aerosols and laser ablation. J Anal At Spectrom 2000, 15, 1125–1131.

[82] de Boer JLM, Ritsema R, Piso S, van Staden H, van den Beld W. Practical and quality-control aspects of multi-element analysis with quadrupole ICP-MS with special attention to urine and whole blood. Anal Bioanal Chem 2004, 379, 872–880.

[83] Castinera MM, Brandt R, Von Bohlen A, Jakubowksi N. Development of a procedure for the multi-element determination of trace elements in wine by ICP-MS. Fresenius J Anal Chem 2001, 370, 553–558.

[84] Bjorn E, Krech W. Non-spectral interference effects in inductively coupled plasma mass spectrometry using direct injection high efficiency and microconcentric nebulisation. J Anal At Spectrom 2001, 16, 4–11.

[85] Sariego-Muniz C, Marchante-Gayon JM, Garcia Alonso JI, Sanz-Medel A. Comparison of electrothermal atomic absorption spectrometry, quadrupole inductively coupled plasma mass spectrometry and double-focusing sector field inductively coupled plasma mass spectrometry for the determination of aluminium in human serum. J Anal At Spectrom 1998, 13, 283–7.

[86] Ruiz Encinar J, Garcia Alonzo JI, Sanz-Medel A, Main S, Turner PJ. A comparison between quadrupole, double focusing, multicollector ICP-MS. Part II. Evaluation of total combined uncertainty in the determination of lead in biological matrices by isotope dilution. J Anal At Spectrom 2001, 16, 322–6.

[87] Rodushkin I, Odman F, Branth S. Multielement analysis of whole blood by high resolution inductively coupled plasma mass spectrometry. Fresenius J Anal Chem 1999, 364(4), 338–46.

[88] Barbaste M, Halicz L, Galy A, Medina B, Emteborg H, Adams FC, Lobinski R. Evaluation of the accuracy of the determination of lead isotope ratios in wine by ICP MS using quadrupole, multicollector magnetic sector and time-of-flight analyzers. Talanta 2001, 54, 307–17.

[89] Sariego-Muniz C, Marchante-Gayon JM, Garcia Alonso JI, Sanz-Medel A. Accurate determination of iron, copper, zinc in human serum by isotope dilution analysis using double focusing ICP-MS. J Anal At Spectrom 1999, 14(9), 1505–10.

[90] Park CJ, Suh JK. Determination of trace elements in rice flour by isotope dilution inductively coupled plasma mass spectrometry. J Anal At Spectrom 1997, 12, 573–577.

[91] Kent AJR, Jacobsen B, Peate DW, Waight TE, Baker JA. Isotope dilution MC-ICP-MS rare earth elements analysis of geochemical reference materials NIST SRM 610, NIST SRM 612, NIST SRM 614, BHVO-2G, BHVO-2, BCR-2G, JB-2, WS-E, W-2, AGV-1 and AGV-2. Geostand Geoanal Res 2004, 28(3), 417–29.

[92] de Bievre P, Isotope-dilution mass spectrometry. Tech Instrum Anal Chem 1994, 15, 169–183.

[93] Heumann KG. Isotope Dilution mass spectrometry (IDMS) of the elements, Mass Spectrom Rev 1992, 11, 41– 67.

[94] Wagner B, Jędral W. Open ablation cell for LA- ICP-MS investigations of historic objects. J Anal At Spectrom 2011, 26, 2058–63.
[95] Diwakar PK, Gonzalez JJ, Harilal SS, Russo RE, Hassanein A. Ultrafast laser ablation ICP MS: role of spot size, laser fluence, and repetition in signal intensity and elemental fractionation. J Anal At Spectrom 2014, 29, 339–46.
[96] Pisonero J, Fernandez B, Gunther D, Critical revision of GD-MS, LA-ICP-MS and SIMS as inorganic mass spectrometric techniques for direct solid analysis, Journal of Analytical Atomic Spectrometry 2009, 24, 1145–60.
[97] Pozebon, Sheffler GL, Dressler VL, Nunes MAG. Review of the applications of laser ablation inductively coupled plasma mass spectrometry to the analysis of biological samples. J Anal At Spectrom 2014, 29, 2204–28.
[98] Wagner B, Nowak A, Bulska E, Hametner K, Günther D. Critical assessment of the elemental composition of corning archeological reference glasses by LA-ICP-MS. Anal Bioanal Chem 2012, 402, 1667–1677.
[99] Durrant SF, Ward NI. Recent biological and environmental applications of laser ablation inductively coupled plasma mass spectrometry (LA-ICP-MS). J Anal At Spectrom 2005, 20, 821–9.
[100] Wagner B, Garboś S, Bulska E, Hulanicki A. Determination of iron and copper in old manuscripts by slurry sampling graphite furnace atomic absorption spectrometry and laser ablation inductively coupled plasma mass spectrometry. Spectrochim Acta Part B 1999, 54, 797–804.
[101] Syta O, Rozum K, Choińska M, Zielińska D, Żukowska Z, Kijowska A, Wagner B. Analytical procedure for characterization of medieval wall-paintings by X-ray fluorescence spectrometry, laser ablation inductively coupled plasma mass spectrometry and Raman spectroscopy. Spectrochim Acta Part B 2014, 101, 140–148.
[102] Cagno S, Cosyns P, Van der Linden V, Schalm O, Izmer A, Deconinck I, Vanhaecke F, Nowak A, Wagner B, Bulska E, Nys K, Janssens K. Composition data of a large collection of black-appearing Roman glass. Open J Archaeometry 2013, 1. DOI: 10.4081/4964.
[103] Orellana FA, Galvez CG, Roldan MT, Garcia-Ruiz C. Applications of laser-ablation-inductively-couples plasma-mass spectrometry in chemical analysis of forensic evidence. Trends Anal Chem 2013, 42, 1–33.
[104] Becker JS, Matush A, Wu B. Bioimaging mass spectrometry of trace elements – recent advance and applications of LA-ICP-MS: A review. Anal Chim Acta 2014, 835, 1–18.
[105] Shortreed M, Bakker E, Kopelman R. Miniature sodium-selective ion-exchange optode with fluorescent pH chromoionophores and tunable dynamic range. Anal Chem 1996, 68, 2656–62.
[106] Bakker E, Diamond D, Lewenstam A, Pretsch E. Ion sensors: current limits and new trends. Anal Chim Acta 1999, 393, 11–18.
[107] Gyurcsanyi RE, Pergel E, Nagy R, Kapui I, Thu Lan BT, Toth K, Bitter I, Lindner E. Direct evidence of ionic fluxes across ion-selective membranes: a scanning electrochemical microscopic and potentiometric study. Anal Chem 2001, 73, 2104–111.
[108] Munoz E, Palmero S. Analysis and speciation of arsenic by stripping potentiometry: A review. Talanta 2005, 65, 613–20.
[109] Pergel E, Gyurcsanyi RE, Toth K, Lindner E. Picomolar detection limits with current-polarized Pb2+ ion-selective membranes. Anal Chem 2001, 73, 4249–53.
[110] Sbhanardakani S, Tayebi L, Farmany A, Cheraghi M. Analysis of trace elements (Cu, Cd and Zn) in the muscle, gill and liver tissues of some fish species using anodic stripping voltammetry. Environ Monit Assess 2013, 184, 6607–6611.
[111] Zhao C, Lui H, Wang L. Simultaneous determination of Pb(II) and Cd(II) using an electrode modified with electropolymerized thiadazole film. Anal Methods 2012, 4, 3586–92.

[112] Ivandini TA, Sato R, Makide Y, Fujishima A, Einaga Y. Electrochemical detection of arsenic (III) using iridium-implanted boron-doped diamond electrodes. Anal Chem 2006, 78, 6291–8.

[113] Alves GMS, Magalhaes JMCS, Salaun P, van der Berg CMG, Soares HMVM. Simultaneous electrochemical determination of arsenic, copper, lead and mercury in unpolluted fresh waters using a vibrating gold microwire electrode. Anal Chim Acta 2011, 703, 1–7.

[114] Sezgin HV, Gokcel HI, Dilgin Y. Adsorptive anodic stropping voltammetric determination of antimony (III) on a glassy carbon electrode using rivastigine as a new chemical receptor. Sens Actuators B Chem 2015, 209, 686–94.

[115] Prasad BB, Jauhari D, Verma A. A dual-ion imprinted polymer embedded in sol-gel matrix for the ultra trace simultaneous analysis of cadmium and copper. Talanta 2014, 120, 398–407.

[116] Santos LB, de Souza MTF, Paulino AT, Garcia EE, Nogami EM, Garcia JC, de Souza NE. Determination of aluminum in botanical samples by adsorptive cathodic stripping voltammetry as Al-8-hydroxyquinoline complex. Microchem J 2014, 112, 50–5.

[117] Locatelli C. Catalytic-adsorptive stripping voltammetric determination of ultra-trace iridium (III). Application to fresh- and sea-water. Talanta 2011, 85, 546–50.

[118] Salaun P, Gibbon-Walsh K, van den Berg CMG. Beyond the hydrogen wave: new frontier in the detection of trace elements by stripping voltammetry. Anal Chem 2011, 83, 3848–56.

[119] Aouarram A, Galindo-Riano MD, Garcia-Vargas M, Stitou M, El Yousfi F, Espaa-Bellido E. An efficient approach to designing and optimizing the analysis of Ni(II) by AdCSV in seawater. Talanta 2010, 82, 1749–56.

[120] Wang J. Electrochemical detection for microscale analytical systems: A review. Talanta 2002, 56, 223–31.

[121] Monticelli D, Laglera LM, Caprara S. Miniaturization in voltammetry: Ultratrace element analysis and speciation with twenty-fold sample size reduction. Talanta 2014, 128, 273–7.

[122] Strange RW, Feiters MC. Biological X-ray absorption spectroscopy (BioXAS): a valuable tool for the study of trace elements in the life sciences. Curr Opin Struct Biol 2008, 18, 609–16.

[123] Matsumoto E, Simabuco SM, Perez CA, Nascimento VF. Atmospheric particulate analysis by synchrotron radiation total reflection (SR-TXRF). X-Ray Spectrom 2002, 31, 136–14.

Part II: **Matrices – Selected applications**

Emilia Vassileva and Petko Mandjukov

3 Trace Elements in the Environment

Abstract: The chapter gives concise summary on the trace elements in the environment. Information covering essentiality, toxicity, phytotoxicity, chemical species, main emission sources, as well as state diagrams for aquatic media is provided for the most frequently studied trace elements.

Basic descriptions on some general characteristics and analytical aspects for some of the extensively investigated environmental matrices like natural waters, soils, sediments and air are presented. Subject of the chapter is also the environmental monitoring of trace elements in different environmental media.

An overview of the methods applied for element determination and species separation, illustrated by large number of contemporary literature sources on environmental trace and speciation analysis is also provided. As the essentiality, toxicity or phytotoxicity usually strongly depends on the chemical form in the nature the significance of the speciation analysis is underlined.

Throughout the chapter the requirements for obtaining reliable trace analysis data, including both the scientific problems relating to trace analysis and all the relevant quality assurance issues are presented.

In addition, some metrological aspects of environmental monitoring, such as uncertainty evaluation, method validation, metrological traceability and proper use of certified reference materials, are discussed and illustrated by cases of study.

Keywords: trace elements, environment, environmental monitoring, trace elements analysis, environmental matrices, speciation analysis, metrological aspects, certified reference materials, uncertainty, traceability

3.1 General aspects of environmental analysis

The term "trace element" has no precise definition in the earth sciences because the concentration of an element in a given phase can be so low that it is considered as trace element, whereas the same element can constitute a main part of another phase. In geochemistry, the most frequently used definition is: trace elements are chemical elements whose concentration in the Earth's crust is less than 0.1% (m/m).

Only 10 elements (oxygen, silicon, iron, aluminum, calcium, potassium, sodium, magnesium, titanium, and hydrogen) of the naturally occurring 89 account for over 99% (m/m) of the Earth's crust. The remaining 79 elements (including the inert gases), which form less than 0.5% (m/m), are known as trace elements. They do not play a fundamental role in the makeup of the Earth's crust but their significance in economy, ecology, agriculture, medicine, toxicology, and many other areas is completely disproportional to their low crustal abundances.

https://doi.org/10.1515/9783110366730-004

Trace elements are of the most critical contaminants in the environment and is present in water, soils, sediments, and air usually at trace levels. However, several human activities (e.g., mining, industry, and sludge dumping) have increased the natural concentration of trace elements to the level of severely contaminated environments.

Soils and sediments, in particular, play an important role in the biogeochemical cycle of trace elements, acting as a sink and a source of the elements for biota and humans. Some of the trace elements are chemically very reactive in the environment, which results in their mobility and bioavailability to living organisms. Due to the effects of bioaccumulation and biomagnification, the levels of trace elements, including toxic species, might increase in plants grown up on polluted soils to high and even dangerous levels. Thus, in the food chain, significant amounts of toxins can be introduced. Their levels are permanently increasing from lower to higher hierarchical levels. People can be exposed to high levels of toxic elements directly by inhalation, through skin, or via ingestion of contaminated food.

Recent industrial development has caused substantial enrichment in basic environmental media such as soil, vegetation, water, and air, resulting in geochemical neoanomalies that may contain up to 1,000 times the normal crustal trace element content. The normal crustal trace element concentration is usually described as the “background concentration.” The trace elements concentrated in neoanomalies are often metals, especially heavy metals (density >6 g/cm^3). Some of them are essential for life processes in trace quantities, but in higher concentrations are all toxic to organisms. Metals and nonmetals enter the environment, and the extent of their emission will depend on the nature and efficiency of the particular industrial process. Considerable neoanomalies of many trace elements are found around mines, smelters, and industrial complexes, along roads and in most urban areas.

At present, the natural geochemical cycles of some trace elements are considerably disrupted by human activities. Contamination of the environment refers to the tertiary dispersion of chemical substances from localized sources into the environment, caused by human activities. However, the natural cycles of some elements have been significantly disturbed by the anthropogenic impacts of recent years.

Some of the chemicals entering the environment have direct and/or indirect influences on life forms and can cause serious damage. Toxic metals such as mercury, cadmium, arsenic, and lead tend to accumulate in certain reservoirs (soil, sediments, etc.) from which they may be released by various processes of remobilization. They may also change form or speciation and become available to the biological food chain, thereby affecting life, including human life, by causing chronic and acute disorders.

Their concentrations in different ecological systems and their elements are varying in wide range. Ecological importance of trace elements is determined by their direct or indirect influence on the natural ecosystems and human.

Analysis of trace elements in various environmental samples is a challenging task. The national and international legislations require control of the toxic elements content in different environmental media at permanently decreasing levels. Ecological analyses are complicated at every step of the study. The objects are typically inhomogeneous with complex sample matrices. The correct definition of analytical problem, representative sampling, and sample preparation, analysis and data treatment are not trivial tasks. Very often, especially in fluidic media, the trace element levels are dynamic variables.

Both the properties of the analyte species and characteristics of the environmental media, objects of the study, should be considered in planning and performing the environmental analysis of trace elements.

3.2 Trace elements (general information, ecological importance, common species in the nature)

Trace elements in the nature are originated both by natural and anthropogenic sources. The concentrations of trace elements in different ecological systems and their species are varying in wide range. Ecological importance of trace elements is determined by their direct or indirect influence on the biological systems plants, animals, and humans.

The biological impact of the trace elements can be detected on different levels:

- Molecular and subcellular level – effects on bio-macromolecules: DNA, RNA, key proteins, enzymes, hormones, metallothioneins, etc.
- Individual level – effects related to individual organism: growth rate, organs functioning, behavioral responses, etc.
- Entire population level – effects related to genetics, reproduction of species, morbidity, mortality, etc.

As a general trend, effects on lower level appear after shorter time and are more clearly and definitely detected. Effects on individual level are disturbed by the natural variability of the living organisms and appearance of atypical symptoms. Besides this, hypersensitivity or hyper-resistivity to insufficiency or toxic impacts might be observed as well. The impacts affecting entire population are most slowly manifested and difficult for prediction and evaluation. The consequences are heavy and often irreversible. Besides the response time, increasing of organizational level leads to more and more difficult linkage to specific chemical species. These effects are strongly dependent on the bioavailability of the elements at different levels of the ecosystem. Typical diagram of the dependence of the effect on the population of certain biological species from the content of the element is presented in Figure 3.1.

Essentiality. Major criterion for essentiality of an element is the appearance of different abnormalities and malfunctions in living organisms in case of its deficiency or absence. Three groups of elements can be clearly distinguished:

- Essential elements: Ca, Co, Cr(III), Cu, Fe, Mg, Mn, Mo, Se, Zn, and so on. For these elements, very clear insufficiency effects and relatively wide acceptable range are typical. At higher concentration levels toxic effects are also observed.
- Elements with possible beneficial effects: As, B, Ni, Si, V, and so on. For the elements belonging to this group the insufficiency effects exist but are not very

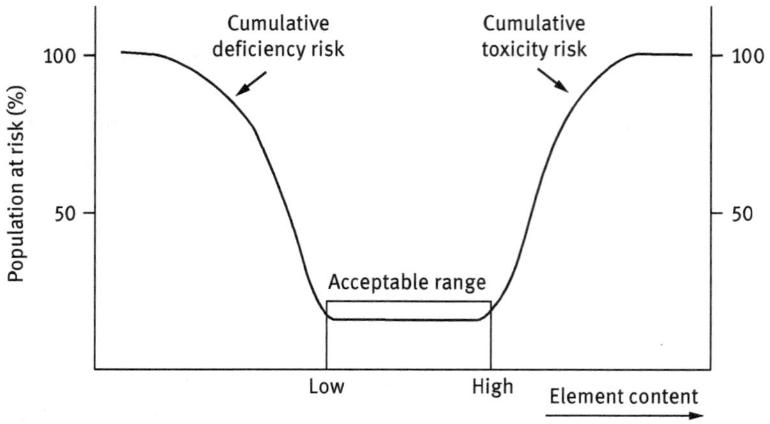

Figure 3.1: Typical diagram of the dependence in the population risk from the content of trace elements.

significant and sometimes uncertain. Dominant are the toxic effects in case of excessive amounts. The acceptable range is usually not wide.

– Elements without any beneficial effects: Al, As, Be, Cd, Hg, Pb, Sn, Tl, and so on. For the elements belonging to this group the acceptable range of content (Figure 3.1) is quite narrow and starting from 0. Besides the typical highly toxic elements, to this group belong also some elements existing in chemically inert forms and practically without biological impact.

Practically, for all trace elements toxic effects can be observed at excessive uptake. Generally, the essentiality and toxicity are substantially different for the various biological species and are dependent on the chemical form of the element. Some of the biological effects, mostly toxic, of the trace elements can be explained by their ability to initiate free radical reactions or the decomposition of peroxides and other unstable molecules, allowing the propagation of free radical reactions in the organism (Figure 3.1).

Toxicity. Two are the major targets for the toxic action of the trace metals.

Enzymes: Most of the toxic metals are acting as enzyme inhibitors by one of the following mechanisms:

– Metal ion–enzyme interactions. Usually it is due to bonding of metal ion to the –SH from the sulfhydryl groups in the enzyme molecules
– Replacing of the essential metal ion (cofactor) in the enzyme
– Blocking the enzyme synthesis in the organism and causing in such a way enzyme deficiency

Subcellular organelles: Some metal ions or organometallic species are able to penetrate the cell membranes, to form associates with the macromolecules from the subcellular organelles and thus causes various malfunctions. It is important to note that the toxicity of the metals as well as the mechanism of their action is strongly dependent on their chemical forms. That makes the speciation analysis a high-priority analytical task [1–4].

Phytotoxicity. Phytotoxicity is the capacity of a compound (such as a plant protection product) to cause temporary or long-lasting damage to plants [5]. The question of phytotoxicity of chemical species is rather important since the plants are in the basis of trophic pyramid. Many trace metals are acting as a phytotoxins for different plant organisms. Studies of such phenomena are complicated by typically slow response and differences in the reactions of various organisms. Mainly studies are orientated to agricultural plants and species important for biomonitoring purposes. Studies are carried out usually under controlled conditions far from the natural.

Across a range of plant species and experimental conditions, the phytotoxicity of some of the trace metals followed the trend (from most to least toxic):

Pb ≈ Hg > Cu > Cd ≈ As > Co ≈ Ni ≈ Zn > Mn, with median toxic concentrations (μM): 0.30 (Pb), 0.47 (Hg), 2.0 (Cu), 5.0 (Cd), 9.0 (As), 17 (Co), 19 (Ni), 25 (Zn), and 46 (Mn) [6, 7].

At molecular level the phytotoxic trace metals are acting, in most of the cases, blocking of sulfhydryl groups in proteins or replacing essential metal ions.

Physicochemical properties of the trace element species. Besides their direct biological impact, the ecological importance of trace metal species is determined by the number of physicochemical properties controlling their behavior in the environment. The mobility and bioavailability of the chemical species are dependent on their oxidation state, solubility, and electric charge of ions and complexes. The equilibrium between particular chemical forms of elements is controlled by redox potential and pH value in the water media.

State diagrams E/pH. The state diagrams E/pH (Pourbaix diagrams) are convenient graphical representations of the equilibriums between species of an element in aqueous media controlled by redox potential (E) and pH. Depending on the values of E and pH, every solution corresponds to a point of the diagram. This is an easy way to obtain information about the dominant chemical form of an element under the specified conditions, as well as to evaluate the potential risks for change of the chemical form during the sample preparation. On the other hand, from the diagrams can be easily revealed information about the degree of protonation of anion forms and charges of the complexes formed. These are questions of crucial importance planning the separation and preconcentration procedures needed in speciation analysis. Each line in the diagram corresponds to the points (in E/pH coordinates) in which the concentrations of the neighbor chemical forms are equal. Dashed lines in the diagrams correspond to the limits of stability of the water. The state diagrams E/pH are entirely theoretical and might vary significantly depending on the chemical forms considered as substantial for the studied system. Minor variations due to differences in thermochemical constants used are possible as well [8].

3.2.1 General information on trace elements

Only most frequently examined elements are considered.

3.2.1.1 Arsenic

Application and As emission sources. Arsenic is a highly abundant element in the nature. Industrial use of the element is mainly in semiconductor technologies as GaAs or n-type dopant of silicon- and germanium-based semiconductors. In metallic form it is used as a strengthening component of Cu and especially Pb alloys (in car accumulators). Due to its toxicity, As is used also in production of various pesticides, herbicides, insecticides, and wood preservation reagents. In some cases it is released in the nature in significant amounts during metal ore treatment (as accompanying element) and coal combustion. Average As abundances in different environments are as follows: lithosphere – 2,100 μg/kg, ocean surface – 2.3 μg/kg and in freshwaters 1 μg/kg. State diagram E/pH for the most common inorganic species of As [8] is presented in Figure 3.2.

Easy inter-conversion between forms As(III) and As(V) should be considered. Special attention to sample preparation for speciation analysis is needed.

Essentiality. The essentiality of As is not definitely confirmed. It is supposed to participate in the metabolism of methionine. Deficiency of As might cause vascular diseases and nervous system disorder. A few bacteria species are reported to be able to use arsenic compounds as respiratory metabolites. In some cases, As compounds are used also for medical purposes.

Toxicity. Usually the As species are assimilated by inhalation or via the gastrointestinal route. Toxic effects can be attributed to affinity of As to S and ability to form complexes with coenzymes. As species are inhibitors for the synthesis of adenosine triphosphate. Carcinogenic, mutagenic, and genotoxic effects are reported for some chemical forms. Generally, the inorganic As species are more toxic than organometallics. AsB and As proteins are reported as not toxic forms [1].

Phytotoxicity. As is observed to cause decrease of chlorophyll *a* and *b* contents and consecutively to suppress the plant growth and vegetation. Highly different phytotoxicity of As species can be attributed to the differences in ability of chemical forms to penetrate the cell membranes [9–11].

Chemical species of As. As(III), As(V) – inorganic compounds, MMA(III) – $CH_3As(OH)_2$/monomethylarsonous acid, MMA(V) – $CH_3AsO(OH)_2$/monomethylarsonic acid, DMA(III) – $(CH_3)_2AsOH$/dimethylarsinous acid, DMA(V) – $(CH_3)_2AsO(OH$/dimethylarsinic acid, AsB – arsenobetaine, AsC – arsenocholine, TMAO – $(CH_3)_3AsO$/trimethylarsine oxide, Me_4As^+ – $(CH_3)_4As^+$/tetramethylarsonium ion, AS – arsenosugars, PhAs – phenylarsonic acid, PhAsO – phenylarsine oxide, DPhAs – diphenylarsinic acid.

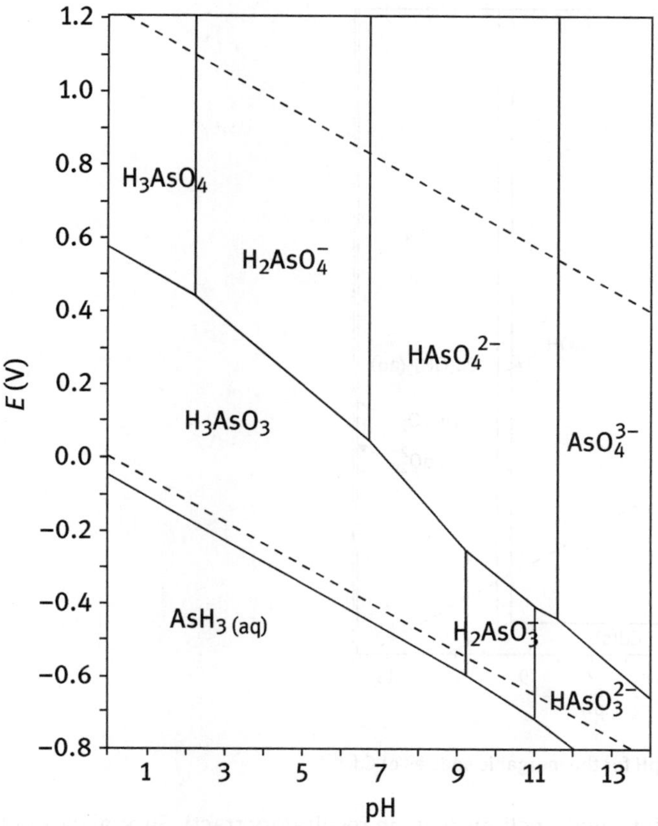

Figure 3.2: State diagram E/pH for the most common inorganic species of As.

3.2.1.2 Cadmium

Application and Cd emission sources. The most significant industrial application of cadmium is in production of batteries, and predominantly in rechargeable nickel–cadmium batteries. Electroplating with Cd is used for protection from corrosion of various constructions and its elements (including aircrafts). Specific application of the metal is in nuclear reactors as component of alloys used as barrier controlling neutrons. It is also used in production of paints and pigments. Average Cd abundances in different environments are as follows: lithosphere – 150 μg/kg and ocean surface – 0.05 μg/kg. State diagram E/pH for the most common inorganic species of Cd [8] is presented in Figure 3.3. Change in oxidation state during sample preparation is not probable.

Essentiality. Essentiality of cadmium is not proved for higher organisms. Cadmium-dependent carbonic anhydrase has been reported for some marine diatoms.

Toxicity. The usual ways for absorption of Cd in the body are inhalation of its dusts and volatile species or ingestion. Cadmium is a poison highly accumulated in the liver

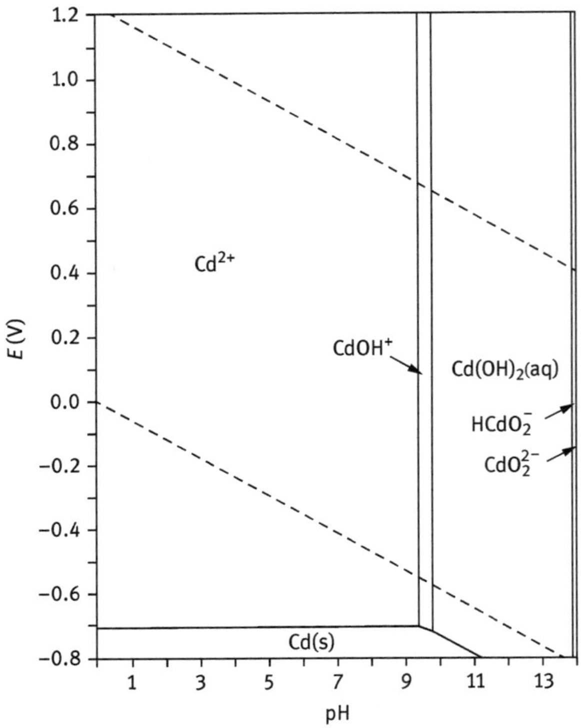

Figure 3.3: State diagram E/pH for the inorganic species of Cd.

and kidneys causing various malfunctions (e.g., in respiratory tract). Several cases of human death caused by Cd intoxication are reported. The most probable explanation of Cd toxicity is in its ability to bound sulfhydryl (–SH) groups, and thus, to inhibit the number of enzymes [1].

Phytotoxicity. Cd contamination in plants causes seed germination inhibition. Synergetic effects with As and Pb are reported [11].

Chemical species of Cd. Cd(II), Cd complexes with natural ligands.

3.2.1.3 Mercury

Application and Hg emission sources. Mercury is used primarily for the manufacture of industrial chemicals or for electrical and electronic applications. It is used in some thermometers, especially ones which are used to measure high temperatures. A still significant amount is used as gaseous mercury in fluorescent lamps. Industrial applications of Hg tend to decrease due to its very high toxicity. The most significant Hg emission source is the coal combustion (in power plants). Average Hg abundances in different environments are as follows: lithosphere – 1 µg/kg, ocean surface – 0.05 µg/kg, and surface freshwater 0.07 µg/kg. State diagram E/pH for the most common inorganic species of Hg [8] is presented in Figure 3.4. As shown in Figure 3.4, Hg can be easily reduced to elemental state.

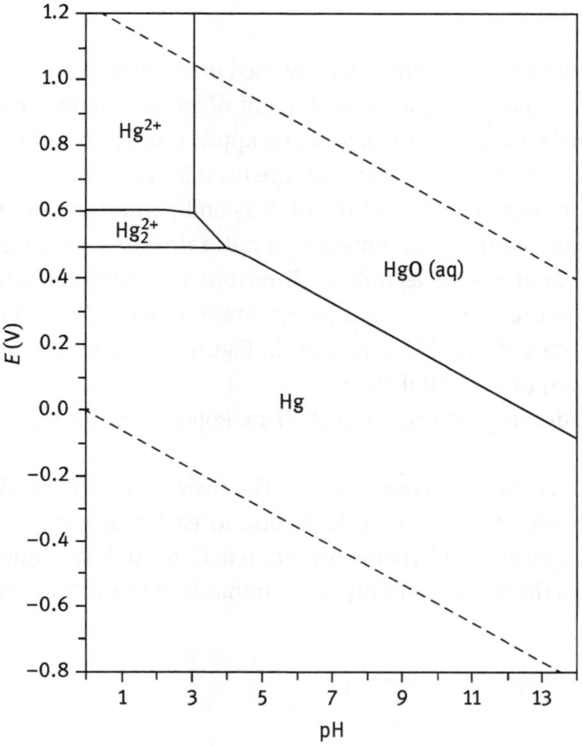

Figure 3.4: State diagram E/pH for the inorganic species of Hg.

Essentiality. Mercury has no positive biological role.

Toxicity. All chemical forms of Hg (elemental Hg, inorganic, and organometallic compounds) are highly toxic. MeHg is the most toxic among the Hg species studied. Hg penetrates the skin, and the inhalation risk is very high. Toxicity of inorganic species depends strongly on solubility of its compounds. Mercury affects the central nervous system (Minamata disease); causes damage to the brain, liver, kidney, lungs; and attacks hemoglobin. It is being accumulated in the kidneys. Hg in all forms might cause death. In the red blood cells the elemental Hg is easily oxidized to Hg^{2+} which has affinity to the sulfhydryl (–SH) groups. It forms stable bonds with proteins and large number of enzymes. Methylated Hg forms are more dangerous since the penetration through the lipid membranes of the cells is easier and accumulation factor is higher than for inorganic forms [1].

Phytotoxicity. The possible mechanisms of phytotoxic action of Hg species comprise the blocking of important molecules (enzymes and polynucleotides), the transport of essential ions, displacement or substitution of metal ions from molecules (e.g., Mg from chlorophyll). Hg causes also DNA changes [12].

Chemical species of Hg. Hg, Hg_2^{2+}, Hg^{2+}, MeHg – methyl-Hg, EtHg – ethyl-Hg, PhHg – phenyl-Hg.

3.2.1.4 Chromium

Application and Cr emission sources. Chromium is widely used in different branches of industry. In metallurgy it finds application as a component of various alloys and for electroplating of metallic surfaces. Besides constructive applications, Cr and its alloys are also used as industrial catalysts. Cr compounds are used in wood preservation, in leather treatment, and for production of different dyes and pigments. These industrial branches are significant source of Cr emission in natural waters. Average Cr abundances in different environments are as follows: lithosphere – 15,000 μg/kg, ocean surface – 0.6 μg/kg, and surface freshwater 1 μg/kg. State diagram *E*/pH for the most common inorganic species of Cr [8] is presented in Figure 3.5. As shown in Figure 3.5, Cr can be easily reduced to elemental state.

Essentiality. Cr(III) participates in the glucose metabolism. Potentiates the action of the enzyme insulin

Toxicity. Cr shows affinity to O atoms in biomolecules. The toxicity of Cr(III) and Cr(II) is relatively low. Cr(VI) is significantly more toxic due to easier penetration through the cell membranes. It causes skin hypersensitivity reactions and lung and kidney damages. Cr(VI) enhances the binding of polycyclic aromatic hydrocarbons to

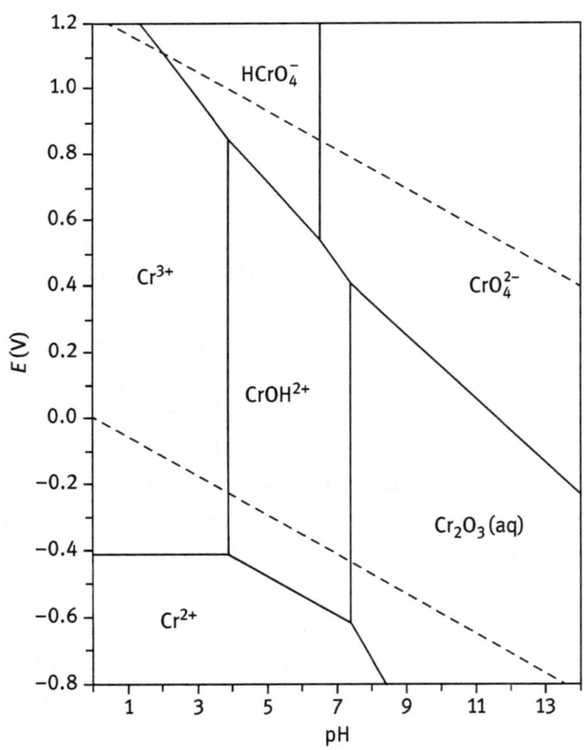

Figure 3.5: State diagram *E*/pH for the inorganic species of Cr.

DNA. Chromium shows the carcinogenic activity as well. The probable mechanism includes intracellular reduction of Cr(VI) to Cr(III) which can bind with nucleic acid and thus initiate carcinogenesis [1].

Phytotoxicity. Chromium is toxic to plants even at low concentrations, and the hexavalent form is reported to be more toxic. Phytotoxicity results in inhibition of seed germination, degradation of nutrient balance, antioxidant enzymes. Cr species are inducing oxidative damages in plant cells affecting growth, water balance, pigmentation, and so on [13].

Chemical species of Cr. Cr(III), Cr(VI).

3.2.1.5 Selenium

Application and Se emission sources. Selenite and selenite salts found application in glass production. These compounds give red color to the glass and can be used both for colored glass production and for neutralization of green and yellow colors appearing due to impurities. In copper alloys, selenium in combination with bismuth is recently used to replace more toxic lead. Various selenides are widely used in solar cells technologies. Average Se abundances in different environments are as follows: lithosphere – 450 μg/kg, ocean surface – 0.05 μg/kg, and 0.2 μg/kg in the surface freshwater. State diagram E/pH for the most common inorganic species of Se [8] is presented in Figure 3.6.

Essentiality. At trace levels, Se is component of the amino acids selenocysteine and selenomethionine. Selenium is a cofactor in some enzymes, for example, glutathione peroxidases, iodothyronine deiodinases, and thioredoxin reductases.

Toxicity. Inorganic Se in excessive amounts shows significant toxicity. The target organs are the respiratory tract, liver, kidneys, blood, skin, and eyes. The symptom of Se poisoning is a garlic-like odor in the breath and sweat. The molecular mechanism of Se toxic action is not well established. Se is supposed to be potential inhibitor of the enzyme squalene monooxygenase participating in the cholesterol biosynthesis [1].

Phytotoxicity. Most commonly described effects are inhibition of plant growth, chlorosis, and fruit/seed abortion. Indirectly Se affects plant species by killing the bees and other insects participating in pollination [14, 15].

Chemical species of Se. Se(IV), Se(VI) and Se-containing amino acids.

3.2.1.6 Antimony

Application and Sb emission sources. The main industrial application of antimony is in production of flame-proof materials, fiberglass composites, and polyester resins. In metallurgy, Sb is used in alloys with lead, increasing its hardness and mechanical strength. For most applications involving lead, varying amounts of antimony are used as alloying metal (e.g., in lead acid accumulators). Sb is used also as a catalyst and stabilizer in polymer production, in semiconductor technologies as n-type dopant, in medicine as a component of some drugs, and so on. Average Sb abundances in different environments are as follows: lithosphere – 200 μg/kg, ocean surface – 0.05 μg/kg,

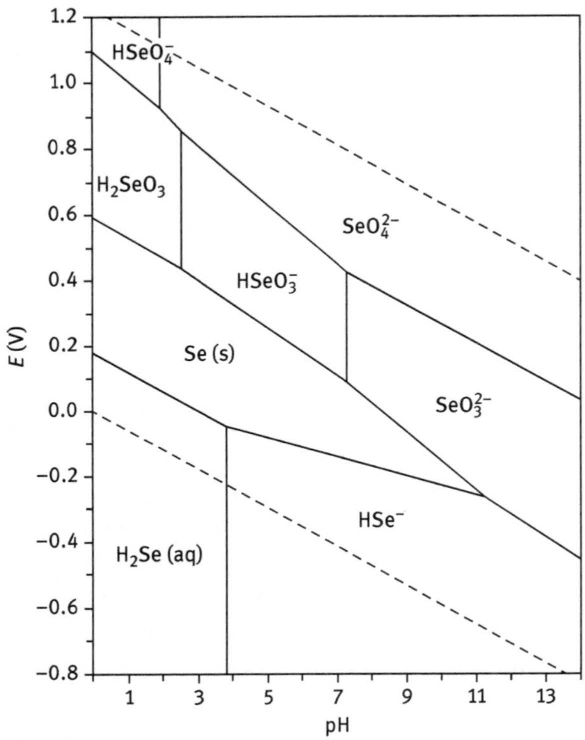

Figure 3.6: State diagram E/pH for the inorganic species of Se.

and 2 µg/kg in surface freshwater. State diagram E/pH for the most common inorganic species of Sb [8] is presented in Figure 3.7.

Essentiality. Essentiality of Sb is not confirmed.

Toxicity. Sb is a relatively low toxic element (significantly less than As). The primary intoxication routes for Sb are inhalation of dust or fumes, skin absorption, or contaminated food. Sb causes weight loss, loss of hair, and malfunctions of heart, liver, and kidney. Sb(III) is claimed to be ca. 10 times more toxic than the prevailing in the nature form Sb(V).

Phytotoxicity. Very few studies on Sb phytotoxicity are published. Suppressing of the growth of root and the activity of α-amylase are reported. Negative influence on the plant growth was observed as well [16].

Chemical species of Sb. Sb(III), Sb(V), methylated forms (e.g., $TMSbCl_2$), Sb complexes with natural ligands.

3.2.1.7 Nickel

Application and Ni emission sources. Nickel finds wide application in metallurgy as a component of steels and nonferrous alloys and for electroplating of metal surfaces. Only about 6% of the world's production of Ni is used for nonmetallurgical

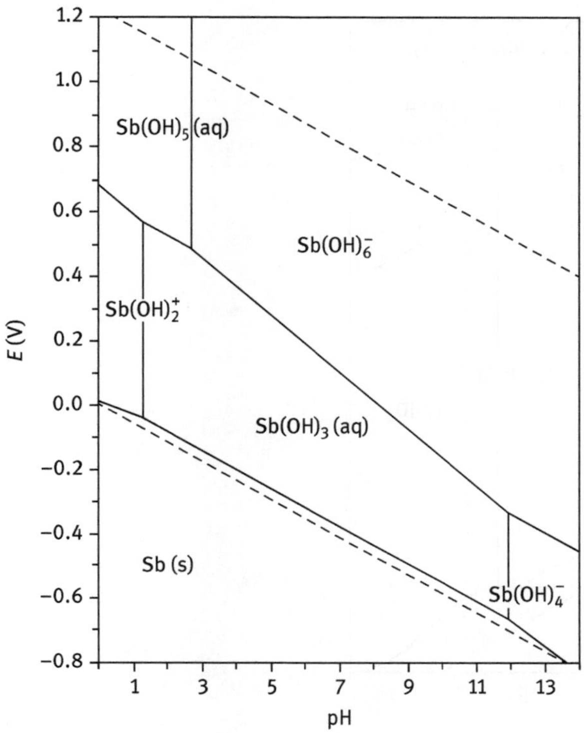

Figure 3.7: State diagram E/pH for the inorganic species of Sb.

applications. Nickel foam or nickel mesh is used in gas diffusion electrodes for alkaline fuel cells. Nickel is a naturally magnetostrictive material, that is, in the presence of a magnetic field the material undergoes a minor change in length. Average Ni abundances in different environments are as follows: lithosphere – 2,000 μg/kg, ocean surface – 90 μg/kg, and 0.3 μg/kg in surface freshwater. State diagram E/pH for the most common inorganic species of Ni [8] is presented in Figure 3.8.

Essentiality. Possibly Ni is cofactor for some enzymes (e.g., urease). It also participates in the folate metabolism.

Toxicity. Ingestion of Ni compounds can cause hyperglycemia, depression of the central nervous system, myocardial weakness, and kidney damage.Ni induces oxidative stress as well.

Phytotoxicity. Ni induces a significant reduction in germination and fresh biomass of seedlings [17, 18].

Chemical species. Ni(II), Ni complexes with natural ligands.

3.2.1.8 Cobalt

Application and Co emission sources. Cobalt-based superalloys consume most of the produced cobalt. The temperature stability of these alloys makes them suitable for use

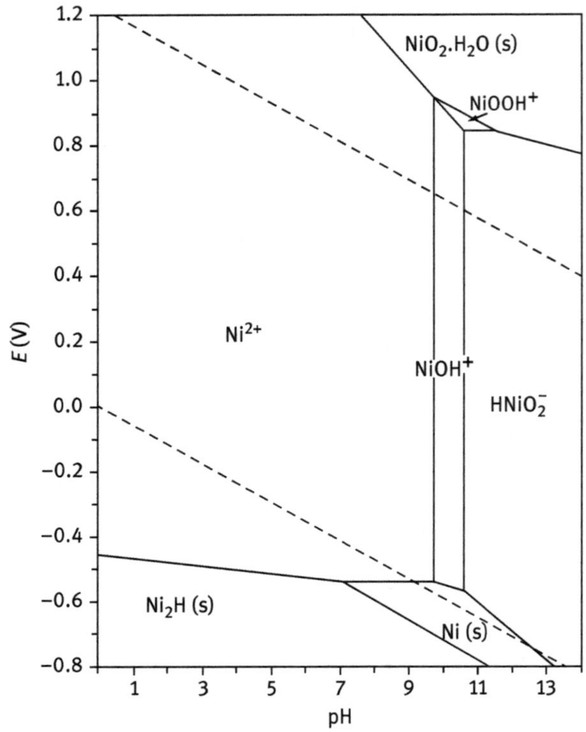

Figure 3.8: State diagram E/pH for the inorganic species of Ni.

in turbine blades for gas turbines and jet aircraft engines. Mixed lithium cobalt oxide ($LiCoO_2$) is widely used in lithium ion battery cathodes. Cobalt is used as a catalyst and component of pigments and colored glasses. Radioactive isotopes ^{60}Co and ^{57}Co have specific applications as radiation source and radiotracers. Average Co abundances in different environments are as follows: lithosphere – 30,000 µg/kg, ocean surface – 0.08 µg/kg, and 0.2 µg/kg in surface freshwater. State diagram E/pH for the most common inorganic species of Co [8] is presented in Figure 3.9.

Essentiality. Co is part of vitamin B_{12} (also known as cobalamin).

Toxicity. Excessive intake of this element may result in polycythemia (overproduction of erythrocytes) and heart problems.

Phytotoxicity. Limited data for relatively low phytotoxicity of Co exists in literature [6].

Chemical species of Co. Co(II), Co complexes with natural ligands.

3.2.1.9 Lead

Application and Pb emission sources. Generally, the industrial applications of Pb tend to decrease, where possible, due to its significant toxicity. As a high-density material, lead is used as a ballast weights. Lead and lead glass are used as a protection

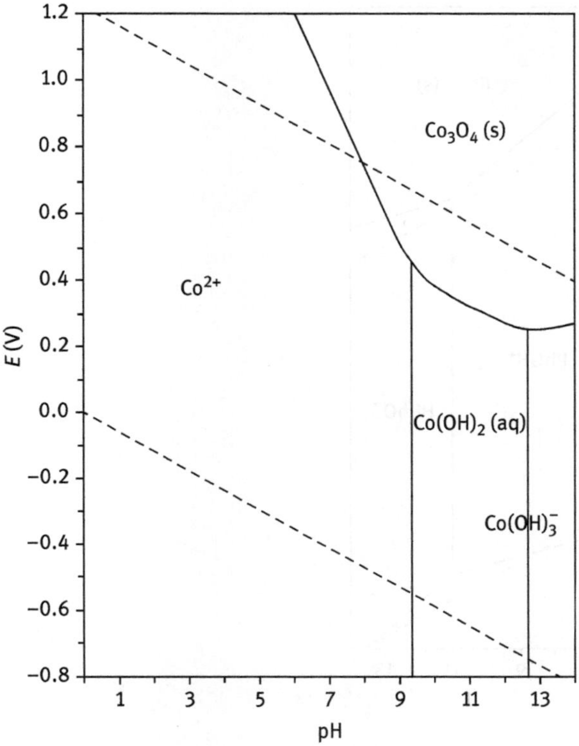

Figure 3.9: State diagram E/pH for the inorganic species of Co.

against radiation applicable to low-intensity radiation. Major application at the moment is in Pb acid batteries, widely used as automobile electric source. Average Pb abundances in different environments are as follows: lithosphere – 10,000 µg/kg, ocean surface – 0.03 µg/kg, and 3 µg/kg in surface freshwater. State diagram E/pH for the most common inorganic species of Pb [8] is presented in Figure 3.10.

Essentiality. Essentiality is not confirmed for lead.

Toxicity. Usually Pb is accepted by the human's body with food and water or inhaled in volatile forms. The metal may affect the central and peripheral nervous system, causes anemia, and damages the kidney. In the human body, lead inhibits porphobilinogen synthase and ferrochelatase, preventing both porphobilinogen formation and the incorporation of iron into protoporphyrin IX, the final step in heme synthesis. This causes ineffective heme synthesis and subsequent microcytic anemia [1].

Phytotoxicity. Pb is known to affect negatively almost all processes related to plant life cycle: seed germination rate, seedling growth, dry mass of roots and shoots, photosynthesis, plant water status, mineral nutrition, and enzymatic activities. In a competitive way, Pb reduces uptake and transport of the nutrients and essential metal ions (Ca, Fe, Mg) [6, 19, 20].

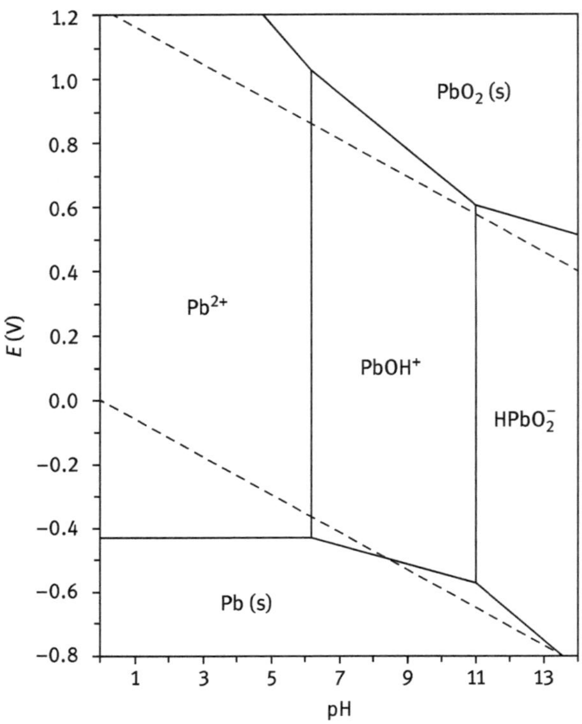

Figure 3.10: State diagram E/pH for the inorganic species of Pb.

Chemical species of Pb. Pb^{2+}, Pb^{+}, DML – dimethyl Pb, TML – trimethyl Pb, DEL – diethyl Pb , TEL – triethyl Pb, TetrEL – tetraethyl Pb, MEL – methylethyl Pb.

3.2.1.10 Tin

Application and Sn emission sources. Tin is mainly used as a component of soldering alloys. As a metal, Sn is used for plating of steel and Cu alloys and component of special alloys (bronze). It is used in Li-ion batteries and in organic synthesis. Organotin compounds are used as polymer stabilizers and biocide component of paints and protective covers (e.g., paint for ships). These are the most toxic and dangerous forms of Sn emitted in the nature. Average Sn abundances in different environments are as follows: lithosphere – 2,200 µg/kg, ocean surface – 0.01 µg/kg, and 0.06 µg/kg in surface freshwater.

State diagram E/pH for the most common inorganic species of Sn [8] is presented in Figure 3.11.

Essentiality. Essentiality is not confirmed for Sn.

Toxicity. Inorganic forms of Sn are low toxic. Cases of poisoning from tin metal, its oxides, and its salts are "almost unknown." On the other hand, certain organotin compounds are almost as toxic as cyanide. Methylated forms are moderately toxic by all routes of exposure. The symptoms include headache, muscle weakness, and paralysis.

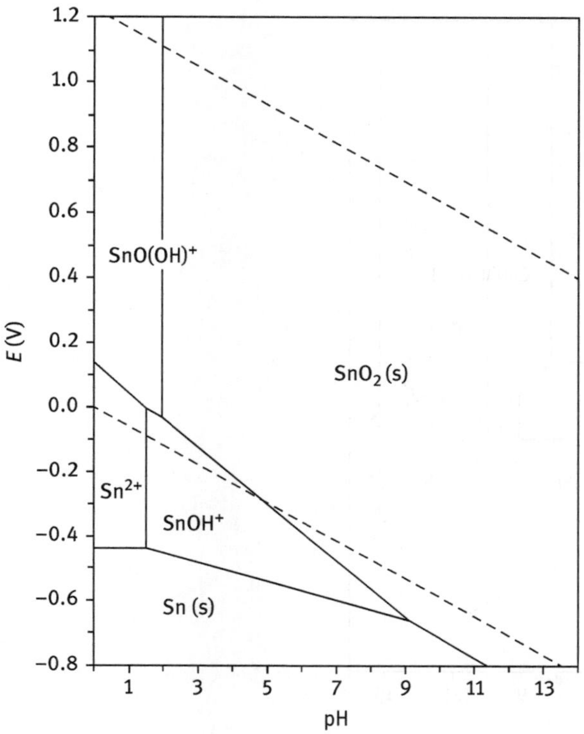

Figure 3.11: State diagram E/pH for the inorganic species of Sn.

Phytotoxicity. Limited information exists. Phytotoxic effects are related to oxidative stress. Tin is similar to Pb [20].

Chemical species of Sn. Sn(IV) – Sn^{4+}, MBSn – monobutyltin, DBSn – dibutyltin, TBSn – tributyltin.

3.2.1.11 Copper

Application and Cu emission sources. Copper meets extensive application in various fields of human activity. Metallic copper is mainly used as conducting material for cables and wires in various electric devices and circuits. It is also used for various alloys and in architecture for covering of buildings and constructions and as a decoration. Copper, its alloys, and compounds have wide antibiofouling, antimicrobial, and generally, biostatic applications. Average Cu abundances in different environments are as follows: lithosphere – 68,000 µg/kg, ocean surface – 3 µg/kg, and 6 µg/kg in surface freshwater. State diagram E/pH for the most common inorganic species of Cu [8] is presented in Figure 3.12.

Essentiality. Cu is involved in the function of some enzymes, for example, cytochrome c oxidase and superoxide dismutase

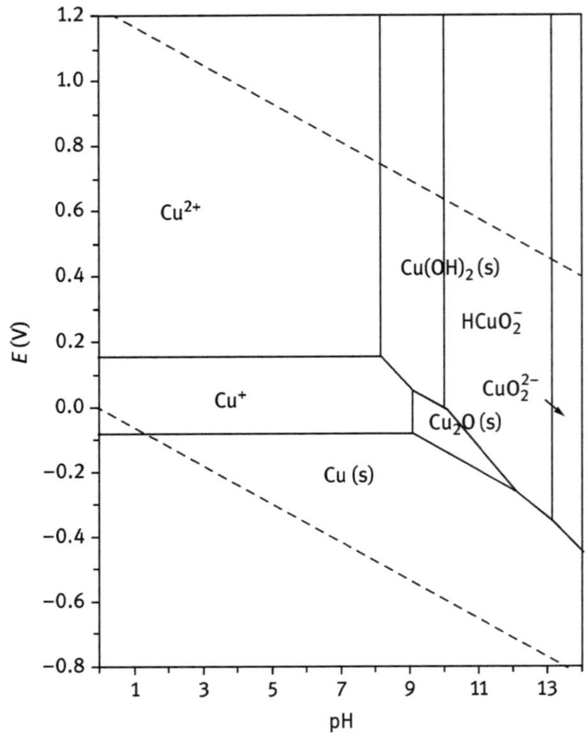

Figure 3.12: State diagram E/pH for the inorganic species of Cu.

Toxicity. Toxicity of elemental Cu is very low. Skin contact results in dermatitis. Many salts of Cu(II) are toxic. Typical symptoms are dry throat, muscle ache, chills, and metal fever.

Phytotoxicity. Cu is an essential element involved in many physiological processes in plants. It acts as a structural element in regulatory proteins and participates in photosynthetic electron transport, mitochondrial respiration, oxidative stress response, cell wall metabolism, and hormone signaling. Plant toxicity is reported at excessive levels. Cu is known to cause chlorosis and shows antagonism with Fe [21, 22].

Chemical species of Cu. Cu(II) and Cu(I), which is extremely unstable under the natural conditions, Cu complexes with natural ligands.

3.2.1.12 Tellurium

Application and Te emission sources. The largest consumer of tellurium is metallurgy, where it is used in iron, copper, and lead alloys. Recent and permanently increasing application of Te is in cadmium telluride (CdTe) solar panels. Organotellurium compounds are used as precursors for metalorganic vapor phase epitaxy in production of semiconductors. Average Te abundances in the lithosphere is 1 µg/kg. State diagram E/pH for the most common inorganic species of Te [8] is presented in Figure 3.13.

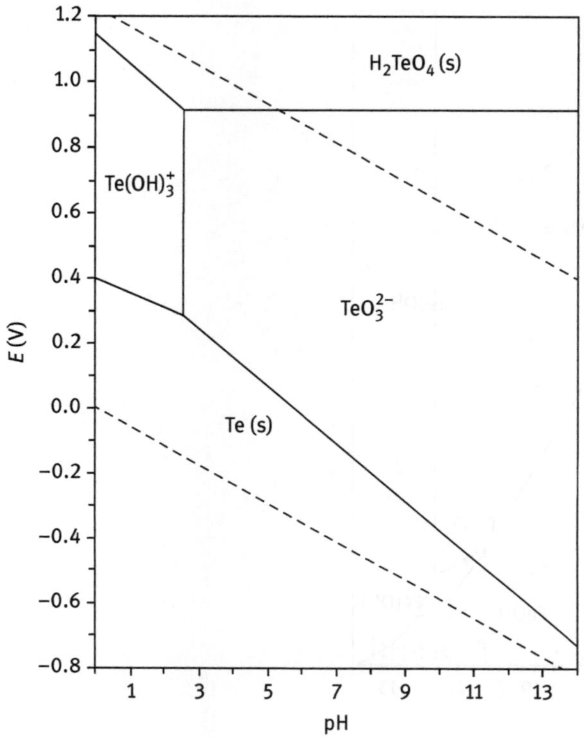

Figure 3.13: State diagram E/pH for the inorganic species of Te.

Toxicity. Tellurium shows low toxicity in elemental state. Clinical symptoms for intoxication are similar for most of the Te salts, including headache, metallic taste, dry throat, loss of appetite, nausea, tremors, and convulsions. Exposure to Te compounds can generate garlic-like odor in breath, sweat, and urine due to the dimethyl telluride formed in the body [1].

Phytotoxicity and essentiality. Not confirmed.

Chemical species of Te. Te(IV) and Te(VI).

3.2.1.13 Iron

Application and Fe emission sources. Iron is extensively used in metallurgy and machinery. Average Fe abundances in different environments are as follows: lithosphere – 63,000,000 µg/kg, ocean surface – 3 µg/kg, and 670 µg/kg in surface freshwater. State diagram E/pH for the most common inorganic species of Fe [8] is presented in Figure 3.14.

Essentiality. Fe is an integral part of the hemoglobin. It participates in some enzymes as well as in electron transfer processes.

Toxicity. Fe is an essential element with very low toxicity at high concentrations.

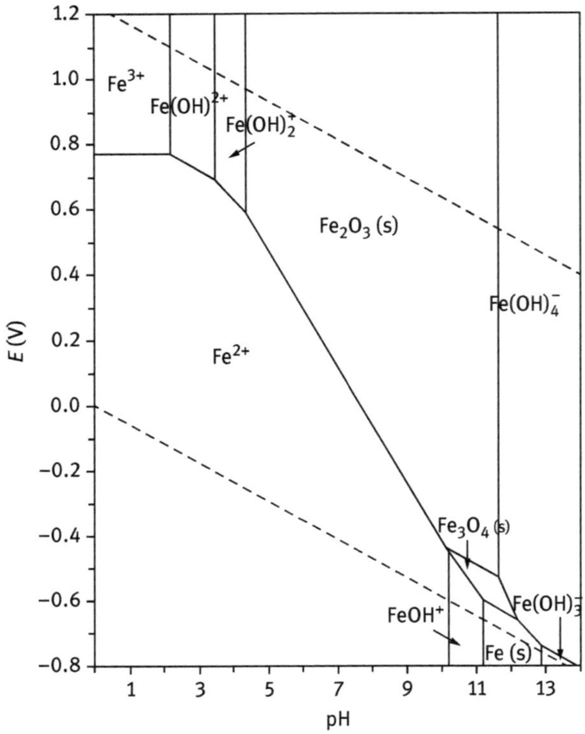

Figure 3.14: State diagram E/pH for the inorganic species of Fe.

Phytotoxicity. At high concentration levels, Fe inhibits root elongation and seed germination [23].

Chemical species of Fe. Fe(II), Fe(III), Fe complexes with natural ligands.

3.2.1.14 Aluminum

Application and Al emission sources. Aluminum is extensively used for production of nonferrous light alloys. Average Fe abundances in different environments are as follows: lithosphere – 82,000,000 µg/kg, ocean surface – 5 µg/kg, and 400 µg/kg in surface freshwater. State diagram E/pH for the most common inorganic species of Al [8] is presented in Figure 3.15.

Essentiality. Essentiality of Al is not confirmed.

Toxicity. Al is toxic for human and animals at very high concentration levels.It has controversially been reported as a factor in Alzheimer's disease.

Phytotoxicity. The phytotoxic effect of soil acidification (pH $\leq$ 5.5) is usually caused by transition of Al in moveable form and subsequent excessive uptake by the plants. Affects mainly roots, but damages also the upper parts (including stem, leaves, and fruits). In addition, Al toxicity triggers an increase in reactive oxygen species. The oxidative stress damages the chloroplasts and inhibits the photosynthetic processes [24].

Chemical species of Al. Al(III), Al complexes with natural ligands.

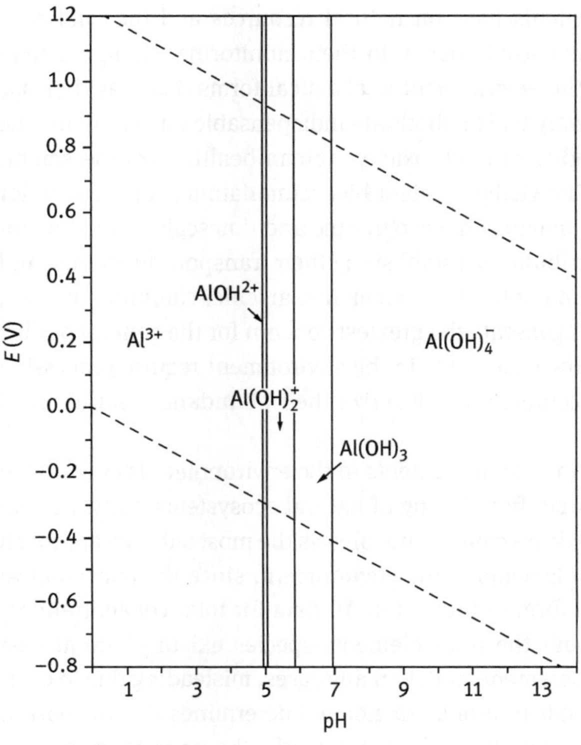

Figure 3.15: State diagram *E*/pH for the inorganic species of Al.

3.3 Environmental monitoring

Trace elements are one of the most critical contaminants in the environment [25] and are present in water, soils, sediments, and air usually at trace levels. However, several human activities (e.g., mining, industry, and sludge dumping) have increased the natural concentration of trace elements and led to severely contaminated environments [26].

Despite efforts to reduce their emissions, atmospheric, terrestrial, aquatic, and biotic compartments are very often affected. This situation is enhanced by some trace elements being particularly reactive in the environment, sifting rapidly between the four interconnected compartments in their biogeochemical cycle [27]. Soils and sediments, in particular, play an important role in the biogeochemical cycle of several trace elements, acting as a sink and a source of the element for biota and humans [28, 29]. Some trace elements can bioaccumulate and biomagnify in biota or affect humans by direct exposure or through food consumption [30, 31].

Monitoring and identification of trace element hotspots in the environment are therefore essential to maintain the health and the proper functioning of natural ecosystems, and to assure the integrity of the food chain and the health of humans and animals that directly or indirectly benefit from these systems.

The great impact trace elements have on natural resources and the quality of the environment justifies the increased interest in their monitoring. Human activities release trace elements into the environment in chemical forms that may increase their bioavailability. Powerful analytical methods are indispensable for evaluating the impact of trace elements in studies of food toxicity, human health risk assessment, and environmental pollution. Knowledge of plant bioaccumulators is also useful for evaluating the effects of trace elements on a geographic and timescale, providing information on trace element distributions, establishing their transport directions, and localizing the emission sites. Among the trace elements, arsenic, cadmium, copper, mercury, and lead are those that generate the greatest concern for the general public. The monitoring and control of trace elements in the environment require processing large numbers of samples to accurately characterize their abundance and to reach reliable conclusions.

Monitoring and identification of toxic elements in the environment is essential to maintain the health and the proper functioning of natural ecosystems. However, determining the total content of toxic elements is not always the most suitable approach to estimate the hazard of trace elements in the environment, since the toxicological effects depend on the chemical form of the metal. All data for total concentrations, without proper information about the trace elements species existing, are at best insufficient for adequate risk assessment, and, in many cases, misleading due to overestimation. In addition, the chemical form of an element determines its transport in and among the environmental compartments. Consequently, the importance of analytical methods for speciation analysis increased in recent years, enhancing the need for validated methods for identification and quantification of trace element fractions.

Generally defined, monitoring refers to any activity in which information about the status of the physical, chemical, and biological characteristics of the environment is collected over time. For the aquatic environment, monitoring usually includes documenting the chemical status of the water column and sediment and the biological status of the aquatic animals associated with them. A long-term assessment and evaluation of monitoring data can help to identify historical trends in the status of a given area. Long-term environmental monitoring involves ongoing data collection in the receiving aquatic environment. These programs are designed to evaluate continuously trends and status in the environment, in terms of the health of important ecosystems.

Environmental monitoring is used in the preparation of environmental impact assessments, as well as in many circumstances in which human activities carry a risk of harmful effects on the natural environment. Assessment of monitoring data helps to identify trends in environment quality and the effectiveness of management measures. For example, it helps to determine whether environment quality standards are being met and whether these standards are adequate to protect environmental resources. According to the Environmental Protection Agency's guidance the monitoring strategy should incorporate or reference the elements listed below [32].

1) ***Monitoring strategy*** – a long-term and detailed implementation plan, which should not exceed 10 years.

2) ***Monitoring objectives*** – these are critical to the design of a monitoring program that is efficient and effective in generating data that serve management decision needs. The objectives should reflect the goals and requirements of the relevant state law.
3) ***Monitoring design*** – an approach and rationale for the selection of a mix of monitoring designs and sample sites that best serves the monitoring objectives. The strategy should be comprehensive with a goal of assessing all waters on a periodic basis. The design should produce scientifically valid data that meet the needs of resource managers and decision-makers.

There are several questions to be answered in the detailed planning of an environmental monitoring program.
- Which sample strategy to be applied?
- Which contaminants are to be measured?
- Which species are to be selected?
- Where are the samples to be collected?
- Which sample design to be selected?
- When is sampling to be done and how frequently?

A sound, science-based decision is based on accurate information. To generating accurate information about the level of contamination in the environment, one should consider the following:
- The appropriateness and accuracy of the sample collection and handling method
- The effect of measurement uncertainty
- The quality and appropriateness of the laboratory analysis
- The representativeness of the data with respect to the objective of the study

4) ***Quality assurance*** – quality management plans and quality assurance program/project plans should be established, maintained, and peer reviewed to ensure the scientific validity of monitoring and laboratory activities, and to ensure that reporting requirements are met. Quality assurance is an important component of the major monitoring programs relied upon by environment protection programs. It is important to ensure that the data generated by monitoring and used to support decision-making in environment protection programs are valid and appropriate. All monitoring programs should have written quality assurance plans that address how the quality of data is assured.

It should be stressed, however, that the acquisition of reliable and relevant data of suitable quality for contaminants in the environment is not solely dependent on the production of accurate analytical measurements. The overall quality of data is also dependent on three other factors, which are

- a meaningful sampling program;
- a suitable storage and pretreatment procedure for samples following collection and prior to analysis;
- a data assessment procedure.

Reliable data are the result of a chain of actions that starts with the proper definition of the problem to be solved. This should lead to a clear definition of the parameters concerned. It includes the target samples to be selected, proper sampling strategy, and sampling techniques to be applied. When adequate samples have been analyzed, the reliable data should be produced by applying the basic principles of the science of measurements, metrology in chemistry, and reported in such a manner that decision-makers can draw conclusions and take actions with all possible assurance. Environmental Quality Standards (EQS) have been developed and although there are no global values, many countries have their own standards which are used to assess pollution levels in the environment. EQS values vary from country to country and are often incomplete. Metal speciation directly impacts on toxicity, but this is often ignored. Despite these omissions, EQS are nevertheless invaluable in the interpretation of monitoring data and the protection of the environment.

3.4 Determination of trace elements in environmental samples

Monitoring and identification of trace element content in environmental samples and trace element hotspots in the environment are essential to maintain the health and the proper functioning of natural ecosystems, to assure the integrity of the food chain, the health of humans and animals that directly or indirectly benefit from the natural ecosystems. Establishing easyto-use protocols for analysis is key to successful assessment of risk from the presence of toxic trace elements in the environment. Sampling is the starting point and a major part of any monitoring program. Sampling design specifies the number, type, and location (spatial and/or temporal) of sampling units to be selected for measurement. A well-developed sampling design plays a critical role in ensuring that data are sufficient to draw the conclusions needed. Inadequate or inconsistent sampling procedures produce data of questionable quality, which may lead to erroneous interpretation and conclusions.

Most analytical testing constitutes three steps: sample preparation, decomposition, and determination.

Sample representivity is requisite to valuable chemical analysis. Following proper sample collection that ensures adequate characterization of the sample entity, sample preparation step must reduce the sample volume to a size suitable for analysis and in parallel preserves the initial signature of the larger sample. The volume of subsample and grain size needed for a statistically valid chemical analysis should be carefully considered.

Sample decomposition can be total, partial, or selective given the requirements of the analysis. Mixtures of strong mineral acids such as hydrofluoric-perchloric-nitric-hydrochloric acids will digest all minerals within the sample permitting total determination of trace elements. Strong acids will dissolve some minerals while leaving others intact for a partial determination of trace elements. Liberation of the trace metals may be studied with a selective or sequential digestion scheme. A sample can be sequentially leached with water, ion exchange reagents, oxidizing or reducing agents, and weak or dilute acids. The bioavailability or the rate of loading of trace metals in the environment can be quantified.

However, the matrices of environmental samples are complex and challenging due to the presence of high amounts of soluble solid substances (i.e., salts of Ca, K, Na, and Mg, chlorides, phosphates, and silicates). Due to high content of dissolved solids, the direct analysis commonly gives many difficulties in sample introduction and nonspectral (matrix effects mainly) and spectral interferences in measurements by flame atomic absorption spectrometry (FAAS), graphite furnace atomic absorption spectrometry (GF-AAS), inductively coupled plasma optical emission spectrometry (ICP-OES), or inductively coupled plasma mass spectrometry (ICP-MS). As a remedy, most of environment samples have to be mineralized before the analysis to destroy the organic matter or diluted to decrease the content of concomitant substances. However, digestion procedures, although very effective in removing the organic matrix of samples, commonly require use of large amounts of reagents that can be an additional source of contamination of the samples being analyzed.

State-of-the-art multielement trace element analyses of environment samples are instrumental determinations measuring mass, light emission/absorption, or gamma radiation properties of atoms within the sample. ICP-OES and ICP-MS are particularly helpful for measurements of trace and ultra-trace concentrations of elements, where high detection power is required. INAA (Instrumental Neutron Activation Analysis) is an example of a nonspectrochemical technique that is successfully used for the multielemental analysis of environment samples. Several review papers giving numerous examples on the determination of trace elements in different environment sample matrices are given in Annex. More details on the nature of several environment matrices and the analytical aspects of determination of trace elements in them are given in Section 3.5.

3.5 Environmental matrices – general characteristics and analytical aspects

3.5.1 Natural waters

Monitoring and identification of toxic elements in the environment are essential to maintain the health and the proper functioning of natural ecosystems. However, determining the total content of toxic elements is not always the most suitable approach to estimate the hazard of trace elements in the environment, since the toxicological effects depend on the chemical form of the metal. All data for total concentrations, without proper information about the trace element species existing, are at best insufficient for adequate risk assessment, and, in many cases, misleading due to overestimation. Distribution, mobility, bioavailability, and toxicity of trace elements depend also on the form in which element exists. Full understanding and prediction of chemical behavior of an element in the environment is possible only by identification of all forms in which that element can be found under different environmental conditions.

In addition, the chemical form of an element determines its transport in and among the environmental compartments. Consequently, the importance of analytical methods for speciation analysis increased in recent years, enhancing the need for validated methods for identification and quantification of trace element speciations. More details on speciation studies for trace elements in the environmental samples are given in Section 3.6.

The chemical composition of all natural waters consists of dissolved inorganic salts, suspended inorganic or organic material, and dissolved gasses at different concentration levels. Practically all natural water samples contain Ca, Mg, Na, K as simple inorganic cations and CO_3^{2-} (usually protonated at some extend), SO_4^{2-}, and Cl^- anions. The trace elements exist in natural waters mainly in the form of soluble inorganic compounds, complexes with inorganic or organic ligands, metal-organic compounds, or as suspended material. Basic characteristics of natural waters, directly influencing chemical forms of trace elements and analytical properties of natural waters as a matrix, are acidity (pH), redox potential, and total dissolved solids (TDS). The ecological importance of natural waters is quite high since they are media for migration of chemical species. Besides this, most of the chemical transformations of trace elements and their uptake in the biosphere are taking place also in aquatic media.

Acidity. In most of the cases, the acidity of natural waters is determined by dissolved gasses (usually acidic, like CO_2 and SO_2) or equilibrium involving minerals being in contact with the waters. It is influenced also by living forms producing in water O_2 or CO_2. The pH value is controlled by protolithic species among the main components of the water. Typical pH values for natural waters are between 5 and 9.5. The ocean water has slightly basic character (pH between 7.5 and 8.4), while fresh surface waters are slightly acidic to neutral (pH between 5 and 7.5–8). Rarely, extremely high or low values can also be observed. For example, the active volcanoes with high-temperature magma (over 1,000°C) are frequently releasing gases like HCl, HF, and so on which can decrease pH of natural waters in the region to 1 and below. On the contrary, waters of lakes situated in regions rich of Ca and Mg carbonates are often basic ($pH \geq 10$). Acidity of natural waters is crucially important for living organisms as well as for chemical forms of the trace elements. It is important to note that pH is temporally unstable characteristic. After sampling, it can be changed even if the sample is not preserved by addition of acids or bases. The reason for changes is absorption or release of gases, or change in microbiological conditions.

Redox potential. The redox potential is controlled by equilibrium involving usually one pair of oxidant/reductant present in the system. The components are among the main components. Trace components practically cannot change the redox potential of the system and the ratios between their oxidant and reductant forms are such to be in equilibrium with all the rest pairs. For aerated surface waters, redox potential is controlled by the dissolved oxygen (O_2 / $2H_2O$) involved in the following equilibrium:

$$O_2 + 4H^+ + 4e^- \leftrightarrows 2H_2O, \text{ with standard electrode potential } E_0 = 1.23\text{ V}.$$

This equation, under standard conditions, corresponds to the upper limit of stability of water presented on the E/pH state diagram as dashed line. The redox potentials of natural waters isolated from the atmosphere environments are determined by various equilibria involving biota decay products (e.g., SO_4^{2-}/HS^-, CO_2/CH_4, etc.) with low standard potentials. Redox potential is even less stable than pH. It should be measured at sampling sites using Pt and reference electrodes.

Figure 3.15 presents the state diagram E/pH representing typical regions for various types of natural waters. Variations are possible due to anthropogenic influences or seasonal and geochemical variations.

Total dissolved solids. One of the basic characteristics of natural waters is the TDS usually expressed in unit mg/l. Values are determined either gravimetrically, after drying at 110–120°C, or as a sum of main components expressed in corresponding units. Considering the TDS value, natural waters are classified in four groups: fresh, brackish, saline, and brine waters.

Freshwaters. (TDS<1,000 mg/l)

To this group belongs atmosphere waters (including all kinds of precipitations), most of the surface waters, glacial, and some of the groundwaters. Freshwaters are totally about 2.6–2.75% of the hydrosphere. Main components of the freshwaters are cations: Ca^{2+} and Mg^{2+} and anions: CO_3^{2-} (HCO_3^-) and SO_4^{2-}. In regions close to the seas and oceans, dominant ions are Na^+ and Cl^-. The freshwaters usually contain low concentrations of dissolved organic material. Limited amount of organics and low-soluble inorganic material might present in suspended form. Usually freshwater samples are filtered at the sampling sites, which improves the sample stability. In some studies, suspended material is analyzed separately from the water after proper sample preparation. Due to the low content of dissolved matter, freshwaters are one of the simplest from analytical point of view matrices. Conductometric determination of TDS is applicable to freshwaters. For natural waters from other groups, the conductivity becomes too high and this method becomes inapplicable (Figure 3.16).

Saline waters (TDS 30,000–50,000 mg/l). This group contains ca. 97.5% of the hydrosphere. Waters of all oceans and most of the seas on the planet belongs to this group. Solubility of the most of Ca and Mg minerals is rather low and such salinity might be reached only on the basis of highly soluble salts such as NaCl. Significant changes in the dissolved oxygen concentration and redox potential of the water occurs in depth. Usually it is connected with changes in redox state. Due to the mass of the

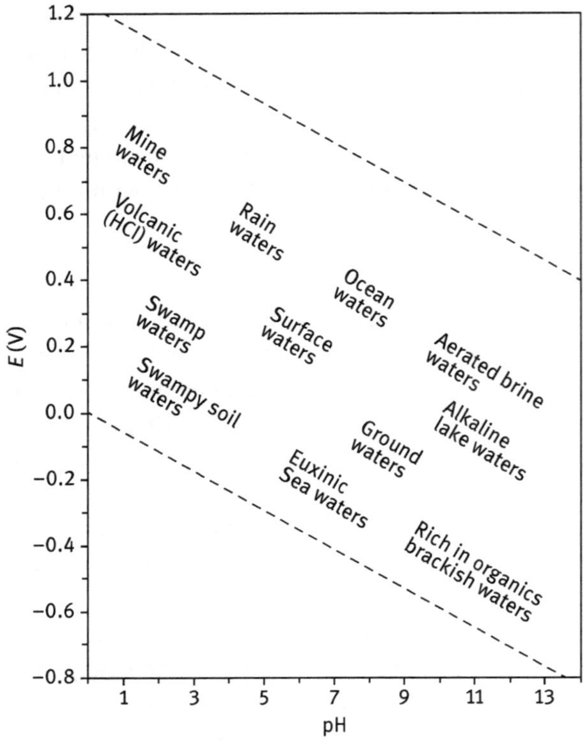

Figure 3.16: Disposition of some types of natural waters on state diagram *E*/pH. Typical values for *E* and pH are taken into account. Deviations are possible due to anthropogenic influences or peculiar local geochemical conditions.

ocean and seawaters the average physicochemical parameters and chemical composition (including trace components) are extremely stable. However, the changes are practically irreversible. Analysis of seawaters usually requires decrease of salt content.

Brines and evaporites (TDS > 50,000 mg/l). To brine waters belong surface and underground basins in which sedimentation of solid minerals takes place. This is a very rare type of natural waters and seldom analyzed. Strong dilution is required.

Generally, the natural waters of all kinds are highly homogeneous samples after filtering the suspended matter. For freshwaters, preconcentration might be required due to the low levels of trace elements, while for waters with higher TDS procedures for decreasing of matrix content are usually needed. Possible Cl interferences should be considered.

General analytical aspects on the determination of trace elements in the natural waters. Freshwater is an essential natural resource for the survival of all ecosystems. From the viewpoints of environmental management and protection from hazardous contamination, trace elements, which cause serious pollution or toxicological problems, are a worldwide concern. As a rule, metals in the environment preferably form adducts with organic species such as naturally occurring complexing agents or ligands released from human activity. The majority of the organic species that persist in natural aquatic systems are humic substances (HSs). Ubiquitous HSs are produced by the degradation of biomass and are heterogeneous with respect to their molecular weight and chemical structure. Because the characters of metals are altered by the presence of such natural organic matters, elemental speciation information in evaluating trace heavy metal status is particularly crucial today instead of determining the total metal concentration [33]. Due to the different toxicities and bioavailabilities of trace metal species present in natural waters, the assessment of their distribution over distinct molecular binding chemical forms is very important for the elucidation of their role in the environment and biological interactions with living organisms. All element forms play a crucial role in assessing the health hazard and toxicity of trace element metals in the ecosystem (e.g., ionic forms of a number of trace elements have been recognized to be much more toxic to biota than the corresponding colloidal or particulate forms). What makes speciation analysis so difficult even now is the very low concentration of metals, usually found at the level of several nanograms per liter, and the presence of a highly saline matrix, which results in deterioration in detection power of the analytical techniques used. Furthermore, trace elements in natural waters are present in the form of a wide range of species varying in particle size and chemical nature, for example, free hydrated ions, dissolved inorganic and organic complexes of different stabilities, and metals associated with colloidal or particle matter. A large number of individual species make it impossible in practice to determine speciation that is in line with the International Union of Pure and Applied Chemistry (IUPAC) recommendations about the distribution of an element among defined chemical species in the system under consideration [33]. For the reasons above, alternative approach has been proved namely to identify the fractions (classes) of the metals that exist in the distinct groupings and to determine the sum of their concentrations in each class using particular techniques, rather than providing a detailed speciation analysis by exact definition of the molecular forms of metals and the structure of ligands complexing the metal ion [34, 35]. To obtain the total concentrations of dissolved metals, the development of an efficient decomposition procedure for the degradation of metal-organic complexes is driven by the need to gain detailed information on different aqueous metal species. Generally, decomposition procedures are carried out using batch-wise wet digestion methods such as microwave digestion [36, 37] and ultrasound-assisted digestion [38, 39]. Although wet digestion is currently the prevalent method, it suffers from contamination associated with reagent addition, analyte loss resulting from the severe operating conditions, and difficult adaptation to flow injection operation. Therefore, the approach of using ultraviolet (UV) irradiation for the decomposition of dissolved metal–organic complexes has also been developed [40, 41]. To improve the performance in terms of the decomposition efficiency, intense investigations have been undertaken on the photochemical digestion of organic substances with the aid of a photocatalyst [42]. Various instrumental techniques, including GF-AAS [43], ICP-OES [44], FAAS [45], and ICP-MS, have been applied for the heavy metal determination [46]. Since trace element concentration level in natural waters is very low and the complexity of matrices is a problem, preconcentration techniques are often required. Different procedures such as liquid–liquid extraction [47], chemical precipitation [48], ion exchange [49], and solid-phase extraction (SPE) have been developed for the extraction and preconcentration of trace element metals in natural matrices [50]. Among the above-mentioned methods, the most commonly used technique for the preconcentration of heavy metal ions from environmental samples is SPE. Its common application is due to its simplicity, rapidity, minimal cost, and low consumption of reagents. By the advent of SPE, diverse sorbents such as carbon nanotubes [51] and magnetic nanoparticles [52, 53] have been applied.

Seawater analysis represents the combination of low elemental concentration and a complex matrix due to the high salt concentration [50, 51].

Modern, high-resolution, magnetic sector ICP-MS instruments have become available to meet this need for a multielement, highly sensitive method with a wide dynamic range. Seawater, however, is a complex matrix with high concentrations of major ions that can interfere with trace metal analysis. Most analysts consider that it is essential to extract and separate the trace metals of interest from the original matrix. Trace metals generally exist at such low concentrations in seawater that a concentration step prior to detection is also often necessary. The avoidance of sample contamination and proper determination of analytical blanks are also critical steps taken in the analysis of trace metals.

3.5.2 Soils

Soils are complex mixtures containing mineral components, organic matter, groundwaters, and gases. The chemical composition of the soils might vary in very wide range depending on the geographic region, climate, and depth of the sampling. Various processes are involved in the soil formation and being responsible for the chemical composition. Historically, the initial soil formation was started ca. 500 million years ago and was related to migration of macroscopic living organisms to the surface of continents and islands. They were the primary source of organic matter for protosoils.

Mineral components. Major source of mineral components in the soils is the parent rocks undergoing physical and chemical transformations, translocations, adsorption, and release of the components in contact with groundwaters as well as an influence of plants. General term of all these processes is weathering of the rocks. As average, soils contain 45–47% (v/v) inorganic mineral components with various porosity and particle size. Considering particle size, three types of mineral components are present in the soils: sand (2–0.05 mm), silt (0.05–0.002 mm), and clay (<0.002 mm). As a chemical composition it consists of main components (expressed as oxides) such as SiO_2, Al_2O_3, Fe_2O_3, and minor components TiO_2, MnO, P_2O_5, CaO, MgO, and so on. The mineral composition varies significantly depending on the bed rock type and the depth of sampling. The main components are always present in the samples, although varying in ratios, while the minor ones can be completely different. Besides this, some trace elements can be included as isomorphic substitutes of ions or occluded in the gaps of crystal lattice.

Organic components. As average, soils contain ca. 5% (v/v) organic components. The source of organic matter is degradation of animal and plant residuals. Amount of organic compounds in soil depends on the climatic conditions, landscape, and biota development in the region. The organic matter in soils is a complex mixture of organic compounds within the range from plant and animal tissues to relatively stable product of their physical, chemical, and biochemical degradation – humus. Humus is also a complex mixture of carbohydrates, lignins, resins, waxes, and organic acids (mainly humic and fulvic acids). Presence of large number of O and N donor atoms in the molecules provides the possibility for chelate complex formation with transition metals. In this form, trace elements are able to migrate in the stream of groundwaters.

Liquids. Soils contain normally about 25% (v/v) liquid substances. This value depends strongly on local climatic conditions and weather situation. The biggest fraction of the liquids is capillary moisture. Soil liquids contain mainly soluble minerals and minor amount of organic substances.

The soil moisture is an integral part of the soil and generally is not subject of particular analysis. Due to the high variability of the moisture content, it is important to determine its value in order to recalculate results for trace elements to comparable state. Most frequently, the moisture is determined gravimetrically after heating at 105°C. Use of higher temperature might produce irreproducible destruction of the organic components. Reduction of moisture content from sample mass is needed in order to guarantee traceability and comparability of the obtained results.

Gases. Gases in the soil are about 25% (v/v). Most of all these are components of atmosphere filling the pores of the soil. Partly soil gases are produced by decomposition of the biomass in the soil (CO_2, CH_4, etc.) and metabolism of microorganisms living in soil (N_2, N_2O, CO_2, etc.). Thus the CO_2 concentration in soil pores might be 10–100 times higher than in the atmosphere. The case with O_2 is opposite. In highly porous (sandy) soils, the gas exchange rate is higher and deviations in composition from the atmosphere are lower. Composition of the soil atmosphere also varies depending on the depth, climate, and season. Soil gases are crucially important for plant development, but are not an object of interests in trace analysis.

Analytical aspects on the determination of trace elements in the soils. Nowadays, the knowledge of toxic element content in soil is considered an important aspect in environmental research, as it is one of the most important recipients of trace elements that can be originated from anthropogenic activities [54–56]. Trace elements in soils can be either included in the mineral particles (mainly silicates) or exist as a mobile species in the form of simple inorganic ions, organometallic substances, and complexes with inorganic or organic ligands. The trace elements included in mineral's crystal lattice are generally not absorbable for plants. Targets for trace element determinations in soil are either total content, or only mobile forms that can be assimilated by the plants and thus introduced in the food chain. Sample preparation technique is essential in both cases. Trace elements in soils may exist in different chemical forms or ways of binding. In unpolluted soils or sediment, they are commonly bound to silicates and primary minerals forming relatively immobile species.

However, trace element determination in soil is still a challenging task because soils with high inorganic fraction are difficult to bring into solution and a time-consuming and not always simple sample preparation procedure is generally required.

The direct determination of these analytes in soil has been carried out using laser-induced breakdown spectroscopy [57, 58], X-ray fluorescence (XRF) [59], laser ablation (LA) or electrothermal vaporization (ETV) coupled to ICP-MS [60], and solid sampling graphite furnace atomic absorption spectrometry (SS-GF-AAS) [61, 62]. In spite of the relatively good performance of these techniques, problems related to calibration, necessity of certified reference materials (CRMs), as well as elemental and isotope fractionation have been still reported [63]. Moreover, matrix effects can affect the accuracy, and problems related to sample homogeneity are known to affect the precision for some of these techniques. Analytical techniques such as ICP-MS [64], ICP-OES [65], and AAS [66] are among the most common spectroscopic techniques used for the determination of toxic elements in soil. In the case of mercury, cold vapor generation coupled to AAS [67] or atomic fluorescence spectrometry (AFS) is currently used [68].

In general, procedures for soil digestion for ICP-MS, ICP-OES, and AAS determination involve dry ashing [69], alkaline fusion [70], microwave-assisted wet digestion [71], and microwave-assisted extraction in closed vessels [72]. Sample digestion by using strong acids – concentrated HNO_3, aqua regia, or other mixtures – might provide total destruction of organic part of the soil. However, trace elements included in silicate fraction can be only partly leached. For determination of total content of trace elements, an addition of HF is required to provide complete dissolution of minerals in soil. For mobile forms determination, the use of concentrated strong acids and their mixtures is not applicable. Solutions imitating fluids in plant roots (e.g., slightly acidic buffers with/without addition of organic substances) are usually used.

3.5.3 Sediments

As a matrix, the sediments are rather close to soils. The significant differences in composition are related to the processes of formation of sediments. There are four basic types of sediments according to their genesis.

Lithogenous (terrigenous) sediments. Lithogenous sediments consist of various in size and composition soil and lithosphere particles. They are produced in the process of soil and rock erosion and transported by wind, surface flows and glacial ice movements. Lithogenous particles are dominant parts of the freshwater and seashore sediments. Depending on their size, their content decreases when separated from the shore. The finest particles are carried out by the streams far from the input point. The composition of this type of sediments is very close to the soils with a general exception of decreased amount of humus, which is remaining mainly in the water phase.

Biogenic sediments. Biogenic sediments are mainly originated by plankton remains. Contributions of macroorganisms are considerably lower and correspond to the biomass distributions between various aquatic species. Two are the basic types of biogenic sediments: calcareous and siliceous, consisting of $CaCO_3$ and hydrated SiO_2, respectively, depending on the types of plankton. Deep ocean sediments are usually rich in such biogenic material – up to 30% or more, while the content in seashore sediments and especially in freshwater sediments is significantly lower.

Hydrogenous sediments. Hydrogenous sediments are formed by precipitation of minerals from seawaters. Seawater contains low concentrations of metal ions and anions that normally do not exceed the solubility of minerals which potentially can be originated. However, due to steep change in the seawater properties, sedimentation conditions might be reached. Zones where such sudden change in conditions takes place are known as geochemical barriers. Usually deposition of minerals from natural waters takes place around the geochemical barriers. Usual reasons for appearance of geochemical barriers are evaporation, decrease of the temperature or increase of the concentration of a precipitating component caused by streams, change in redox potential or pH, and adsorption on minerals already existing on the ocean floor.

Accidental impacts. The main natural sources of accidental impacts to sediments are volcano eruptions and cosmogenous particles. Some anthropogenic impact has occasional nature as well.

Volcano eruptions are accidental events introducing, in some cases, vast amount of magma and volcanic dust particles in the atmosphere. For instance, the eruption of Mt. Pinatubo, Philippines (1991), has ejected 10^{10} t (or ca. 10 km^3) volcanic particles in the atmosphere. Such events are affecting the surface of the entire planet and the composition of soils and sediments, formed in that period, as well.

Cosmogenous sediments are extraterrestrial in nature, consisting generally of micrometeorites formed by impacts of larger bodies of space material destructed in the Earth atmosphere. Specific component of these sediment fractions is tektites – glassy structure particles formed either by heating in the atmosphere of extraterrestrial material or, more frequently, lithosphere debris ejected during extraterrestrial impacts. The amount of outer space material reaching the Earth surface daily is significant.

Chemical composition of sediments, in general, is very close to those of soils. Frequently, considering environmental trace analysis and speciation, soils and sediments are objects of simultaneous study using the same or very similar sample preparation and analytical techniques. Despite the closeness in composition, several substantial differences should be underlined. Due to the underwater formation, the gas content of the sediments is always considerably lower than in soils. Organic material content is much lower as well. On the other hand, the sediments are formed as a dense layer of material deposited in different times, under conditions ensuring far lower intensity of erosion, as compared to the soils and lithosphere in contact with the atmosphere. This provides unique opportunity to study indirectly processes taking place in history. A depth profile distributions of certain elements allows detecting and dating of substantial events, both in human's history and geological history of the planet.

General analytical aspects on the determination of trace elements in sediments. Analytical determination of trace element concentrations and distribution in marine environment leads to better understanding of their behavior in the aquatic environment and is important for detecting sources of pollution. Sediments are important carriers of trace metals in the hydrological cycle and because metals are partitioned with the surrounding waters, they reflect the quality of an aquatic system. The primary mechanisms of deposition of trace elements found in the marine environment are the deposition from the atmosphere, industrial and agricultural discharges, and storm water runoff. Intense sedimentation within estuarine and marine environments traps trace elements within fine-grained particles which then precipitate and filter heavy metals out of the immediate biosphere. The intense sedimentation concentrates trace element metals and helps to limit their environmental impact. Thus, geochemical characteristics of the sediments can be used to infer the sources of pollution. Because of their large adsorption capabilities, fine-grained sediments represent a major repository for trace metal and a record of the temporal changes in contamination. Thus, they can be used for historical reconstruction. Vertical profiles of pollutant species in sediment cores have been commonly used as "pollution records" over the last decades. Over the last decades, the study of sediment cores has shown to be an excellent tool for establishing the effects of anthropogenic and natural processes on depositional environments. A number of recent pieces of work have used sediment profiles to describe the contamination history of different environments [73, 74].

Due to their stability within the sedimentary column, most of the trace metals can leave their fingerprint in sediments. Metals enter the environment and oceans by two means: natural processes (including erosion of ore-bearing rocks, wind-blown dust, volcanic activity, and forest fires) and processes derived from human activities by means of atmospheric deposition, rivers, and direct discharges or dumping. For some metals, natural and anthropogenic inputs are of the same order (for sample Hg and Cd), whereas for others (e.g., Pb) inputs due to human activities dwarf natural inputs [74, 75].

The trend of trace elements to accumulate in sediments is due to adsorption, coprecipitation and metal complexation in the surface layers of the sediment particles. Four types of metal fractions can be distinguished: (1) the exchangeable fraction, which is located in the ion-exchange sites, and it is freely available to take part in chemical reactions; (2) the reducible fraction, consisting of complex oxides and hydroxides, which is soluble under reducing conditions; (3) the oxidizable fraction, which is formed with the organic matter and sulfides, and it is available under oxidizing conditions; and (4) the residual fraction, which is introduced into the environment in the crystalline form by geological processes and it is generally not available [76, 77].

Sediments act as metal reservoirs and, for that reason, the analysis of metals in sediments allows us to detect pollution in the marine environment and provides information about the ecosystem health.

A critical point in the analysis of soil and sediment samples is the sample preparation step, which can largely influence the obtained results. Results from various studies are only comparable, when identical sample dissolution steps are applied for investigated elements.

The two principal types of analysis include the analysis of total trace elements and analyses associated with various matrix components (speciation analysis). The metals incorporated in the crystal lattice of quartz particles of soil and sediments are of no consequence to food chains, and plant growths are seldom of concern. In order to achieve a complete removal of the pseudo-total metal content, the exchangeable fraction mentioned earlier, acids such as HCl and HNO_3 are used; they can remove this fraction of metals avoiding its incorporation into the mineral structure of the sediments joined to silicates, also called the residual fraction [77]. To remove this residual fraction, which would enable the determination of the total metal content, HF must also be used. This methodology can be combined with the application of microwave energy, a process known as microwave-assisted digestion, which provides a faster heating alternative, allowing shorter radiation times [78, 79].

A new green method for the extraction of the pseudo-total content of the heavy metals such as Ni, Cu, Cr, Pb, and Cd from marine sediments using a mixture of biodegradable micellar media (SDS, Triton X-100) as extractants and GF-AAS for their determination has been proposed [80]. The use of surfactants as extractants allows adapting the technique of microwave-assisted digestion to completely avoid the use of toxic and corrosive acids, a method known as microwave-assisted micellar extraction [81].

Several studies have been reported good results using slurry, with recoveries comparable to those obtained with other pretreatments such as acid digestions or solid sampling when all these treatments have been applied to the same samples [81].

A harmonized protocol (ethylenediaminetetraacetic acid, diethylenetriaminepentaacetic acid, acetic acid as single extraction procedures, and the three-step sequential extraction procedure) for determination of forms or phases of elements (e.g., “bioavailable” forms of elements) in soils or sediments had been validated in the 1980s, and reproducibility and the repeatability of the measurements evaluated. Further investigations have led to slight modifications of the three-step sequential extraction scheme [77] and a considerable improvement in its reproducibility.

GF-AAS [81–83] is widely used for the determination of trace and ultra-trace element analysis of sediments; FAAS and ICP-OES are applied for the determination of minor and major elements [84].

ICP-MS is the most sensitive, multielemental technique, but it requires sample digestion, unless ETV or LA is used. ICP-MS allows for the simultaneous determination of both elemental concentration and isotope ratios. Polyatomic interferences generated from the sample itself or from the medium used for sample dissolution pose a serious problem in ICP-MS analysis of sediment samples [85, 86].

Direct analytical techniques are obviously preferable over those that require significant sample preparation, particularly in the case of complex samples, such as marine sediments. SS-GF-AAS has been shown to be a powerful technique that, after careful optimization of the conditions, provides accurate results and often allows calibration against aqueous standards [87]. The main advantages of this technique are high sensitivity, since no dilution is involved; minimum risk of contamination; high sample throughput; and no use of reagents. The only disadvantage, which is common to all direct sampling techniques, is the relatively high uncertainty, caused by the heterogeneity of natural samples and the small amount of sample, typically lower than 1 mg, which is introduced into the atomizer [88, 89].

3.5.4 Air, airborne particles

Typically trace elements form very limited number of gas-phase compounds. They are usually not stable and in atmosphere undergo quick transformation to suspended particles. Depending on their origin, the size and composition of air particles might vary in a wide range. Two types of sources of particulate matter are clearly distinguished: primary and secondary sources.

Primary sources. Primary sources are those that provide direct impact of particulate matter in the atmosphere. Naturally originated particles in the atmosphere include the following:

Mineral and soil weathering products blown by wind. Their composition is varying widely and is determined by the geochemistry of the origin area. Permanent main components are metal oxides (dominant elements – Si, Al, Fe, etc.) and minor amounts of organic material (humus and humic acids).

Second largest contributor is dried in the air sea aerosols. Composition of sea aerosols is identical with seawater. The main component is NaCl with minor amounts of Mg^{2+}, Ca^{2+}, $SO_4{}^{2-}$, and so on. They are distributed around the sea and ocean shores, however, can be carried by winds (sometimes over 200 km) within the continents.

Forest and grassland fires are emitting in the atmospheric carbon particles and organic materials – products of incomplete burning of the plant and tree tissues.

Biological sources (pollens and various microorganisms) are seasonal factors.

Volcanic dust is an occasional source ejecting in the atmosphere enormous amounts of solid material. Composition corresponds to the average abundance of elements in lithosphere – predominantly oxides of Si, Al, and Fe. Typical for volcanogenic particles is their glassy structure.

Major anthropogenic primary sources of particles include:

Industrial and urban dust. Despite the regulations and trend to decrease such emissions (especially in developed countries), this remains a rather significant source.

The composition of these particulates is extremely diverse and remains one of the major sources of toxic trace elements in the atmosphere.

Products of incomplete combustion of fossil fuels. Specific for these aerosols is their organic nature including liquids in many cases and increased content of platinum group metals (originated from car catalysts).

Secondary sources. Secondary sources are sources of aerosols which are emitting initially gaseous substances, which later undergo oxidation (e.g., H_2S oxidized to solid sulfur particles), condensation, or reaction with other gases producing solid salt particles (e.g., NH_3, NO_x, SO_2, etc.).

Particle sizes. Typical sizes of airborne particulates are between 1 nm and 100 µm (as averaged diameter). Larger particle size leads to faster sedimentation, limited uptake, and significantly lower health risk. Two fractions are major objects of study: PM_{10} (the fraction of particulates in air of size <10 µm) and $PM_{2.5}$ (particles of size <2.5 µm). The particles between 1 and 5 µm are related to the highest health risk and the most effective uptake of toxic components. Over 5 µm the particles are effectively stopped in oropharynx and not reaching lungs. Below 1 µm is very likely to be exhaled by tidal breathing.

Analytical aspects of trace element analysis of air and air particles. Trace analysis of air samples usually include filtering of the air to select fraction of particulates, sample treatment (very similar to those for soils and sediments), eventually separation of chemical species and determination.

The concentrations and size distributions of trace metals in the air are governed by the nature of emissions into the atmosphere as well as by the rates of wet and dry deposition, cloud processing, chemical transformations, and the exchange of air between the boundary layer and free troposphere [90].

One of the main sources of atmospheric contamination in large cities comes from atmospheric particulate matter and gases derived from anthropogenic activities such as traffic and industry [91], which can impact on rural and ecological interest areas modifying the regional background concentrations.

The size distribution, concentration, component, and toxicity of PM are important factors that can greatly affect human health problems related to exposure to airborne PM. According to health experts, PM_{10} is known to affect human thoracic health and is thus considered a thoracic fraction, and $PM_{4.0}$ is known to affect the respiratory tract and is thus a respirable fraction [92]. However, PM_{10} and $PM_{2.5}$ alveolar fractions are usually selected as the monitoring parameters for evaluations of air quality. Air pollution by PM is considered a serious environmental issue due to the presence in the atmosphere of hazardous materials such as toxic trace metals [93]. Metals associated with respirable particles have been shown to increase lung and cardiopulmonary injuries in humans. Trace metals are found in almost all atmospheric aerosol size fractions and, in general, fine PM carries a higher burden of heavy toxic metals than does coarse PM [94]. Natural emissions (from crustal minerals, forest fires, and oceans) and industrial processes (fuel combustion, waste incineration, mining, and automobile exhaust) are the principal sources of the metals in ambient air.

The measurement of trace element concentrations in aerosols is of great interest for the assessment of human population exposure and for a wide set of environmental studies. The airborne particles are generally collected by filtration of a known volume of air pulled for a preset time. Choice of filters is critical with respect to particle size, collection efficiency, and likely contamination. Concentrations of some trace elements in aerosols have been reported for a large variety of locations all over the world [95, 96]. These studies have been favored by the recent spreading of ICP-MS technique, allowing for the simultaneous and multielemental determination of a wide set of target analytes at trace and ultra-trace concentrations [97–99]. Another multielement methods for analysis of the filters without digestion are XRF [100] and INAA [101]. Filters can be directly introduced into a plasma or a GF atomizer, but elevated background problems, calibration difficulties, weighing errors, and sample representativity are problematic factors for obtaining reliable measurement results [102].

There are many studies of trace elements in air filter samples based on open vessel digestion procedures with different reagents such as aqua regia, HF + HNO_3, and H_2SO_4 + H_2O_2 [72], but the most satisfactory results are obtained using microwave-assisted digestion procedures [103]. For all elements with fractions are associated with silicates, many studies proposed reagent mixtures with HF. The use of HF in the filter digestions requires the use of H_3BO_3 to redissolve fluoride precipitates [104]. The removal of the silicate matrix is essential to increase the stability of a sample solution and to avoid the precipitation of the silicate. Loss of certain elements (Si, B, As, Be, Sb) as volatile fluorides is known limitation, when HF is used for the sample digestion. For this reason, many other studies use digestion procedures based on methodology using only HNO_3 [104].

3.6 Speciation analysis in the environment samples

According to the IUPAC recommendations – 2000 [34, 105], the following basic definitions related to the speciation of trace elements should be considered.

Chemical species. Specific form of an element defined as to isotopic composition, electronic or oxidation state, and/or complex or molecular structure.

Speciation analysis in the analytical activities of identifying and/or measuring the quantities of one or more individual chemical species in a sample. ***Speciation of an element*** is the distribution of an element among defined chemical species in a system.

The term speciation is different from fractionation which describes the process of classification of an analyte or a group of analytes from a certain sample according to physical or chemical properties.

In most of the cases, the speciation analysis methods are based on hyphenated techniques combining separation and detection method. The term hyphenated techniques refers to online combinations of chromatographic or electrophoretic separation technique with a sensitive and element-specific detector (atomic absorption, atomic emission, or mass spectrometer) [106–108]. The choice of detector component becomes critical when the concentration of the analyte species in the sample is very low and when the detection limits required are in the low concentration range. The separation component of the coupled system becomes of particular importance when the target species have similar properties.

For separation purposes, various types of chromatography are used most frequently. Below are presented some important definitions related to separation techniques and acronyms used further.

Chromatography is a physical method of separation in which the components to be separated are distributed between two phases, one of which is stationary phase while the other the mobile phase.The stationary phase is one of the two phases forming a chromatographic system. It may be a solid, a gel, or a liquid [109]. If it is a liquid, it may be distributed on a solid. This solid may or may not contribute to the separation process. The liquid may also be chemically bonded to the solid (bonded phase) or immobilized onto it (immobilized phase). A fluid percolates through or along the stationary bed in a definite direction. It may be a liquid chromatography (LC) or a gas chromatography or a supercritical fluid chromatography. The possibility for changes of the stationary and mobile phases, size and construction of the columns, as well as the experimental conditions with respect to the sample composition provides unique flexibility and makes the chromatography the most widely used separation technique in the speciation analysis (Figure 3.17).

3.6.1 Types of chromatographic techniques used in speciation analysis

Gas Chromatography is a separation technique in which the mobile phase is a gas. GC separation is always carried out in a column. *LC* is a separation technique in

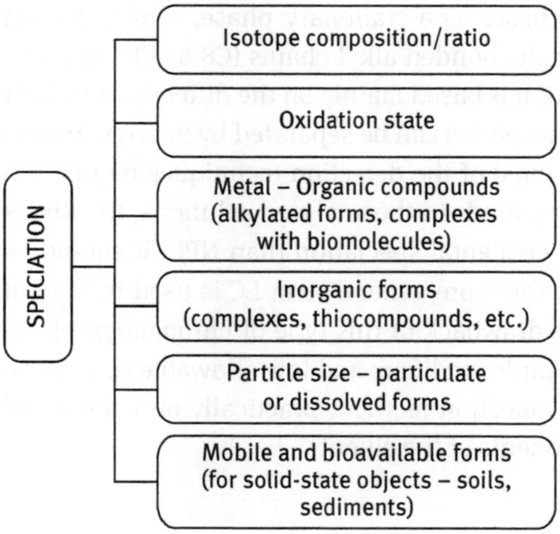

Figure 3.17: Possible distribution of an element among defined chemical species in a system.

which the mobile phase is a liquid. LC can be carried out either in a column or on a plane. At present the stationary phase for LC, in most of the cases, consists of very small particles providing better separation of the sample components, but requiring a relatively high inlet pressure. This variation is usually described by the term *high-performance liquid chromatography* (HPLC). At present this is the most frequently used variation of LC.

Further chromatographic methods used in the speciation studies can be divided into following groups:

– *Normal-phase chromatography* (NPC)

  IUPAC definition: An elution procedure in which the stationary phase is more polar than the mobile phase.

  Usually modified microporous silica gel and organic polymer materials with cyano- and amino-functional groups immobilized on the surface are used as a stationary phase. The elution procedure is carried out using reagents less polar than the stationary phase. The method can be used to separate analytes with different polarity as the most polar ones have highest retention times and are eluted last. High content of organic solvents is not tolerable for the most of commonly used detection methods, and thus, the NPC is relatively infrequently used in trace element speciation. On the other hand, the higher retention times for polar analytes (normal case for trace element forms) increase the separation time. The term "normal phase" is used in LC to emphasize the contrast to the Reserved Phase Chromatography (RPC).

– *Reversed-phase chromatography*

  IUPAC definition: An elution procedure used in LC in which the mobile phase is significantly more polar than the stationary phase, for example, a microporous silica-based material with chemically bonded alkyl chains. The term "reverse phase" is an incorrect expression to be avoided.

  An elution procedure used in LC in which the mobile phase is significantly more polar than the stationary phase. As a stationary phase, a microporous silica-based material with chemically bonded alkyl chains (C8 or C18) is typically used. Analyte species separation is based mainly on the differences in their polarity. Large number of chemical species can be separated by minor variation of the eluent composition. Since most of the detection techniques require relatively low content of organic compounds in the analyzed solutions, the RPC is recently more frequently used in elemental speciation than NPC. It should be noted that in many publications, the more general term LC is used instead of the more detailed RPC. The main drawback of this type of chromatography is the limited number of suitable organic modifiers and low allowable concentrations. When ICP-MS is used as a detection method, practically only methanol and ethanol are acceptable components of the eluent.

- *Ion pair – RPC* (IP-RPC)

  Variation of RPC is used for better separation of ionic or easily ionizable species. Introducing minor addition (ion-pairing reagent) to the organic modifiers in the eluent for RPC provides additional versatility in separation of ionic analyte species by changing their size and mobility. Typically ion-pairing agents contain both polar and nonpolar fragments. The IP-RPC permits determination of charged and uncharged chemical species within single chromatographic run.
- *Affinity chromatography* (AC)

  IUPAC definition: An expression characterizing the particular variant of chromatography in which the unique biological specificity of the analyte and ligand interaction is utilized for the separation.

  This method is used for the separation of chemical and biochemical mixtures based on a highly specific interaction. Usually this technique uses a biologically related agent as a stationary phase for the purification or analysis of sample components.The stationary phase in AC has specific structure and consists of three parts:

  Matrix: The matrix is an inert support to which a ligand can be directly or indirectly coupled. The main requirement to the material for the matrix is to be an inert material providing extremely low nonspecific adsorption. It is essential since the AC is based on specific interactions.

  Ligand: The ligand is the chemical form binding reversibly specific molecule or group of molecules, enabling separation (or purification) by AC. The selection of the ligand for AC is influenced by two factors: the ligand must exhibit specific and reversible binding affinity for the target substance(s) and it must have chemically modifiable groups allowing to be attached to the matrix without destroying binding activity. Elution in AC is based on change of the physicochemical parameters in the mobile phase or involving different reagents. Linear gradient elution or stepwise changes of eluent composition are frequently used in AC to provide efficient separation of the analyte species.
- *Ion-exchange chromatography* (IEC)

  IUPAC definition: Chromatography in which the separation of the ions is based mainly on differences in the ion-exchange affinities of the sample components. Present-day IEC on small-particle high-efficiency columns and usually utilizing conductometric or spectroscopic detectors is often referred to as ion chromatography (IC).

Separation is based mainly on differences in the ion-exchange affinities of the sample components. Depending on the charge of the ions being exchanged between stationary and mobile phases two types of IEC are used. The anion-exchange separation is based on adsorption of anions on positively charged cites on the stationary phase and subsequent elution during which the adsorbed ions are being replaced by anions (usually OH^-) from the eluent. In cation-exchange the stationary phase contains negatively charged sites and adsorbed analyte species are replaced most frequently by oxonium ions (H_3O^+) from the eluent. In both cases, the analyte species distribution between stationary and mobile phases is highly dependent on the pH of the sample and eluent solutions. The packing material for IEC columns consists of beads of cross-linked styrene and divinylbenzene. Charged functional groups bonded on the surface of stationary phase are usually sulfonic and carboxylic acids for cation exchange and primary or quaternary amines for anion exchange. The method is suitable for large number of inorganic or element-organic ion species and charged complexes. Normally as a mobile phase, aqueous buffer solutions are used frequently with minor addition of organic modifiers – polar substances, such as methanol or acetonitrile.

At present IEC on small-particle high-efficiency columns is often referred as *ion chromatography*. IEC employing conductometric or detectors might be used as a direct method for speciation analysis. However, the sensitivity is relatively low. The state diagram E versus pH could be potentially a powerful tool for checking applicability of the method for separation of certain analyte species as well as the approximate optimal conditions.

Capillary electrophoresis (CE) is a family of electrokinetic separation methods performed in submillimeter capillaries and in micro- and nanofluidic channels carried out in external electric field. The separation is based on differences in electrophoretic mobility of the analyte species controlled by their electric charges and particle sizes. Most frequently, CE refers to capillary zone electrophoresis, but other electrophoretic techniques including capillary gel electrophoresis, capillary isoelectric focusing, capillary isotachophoresis, and the micellar electrokinetic chromatography belong also to this class of methods. In CE methods, analytes migrate through electrolyte solutions under the influence of an electric field. Capillary electrophoresis has many well-known advantages, including (i) low sample volume requirements; (ii) high separation efficiency; (iii) the ability to separate positive, neutral, and negatively charged species in a single run; (iv) relatively short analysis times (when optimized); and (v) low matrix/solvent amount. The reduced matrix amount makes the capillary electrophoresis especially a suitable separation technique for coupling with ICP-MS detection device.

Speciation analysis time can be reduced considerably by the application of online hyphenated techniques. Considerable work has been devoted to their performance enhancement. Detection limits can be improved by the application of new sample introduction systems such as thermospray or ultrasonic nebulization, ETV, or direct coupling of GC to ICP-MS. However, the preference for a highly sensitive discrete atomization technique such as ETV, ICP-MS, or GF-AAS may be the reason for choosing an off-line method of coupling.

Table 3.1 summarizes the selection of a hyphenated technique for speciation analysis of environmental samples as a function of the application objective.

Table 3.1: Selected applications for speciation analysis in environment samples.

Separation method	Analyte/chemical forms	Matrix	Detection method	Comments	References
Flow injection – capillary microextraction	**As**/As(III), As(V) **Cr**/Cr(III), Cr(VI)	Natural waters	ICP-MS	Procedure optimized. Interferences from matrix ions investigated	[110]
Microcolumn HPLC	**As**/As(III), As(V), MMA, DMA **Se**/Se(IV), Se(VI)	Natural waters, high salinity waters	ICP-MS	Validation by spiked natural waters. Comparison of external calibration and standard addition methods	[111]
HPLC	**As**/AS	Biota (algae)	ICP-MS	Validation by CRMs analysis	[112]
IC	**As**/As(III), As(V), MMA, DMA, AB	Groundwaters	ICP-MS	Gradient elution. LODs 5 and 10 ng/l for AB	[113]
IC	**As**/As(III), As(V); **Se**/Se(IV), Se(VI)	Natural waters and polluted soils	ICP-MS	Microconcentric nebulizer. LODs 1 µg/l for As and 4 µg/l for Se	[114]
Microcolumn LC	**Hg**/Hg(II), $MeHg^+$, $EtHg^+$, $PhHg^+$	Natural water	ICP-MS	LODs from 8 ng/l for $EtHg^+$ to 32 ng/l for $PhHg^+$	[115]
CE	**As**/10 species separated and determined	Groundwaters, plants	ICP-MS	LODs 09–3 ng/g. Recoveries estimated by spiking of mixed standards CE-ESI-MS as an interface to ICP-MS	[116]
Sequential extraction	**Hg**/total Hg, elemental, MeHg, organically and sulfur bonded	Sediments	CV-GF-AAS	Bioavailable Hg determined	[117]
GC	**Sn**/total Sn, TBSn, DBSn, MBSn	Antarctic marine sediments	GC-specific detectors	Results reported. Sample storage – in distilled water and frozen. Sample preparation – freeze-dried,ethylation with Na-tetraethyl-borate, adsorption, and thermal desorption	[118]
LC	**As**/As(III), As(V), MMAs, DMAs	Biological samples	ICP-MS	Comparison of different sample introduction methods	[119]

(*Continued*)

Table 3.1: (Continued)

Separation method	Analyte/chemical forms	Matrix	Detection method	Comments	References
Selective hydride generation	**As**/As(III), total inorganic As, MMAs, DMAs, TMAO	Natural waters, CRMs	AFS	Cryogen trapping. LODs –0.44, 0.74, 0.15, 0.17, and 0.67 ng/l for As(III), total inorganic As, MMAs, DMAs, and TMAO, respectively	[120]
Selective hydride generation	**As**/As(III), total inorganic As, MMAs, DMAs	Natural waters, biological samples, CRMs	ICP-MS	LODs – 0.06, 3.4, 0.14, 0.10 ng/l for As(III), total inorganic As, MMAs, and DMAs, respectively. Cryogen trapping	[121]
Micro-LC	**Hg**/Hg(II), MeHg, EtHg, PhHg	Natural waters	ICP-MS	Alternative use of two eluents with different concentrations. Comparison with other methods	[115]
IC-HG	**As**/As(III), As(V), thioarsenite, mono-, di-, tri-thioarsenate	Sulfidic natural waters	AFS – detector	Standard material synthesis and preparation discussed	[122]
Dispersive SPME	**Se**/Se(IV), Se(VI)	Natural waters, biological samples	ED-XRF	Graphene with APDC as chelatimg agent. Se(IV)–APDC complex. Se(VI) is determined as a difference of results obtained with methods containing pre-reduction step and direct one	[123]
HPLC	**As**/As(III), MMA, As(V), DMA, AsB, TMAO	Landfill leachate	ICP-MS	Validation of the method presented. Comparison between external calibration and standard addition proves no interferences	[124]
SPME-GC	**Hg**/Hg^{2+}, MeHg	Seawater	ICP-MS	Derivatizing agents: $NaBEt_4$ and $NaBPr_4$ LODs below 1 ng/l	[125]
Thermodesorption	**Hg**/Hg, $HgCl_2$, Hg–Fe_2O_3, Hg–humic acid, HgS, HgO	Soils and sediments	AMA, AAS	In most of the cases complete separation seems impossible. Synthetic reference materials and CRMs used.	[126]

Thermodesorption	**Hg**/$HgCl_2$, Hg–humic acid, HgS	Soils and sediments	AMA, AAS	Incomplete separation. Synthetic reference materials and CRMs used	[127]
Thermodesorption	**Hg**/$HgCl_2$, MeHg, HgS	Lake sediments, plankton	AAS	Validation performed. Incomplete separation. Solid-state dilution using graphite, SiO, and Al_2O_3	[128]
SPME–multicapillary GC	**Hg**/Hg^{2+}, MeHg, PrHg	Antarctica ice, sediment CRMs	ICP-MS (TOF)	Hg derivatization by $NaBEt_4$, $NaBPr_4$, and $NaBPh_4$. Precision checked by spikes	[129]
Sequential selective extraction	**Hg**/HgS, $HgCl_2$, Hg, Hg_2Cl_2, HgSe, HgO, Hg(II) adsorbed on goethite, Hg–humic acid, and Hg–Au amalgam	Sediments and soils	CV-AFS after Au amalgamation	Fractions: F1 – H_2O; F2 – pH = 2 CH_3COOH + HCl; F3 – KOH; F4 – HNO_3; F5 – aqua regia Reduction with $SnCl_2$ of each fraction	[130]
Extraction, speciated isotope dilution	**Hg**/MeHg, total Hg	Natural waters – from fresh to hypersaline lake waters, sediments	ICP-MS, CV-AFS	Estimation of bioavailable Hg, migration studied	[131]
HPLC	**As**/As(III), As(V)	High Fe and Mn groundwaters	Voltammetry: CSV – As(III), ASV – total inorganic As	Direct speciation method. As(III) determined by CSV immediately after sampling, EDTA added to eliminate Fe and Mn interferences. HPLC-ICP-MS used as a reference method.	[132]
Magnetic SPE	**As**/As(III), As(V)	Water samples	Chemiluminescence	Method validation. Recovery: 93.3–106.7% for spiked samples	[133]
SFE	Cd, Co, Cu, Ni, Pb **Cu**/speciation	Rainwater, seawater	Voltammetry – CSV	Direct speciation. Miniaturization – 20-fold reduced sample size (6 ml)	[134]
Selective extraction	**As**/As(III), As(V)	Water samples	Spectrophotometry As(V) (Mo – blue)	Total As – after As(III) oxidation with $KBrO_3$, FIA, P(V) interference. Recovery evaluated by spikes	[135]

(Continued)

Table 3.1: (Continued)

Separation method	Analyte/chemical forms	Matrix	Detection method	Comments	References
LC (RPC)	**Hg**/Hg(II), MeHg, EtHg, PhHg	CRMs – biological samples	Online chemical vapor generation – AFS	LODs 16–20 pg (as Hg) for all species. Validation by use of CRMs	[136]
HPLC	**Hg**/Hg(II), MeHg	Filtered seawater and particulate phase	ICP-MS	LODs 0.07 and 0.02 ng/l for Hg (II) and MeHg. Validation by CRM.	[137]
HPLC	**As**/As(III), As(V), phenyl arsenicals	Groundwater	ICP-MS	Contamination by degradation products of As-containing chemical warfare agents	[138]
Selective hydride generation	**As**/As(III), As(V)	River water and synthetic samples	CCD spectrophotometric detection	FIA – hydride generation – pervaporation – hydride trapping in acidified $KMnO_4$ soln. – detection at 528 nm. LODs – 22 µg/l for As(III) and 51 µg/l for total As	[139]
HPLC	**As**/As(III), As(V) **Se**/Se(IV), Se(VI) **Cr**/Cr(III), Cr(VI)	Wildfire ash and soil leachates	ICP-MS	Dynamic reaction cell used. Leachates with deionized water and simulated extracellular lung fluid	[140]
LC	**Hg**/Hg(II), MeHg	Water samples	CV-AFS	$SnCl_2$ reduction. LOD for MeHg 40 pg/l (expressed as Hg)	[141]
Selective hydride generation	**As**/As(III), As(V), total As	Mushrooms	AFS	Speciation by use of proportional equations corresponding to different analytical conditions	[142]
HPLC	**As**/As(III), MMA, As(V), DMA	Rice	HG-AFS	Method validation. LODs 0.020 µg/g for As(III) and 0.025 µg/g for other species. Linearity up to 15 µg/l for the final solution	[143]
LC/IC	**As**/As(III), As(V), MMA, DMA, AC, TMAO, AB, AS	Seafood from Aegean Sea	HG-AFS Photo-oxidation step included before or after HG	Used anion and cation exchange columns for separation of different groups of As species	[144]

Adsorption on Zr(IV) modified silica gel	**Cr**/Cr(III), Cr(VI)	Natural waters	Flame AAS	FIA preconcentration and separation. LODs for Cr(III) and Cr(VI) 1.9 and 2.3 µg/	[145]
GC	**Hg**/MeHg	Whale tissue, CRMs (marine samples)	ICP-MS compared with AFS	Both detection techniques validated and suitable. LODs (as Hg) 0.25 and 0.9 pg for AFS and ICP-MS	[146]
LC	**Hg**/MeHg	Biota sample	Speciated ID ICP-MS	Nine MeHg extraction systems compared in terms of efficiency, recovery, and risk for transformations during the sample preparation	[147]
GC	**Hg**/inorganic Hg, MeHg	Freshwater samples	ID-ICP-MS	Hg volatilization by ethylation with $NaBEt_4$	[148]
GC	**Hg**/MeHg, EtHg	Seafood	ICP-MS	LODs for MeHg and EtHg 0.5 and 1.0 pg (as Hg)	[149]
HPLC	**Hg**/Hg(II), MeHg, EtHg **Pb**/Pb(II), TMePb, TEtPb	Fish samples	ICP-MS	Identification of chromatogram peaks performed by analyte addition experiments	[150]
HPLC	**As**/AC, AB	Water and fish samples	AFS (ultrasonic nebulizer); thermospray MS detector	LODs 13.6 and 11.2 pg/l for AC and AB	[151]
HPLC	**As**/As(III), As(V), MMA, DMA	Rice grains	ICP-MS	Extraction with 0.29 M HNO_3 at 95°C for 90 min comparison between different extraction systems	[152]
HPLC	**As**/As(III), As(V), MMA, DMA	Biota sample	ICP-MS	Several extraction procedures studied and compared	[153]
LC	**Hg**/Hg(II), MeHg	Fish tissue	ICP-MS	Bi(III) used as internal standard. LODs 0.8 and 0.7 µg/l for Hg(II) and MeHg	[154]

(Continued)

Table 3.1: (Continued)

Separation method	Analyte/chemical forms	Matrix	Detection method	Comments	References
Capillary GC	**Hg**/Hg(II), MeHg	Sediments	SS ID ICP-MS	Volatilization by ethylation ($NaBEt_4$) or propylation ($NaBPr_4$)	[155]
Sequential extraction	**Hg**/Hg– total, Hg– inorganic, Hg– organic	Fish tissue, CRMs	CV – Au amalgamation – GF-AAS	Hg amalgamation on Au gauze inserted in the atomizer. Collection at 80°C, release at 650°C	[156]
GC	**Hg**/Hg(II), MeHg, DMHg	Artificial water samples	SS-ID-ICP-MS	Application of reduced plasma power	[157]
IC	**As**/As(III), As(V), MMA, DMA, AB **Se**/Se(IV), Se(VI)	Groundwater, soil extracts	ICP-MS	Soft soil leaching conditions to prevent changes in analyte species. Peaks identified by species addition	[158]
IC	**Fe, Ni, Cu, Zn, Cd, Hg, Pb** – ionic species	Estuarine and coastal waters	Voltammetry	Data used in the model: pH, salinity, dissolved organic compounds and trace element concentrations. Comparison of experimental data with predicted distribution based on theoretical model	[159]
Ion exchange resins	**As**/As(III), As(V), MMA, DMA	Natural water samples	ICP-MS	Three types of resins used. Sorbent capacity evaluated	[160]
HPLC	**Cr**/ Cr(III), Cr(VI)	Industrial solutions	ICP-MS	Molecular ion interferences studied and reduced by variation of resolution and plasma temperature	[161]
IC	**Hg**/Hg(II), MeHg, EtHg, PhHg	Seawater, marine fish tissue	ICP-MS	LODs: 0.019, 0.027, 0.031 and 0.022 µg/l for Hg(II), MeHg, EtHg and PhHg. Validation with CRMs	[162]
IC	**Cr**/ Cr(III), Cr(VI)	Contaminated waters	ICP-MS	Several column configurations tested. Cr(III) determined as anionic complex. LODs below 0.2 µg/l for both species	[163]
Sequential chemical extraction	**Hg**/Hg(II), HgS, HgO, elemental Hg	Mine wastes, soil, river sediments, river water suspended particulate matter	SEM-EDS, XANES	Separate particles studied. Low sensitivity of XANES (≥10% fraction can be determined)	[164]

HPLC	**Hg**/Hg(I), Hg(II), MeHg, EtHg	Natural waters: fresh, drinking and sea	ICP-MS	Ion exchange online preconcentration column prior to HPLC separation. LODs: 0.015, 0.010, 0.009, and 0.016 ng/l for Hg(I), MeHg, EtHg, and Hg(II)	[165]
HPLC	**Sb**/Sb(III), Sb(V), $TMSbCl_2$	Uncontaminated freshwater samples	ICP-MS	Ultrasonic nebulizer. Decreasing USN heating temp. to 80°C provides ≈25% higher signals. LODs 14, 12, 9 ng/l for Sb(III), Sb(V), and $TMSbCl_2$	[166]
HPLC	**As**/As(III), As(V), MMA, DMA, AsB, AsC, TMAO, AsS, TMA, total As	Freshwater and biota from As-rich Hayakawa River (Japan)	ICP-MS	Water samples – cryopreserved at −84°C, biota samples – rinsed with water and freeze-dried	[167]
HPLC	**Hg**/Hg(I), Hg(II), MeHg, EtHg	Soil samples	ICP-MS, AMA	DGT is used for mobile forms and Hg uptake by plants	[168]
HPLC	**As**/As(III), As(V), MMA, DMA **Se**/Se(IV), Se(VI) **Cr**/Cr(VI)	Surface water samples	ICP-MS	Anion exchange column, gradient elution Samples stored in polyethylene at 4°C, no conservation. LODs: 0.006, 0.02, 0.4, 0.05, 1, 1, and 0.2 µg/l for As(III), As(V), MMA, DMA, Se(IV), Se(VI), and Cr(VI)	[169]
GC	**Sn**/MBT, DBT, TBT	Seawater and harbor sediments	ICP-MS	Volatilization by ethylation. Trend in changes of Sn species concentration	[170]
GC	**Sn**/MBT, DBT, TBT	Freshwater sediments and biota	ICP-MS, species specific isotope dilution.	ICP-MS optimization performed by addition of Xe to the plasma Ar. LODs (as Sn) 5.6, 6.6. and 3.4 ng/kg for MBT, DBT. and TBT	[171]
SPE with carbon nanotubes	**As**/As(III), As(V) **Se**/Se(IV), Se(VI) **Cr**/Cr(III), Cr(VI)	Natural waters (rain, river, lake)	ICP-MS	As(V), Se(VI). and Cr(VI) adsorbed at pH ≈ 2.2. Total As, Se, and Cr determined after oxidation with $KMnO_4$. Lower oxidation states determined as a difference	[172]

(Continued)

Table 3.1: (Continued)

Separation method	Analyte/chemical forms	Matrix	Detection method	Comments	References
Liquid–liquid microextraction and HPLC	**Hg**/Hg(II), MeHg, EtHg	Sediment samples	CV-AFS	Extraction agent L-cysteine; reducing agent $NaBH_4$ LODs 0.029, 0.028. and 0.057 µg/kg for Hg(II), MeHg. and EtHg	[173]
IC	**As**/As(III), As(V), total dissolved As	Groundwater	ICP-MS	Cl^-interferences studied	[174]
GC	**Pb**/DML, TML, DEL, TEL, MEL	Airborne particulate matter	ICP-MS	Sampling of particulate matter. TetrEL used as internal standard	[175]
Single-drop microextraction	**Sn**/MBT, DBT, TBT	Environmental samples	GC-ICP-MS	Headspace sampling. Tri-propyl-Sn used as internal standard	[176]
HPLC	**As**/As(III), As(V), AsB	CRMs	NAA with internal standard Au	Two CRMs studied. About 20% bias for As in both CRMs	[177]
Gas-diffusion flow injection	**As**/As(III)	Natural waters	Chemiluminescence detection	HG using $NaBH_4$. Applicable to Sb(III)	[178]
HPLC	**As**/As(III), As(V) oxo- and thio-anions	Sulfidic waters	ICP-MS	Following equilibrium reactions are considered: (i) redox reactions, (ii) sulfidation, (iii) protolysis, (iv) solubility	[179]

Note: ICP-MS, inductively coupled plasma mass spectrometry;HPLC, high-performance liquid chromatography; MMA, monomethylarsonous acid; DMA, dimethylarsinous acid; IC, ion chromatography; CRM, certified reference materials; LOD, limit of detection; AB, arsenobetaine; TMAO, trimethylarsine oxide; LC, liquid chromatography; CE, capillary electrophoresis; ESI-MS, electrospray ionization-mass spectrometry; ED-XRF, energy-dispersive X-ray fluorescence; RPC, reversed-phase chromatography; SPE, solid-phase extraction; EDTA, ethylenediaminetetraacetic acid; SFE, supercritical fluid extraction; CSV, comma separated variable; TOF, time of flight; GF-AAS, atomic absorption spectrometry with graphite furnace; GC, gas chromatography; MBSn, monobutyltin; DBSn, dibutyltin; TBSn, tributyltin; AFS, atomic fluorescence spectrometry; TMAO, trimethylarsine oxide; AS, arsenosugars; IC-HG, ion chromatography hydride generation; SPME, solid-phase microextraction; APDC, ammonium pyrrolidinedithiocarbamate; AMA, advanced mercury analyzer; CV-AFS, cold vapor atomic fluorescence spectrometry; CCD, charge coupled device; SS ID, species-specific isotope dilution; SEM-EDS, scanning electron microscopy/energy-dispersive spectrum; XANES, X-ray absorption near-edge spectroscopy; MBT, monobutyl tin; DBT, dibutyl tin; TBT, tributyl tin; HG-AFS, hydride-generation atomic fluorescence spectrometry; FIA

3.7 Isotopic analysis

3.7.1 Use the isotope ratios as a tool for investigation of environmental pollution

The isotopic composition of different elements is widely used as a clear indicator for the sources of origin and pollution provenance studies. Identifying pollution sources and tracking the transport pathways of toxic elements are of fundamental interest, in terms of the "polluter pays" principle, that is, now generally applied by environmental legislation. Isotopic fingerprinting can serve as a powerful tool in addressing these needs. Stable isotopic ratios have been used for decades for investigating biogeochemical cycling of trace metals and for tracing natural and anthropogenic sources of toxic elements in the environment.

The term isotope is referred to those atoms that differ from other only in their number of neutrons. Generally speaking, it can be stated that isotopic composition of an element is invariant in nature but, however, there are various reasons why this composition can be modified:

Radiogenic nuclides are produced as a result of natural decay of long-lived radionuclide. The additional production of these nuclides leads to a temporal modification in the isotopic composition [180].

As a result of their difference in mass, isotope of the same elements participate to physical and chemical processes with slightly different efficiencies. This phenomenon is caused by thermodynamic and kinetic effects and is well characterized for light elements that show big differences in mass between their isotopes [180]. For an increasing number of elements there is not a linear correlation between the magnitude of the fractionation effect and the difference in mass between the isotopes considered. This phenomenon is attributed to a slight difference in the interactions between nucleus and surrounding electrons, which results in mass-independent fractionation in chemical reactions [180].

The natural variations typically are the result of light and heavy isotopes of an element portioning differently in chemical reaction or phase transformation products and reactants. As such, the stable isotopic variations of elements provide new information, in addition to concentration and speciation measurements.

The capabilities of ICP-MS for performing precise isotope ratio measurements are nowadays well known. Instrumental progress in this field has permitted to obtain highly reliable isotope ratio data useful for many environmental applications.

In this particular field, one of the most investigated analytes is Pb. Pb isotopic analysis has been proved to be a very efficient tool for tracking sources of local and global Pb pollution. Lead is a nonessential toxic element and at high levels of human exposure, it causes damage to almost all organs and organ system. The isotopic composition of lead is variable in nature as only one (^{204}Pb) of the four stable isotopes of this element (^{204}Pb, ^{206}Pb, ^{207}Pb, and ^{208}Pb) is not the end product of the decay of thorium and uranium isotopes. Hence, although primordial lead had a fixed isotopic composition, the amounts of ^{206}Pb, ^{207}Pb, and ^{208}Pb have increased over time relative to ^{204}Pb [180]. Taking this into account, the use of variations in stable lead isotope ratios has become a well-established diagnostic technique to characterize sources of lead contamination.

Lead isotopic analysis of different environmental matrices (atmospheric aerosols, lichens, tree rings, peat deposits, lake, stream, marine sediments, soils, etc.) has been used in environmental and climate change studies and the analysis of Pb isotopes using mass spectrometry techniques has been successfully applied to investigate the transport of anthropogenic pollution as well as for tracing the sources of Pb pollution [181–184].

A substantial amount of applications reported in the literature aim at provenance determination. The isotopic composition of elements such as Sr (soil) and Pb (mining area) varies substantially from one location to another as some of their isotopes are radiogenic. As a result, isotopic analysis of these elements may provide insight into their origin [185].

In contrast to Pb, Hg isotopes are nonradiogenic and isotopic variability is caused by fractionation during biogeochemical processes.

Mercury has seven stable isotopes (196, 198, 199, 200, 201, 202, and 204 amu), and the isotopic composition of Hg in environmental matrices can now be determined with high precision [186–189]. Stable isotopes of mercury are becoming a useful tool to characterize sources and identify biogeochemical transformations of Hg in the environment. Stable isotopes of mercury have been employed to trace geogenic, urban, industrial, and atmospheric sources of Hg in the environment [189, 190] and to estimate the relative contribution of different Hg sources.

Analysis of a sediment core from the Niagara Basin of Lake Ontario [191] yielded the first recorded evidence for systematic variations in the stable isotope composition of mercury (Hg) in natural environments on the Earth. The sediments comprised younger strata enriched in Hg by recent pollution overlying older strata containing background Hg only. Several isotopic ratios of Hg varied significantly with the Mn/Fe ratio of the extractable readily reduced oxyhydroxide fraction of the sediment and with the concentrations of extractable Mn or Fe fractions, or both, in the sediment, and the zone of recent Hg pollution gave strikingly different results than the zone of background Hg. The results of the study research suggest fractionation of Hg isotopes by natural processes, including Hg methylation, with effects linked to temporal variations in the oxidation–reduction potential of the sediments. These findings show that Hg isotopes could provide valuable information about the sources and biogeochemical cycling of natural and anthropogenic Hg.

Recent advances in stable isotopic techniques of Hg using multicollector inductively coupled plasma mass spectrometry (MC-ICP-MS) offer promising methods for delineating sources. Isotopic signatures of Hg in environmental samples obtained by MC-ICP-MS have been reported to differ significantly [192]. The potential of Hg isotopes to distinguish sources of contamination has therefore been investigated. However, many biotic and abiotic processes undergone by Hg compounds in the environment can cause significant shifts of isotopic signatures [193]. These shifts, referred to as "isotopic fractionation," must be thoroughly understood during interpretation of isotopic signatures. In summary, real-time monitoring and marine cycling research studies are continually emerging that better define sources and fate of Hg. While it is difficult to truly "separate" anthropogenic from natural sources in samples, cumulative evidence in studies allows for source attribution and hypotheses about important sites of transformation and bioaccumulation.

Emerging technologies for atmospheric Hg speciation and for Hg analyses through MC-ICP-MS show great promise, but are subject to interpretation based on current limits of knowledge of transformation and fractionation reactions. Clearly, studies with proper ancillary measurements enhance the ability to differentiate between natural and anthropogenic Hg.

Boron also shows a broad range of isotopic variation in the nature. In environmental sciences, boron isotopes can be used to gain information about the source of contaminants in groundwater and freshwater systems [194]. The reason for the high variability of $\delta^{11}B$ in nature is the isotopic fractionation within the global geochemical cycle of boron. This mass-dependent isotopic fractionation is mainly driven by different ground energies of ^{10}B and ^{11}B in the two boron species $B(OH)_3$ and $B(OH)_4$ in aqueous solutions.

To obtain the best results, ratios should be used in conjunction with other data, such as concentrations and other elemental compositions, and assisted with tools such as geospatial analysis, when applied to identify the sources and transport pathways of trace elements in the environment. Increased availability of instrument and development of new analytical instrument requiring minimum sample preparation are expected to make this method more prevalent in trace elements pollution investigation in the future.

Recently, also other elements became the subject of investigation in terms of fingerprinting of environmental pollution, mainly because of the very good isotope ratio precision that can be obtained by means of MC-ICP-MS. By applying this instrumentation, it was shown that also elements such as Cd [195], Zn [196], and Fe [197] show fractionation effects, and hence, could be a useful tool for environmental studies. Zn isotopes were mainly studied as tracers of biogeochemical and chemical processes but recently also the potential of Zn isotopic analysis for source assessment in environmental samples has been described [198].

3.8 Metrological aspects of environmental monitoring of trace elements

The protection of the environment is among the highest priorities worldwide. The identification of environmental pollution is based on monitoring campaigns that periodically assess the quality of our water, soil, and air. The reliability and comparability of the analytical data produced in this context are a crucial point for management of the environment in general, for example, taking decisions and meaningful actions in remediation policies. Good laboratory practice associated with quality assurance/quality control should be essential components of the analytical process for generation of environmental data.

3.8.1 Use of reference materials

Quality control procedures are commonly based on the analysis of CRMs. CRMs are the anchor points for comparability of measurement results, in both space and time, ensuring their traceability to a common reference. These materials are essential tool in laboratories to meet the requirements of the quality assurance system.

According to the International Vocabulary of Metrology-Basic and General Concepts and Associated Terms [199], the definitions for reference material and CRM are as follows:

3.8.1.1 Reference material

Definition. It is a material that is sufficiently homogeneous and stable with reference to specified properties, which has been established to be fit for its intended use in measurement or in examination of nominal properties.

NOTE 1 Examination of a nominal property provides a nominal property value and associated uncertainty. This uncertainty is not a measurement uncertainty.

NOTE 2 Reference materials with or without assigned quantity values can be used for measurement precision control whereas only reference materials with assigned quantity values can be used for calibration or measurement trueness control.

NOTE 3 "Reference material" comprises materials embodying quantities as well as nominal properties.

3.8.1.2 Certified reference material

Definition. Reference material accompanied by documentation is sued by an authoritative body and providing one or more specified property values with associated uncertainties and traceabilities using valid procedures.

NOTE 1 "Documentation" is given in the form of a "certificate".

NOTE 2 Procedures for the production and certification of CRMs are given, for example, in ISO Guides 34 and 35.

Reference materials can be artificially (synthetically) prepared or obtained from materials with natural origin. When analyzing natural matrices, the matrix itself can cause substantial bias in the analytical result.

"Matrix reference materials" realize the measurement in a specific matrix, often produced from a natural sample.

Reference materials are used for validation of analytical methods, for the estimation of measurement uncertainties, and for establishing traceability of measurement results to internationally agreed stated references. As such they are cornerstones for the proper implementation of relevant regulations in national and international legislations, for quality control and laboratory accreditation. By using the reference materials for calibration and method validation, laboratories can demonstrate that their measurement results are traceable, or in other words that they are globally comparable.

A number of metrological and international institutes provide CRMs, such as the National Institute of Standards and Technology in the USA, the National Research Council in Canada, the International Atomic Energy Agency, the Institute for Reference Materials and Measurements in Belgium, the Federal Institute for Materials Research and Testing, and National Metrology Institute of Japan, Japan.

Since the production of CRMs is a very complex and laborious process, the increasing demand of reference materials cannot be satisfied.

The scope of this chapter is therefore to present an overview of existing matrix CRMs, related to the environmental monitoring needs. Our survey concerns available matrix CRMs for trace elements in water, biota sediment, and soils. They are shown in Table 3.2.

3.8.2 Measurement uncertainty in the environment analysis: study case

According to VIM [199], measurement uncertainty is a nonnegative parameter characterizing the dispersion of the quantity values being attributed to a measurand, based on the information used.

The word uncertainty means doubt and uncertainty of a measurement result is a parameter that characterizes the dispersion of the values that could reasonably be attributed to the measurand. Uncertainty estimation is not meant to bring unnecessary complication but, instead, it is essential to the evaluation of the quality of a measurement result. It is not necessarily a difficult task, and it must remain fit for purpose and proportionate to the objective of the measurement. What must be achieved in priority is sufficient understanding of the measurement process to ensure the declaration of accurate results within realistic uncertainty boundaries given the scientific, economical, and legal context of the measurements.

Every analytical measurement, upon which important decision are often made, has an associated uncertainty, resulting from errors arising in the various stages of sampling and analysis, which characterizes the range of values within which the “true” values asserted to lie [200]. The estimation of measurement uncertainty associated with an analytical measurement is increasingly being recognized as an essential part of the measurement process, because it facilitates and improves comparison of analytical results and is a requirement for the ISO accredited methods [201]. In order to estimate the uncertainty associated with an analytical measurement a clear statement of the measurand, and the quantities used to derive it, is required. The formulation of an uncertainty budget can be used as a method development tool, allowing the analyst

Table 3.2: Selected certified reference materials (CRMs) for trace elements in environment matrices.

Analyte	CRM	Matrix	Producer	References
Trace elements	SRM 1974b	Mussel tissue (*Mytilus edulis*)	NIST	http://www.nist.gov
Trace elements	SRM 1947	Mussel tissue (*Mytilus edulis*)	NIST	http://www.nist.gov
Trace elements	BCR-414	Plankton	JRC-IRMM	https://irmm.jrc.ec.europa.eu
Trace elements	ERM-CE278	Mussel tissue (*Mytilus edulis*)	JRC-IRMM	https://irmm.jrc.ec.europa.eu
Trace elements	DOLT-4	Dogfish liver	NRC-CNRC	http://www.nrc-cnrc.gc.ca
Trace elements	DORM-3	Fish protein	NRC-CNRC	http://www.nrc-cnrc.gc.ca
Trace elements	SRM 1566b	Oyster tissue	NIST	http://www.nist.gov
Trace elements and methylmercury	SRM 2976	Mussel tissue	NIST	http://www.nist.gov
Trace elements	LUTS-1	Non-defatted lobster hepatopancreas	NRC-CNRC	http://www.nrc-cnrc.gc.ca
Trace elements	TORT-2	Lobster hepatopancreas	NRC-CNRC	http://www.nrc-cnrc.gc.ca
Trace elements and methylmercury	IAEA 452	Scallop	IAEA	http://www.iaea.org
Trace elements and methylmercury	IAEA 461	Clams	IAEA	http://www.iaea.org
Trace elements, arsenobetaine. and methylmercury	NMIJ CRM 7402-a	Cod fish	NMIJ	https://www.nmij.jp
Trace elements	NMIJ CRM 7405-a	Tissue	NMIJ	https://www.nmij.jp
Mercury and its compounds	BCR 463	Tuna fish	JRC-IRMM	https://irmm.jrc.ec.europa.eu
Mercury and its compounds	ERM-CE464	Tuna fish	JRC-IRMM	https://irmm.jrc.ec.europa.eu
Trace elements and methylmercury	NMIJ CRM 7403-a	Swordfish	NMIJ	https://www.nmij.jp
Trace elements	FEBS-1	Otolith	NRC-CNRC	http://www.nrc-cnrc.gc.ca
Trace, major, and minor elements	IAEA-392	Algae	IAEA	http://www.iaea.org
Methylmercury, total mercury, arsenic, and iron	SRM 1946	Lake superior fish tissue	NIST	http://www.nist.gov
Trace elements	IAEA-413	Algae	IAEA	http://www.iaea.org
Trace elements	ERM CD200	Seaweed	JRC-IRMM	https://irmm.jrc.ec.europa.eu
Trace elements	NIES No.3	*Chlorella* (algae)	NIES	http://www.comar.bam.de
Trace elements	BCR-060	Aquatic plant (*Lagarosiphon major*)	JRC-IRMM	https://irmm.jrc.ec.europa.eu

Trace elements	BCR-279	Sea lettuce (*Ulva lactuca*)	JRC-IRMM	https://irmm.jrc.ec.europa.eu
Trace elements	BCR-670	Aquatic plant – Duck weed (*Lemna minor*)	JRC-IRMM	https://irmm.jrc.ec.europa.eu
Trace elements and arsenic compounds	NMIJ CRM 7405-a	Seaweed (Hijiki)	NMIJ	https://www.nmij.jp
Chromium	BCR-596	Aquatic plant	JRC-IRMM	https://irmm.jrc.ec.europa.eu
Trace elements	IAEA-433	Marine sediment	IAEA	http://www.iaea.org
Trace elements	IAEA-456	Marine sediment	IAEA	http://www.iaea.org
Trace elements	IAEA-457	Marine sediment	IAEA	http://www.iaea.org
Trace elements	IAEA-458	Marine sediment	IAEA	http://www.iaea.org
Trace elements	BCR-667	Estuarine sediment	JRC-IRMM	https://irmm.jrc.ec.europa.eu
Extractable trace elements	BCR-701	Lake sediment	JRC-IRMM	https://irmm.jrc.ec.europa.eu
Trace metals and other constituents	HISS-1	Marine sediment	NRC-CNRC	http://www.nrc-cnrc.gc.ca
Trace elements	MESS-3	Marine sediment	NRC-CNRC	http://www.nrc-cnrc.gc.ca
Trace elements	PACS-2	Marine sediment	NRC-CNRC	http://www.nrc-cnrc.gc.ca
Trace elements	IAEA-SL-1	Lake sediment	IAEA	http://www.iaea.org
Trace elements	IAEA-SL-3	Lake sediment	IAEA	http://www.iaea.org
Trace elements	IAEA-405	Estuarine sediment	IAEA	http://www.iaea.org
Trace elements	NMIJ CRM 7302-a	Marine sediment	NMIJ	https://www.nmij.jp
Trace elements	NMIJ CRM 7303-a	Lake sediment	NMIJ	https://www.nmij.jp
Trace elements	SRM 1646a	Estuarine sediment	NIST	http://www.nist.gov
Trace elements	SRM 1944	New York/New Jersey waterway sediment	NIST	http://www.nist.gov
Trace elements	BCR-320R	Channel sediment	JRC-IRMM	https://irmm.jrc.ec.europa.eu
Trace elements	BCR 277R	Estuarine sediment	JRC-IRMM	https://irmm.jrc.ec.europa.eu
Inorganics	SRM 2702	Marine sediment	NIST	http://www.nist.gov/srm/index.cfm

(*Continued*)

Table 3.2: (Continued)

Analyte	CRM	Matrix	Producer	References
Trace elements	LGC6137	Estuarine sediment	LGC	http://www.comar.bam.de
Trace elements	LGC6187	River sediment	LGC	http://www.comar.bam.de
Trace elements	LGC6189	River sediment	LGC	http://www.comar.bam.de
Trace elements	ERM-CC-020	Estuarine sediment	JRC-IRMM	https://irmm.jrc.ec.europa.eu
Trace elements	SRM 2586	Soil (containing lead from paint)	NIST	http://www.nist.gov
Trace elements	SRM 2587	Soil (containing lead from paint)	NIST	http://www.nist.gov
Trace elements	NWWQB-1	Lake sediment	LGC	http://www.lgc.co.uk
Trace elements	BCR-280R	Lake sediment	JRC-IRMM	https://irmm.jrc.ec.europa.eu
Trace metals	RTC-CRM015-50G	Freshwater sediment	LGC	http://www.lgc.co.uk
Trace metals	RTC-CRM016-50G	Freshwater sediment	LGC	http://www.lgc.co.uk
Trace elements and methylmercury	IAEA 158	Sediment	IAEA	http://www.iaea.org
Total and methylmercury	ERM-CC-580	Estuarine sediment	JRC-IRMM	https://irmm.jrc.ec.europa.eu
Trace elements	NCS ZC76308	Water	NCS	http://www.lgc.co.uk
Trace elements	ERM-CA022	Soft drinking water	LGC	http://www.lgc.co.uk
Trace elements	CASS-5	Nearshore seawater	NRC-CNRC	http://www.nrc-cnrc.gc.ca
Trace elements	NASS-6	Seawater	NRC-CNRC	http://www.nrc-cnrc.gc.ca
Trace elements	SLEW-3	Estuarine water	NRC-CNRC	http://www.nrc-cnrc.gc.ca
Trace elements	SLRS-5	Riverine water	NRC-CNRC	http://www.nrc-cnrc.gc.ca
Trace elements	LGC6016	Estuarine water	LGC	http://www.comar.bam.de
Trace elements	LGC6017	Rainwater	LGC	http://www.comar.bam.de
Trace elements	LGC6019	River water	LGC	http://www.comar.bam.de
Trace elements	ERM-CA011	Hard drinking water	LGC	http://www.comar.bam.de
Trace elements	SRM 1640a	Natural water	NIST	http://www.nist.gov
Trace elements	SRM 1643e	Water	NIST	http://www.nist.gov
Trace elements	SRM 1643d	Water	NIST	http://www.nist.gov
Trace elements	BCR-505	Estuarine water	JRC-IRMM	https://irmm.jrc.ec.europa.eu

Trace elements	BCR-609	Groundwater	JRC-IRMM	https://irmm.jrc.ec.europa.eu
Trace elements	BCR-610	Groundwater	JRC-IRMM	https://irmm.jrc.ec.europa.eu
Trace elements	NMIJ CRM 7201-a	River water	NMIJ	http://www.comar.bam.de
Trace elements	NMIJ CRM 7202-a	River water	NMIJ	https://www.nmij.jp
Trace elements	BCR-713	Wastewater	JRC-IRMM	https://irmm.jrc.ec.europa.eu
Trace elements	ERM-CA615	Natural groundwater	JRC-IRMM	https://irmm.jrc.ec.europa.eu
Trace elements	ERM-CA713	Wastewater	JRC-IRMM	https://irmm.jrc.ec.europa.eu
Mercury ions	BCR-579	Coastal seawater	JRC-IRMM	https://irmm.jrc.ec.europa.eu
Mercury	ORMS-4	River water	NRC-CNRC	http://www.nrc-cnrc.gc.ca
Mercury	ORMS-5	River water	NRC-CNRC	http://www.nrc-cnrc.gc.ca
Mercury	ERM-CA615	Groundwater	JRC-IRMM	https://irmm.jrc.ec.europa.eu
Mercury	SRM 1641d	Water	NIST	http://www.nist.gov
Mercury	SRM 1641e	Water	NIST	http://www.nist.gov
Metals	NIST-2781	Domestic sludge	NIST	http://www.nist.gov
Leachable and total metals	NIST-2782	Industrial sludge	NIST	http://www.nist.gov
Trace elements	BCR-146R	Sewage sludge (industrial origin)	JRC-IRMM	https://irmm.jrc.ec.europa.eu
Trace elements	BCR-145R	Sewage sludge (mixed origin)	JRC-IRMM	https://irmm.jrc.ec.europa.eu
Extractable metals	LGC6181	Sewage sludge	LGC	http://www.comar.bam.de
Extractable and total metals	ERM-CC136	Sewage sludge	JRC-IRMM	https://irmm.jrc.ec.europa.eu
Chromium	BCR-597	Sewage sludge	JRC-IRMM	https://irmm.jrc.ec.europa.eu
Constituent elements	SRM 1648	Urban particulate matter	NIST	http://www.nist.gov
Constituent elements	SRM 1650	Diesel particulate matter	NIST	http://www.nist.gov
Constituent elements	NIST-2691	Coal fly ash	NIST	http://www.nist.gov

(*Continued*)

Table 3.2: (Continued)

Analyte	CRM	Matrix	Producer	References
Constituent elements	NIST-2689	Coal fly ash	NIST	http://www.nist.gov
Constituent elements	NIST-2690	Coal fly ash	NIST	http://www.nist.gov
Trace elements	NIST-1633C	Coal fly ash	NIST	http://www.nist.gov
Organics and trace metals	NIST-2787	Fine particulate matter	NIST	http://www.nist.gov
Organics and trace metals	NIST-2786	Fine particulate matter	NIST	http://www.nist.gov
Trace elements	BCR-176R	Fly ash	JRC-IRMM	https://irmm.jrc.ec.europa.eu
Trace elements	BCR-038	Fly ash from pulverized coal	JRC-IRMM	https://irmm.jrc.ec.europa.eu
Trace elements	NIST-2584	Indoor dust	NIST	http://www.nist.gov
Respirable silica	NIST-1879A	Respirable cristobalite for XRD	NIST	http://www.nist.gov
Palladium, platinum, and rhodium	BCR-723	Road dust	JRC-IRMM	https://irmm.jrc.ec.europa.eu
Trace elements	ERM-CC690	Calcareous soil	JRC-IRMM	https://irmm.jrc.ec.europa.eu
Trace elements	BCR-142R	Light sandy soil	JRC-IRMM	https://irmm.jrc.ec.europa.eu
Trace elements	ERM-CC141	Loam soil	JRC-IRMM	https://irmm.jrc.ec.europa.eu
Metals	RTC-SQC001-50G	Soil	LGC	http://www.comar.bam.de
Trace and constituent elements	NIST-2710A	Montana soil	NIST	http://www.nist.gov
Trace and constituent elements	NIST-2711A	Montana soil	NIST	http://www.nist.gov

Note: XRD, X-ray diffraction; NIST, National Institute of Standards and Technology, IAEA, the International Atomic Energy Agency; JRC-IRMM, the Institute for Reference Materials and Measurements; NMIJ, National Metrology Institute of Japan; LGC.

to identify which areas of the measurement process provide the major contributions to the overall measurement uncertainty, making it possible to focus on these areas in order to improve the quality of the analytical result. It also allows an analytical method to be monitored in order to determine if the method is under control.

The estimation of combined uncertainty of measurement results can be obtained by applying different approaches, modeling approach recommended by the ISO GUM [201] or single laboratory validation approach [202].

The steps considered to evaluate the measurement uncertainty according to ISO-GUM, applying modeling approach, are as follows:

- define the measurand;
- describe the measurement procedure by the corresponding model equations;
- identify the possible sources of uncertainty;
- estimate all the input quantities;
- evaluate the standard uncertainty of each input quantity;
- estimate the measured value;
- evaluate the combined standard uncertainty of the result;
- evaluate the uncertainty sources contribution/index;
- calculate the expanded uncertainty; and
- report the result with the associated uncertainty.

Uncertainty estimation requires that the measurement process (i.e., from sample preparation to the instrumental and the data evaluation steps) is described entirely under the form of (not necessarily complex) mathematical expressions establishing the functional relationship between a measurand and the input quantities. All sources of uncertainty associated with the analytical procedure are identified and incorporated into the model used. The value for each parameter in the measurement equations, obtained either by an analytical measurement or a mathematical calculation, has an associated standard uncertainty. All knowledge of a specific measurement can be easily implemented in this model. It brings objectivity to the description of the reality as perceived, and it provides the transparency required for the establishment of the traceability of the measurement results.

The equations described in Table 3.3 represent the mathematical model, which is used to calculate MeHg mass fractions in biota sample applying external calibration method and Advance Mercury Analyzer [203]. The value for each parameter in the described equations, obtained either by an analytical measurement or a mathematical calculation, was with an associated standard uncertainty. The uncertainties of all input quantities are found and thereafter combined into the combined standard uncertainty. Bracketing calibration strategies is used in the presented study case. This type of calibration is in theory more advantageous in terms of uncertainty propagation as (i) the measurement cycle is fast and the instrumental drift is minimized and

Table 3.3: Modeling approach for the calculation of combined uncertainty of MeHg mass fraction in biota sample [206].

Preparation of calibration standards

$$C_{D\_i} = C_M \times \frac{m_M}{(m_M+m_{d\_1})_1} \times \frac{m_1}{(m_1+m_{d\_2})_2} \times \cdots \times \frac{m_{(i-1)}}{(m_{(i-1)}+m_{d\_i})_i} \quad (1)$$

Sample bracketing calibration and other corrections

$$C_{\text{meas}} = \frac{C_{D\_(i+1)} \times (A_S - A_{i\text{-corr}} - A_{\text{Blk}}) + C_{D\_i} \times (A_{(i+1)\text{corr}} - A_S + A_{\text{Blk}})}{(A_{(i+1)\text{corr}} - A_{i\text{-corr}})} \quad (2)$$

Absorbance correction

$$A_{\text{corr}} = A_{\text{std}} - A_{\text{I\_Blk}} \quad (3)$$

Recovery calculation

$$\bar{R} = \frac{1}{n} \times \sum_1^n \frac{[C_{\text{CRM}}]_n}{C_{\text{CRM\_cert}}} \quad (4)$$

Final calculation with corrections for recovery and moisture content

$$\bar{C}_S = \frac{1}{p} \times \sum_1^p \frac{C_{\text{meas}} \cdot V_{t_1} \cdot V_{\text{thio}}}{m_S \cdot V_{t_2} \cdot \bar{R}} \times \frac{1}{1-W} \times f \quad (5)$$

Parameter		Index	
C	Mass fraction (mg/kg)	D	Working calibration standard
$\bar{C}$	Mass fraction average (mg MeHg/kg)	M	Stock solution
m	Mass (kg)	i	Dilution steps ($i \geq 2$)
A	Absorbance	S	Sample
$\bar{R}$	Recovery	CRM	Certified reference material
V	Volume (ml)	Std	Calibration standard
W	Moisture content of biota sample (%)	meas	Measured
f	Dilution factor	cert	Certified
		n, p	Number of repeats
		Blk	Procedural blank
		I_Blk	Instrumental background
		Corr	Correction for instrumental background
		t_1, t_2	Added, collected toluene
		thio	Volume of sodium thiosulfate
		aliquot	Measured aliquot thiosulfate

(ii) the effect of instrumental nonlinearity is insignificant. Therefore, the uncertainty coming from the calibration step of the measurement procedure is estimated according to eq. (2) (Table 3.2), which describes the application of bracketing approach. Besides the correction for recovery, the measured mass fractions were also corrected for procedural blank and moisture content.

Typically, the relative expanded uncertainty on the MeHg content in biota sample is found to be 15% (k = 2). In this way, the combined uncertainty statements that arise go far beyond the simple repeatability calculations and reflect our understanding of the measurement process.

The single-laboratory validation approach, contrary to the ISO GUM modeling approach, does not go deeply into the measurement procedure and does not attempt to quantify all uncertainty sources individually. Instead uncertainty sources are quantified in large "batches" via components that take a number of uncertainty sources into account. Most of the data that are used come from validation of the analytical procedure. This type of approach is sometimes called the "top-down" approach and is also known as a Nordtest approach [252]. The equations for the single-laboratory validation approach for the same example are presented in Table 3.4.

Equation (1) includes the two main components of uncertainty budget: u (Rw) which stands for the within laboratory reproducibility component of uncertainty and u (bias) accounting the uncertainty component taking into account possible bias. The within-laboratory reproducibility (intermediate precision) component takes into

Table 3.4: Single laboratory validation approach for the calculation of combined uncertainty of MeHg mass fraction in biota sample [206].

Combined standard uncertainty (as relative)

$$u_c = \sqrt{u_{\mathrm{Rw}}^2 + u_{\mathrm{bias}}^2} \qquad (1)$$

Random effect (as relative)

$$u_{\mathrm{Rw}} = S_{\mathrm{Rw}} \qquad (2)$$

Possible bias (as relative)

$$u_{\mathrm{bias}} = \sqrt{\mathrm{RMS}_{\mathrm{bias}}^2 + u_{\mathrm{ref}}^2} \qquad (3)$$

$$\mathrm{RMS}_{\mathrm{bias}} = \sqrt{\frac{\sum_{i=1}^{n}(\mathrm{bias_i})^2}{n}} \qquad (4)$$

$$\mathrm{bias}_i = \frac{C_{\mathrm{lab}\_i} - C_{\mathrm{ref}\_i}}{C_{\mathrm{ref}\_i}} \qquad (5)$$

$$u_{C\_\mathrm{ref}} = \sqrt{\frac{\sum_{i=1}^{p}(u_{C\_\mathrm{ref}\_i})^2}{n}} \qquad (6)$$

Parameter		Index	
u	Standard uncertainty	lab	Laboratory
S	Standard deviation	n	Number of CRM used
RMS	Root mean square	R_w	Within laboratory reproducibility
C	Mass fraction (mg/kg)	ref	Certified reference material

account all uncertainty sources that are random in the long term. The bias component takes into account the systematic effects that cause long-term bias (but not those that just cause bias within a given day). The long-term bias can be regarded as a sum of procedure bias (bias inherent in the nature of the procedure) and laboratory bias (bias caused by the way how the procedure is implemented in the laboratory).

The relative expanded uncertainty calculated with validation data was found to be 20% (k = 2), which is consistent with the one calculated combined uncertainty by the modeling approach.

Validation. According to the ISO-17025 guidelines, validation is "the confirmation by examination and the provision of objective evidence that the particular requirements for a specific intended use are fulfilled" [200]. The factors influencing the final results were systematically assessed as follows: different experiments have to be set up in order to investigate the performance characteristics of the method in terms of linearity, repeatability, intermediate precision, limit of detection, limit of quantification, trueness and stability of the extracts, procedural and instrumental blanks.

Alternative approach for the method validation indicated in the ISO 17025 [200] is the participation in inter-laboratory comparisons and comparison of obtained results with results from independent method.

3.8.3 Practical aspects on the traceability of trace elements in environmental samples

The concept of metrological traceability underpins comparability of measurement results. According to VIM [199] traceability is "property of a measurement result whereby the result can be related to a reference through a documented unbroken chain of calibrations, each contributing to the measurement uncertainty." A principal requirement exists in ISO 17025 for laboratories to produce measurements that are traceable to a common system of measurement, International System of Units (SI) in this case, to ensure comparability of measurement results. A typical chemical measurement involves a number of steps as illustrated in Table 3.2. The way to demonstrate traceability is to use an uncertainty budget, where all the parameters influencing the final result are systematically assessed. Key steps in the attainment of traceability were as follows:

1. The analytical method used should be properly selected and validated, both in terms of matrix composition and analyte concentration.
2. The unbroken chain linking the trace element mass fraction to SI unit should be described with the mathematical model. Table 3.2 presents the mathematical model of the analytical procedure, which is completely understood. This model, together with subcalculations or references to certified values, relates each of the input quantity to SI units of the mol or the kilogram.
3. The use of CRM for calibration and bias (recovery) estimation is the way to link trace element mass fraction to the common system of reference SI. CRM with similar matrix should be used also in the validation of sample preparation step.

Due to the extreme diversity of analytical problems related to determination of trace elements in environmental samples, it is impossible to reveal universal rules and approaches for their solving. Despite the significant number of publications on this topic issued during the last decades, the environmental trace analysis remains extremely dynamically developing branch of the analytical science. Even minor variations in the matrix composition or structure might cause problems in method transfer and application. All approaches should be adapted to the particular object of investigation with respect to the specific sample properties, analyte to be determined, applicable methods, and personal experience of the operator. Development and application of reliable and validated analytical methods providing traceable results related to environmental objects remain a challenging analytical task.

References

[1] Patnaik P. A, Comprehensive Guide to the Hazardous Properties of Chemical Substances. A John Wiley & Sons, Inc., New Jercey, USA, 2007.

[2] Bini C, Bech J, eds. PHEs, Environment and Human Health. Potentially Harmful Elements in the Environment and the Impact on Human Health. Springer-Verlag Berlin Heidelberg New York, 2014.

[3] Prasad M, ed. Trace Elements as Contaminants and Nutrients. Consequences in Ecosystems and Human Health. John Wiley & Sons, Inc., New Jercey, USA, 2008.

[4] Cesar G, Fraga C. Relevance, essentiality and toxicity of trace elements in human health. Mol Aspect Med 2005, 26, 235–44.

[5] European and Mediterranean Plant Protection Organization. PP 1/135 (4) Phytotoxicity assessment Bulletin OEPP/EPPO Bulletin, 2014, 44, 265–73.

[6] Kopittke PM, Pax F, Blamey C, Asher CJ, Neal W. Menzies, trace metal phytotoxicity in solution culture: A review. J Exp Bot 2010, 61, 945–54.

[7] McGeer J, Henningsen G, Lanno R, Fisher N, Sappington K, Drexler J; Contributor: Michael Beringer. Issue Paper on the Bioavailability and Bioaccumulation of Metals, U.S. EPA, 2004 (http://www2.epa.gov/sites/production/files/2014-11/documents/bio_final.pdf.)

[8] Takeno N. Atlas of Eh-pH diagrams. Intercomparison of thermodynamic databases. National Geological Survey of Japan Open File Report No. 419, 2005.

[9] Rahman MA, Hasegawa H, Rahman M, Islam MN, Miah M, Tasmen T. Effect of arsenic on photosynthesis, growth and yield of five widely cultivated rice (Oryza sativa L.) varieties in Bangladesh. Chemosphere 2007, 67, 1072–9.

[10] Carbonell A, Aarab M, DeLaune R, Gambrell R, Patrick WH Jr. Arsenic in wetland vegetation: Availability, phytotoxicity, uptake and effects on plant growth and nutrition. Sci Total Environ 1998, 217, 189–99.

[11] Luan H, Cao H, Yan B. Individual and combined phytotoxic effects of cadmium, lead and arsenic on soybean in Phaeozem. Plant Soil Environ 2008, 54, 403–11.

[12] Azevedo R, Rodriguez E. Phytotoxicity of mercury in plants: A review. J Bot 2012, 3456–62.

[13] Panda S, Choudhury S. Chromium stress in plants. Braz J Plant Physiol 2005, 17, 95–102.

[14] Lyons G, Stangoulis J, Graham R. Tolerance of wheat (Triticum aestivum L.) to high soil and solution selenium levels. Plant Soil 2005, 270, 179–88.

[15] Hladun K, Parker D, Tran K, Jrumble J. Effects of selenium accumulation on phytotoxicity, herbivory, and pollination ecology in radish (Raphanus sativus L.). Environ Pollution 2013, 172, 70–5.

[16] Mengchang H, Yang J. Effects of different forms of antimony on rice during the period of germination and growth and antimony concentration in rice tissue. Sci Tot Environ 1999, 243–244, 149–55.

[17] Shabnam I, Seema M. Phytotoxicity of nickel and its accumulation in tissues of three Vigna species at their early growth stages. J Appl Bot Food Qual 2011, 84, 223–8.

[18] Rooney C, Zhao J, McGrath S. Phytotoxicity of nickel in a range of European soils: Influence of soil properties, Ni solubility and speciation. Environ Pollut 2007, 145, 596–605.

[19] Pinho S, Ladeiro B. Phytotoxicity by lead as heavy metal focus on oxidative stress. J Bot 2012, Article ID 369572, 10 pages; doi:10.1155/2012/369572.

[20] Murkowski A, Skorska E. Comparison of phytotoxicity of lead and tin organic compounds by means of luminescence methods. Acta Agrophys 2008, 11, 131–40.

[21] Michaud A, Bravin M, Galleguillos M, Hinsinger P. Copper uptake and phytotoxicity as assessed in situ for durum wheat (Triticum turgidum durum L.) cultivated in Cu-contaminated, former vineyard soils. Plant Soil 2007, 298, 99–111.

[22] Deo B, Nayak P. Study of copper phytotoxicity on in vitro culture of Musa acuminata cv. 'Bantala'. J Agric Biotechnol Sustainable Dev 2011, 3, 136–40.

[23] Bartakova I, Kummerova M, Mandl M, Pospisil M. Phytotoxicity of iron in relation to its solubility conditions and the effect of ionic strength. Plant Soil 2001, 235, 45–51.

[24] Meriño-Gergichevich C, Alberdi M, Ivanov A, Reyes-Díaz M. Al3+- Ca2+ interaction in plants growing in acid soils: Al-phytotoxicity response to calcareous amendments. J Soil Sci Plant Nutr 2010, 10, 217–43.

[25] Mearns A, Reish D, Oshida P, Ginn T, Rempel-Hester M. Effects of pollution on marine organisms (Review). Water Environ Res 2011, 83, 1789–52.

[26] Moore J, Ramamoorthy S, Heavy Metals in Natural Waters: Applied Monitoring and Impact Assessment. Springer, New York, 1984.

[27] Selin N. Global biogeochemical cycling of mercury: A review. Ann Rev Environ Resour, 2009, 34, 43–63.

[28] Gerbersdorf S, Cimatoribus C, Class H, Engesser K, Helbich S, Hollert H, Lange C, Kranert M, Metzger J, Nowak W. Anthropogenic Trace Compounds (ATCs) in aquatic habitats – Research needs on sources, fate, detection and toxicity to ensure timely elimination strategies and risk management. Environ Int 2015, 79, 85–105.

[29] Morel F, Milligan A, Saito M. Marine Bioinorganic Chemistry: The Role of Trace Metals in the Oceanic Cycles of Major Nutrients. Treatise on Geochemistry: Second Edition, Elsevier, Amsterdam, Netherlands, 2013, 8, 123–50.

[30] Avigliano E, Schenone NF, Volpedo AV, Goessler W, Cirelli AF. Heavy metals and trace elements in muscle of silverside (Odontesthes bonariensis) and water from different environments (Argentina): Aquatic pollution and consumption effect approach. Sci Total Environ 2015, 506–507, 102–8.

[31] Mergler D, Anderson HA, Chan LHM, Mahaffey KR, Murray M, Sakamoto M, Stern AH. Methylmercury exposure and the health effects in humans: A worldwide concern. Ambio 2007, 36, 3–11.

[32] USEPA. Guidance on choosing a sampling design for environmental data collection. EPA/240/R-02/005, 2002 (http://www.epa.gov/QUALITY/qs-docs/g5s-final.pdf).

[33] Pohl P. Application of ion-exchange resins to the fractionation of metals in water. Trends Anal Chem 2006, 25(1), 31–43.

[34] Templeton DM, Ariese F, Cornelis R, Danielsson L-G, Muntau H, Van Leeuwen HP, Lobinski R. Guidelines for terms related to chemical speciation and fractionation of elements. Definitions, structural aspects and methodological approaches. (IUPAC Recommendations 2000). Pure Appl Chem 2000, 72, 1453–70.

[35] Caruso JA, Klaue B, Michalke B, Rocke DM. Group assessment: Elemental speciation. Ecotoxicol Environ Safety 2003, 56, 32–44.
[36] Tuzen M, Saygi KO, Soylak M. Solid phase extraction of heavy metal ions in environmental samples on multiwalled carbon nanotubes. J Hazard Mater 2008, 152, 632–9.
[37] Taylor VF, Toms A, Longerich HP. Acid digestion of geological and environmental samples using open-vessel focused microwave digestion. Anal Bioanal Chem 2002, 372, 360–5.
[38] Pontes FVM, Mendesa BAO, De Souza EMF, Ferreira FN, Da Silva LID, Carneiro MC, Monteiro MIC, De Almeidaa MD, Neto AA, Vaitsman DS. Determination of metals in coal fly ashes using ultrasound-assisted digestion followed by inductively coupled plasma optical emission spectrometry. Anal Chim Acta 2010, 659, 55–9.
[39] Capelo-Martinez JL, Ximenez-Embun P, Madrid Y, Camara C. Advanced oxidation processes for sample treatment in atomic spectrometry. Trends Anal Chem 2004, 23, 331–40.
[40] Kailasa SK, Wu HF. Inorganic contaminants: Sample preparation approaches. In: Comprehensive Sampling and Sample Preparation, Ed. Janusz Pawliszyn, Elsevier, Amsterdam, Netherlands, 2012, 3, 743–82.
[41] Achterberg P, Braungardt CB, Sandford RC, Worsfold PJ. UV digestion of seawater samples prior to the determination of copper using flow injection with chemiluminescence detection. Anal Chim Acta 2001, 440, 27–36.
[42] Point ED, Bareille G, Pinaly H, Belin C, Donard OFX. Multielemental speciation of trace elements in estuarine waters with automated on-site UV photolysis and resin chelation coupled to inductively coupled plasma mass spectrometry. Talanta 2007, 72, 1207–16.
[43] Danielsson LG, Magnusson B, Westerlund S. Improved metal extraction procedure for determination of trace-metals in seawater by atomic-absorption spectrometry with electrothermal atomization. Anal Chim Acta 1978, 98, 47–57.
[44] Krishna AK, Mohan KR, Murthy NN. Determination of heavy metals in soil, sediment, and rock by inductively coupled plasma optical emission spectrometry: Microwave-assisted acid digestion versus open acid digestion technique. Atom Spectrosc 2009, 30, 75–81.
[45] Bagheri H, Afkhami A, Saber-Tehrani M, Khoshsafar H. Preparation and characterization of magnetic nanocomposite of Schiff base/silica/magnetite as a preconcentration phase for the trace determination of heavy metal ions in water, food and biological samples using atomic absorption spectrometry. Talanta 2012, 97, 87–95.
[46] Milne A, Landing W, Bizimis M, Morton P. Determination of Mn, Fe, Co, Ni, Cu, Zn, Cd and Pb in seawater using high resolution magnetic sector inductively coupled mass spectrometry (HR-ICP-MS). Anal Chim Acta 2010, 665, 200–7.
[47] Ndung'u K, Franks RP, Bruland KW, Flegal AR. Organic complexation and total dissolved trace metal analysis in estuarine waters: Comparison of solvent-extraction graphite furnace atomic absorption spectrometric and chelating resin flow injection inductively coupled plasma-mass spectrometric analysis. Anal Chim Acta 2003, 481, 127–38.
[48] Wu J, Boyle E. Determination of iron in seawater by high-resolution isotope dilution inductively coupled plasma mass spectrometry after Mg(OH)2 co-precipitation. Anal Chim Acta 1998, 367, 183–91.
[49] Bayon G, Birot D, Bollinger C, Barrat JA. Multi-element determination of trace elements in natural water reference materials by ICP-SFMS after Tm addition and iron co-precipitation. Geostand Geoanal Res 2011, 35, 145–53.
[50] Biller DV, Bruland K. Analysis of Mn, Fe, Co, Ni, Cu, Zn, Cd, and Pb in seawater using the Nobias-chelate PA1 resin and magnetic sector inductively coupled plasma mass spectrometry (ICP-MS). Marine Chemi 2012, 130–131, 12–20.
[51] Lee J, Boyle E, Echegoyen-Sanz Y, Fitzsimmons J, Zhang R, Kayser R. Analysis of trace metals (Cu, Cd, Pb, and Fe) in seawater using single batch Nitrilotriacetate resin extraction and

isotope dilution inductively coupled plasma mass spectrometry. Anal Chim Acta 2011, 686, 93–101.
[52] Rahmi D, Takasaki Y, Zhu Y, Kobayashi H, Konagaya S, Haraguchi H, Umemura T. Preparation of monolithic chelating adsorbent inside a syringe filter tip for solid phase microextraction of trace elements in natural water prior to their determination by ICP-MS. Talanta 2010, 81, 1438–45.
[53] Tuzen M, Saygi KO, Usta C, Soylak M. Pseudomonas aeruginosa immobilized multiwalled carbon nanotubes as biosorbent for heavy metal ions. Bioresour Technol 2008, 99, 1563–70.
[54] Komarek M, Vanek A, Ettler V. Chemical stabilization of metals and arsenic in contaminated soils using oxides: A review. Environ Pollut 2013, 172, 9–22.
[55] Fernandez-Turiel JL, Acenolaza P, Medina ME, Lorens JF, Sardim F. Assessment of a smelter impact area using surface soils and plants. Environ Geochem Health 2001, 23, 65–78.
[56] Rauret G. Extraction procedures for the determination of heavy metals in contaminated soil and sediment. Talanta 1998, 46, 449–55.
[57] Popov MA, Colao F, Fantoni R. Spatial confinement of laser-induced plasma to enhance LIBS sensitivity for trace elements determination in soils. J Anal At Spectrom 2010, 25, 837–48.
[58] Santos Jr D, Nunes LC, Trevizan LC, Godoi Q, Leme FO, Braga JWB, Krug FJ. Evaluation of laser induced breakdown spectroscopy for cadmium determination in soils. Spectrochim Acta, Part B 2009, 64, 1073–8.
[59] Guerra MBB, Schaefer CEGR, Rosa PF, Simas FNB, Pereira TTC, Pereira-Filho ER. Heavy metals contamination in century-old manmade technosols of Hope Bay, Antarctic Peninsula. Water Air Soil Pollut 2011, 222, 91–102.
[60] Aramendía M, Resano M, Vanhaecke F. Electrothermal vaporization-inductively coupled plasma-mass spectrometry: A versatile tool for tackling challenging samples: A critical review. Anal Chim Acta 2009, 648, 23–44.
[61] Savio M, Cerutti S, Martinez LD, Smichowski P, Gil RA. Study of matrix effects and spectral interferences in the determination of lead in sediments, sludges and soils by SR-ETAAS using slurry sampling. Talanta 2010, 82, 523–7.
[62] Vignola F, Borges DLG, Curtius AJ, Welz B, Becker-Ross H. Simultaneous determination of Cd and Fe in sewage sludge by high-resolution continuum source electrothermal atomic absorption spectrometry with slurry sampling. Microchem J 2010, 95, 333–6.
[63] Peter T, Mária Ž. A study of the direct determination of Cd, Cr, Cu, Pb and Zn in certified reference materials of soils by solid sampling electrothermal atomic absorption spectrometry. Spectrochim Acta, Part B 2011, 66, 93–7.
[64] D'Ilio S, Violante N, Majorani C, Petrucci F. Dynamic reaction cell ICP-MS for determination of total As, Cr, Se and V in complex matrices: Still a challenge?: A review. Anal Chim Acta 2011, 698, 6–13.
[65] Bettinelli M, Beone GM, Spezia S, Baffi C. Determination of heavy metals in soils and sediments by microwave-assisted digestion and inductively coupled plasma optical emission spectrometry analysis. Anal Chim Acta 2000, 424, 289–96.
[66] Unsal YE, Soylak M, Tuzen M, Hazer B. Determination of lead, copper, and iron in cosmetics, water, soil, and food using polyhydroxybutyrate-B-polydimethyl siloxane preconcentration and flame atomic absorption spectrometry. Anal Lett 2015, 48 1163–79.
[67] Neto JAG, Zara LF, Rocha JC, Santos A, Dakuzaku CS, Nóbrega JA. Determination of mercury in agro-industrial samples by flow-injection cold vapour atomic absorption spectrometry using ion exchange and reductive elution. Talanta 2000, 51, 587–94.
[68] Li Z, Yang X, Guo Y, Li H, Feng Y. Simultaneous determination of arsenic, antimony, bismuth and mercury in geological materials by vapour generation-four-channel non-dispersive atomic fluorescence spectrometry. Talanta 2008, 74, 915–21.

[69] Pinho J, Canario J, Cesario R, Vale C. Rapid acid digestion method with ICP-MS detection for the determination of selenium in dry sediments. Anal Chim Acta 2005, 551, 207–12.
[70] Medved J, Streko V, Kubová J, Polakovièová J. Efficiency of decomposition procedures for the determination of some elements in soils by atomic spectroscopic methods. Fresenius J Anal Chem 1998, 360, 219–24.
[71] Melaku S, Dams R, Moens L. Determination of trace elements in agricultural soil samples by inductively coupled plasma-mass spectrometry: Microwave acid digestion versus aqua regia extraction. Anal Chim Acta 2005, 543, 117–20.
[72] Sastre J, Sahuquillo A, Vidal M, Rauret G. Determination of Cd, Cu, Pb and Zn in environmental samples: Microwave-assisted total digestion versus aqua regia and nitric acid extraction. Anal Chim Acta 2002, 462, 59–72.
[73] Dhivert E, Grosbois C, Rodrigues S, Desmet M. Influence of fluvial environments on sediment archiving processes and temporal pollutant dynamics (Upper Loire River, France). Sci Total Environ 2015, 505, 121–36.
[74] Morelli G, Gasparon M, Fierro D, Hu W, Zawadzki A. Historical trends in trace metal and sediment accumulation in intertidal sediments of Moreton Bay, southeast Queensland, Australia. Chem Geol 2012, 300–301, 152–64.
[75] Xu W, Ni S, Gao Y, Shi Z. Reconstruction of the cadmium contamination history of a river floodplain from Maoniuping mining area (China) by gamma ray spectrometry and inductively coupled plasma mass spectrometry. Spectrosc Lett 2015, 48, 542–52.
[76] Bettiol C, Stievano L, Bertelle M, Delfino F, Argese E. Evaluation of microwave-assisted acid extraction procedures for the determination of metal content and potential bioavailability in sediments. Appl Geochem 2008, 23, 1140–51.
[77] Quevauviller Ph. Operationally-defined extraction procedures for soil and sediment analysis. Part 3: New CRMs for trace-element extractable contents. Trends Anal Chem 2002, 21, 774–85.
[78] Begum Z, Balaram V, Ahmad SM, Satyanarayanan M, Rao TG. Determination of trace and rare earth elements in marine sediment reference materials by ICP-MS: Comparison of open and closed acid digestion methods. At Spectrosc 2007, 28, 41–50.
[79] Hassan NM, Rasmussen PE, Dabek-Zlotorzynska E, Celo V, Chen H. Analysis of environmental samples using microwave-assisted acid digestion and inductively coupled plasma mass spectrometry: Maximizing total element recoveries. Water Air Soil Pollut 2007, 178, 323–34.
[80] Frena M, Quadros D, Castilho I, De Gois J, Borges D, Welz B, Madureira L. A novel extraction-based procedure for the determination of trace elements in estuarine sediment samples by ICP-MS. Microchem J 2014, 117, 1–6.
[81] Sardans J, Montes F, Peñuelas J. Electrothermal atomic absorption spectrometry to determine as, Cd, Cr, Cu, Hg, and Pb in soils and sediments: A review and perspectives (Review). Soil Sediment Contam 2011, 20, 447–91.
[82] Hernández-Martínez A, Padrón-Sanz C, Padrón ME, Ferrera Z, Rodríguez J. Determination of heavy metals in marine sediments using MAME-GFAAS. J Anal At Spectrom 2015, 30, 435–42.
[83] Welz B, Borges D, Lepri F, Vale MG, Heitmann U. High-resolution continuum source electrothermal atomic absorption spectrometry: An analytical and diagnostic tool for trace analysis. Spectrochim Acta, Part B 2006, 62, 873–83.
[84] Kuznetsova AI, Zarubina OV, Sklyarova OA. Determination of trace elements (Ag, B, Ge, Mo, Sn, W, Tl) in reference samples of soils and bottom sediments by atomic emission spectrometry: A traceability and fitness-for-purpose study. Geostand Geoanal Res 2007, 31, 251–9.
[85] Roje V. Multi-elemental analysis of marine sediment reference material MESS-3: One-step microwave digestion and determination by high resolution inductively coupled plasma-mass spectrometry (HR-ICP-MS). Chem Pap 2010, 64, 409–14.

[86] Madinabeitia SGD, Lorda MES, Ibarguchi JIG. Simultaneous determination of major to ultratrace elements in geological samples by fusion-dissolution and inductively coupled plasma mass spectrometry techniques. Anal Chim Acta 2008, 625, 117–30.

[87] Sahuqulloo A, Rauret G, Rehnert A, Muntau H. Solid sample graphite furnace atomic absorption spectroscopy for supporting arsenic determination in sediments following sequential extraction procedure. Anal Chim Acta 2003, 476, 15–24.

[88] Orani AM, Han E, Mandjukov P, Vassileva E. Marine sediments monitoring studies for trace elements with the application of fast temperature programs and solid sampling continuous source high resolution atomic absorption spectroscopy. Spectrochim Acta, Part B 2015, 103–104, 131–43.

[89] Mandjukov P, Orani AM, Han E, Vassileva E. Determination of total mercury for marine environmental monitoring studies by solid sampling continuous source high resolution atomic absorption spectroscopy. Spectrochim Acta, Part B 2015, 103–104, 24–33.

[90] Finlayson-Pitts BJ, Pitts JN. Chemistry of the Upper and Lower Atmosphere: Theory, Experiments and Applications. Academic Press, London, 2000, 969–74.

[91] Baldasano JM, Valera E, Jiménez P. Air quality data from large cities. Sci Total Environ 2003, 307, 141–65.

[92] Fang G-C, Wu Y-S, Chang S-Y, Huang S-H, Rau J-Y. Size distributions of ambient air particles and enrichment factor analyses of metallic elements at Taichung Harbor near the Taiwan Strait. Atmos Res 2006, 81, 320–33.

[93] Fang G-C, Chang S-Y. Atmospheric particulate (PM10 and PM2.5 mass concentration and seasonal variation study in Taiwan area during 2000-2008. Atmos Res 2010, 98, 368–77.

[94] Pére N, Pey J, Querol X, Alastuey A, López JM, Viana M. Partitioning of major and trace components in PM10-PM2.5-PM1 at and urban site in Southern Europe. Atmos Environ 2008, 42, 1677–91.

[95] Artiñano B, Salvador P, Alonso DG, Querol X, Alastuey A. 2003. Anthropogenic and natural influence on the PM10 and PM2.5 aerosol in Madrid (Spain). Analysis of high concentration episodes. Environ Pollut 2003, 125, 456–65.

[96] Mouli GP, Mohan VS, Balaram V, Kumar PM, Reddy JS. A study on trace elemental composition of atmospheric aerosols at a semi-arid urban site using ICP-MS technique. Atmos Environ 2006, 40, 136–46.

[97] Buzika D, Gerbole M, Borowiak A, Tricherini P, Passarella R, Pedroni V. Comparison of voltammetry and inductively coupled plasma-mass spectrometry for the determination of heavy metals in PM10 airborne particulate matter. Atmos Environ 2006, 40, 4703–10.

[98] Moreno T, Querol X, Alastuey A, Viana M, Salvador P, Sánchezde laCampa A, Artiñano B, de la Rosa, J, Gibbons W. Variations in atmospheric PM trace metal content in Spanish towns: Illustrating the chemical complexity of the inorganic urban aerosol cocktail. Atmos Environ 2006, 40, 6791–803.

[99] Liang SX, Wu H, Sun HW. Determination of trace elements in airborne PM10 by inductively coupled plasma mass spectrometry. Int J Environ Sci Technol 2015, 12, 1373–8.

[100] Bukowiecki N, Lienemann P, Zwicky CN, Gehrig R, Baltensperger U. X-ray fluorescence spectrometry for high throughput analysis of atmospheric aerosol samples: The benefits of synchrotron X-rays. Spectrochim Acta, Part B At Spectrosc 2008, 63, 929–38.

[101] Avino P, Capannesi G, Rosada A. Source identification of inorganic airborne particle fraction (PM10) at ultratrace levels by means of INAA short irradiation. Environ Sci Pollut Res 2014, 21, 4527–38.

[102] Shaltout AA, Bornan J, Welz B, Castilho INB, Al Ashkar EA, Gaita SM. Method development for the determination of Cd, Cu, Ni and Pb in PM2.5 particles sampled in industrial and urban

areas of Greater Cairo, Egypt, using high-resolution continuum source graphite furnace atomic absorption spectrometry. Microchem J 2014, 113, 4–9.
[103] Kulkarni P, Chellam S, Flanagan JB, Jayanty RKM. Microwave digestion ICP-MS for elemental analysis in ambient airborne fine particulate matter: Rare earth elements and validation using a filter borne fine particle certified reference material. Anal Chim Acta 2007, 599, 170–6.
[104] Salcedo D, Bernal JP, Pérez-Arvizu O, Lounejeva E. Assessment of sample preparation methods for the analysis of trace elements in airborne particulate matter. J Anal At Spectrom 2014, 29, 753–61.
[105] Lobinski R. Speciation – targets, analytical solutions and markets. Spectrochim Acta, Part B 1998, 53,177–85.
[106] Ettre LS. Nomenclature for chromatography (IUPAC Recommendations 1993). Pure Appl Chem 1993, 65, 819–72.
[107] Sadek PC. Illustrated Pocket Dictionary of Chromatography. John Wiley & Sons Inc., Wiley Interscience, New York, NY, 2004.
[108] Czes J, ed. Encyclopedia of Chromatography, 3rd ed. Taylor & Francis Group, CRC Press, Boca Raton, FL, 2010.
[109] Fritz JS. Analytical Solid Phase Extraction. John Wiley & Sons Inc., Wiley Interscience, New York, NY, 1999.
[110] Wenling H, Zheng F, Bin H. Simultaneous separation and speciation of inorganic As(III)/As(V) and Cr(III)/Cr(VI) in natural waters utilizing capillary microextraction on ordered mesoporous Al2O3 prior to their on-line determination by ICP-MS. J Hazard Mater 2008, 151, 58–64.
[111] Castillo A, Roig-Navarro AF, Pozo OJ. Capabilities of microbore columns coupled to inductively coupled plasma mass spectrometry in speciation of arsenic and selenium. J Chromatogr A 2008, 1202, 132–7.
[112] Todoli J, Grotti M. Fast determination of arsenosugars in algal extracts by narrow bore high-performance liquid chromatography–inductively coupled plasma mass spectrometry. J Chromatogr A 2010, 1217, 7428–33.
[113] Ammann A. Arsenic speciation by gradient anion exchange narrow bore ion chromatography and high resolution inductively coupled plasma mass spectrometry detection. J. Chromatogr. A 2010, 1217, 2111–6.
[114] Woller A, Garraud H, Boisson J, Dorthe AM, Fodor F, Donard OFX. Simultaneous speciation of redox species of arsenic and selenium using an anion-exchange microbore column coupled with a microconcentric nebulizer and an inductively coupled plasma mass spectrometer as detector. J Anal At Spectrom 1998, 13, 141–9.
[115] Castillo A, Roig-Navarro AF, Pozo OJ. Method optimization for the determination of four mercury species by micro-liquid chromatography–inductively coupled plasma mass spectrometry coupling in environmental water samples. Anal Chim Acta 2006, 577, 18–25.
[116] Liu L, He B, Yun Z, Sun J, Jiang G, Speciation analysis of arsenic compounds by capillary electrophoresis on-line coupled with inductively coupled plasma mass spectrometry using a novel interface. J Chromatogr A 2013, 1304, 227–33.
[117] Pinedo-Hernandez J, Marrugo-Negrete J, Diez S. Speciation and bioavailability of mercury in sediments impacted by gold mining in Colombia. Chemosphere 2015, 119, 1289–95.
[118] Negri AP, Hales LT, Battershill C, Wolff C, Webster NS. TBT contamination identified in Antarctic marine sediments. Mar Pollut Bull 2004, 48, 1142–4.
[119] Sun YC, Lee YS, Shiah TL, Lee PL, Tseng WC, Yang MH. Comparative study on conventional and low-flow nebulizers for arsenic speciation by means of microbore liquid chromatography, with inductively coupled plasma mass spectrometry. J Chromatogr A 2003, 1005, 207–13.
[120] Musil S, Matoušek T, Currier JM, Styblo M, Dedeina J. Speciation analysis of arsenic by selective hydride generation-cryotrapping-atomic fluorescence spectrometry with

flame-in-gas-shield atomizer: Achieving extremely low detection limits with inexpensive instrumentation. Anal Chem 2014, 86, 10422–8.
[121] Matousek T, Currier JM, Trojankova N, Saunders RJ, Ishida MC, Gonzalez-Horta C, Musil S, Mester Z, Stýblo M, Dedina J. Selective hydride generation-cryotrapping-ICP-MS for arsenic speciation analysis at picogram levels: Analysis of river and sea water reference materials and human bladder epithelial cells. J Anal At Spectrom 2013, 28, 1456–65.
[122] Keller NS, Stefansson A, Sigfusson B. Determination of arsenic speciation in sulfidic waters by Ion Chromatography Hydride-Generation Atomic Fluorescence Spectrometry (IC-HG-AFS). Talanta 2014, 128, 466–72.
[123] Kocot K, Leardi R, Walczak B, Sitko R. Determination and speciation of trace and ultratrace selenium ions by energy-dispersive X-ray fluorescence spectrometry using graphene as solid adsorbent in dispersive micro-solid phase extraction. Talanta 2015, 134, 360–5.
[124] Ponthieu M, Pinel-Raffaitin P, Le Hecho I, Mazeas L, Amouroux D, Donard OFX, Potin-Gautier M. Speciation analysis of arsenic in landfill leachate. Water Res 2007, 41, 3177–85.
[125] Bravo-Sanchez LR, Encinar JR, Fidalgo Martinez JI, Sanz-Medel A. Mercury speciation analysis in sea water by solid phase microextraction – gas chromatography–inductively coupled plasma mass spectrometry using ethyl and propyl derivatization. Matrix effects evaluation. Spectrochim Acta, Part B 2004, 59, 59–66.
[126] Reis AT, Coelho JP, Rucandio I, Davidson CM, Duarte AC, Pereira E. Thermo-desorption: A valid tool for mercury speciation in soils and sediments. Geoderma 2015, 237–238, 98–104.
[127] Reis AT, Coelho JP, Rodrigues SM, Rocha R, Davidson CM, Duarte AC, Pereira E. Development and validation of a simple thermo-desorption technique for mercury speciation in soils and sediments. Talanta 2012, 99, 363–8.
[128] Shuvaeva OV, Gustaytis MA, Anoshin GN. Mercury speciation in environmental solid samples using thermal release technique with atomic absorption detection. Anal Chim Acta 2008, 621, 148–54.
[129] Jitaru P, Adams FC. Speciation analysis of mercury by solid-phase microextraction and multicapillary gas chromatography hyphenated to inductively coupled plasma–time-of-flight-mass spectrometry. J Chromatogr A 2004, 1055, 197–207.
[130] Bloom NS, Preus E, Katon J, Hiltner M. Selective extractions to assess the biogeochemically relevant fractionation of inorganic mercury in sediments and soils. Anal Chim Acta 2003, 479, 233–48.
[131] Johnson WP, Swanson N, Black B, Rudd A, Carling G, Fernandez DP, Luft J, Van Leeuwen J, Marvin-Di Pasquale M. Total- and methyl-mercury concentrations and methylation rates across the freshwater to hypersaline continuum of the Great Salt Lake. Sci Tot Environ 2015, 511, 489–500.
[132] Gibbon-Walsh K, Salaun P, Uroic M, Feldmann J, McArthur JM, Van den Berg CMG. Voltammetric determination of arsenic in high iron and manganese groundwaters. Talanta 2011, 85, 1404–11.
[133] Abdolmohammad-Zadehn H, Talleb Z. Speciation of As(III)/As(V) in water samples by a magnetic solid phase extraction based on Fe_3O_4/Mg–Al layered double hydroxide nano-hybrid followed by chemiluminescence detection. Talanta 2014, 128, 147–55.
[134] Monticelli D, Laglera LM, Caprara S. Miniaturization in voltammetry: Ultratrace element analysis and speciation with twenty-fold sample size reduction. Talanta 2014, 128, 273–7.
[135] Dasgupta PK, Huang H, Zhang G, Cobb GP. Photometric measurement of trace As(III) and As(V) in drinking water. Talanta 2002, 58, 153–64.
[136] Bramanti E, Lomonte C, Onor M, Zamboni R, D'Ulivo A, Raspi G. Mercury speciation by liquid chromatography coupled with on-line chemical vapour generation and atomic fluorescence spectrometric detection (LC–CVGAFS). Talanta 2005, 66, 762–8.

[137] Cairns WRL, Ranaldo M, Hennebelle R, Turetta C, Capodaglio G, Ferrari CF, Dommergue A, Cescon P, Barbante C. Speciation analysis of mercury in seawater from the lagoon of Venice by on-line pre-concentration HPLC–ICP-MS. Anal Chim Acta 2008, 622, 62–9.
[138] Daus B, Mattusch J, Wennrich R, Weiss H. Analytical investigations of phenyl arsenicals in groundwater. Talanta 2008, 75, 376–9.
[139] Boonjob W, Miro M, Kolev SD. On-line speciation analysis of inorganic arsenic in complex environmental aqueous samples by pervaporation sequential injection analysis. Talanta 2013, 117, 8–13.
[140] Wolf RE, Morman SA, Hageman PL, Hoefen TM, Plumlee GS. Simultaneous speciation of arsenic, selenium, and chromium: Species stability, sample preservation, and analysis of ash and soil leachates. Anal Bioanal Chem 2011, 401, 2733–45.
[141] Brombach C-C, Cornelius C, Chen B, Corns WT, Feldmann J, Krupp EM. Methylmercury in water samples at the pg/L level by online preconcentration liquid chromatography cold vapor-atomic fluorescence spectrometry. Spectrochim Acta, Part B 2015, 105, 103–8.
[142] Gonzalvez A, Llorens A, Cervera ML, Armenta S, De la Guardia M. Non-chromatographic speciation of inorganic arsenic in mushrooms by hydride generation atomic fluorescence spectrometry. Food Chem 2009, 115, 360–4.
[143] Howell KA, Achterberg EP, Braungardt CB, Tappin AD, Worsfold PJ, Turner DR. Voltammetric in situ measurements of trace metals in coastal waters. Trends Anal Chem 2003, 22, 828–35.
[144] Schaeffer R, Soeroes C, Ipolyi I, Fodor P, Thomaidis NS. Determination of arsenic species in seafood samples from the Aegean Sea by liquid chromatography–(photo-oxidation)–hydride generation–atomic fluorescence spectrometry. Anal Chim Acta 2005, 547, 109–18.
[145] Maltez HF, Carasek E. Chromium speciation and preconcentration using zirconium(IV) and zirconium(IV) phosphate chemically immobilized onto silica gel surface using a flow system and FAAS. Talanta 2005, 65, 537–42.
[146] Armstrong HEL, Corns WR, Stockwell PB, O'Connor G, Ebdon L, Evans EH. Comparison of AFS and ICP-MS detection coupled with gas chromatography for the determination of methylmercury in marine samples. Anal Chim Acta 1999, 390, 245–53.
[147] Rahman GMM, Fahrenholz T, 'Skip' Kingston HM. Application of speciated isotope dilution mass spectrometry to evaluate methods for efficiencies, recoveries, and quantification of mercury species transformations in human hair. J Anal At Spectrom 2009, 24, 83–92.
[148] Jackson B, Taylor V, Arthurbaker R, Miller A. Low-level mercury speciation in freshwaters by isotope dilution GC-ICP-MS. Environ Sci Technol 2009, 43, 2463–9.
[149] Li Y, Liu S-J, Jiang D-Q, Jiang Y, Yan X-P. Gas Chromatography-inductively coupled plasma-mass spectrometry for mercury speciation in seafood. Chin J Anal Chem 2008, 36, 793–8.
[150] Chang L-F, Jiang S-J, Sahayam AC. Speciation analysis of mercury and lead in fish samples using liquid chromatography–inductively coupled plasma mass spectrometry. J Chromatogr A 2007, 1176, 143–8.
[151] Mester Z, Fodor P. Analytical system for arsenobetaine and arsenocholine speciation. J Anal At Spectrom 1997, 12, 363–7.
[152] Huang J-H, Ilgen G, Fecher P. Quantitative chemical extraction for arsenic speciation in rice grains. J Anal At Spectrom 2010, 25, 800–2.
[153] D'Amato M, Aureli F, Ciardullo S, Raggi A, Cubadda F. Arsenic speciation in wheat and wheat products using ultrasound- and microwave-assisted extraction and anion exchange chromatography-inductively coupled plasma mass spectrometry. J Anal At Spectrom 2011, 26, 207–13.
[154] Santoyo MM, Figueroa JAL, Wrobel K, Wrobel K. Analytical speciation of mercury in fish tissues by reversed phase liquid chromatography–inductively coupled plasma mass spectrometry with Bi^{3+} as internal standard. Talanta 2009, 79, 706–11.

[155] Martin-Doimeadios RCR, Monperrus M, Krupp E, Amouroux D, Donard OFX. Using speciated isotope dlution with GC-inductively coupled plasma MS to determine and unravel the artificial formation of monomethylmercury in certified reference sediments. Anal Chem 2003, 75, 3202–11.
[156] Duarte FA, Bizzi CA, Antes FG, Dressler VL, Flores EMM. Organic, inorganic and total mercury determination in fish by chemical vapor generation with collection on a gold gauze and electrothermal atomic absorption spectrometry. Spectrochim Acta, Part B 2009, 64, 513–9.
[157] Snell JP, Stewart II, Sturgeon RE, Frech W. Species specific isotope dilution calibration for determination of mercury species by gas chromatography coupled to inductively coupled plasma- or furnace atomisation plasma ionisation-mass spectrometry. J Anal At Spectrom 2000, 15, 1540–5.
[158] Vassileva E, Becker A, Broekaert JAC. Determination of arsenic and selenium species in groundwater and soil extracts by ion chromatography coupled to inductively coupled plasma mass spectrometry. Anal Chim Acta 2001, 442, 135–46.
[159] Stockdale A, Tipping E, Lofts S. Dissolved trace metal speciation in estuarine and coastal waters: Comparison of WHAM/Model VII predictions with analytical results. Environ Toxicol Chem 2015, 34, 53–63.
[160] Issa NB, Rajakovic-Ognjanovic VN, Marinkovic AD, Rajakovic LV. Separation and determination of arsenic species in water by selective exchange and hybrid resins. Anal Chim Acta 2011, 706, 191–8.
[161] Vanhaecke F, Saverwyns S, De Wannemacker G, Moens L, Dams R. Comparison of the application of higher mass resolution and cool plasma conditions to avoid spectral interferences in Cr(III)/Cr(VI) speciation by means of high-performance liquid chromatography – inductively coupled plasma mass spectrometry. Anal Chim Acta 2000, 419, 55–64.
[162] Chen X, Han C, Cheng H, Wang Y, Liu J, Xu Z, Hu L. Rapid speciation analysis of mercury in seawater and marine fish by cation exchange chromatography hyphenated with inductively coupled plasma mass spectrometry. J Chromatogr A 2013, 1314, 86–93.
[163] Chen ZL, Megharaj M, Naidu R. Speciation of chromium in waste water using ion chromatography inductively coupled plasma mass spectrometry. Talanta 2007, 72, 394–400.
[164] Rimondi V, Bardelli F, Benvenuti M, Costagliola P, Gray JE, Lattanzi P. Mercury speciation in the Mt. Amiata mining district (Italy): Interplay between urban activities and mercury contamination. Chem Geol 2014, 380, 110–8.
[165] Cheng H, Wu C, Shen L, Liu J, Xu Z. Online anion exchange column preconcentration and high performance liquid chromatographic separation with inductively coupled plasma mass spectrometry detection for mercury speciation analysis. Anal Chim Acta 2014, 828, 9–16.
[166] Krachler M, Emons H. Speciation analysis of antimony by high-performance liquid chromatography inductively coupled plasma mass spectrometry using ultrasonic nebulization. Anal Chim Acta 2001, 429, 125–33.
[167] Miyashita S, Shimoya M, Kamidate Y, Kuroiwa T, Shikino O, Fujiwara S, Francesconi KA, Kaise T. Rapid determination of arsenic species in freshwater organisms from the arsenic-rich Hayakawa River in Japan using HPLC-ICP-MS. Chemosphere 2009, 75, 1065–73.
[168] Cattani I, Spalla S, Beone GM, Del Re AAM, Boccelli R, Trevisan M. Characterization of mercury species in soils by HPLC–ICP-MS and measurement of fraction removed by diffusive gradient in thin films. Talanta 2008,74, 1520–6.
[169] Martinez-Bravo Y, Roig-Navarro AF, Lopez FJ, Hernandez F. Multielemental determination of arsenic, selenium and chromium(VI) species in water by high-performance liquid chromatography–inductively coupled plasma mass spectrometry. J Chromatogr A 2001, 926, 265–74.

[170] Ritsema R, De Smaele T, Moens L, De Jong AS, Donard OFX. Determination of butyltins in harbor sediment and water by aqueous phase ethylation GC-ICP-MS and hydride generation GC-AAS. Environ Pollut 1998, 99, 271–7.

[171] Uveges M, Abranko L, Fodor P. Optimization of GC–ICPMS system parameters for the determination of butyltin compounds in Hungarian freshwater origin sediment and mussel samples. Talanta 2007, 73, 490–7.

[172] Peng H, Zhang N, He M, Chen B, Hu B. Simultaneous speciation analysis of inorganic arsenic, chromium and selenium in environmental waters by 3-(2-aminoethylamino) propyltrimethoxysilane modified multi-wall carbon nanotubes packed microcolumn solid phase extraction and ICP-MS. Talanta 2015, 131, 266–72.

[173] Leng G, Yin H, Li S, Chen Y, Dan D. Speciation analysis of mercury in sediments using vortex-assisted liquid–liquid microextraction coupled to high-performance liquid chromatography-cold vapor atomic fluorescence spectrometry. Talanta 2012, 99, 631–6.

[174] Gibbon-Walsh K, Salan P, Van den Berg CMG. Arsenic speciation in natural waters by cathodic stripping voltammetry. Anal Chim Acta 2010, 662, 1–8.

[175] Leal-Granadillo IA, Alonso JIG, Sanz-Medel A. Determination of the speciation of organolead compounds in airborne particulate matter by gas chromatography–inductively coupled plasma mass spectrometry. Anal Chim Acta 2000, 423, 21–9.

[176] Xiao Q, Hu B, He M. Speciation of butyltin compounds in environmental and biological samples using headspace single drop microextraction coupled with gas chromatography-inductively coupled plasma mass spectrometry. J Chromatogr A 2008, 1211, 135–41.

[177] Miura T, Chiba K, Kuroiwa T, Narukawa T, Hioki A, Matsue H. Accurate determination of arsenic in arsenobetaine standard solutions of BCR-626 and NMIJ CRM 7901-a by neutron activation analysis coupled with internal standard method. Talanta 2010, 82, 1143–8.

[178] Lomonte C, Currell M, Morrison RJS, McKelvie ID, Kolev SD. Sensitive and ultra-fast determination of arsenic(III) by gas-diffusion flow injection analysis with chemiluminescence detection. Anal Chim Acta 2007, 583, 7–77.

[179] Helz GR, Tossell JA. Thermodynamic model for arsenic speciation in sulfidic waters: A novel use of ab initio computations. Geochim Cosmochim Acta 2008, 72, 4457–68.

[180] Dickin AP. Radiogenic Isotope Geology, 2nd ed. Cambridge, Cambridge University Press, 2005.

[181] Komárek M, Ettler V, Chrastný V, Mihaljevič M. Lead isotopes in environmental sciences: A review. Environ Int 2008, 34, 562–77.

[182] Prohaska T, Wenzel WW, Stingeder G. ICP-MS-based tracing of metal sources and mobility in a soil depth profile via the isotopic variation of Sr and Pb. Int J Mass Spectrom 2005, 50, 242–3.

[183] Sherman LS, Blum JD, Dvonch JT, Gratz LE, Landis MS. The use of Pb, Sr, and Hg isotopes in Great Lakes precipitation as a tool for pollution source attribution. Sci Total Environ 2015, 502, 362–74.

[184] Bazzano A, Grotti M. Determination of lead isotope ratios in environmental matrices by quadrupole ICP-MS working at low sample consumption rates. J Anal At Spectrom 2014, 29, 926–33.

[185] Balcaen L, Moens L, Vanhaecke F. Determination of isotope ratios of metals (and metalloids) by means of inductively coupled plasma-mass spectrometry for provenancing purposes – A review. Spectrochim Acta, Part B 2010, 65, 769–86.

[186] Hintelmann H, Foucher D , Evans RD, Dillon P. Measurement of mercury isotope ratios using multicollector ICP-MS. Geochim Cosmochim Acta 2003, 67, 147–52.

[187] Vanhaecke F, Balcaen L, Malinovsky D. Use of single-collector and multi-collector ICP-mass spectrometry for isotopic analysis. J Anal At Spectrom 2009, 24, 863–86.

[188] Xie QL, Lu SY, Evans D, Dillon P, Hintelman H. High precision Hg isotope analysis of environmental samples using gold trap-MC-ICP-MS. J Anal At Spectrom 2005, 20, 515–22.

[189] Jackson TA, Muir DCG, Vincent WF. Historical variations in the stable isotope composition of mercury in arctic lake sediments. Environ Sci Technol 2004, 38, 2813–21.
[190] Jackson TA. Variations in the isotope composition of mercury in a freshwater sediment sequence and food web. Can J Fish Aquat Sci 2001, 58,185–96.
[191] Feng XB, Foucher D, Hintelmann H, Yan HY, He TR, Qiu GL. Tracing mercury contamination sources in sediments using mercury isotope compositions. Environ Sci Technol 2010, 44, 3363–8.
[192] Jackson T A, Whittle DM, Evans MS, Muir DCG. Evidence for mass-independent and mass-dependent fractionation of the stable isotopes of mercury by natural processes in aquatic ecosystems. Appl Geochem, 2008, 23, 547–71.
[193] Bergquist BA, Blum JD. The odds and evens of mercury isotopes: Applications of mass-dependent and mass-independent isotope fractionation. Elements 2009, 5, 353–57.
[194] Vogl J, Rosner M, Pritzkow W. Development and validation of a single collector SF-ICPMS procedure for the determination of boron isotope ratios in water and food samples. J Anal At Spectrom 2011, 26, 861.
[195] Gao B, Liu Y, Sun K, Liang XR, Peng P, Sheng G, Fu F. Precise determination of cadmium and lead isotopic compositions in river sediments. Anal Chim Acta 2008, 612, 114–20.
[196] Cloquet C, Carignan J, Libourel G. Isotopic composition of Zn and Pb atmospheric depositions in an urban/periurban area of northeastern France. Environ Sci Technol 2006, 40, 6594–600.
[197] Beard B, Johnson CM, Skulan JL, Nealson KH. Application of Fe isotopes to tracing geochemical and biological cycling of Fe. Chem Geol 2003, 195, 87–117.
[198] Cloquet C, Carignan J, Lehmann MF, Vanhaecke F. Variation in the isotopic composition of zinc in the natural environment and the use of zinc isotopes in biogeosciences: A review. Anal Bioanal Chem 2008, 390, 451–63.
[199] International Vocabulary of Metrology – Basic and General Concepts and Associated Terms, VIM, 3rd ed., JCGM 200:2012 (JCGM 200:2008 with minor corrections).
[200] ISO/IEC 17025. General requirements for the competence of testing and calibration laboratories, Geneva, 2005.
[201] Joint Committee for Guides in Metrology (jgcm), Evaluation of measurement data – Guide to the expression of uncertainty in measurement. JGCM 100:2008 (GUM with minor corrections), 2008. http://www.bipm.org/utils/common/documents/jcgm/JCGM_100_2008_E.pdf.
[202] Nordtest. Handbook for Calculation of Measurement Uncertainty in Environmental Laboratories, Nordtest project 1589-02, Technical report 537 ed. 3.1, 2012.
[203] Azemard S, Vassileva E. Determination of methylmercury in marine biota samples with advanced mercury analyser: Method validation. Food Chem 2015, 176, 367–75.
[204] B'Hymer C, Caruso J. Arsenic and its speciation analysis using high-performance liquid chromatography and inductively coupled plasma mass spectrometry. J Chromatogr A 2004, 1045, 1–13.
[205] B'Hymer C, Caruso JA. Selenium speciation analysis using inductively coupled plasma-mass spectrometry. J Chromatogr A 2006, 1114, 1–20.
[206] Grotti M, Terol A, Todoli J. Speciation analysis by small-bore HPLC coupled to ICP-MS. Trends Anal Chem 2014, 61, 92–106.
[207] Bouyssiere B, Szpunar J, Lobinski R. Gas chromatography with inductively coupled plasma mass spectrometric detection in speciation analysis. Spectrochim Acta B 2002, 57, 805–28.
[208] Leopold K, Foulkes M, Worsfold P. Methods for the determination and speciation of mercury in natural waters – A review. Anal Chim Acta 2010, 663, 127–38.
[209] Montes-Bayon M, DeNicola K, Caruso J. Liquid chromatography–inductively coupled plasma mass spectrometry. J Chromatogr A 2003, 1000, 457–76.
[210] Popp M, Hann S, Koellensperger G. Environmental application of elemental speciation analysis based on liquid or gas chromatography hyphenated to inductively coupled plasma mass spectrometry – A review. Anal Chim Acta 2010, 668, 114–29.

[211] Rodriguez-Gonzalez P, Marchante-Gayon JM, Alonso JIG, Sanz-Medel A. Isotope dilution analysis for elemental speciation: A tutorial review. Spectrochim Acta B 2005, 60, 151–207.
[212] Rosen AL, Hieftje GM. Inductively coupled plasma mass spectrometry and electrospray mass spectrometry for speciation analysis: Applications and instrumentation. Spectrochim Acta B 2004, 59, 135–46.
[213] Oliveira RC, Santelli RE. Occurrence and chemical speciation analysis of organotin compounds in the environment – A review. Talanta 2010, 82, 9–24.
[214] Anawar HM. Arsenic speciation in environmental samples by hydride generation and electrothermal atomic absorption spectrometry. Talanta 2012, 88, 30–42.
[215] Guerin T, Astruc A, Astruc M. Speciation of arsenic and selenium compounds by HPLC hyphenated to specific detectors: A review of the main separation techniques. Talanta 1999, 50, 1–24.
[216] Gonzalvez A, Cervera ML, Armenta S, De la Guardia M. A review of non-chromatographic methods for speciation analysis. Anal Chim Acta 2009, 636, 129–57.
[217] Maa J, Sengupta MK, Yuan D, Purnendu K, Dasgupta PK. Speciation and detection of arsenic in aqueous samples: A review of recent progress in non-atomic spectrometric methods. Anal Chim Acta 2014, 831, 1–23.
[218] Gomez V, Callao MP. Chromium determination and speciation since 2000. Trends Anal Chem 2006, 25, 1006–15.
[219] Chena Y-W, Belzile N. High performance liquid chromatography coupled to atomic fluorescence spectrometry for the speciation of the hydride and chemical vapour-forming elements As, Se, Sb and Hg: A critical review. Anal Chim Acta 2010, 671, 9–26.
[220] Munoz E, Palmero S. Analysis and speciation of arsenic by stripping potentiometry: A review. Talanta 2005, 65, 613–20.
[221] Michalke B. The coupling of LC to ICP-MS in element speciation: I. General aspects, Trends Anal Chem 2002, 21, 142–53.
[222] Michalke B. The coupling of LC to ICP-MS in element speciation: II. Recent trends in application. Trends Anal Chem 2002, 21, 154–65.
[223] Sutton K, Sutton RMC, Caruso JA. Inductively coupled plasma mass spectrometric detection for chromatography and capillary electrophoresis. J Chromatogr A 1997, 789, 85–126.
[224] Kannamkumarath SS, Wrobel K, B'Hymer C, Caruso JA. Capillary electrophoresis–inductively coupled plasma-mass spectrometry: An attractive complementary technique for elemental speciation analysis. J Chromatogr A 2002, 975, 245–66.
[225] Sutton KL, Caruso JA. Liquid chromatography–inductively coupled plasma mass spectrometry. J Chromatogr A 1999, 856, 243–58.
[226] De Mora CJ, Harrison RM. The use of physical separation techniques in trace metal speciation studies. Water Res. 1983, 17, 723–33.
[227] Florence TM. The speciation of trace elements in waters. Talanta 1982, 29, 345–64.
[228] Kataoka H, Ishizaki A, Nonaka Y, Saito K. Developments and applications of capillary microextraction techniques: A review. Anal Chim Acta 2009,655, 8–29.
[229] Wuilloud JCA, Wuilloud RG, Vonderheide AP, Caruso JA. Gas chromatography/plasma spectrometry – an important analytical tool for elemental speciation studies. Spectrochim Acta, Part B 2004, 59, 755–92.
[230] Sonke JE, Salters VJM. Capillary electrophoresis–high resolution sector field inductively coupled plasma mass spectrometry. J Chromatogr A 2007, 1159, 63–74.
[231] Ponce de Leon CA, Montes-Bayon M, J.A. Caruso JA. Elemental speciation by chromatographic separation with inductively coupled plasma mass spectrometry detection. J Chromatogr A 2002, 974, 1–21.

[232] Michalke B. Element speciation definitions, analytical methodology, and some examples. Ecotoxicol Environ Saf 2003, 56, 122–39.
[233] Caruso JA, Montes-Bayon M. Elemental speciation studies – new directions for trace metal analysis. Ecotoxicol Environ Saf 2003, 56, 148–63.
[234] Florence TM, Batley GE, Benes P. Chemical speciation in natural waters. Crit Rev Anal Chem 1980, 9, 219–96.
[235] Mester Z, Sturgeon R. Trace element speciation using solid phase microextraction. Spectrochim Acta, Part B 2005, 60, 1243–69.
[236] Negri A, Marshall P. TBT contamination of remote marine environments: Ship groundings and ice-breakers as sources of organotins in the Great Barrier Reef and Antarctica. J Environ Manage 2009, 90, S31–40.
[237] Mays DE, Hussam A. Voltammetric methods for determination and speciation of inorganic arsenic in the environment: A review. Anal Chim Acta 2009, 646, 6–16.
[238] Mester Z, Sturgeon R, Pawliszyn J. Solid phase microextraction as a tool for trace element speciation. Spectrochim Acta, Part B 2001, 56, 233–60.
[239] Unceta N, Seby F, Malherbe J, Donard OFX. Chromium speciation in solid matrices and regulation: A review. Anal Bioanal Chem (DOI 10.1007/s00216-009-3417-1, Published on-line 23.Jan.2010).
[240] Yang T, Chen M-L, Wang J-H. Genetic and chemical modification of cells for selective separation and analysis of heavy metals of biological or environmental significance. Trends Anal Chem 2015, 66, 90–102.
[241] Florence TM, Morrison GM, Stauber JL. Determination of trace element speciation and the role of speciation in aquatic toxicity. Sci Tot Environ 1992, 125, 1–13.
[242] Alvarez-Llamas G, De la Campa MRF, Sanz-Medel A. ICP-MS for specific detection in capillary electrophoresis. Trends Anal Chem 2005, 24, 28–36.
[243] Gonzalez-Toledo E, Compano R, Granados M, Prat MD. Detection techniques in speciation analysis of organotin compounds by liquid chromatography. Trends Anal Chem 2003, 22, 26–33.
[244] Clough R, Harrington CF, Hill SJ, Madrid Y, Tyson JF. Atomic spectrometry update. Elemental speciation: A review. J Anal At Spectrom 2013, 28, 1153–95.
[245] Hu B, Zheng F, He M, Zhang N. Capillary microextraction (CME) and its application to trace elements analysis and their speciation. Anal Chim Acta 2009, 650, 23–32.
[246] Capelo JL, Fernandez C, Pedras B, Santos P, Gonzalez P, Vaz C. Trends in selenium determination/speciation by hyphenated techniques based on AAS or AFS. Talanta 2006, 68, 1442–7.
[247] Buffle J, Tercier-Waeber M-L. Voltammetric environmental trace metal analysis and speciation: From laboratory to in situ measurements. Trends Anal Chem 2005, 24, 172–91.
[248] Das D, Gupta U, Das AK. Recent developments in solid phase extraction in elemental speciation of environmental samples with special reference to aqueous solutions. Trends Anal Chem 2012, 38, 163–71.
[249] Poole CF. New trends in solid-phase extraction. Trends Anal Chem 2003, 22, 362–73.
[250] Leermakers M, Baeyens W, De Gieter M, Smedts B, Meert C, De Bisschop HC, Morabito R, Quevauviller Ph. Toxic arsenic compounds in environmental samples: Speciation and validation. Trends Anal Chem 2006, 25, 1–10.
[251] Gonzalvez A, Armenta S, Cervera ML, De la Guardia M. Non-chromatographic speciation. Trends Anal Chem 2010, 29, 260–8.
[252] Latorre CH, Garcia JB, Martin SG, Crecente RMP. Solid phase extraction for the speciation and preconcentration of inorganic selenium in water samples: A review. Anal Chim Acta 2013, 804, 37–49.

Annex

Reviews on determination of selected trace elements in environment samples.

Analytes	Environmental matrices	Detection methods	Comments	References
As	Water, soil extracts, plants, marine biota, CRMs	ICP-MS	All steps of the analytical procedures discussed	[204]
Se	Water, soil, sediments, plants, marine biota, CRMs	ICP-MS	Biochemistry of Se species. Environmental effects Sample collection, storage, and treatment	[205]
As, Se, Cr, Cd, Hg, Pb, Cu, Zn	Surface water, groundwater, and seawaters, soil leaching, biological samples	ICP-MS	Analytical performance, column recovery. Instrumental issues: interfaces, nebulizers, and spray chambers	[206]
As, Se, Sb, Bi, Cr, Cd, Hg, Te, Pb	Air, natural waters, sols, sediments, biological samples	ICP/MIP-OES, ICP-MS, MIP-MS, GD-MS	Sample preparation techniques (solid-phase microextraction, derivatization, capillary cryofocussing, stir bar sorptive extraction – thermodesorption)	[207]
Hg	Natural waters	ICP/MIP-OES, ICP-MS, ICP-ID-MS, AAS	Sample preparation techniques (solid-phase microextraction, cryotrapping)	[208]
As, Se, Sn, Fe, Zn, Cu	Natural waters, soils, biological materials	ICP-MS, ESI-MS	Need of CRMs discussed	[209]
As, Cr, Sb, Hg, Sn	Natural waters, sediments, soil, biota, extracts, gas	ICP-MS (SS-ID-MS)	Preconcentration and derivatization methods discussed	[210]
Se	Natural waters, soils, sediments, biota	ID-ICP-MS ID-TIMS	Isotope dilution principles discussed	[211]
As, Cr, Cd, Al, Cu, etc.	Applications: environmental, clinical, and bioinorganic chemistry	ICP-MS, ESI-MS ESI-MS-MS	General review	[212]

Analytes	Environmental matrices	Detection methods	Comments	References
Sn	Natural waters, sediments, soil, biota	UV, AAS, ICP-OES, ICP-MS, electrochemical methods	Various extraction techniques used for preconcentration and sample preparation	[213]
As	Waters, rocks, and sediments, biota clinical samples	HG-AAS, GF-AAS, AFS, ICP-OES, ICP-MS	Chemical modifiers for thermal stabilization of As species: Ni, Cu, Pd, Pd + Mg, W, Rh, La. Matrix interferences in hydride generation	[214]
As, Se	Natural waters, biological samples	AAS, ICP-OES, DCP-OES, ICP-MS, AFS	Comparative study of different techniques	[215]
Al, Sb, As, Cr, Pb, Hg, Fe, Se, Te, Sn	Natural waters, sediments, fly ash, soil, plants, biota	UV-VIS, AAS, AFS, ICP-OES, ICP-MS, NAA, stripping voltammetry	Proportional equations (mathematical evaluation	[216]
As	Waters	Colorimetry/UV-VIS; fluorescence; liquid-, gas-phase chemiluminescence; laser-induced breakdown spectroscopy; surface-enhanced Raman spectroscopy; TXRF; attenuated total reflectance – FTIR; potentiometry and conductometry; ICP-MS; NAA; biosensors	Detailed discussion on interferences	[217]
Cr	Waters, airborne particles, soils and sediments, biological samples, industrial samples	UV-VIS, AAS, ICP-OES, ICP-MS, fluorimetry, amperometry, voltammetry, HPLC, IC, GC	Chemometrical techniques. Multivariate calibrations	[218]
As, Se, Sb, Hg	Waters, polluted soil, sediments, biota	HG/CV-AFS	Chromatographic conditions and hydride generation discussed in details.	[219]
As	Polluted river water, seawater, soils, biota	Stripping potentiometry	As(III), As(V) total As determination	[220]

As		ICP-MS	Part 1: Instrumentation, general aspects, LC-ICP-MS coupling	[221]
As, Se, Cd, Sn, I	Plants, polluted soils, biological samples	ICP-MS	Part 2: Applications. Trends	[222]
As, Se, Hg, Sn, Pb, Cd, Cr, Te	Coal fly ash, seawater, sediments, biota, clinical samples	ICP-MS; MIP-OES; ion-spray MS	Application trends	[223]
As, Se, Sb, Te, Cr, Hg, Au, Cd, Zn, Pt, Co	Water, artificial test samples, clinical samples	ICP-MS	Interfaces and nebulizers compared	[224]
Cr, Se, Co, Fe, Zn, Pt, Te, As, Pb, Hg, Sn, V, lanthanides	Wastewater, biota, clinical samples	ICP-MS	Detailed discussion on interferences	[225]
Pb, Cu, Co, Cd, Mn, Ni	Natural waters	ICP-MS	Physical separation techniques presented	[226]
Cu, Pb, Cd, Zn, Fe, Cr, Hg, Sn, Co, Ni, Mn, As, Se, Pu	Natural waters	Anodic stripping voltammetry	Direct speciation methods	[227]
Co, Ni, Cd, Pb, Hg, Cu	Environmental waters, air, wastewater, clinical samples	LC-MS, GC-FID, LC-MS, HPLC-UV, ICP-MS, ICP-OES, etc.		[228]
As, Bi, Cd, Br, Cu, Ge, Fe, Pb, Mn, Hg, Ni, Se, Te, Sn,	Natural waters, soil, sediments, plants, solid waste deposits	ICP-MS, MIP-MS, ICP-OES, MIP-OES, AAS, AFS		[229]
Cr, As, Sb, Cu, Zn, Cd,	Biochemical and environmental samples	HR-ICP-MS, species-specific/unspecific ID	Calibration methods compared	[230]
Sn, Hg, Se As, Ge, Te, Pb	Natural water sediments, biota	ICP-MS, LC-MS, GC-MS, ICP-OES, MIP-OES	All steps of the analytical procedure discussed	[231]
As, Cr, Cu, I, Fe, Hg, Pt, Zn	Soils, sediments, biological samples	ICP-MS	Sampling procedure discussed	[232]

Analytes	Environmental matrices	Detection methods	Comments	References
As	Natural waters, environmental samples	ASV, CSV	Methods allowing direct speciation analysis. Hg, Bi, Cu, Au electrodes described	[233]
As, Se, Sb, Hg, Cu, Cd, Cr, Mn, Al, Pb, Tl, etc.	Natural waters	ASV	Method allowing direct speciation analysis	[234]
Sn, Hg, Se As, Ge, Te Pb,	Natural water sediments, biota	ICP-MS, LC-MS, GC-MS, ICP-OES, MIP-OES	All steps of the analytical procedure discussed	[235]
Sn	Antarctic sea sediments, biota	ICP-MS	TBT contamination in Antarctica region caused by protective paints of ships and ice-breakers. Sampling and ecological problems discussed	[236]
As	Natural waters, environmental samples	ASV, CSV	Methods allowing direct speciation analysis. Hg, Bi, Cu, Au electrodes described	[237]
Hg, Sn, As, Se, Pb	Natural waters, sediments, soil, fish tissue	MS, GC-MS, ICP-MS	Volatilization by ethylation, hydride generation, headspace	[238]
Cr	Soil, sediments, water, airborne particulates	ICP-MS, ICP-OES, AAS	Interconversion Cr(III)–Cr(VI) during sample preparation discussed	[239]
Fe, Co, Ni, Cd, Pb, Cr, Cu, Zn, Pb, Ag, Se, As	Water samples	AAS, ICP-OES	Natural biospeciation and biomonitoring	[240]
Zn, Cd, Pd, Cr, As, Hg, Cu	Water samples, seawater	ASV	Direct speciation. Toxicity of chemical forms discussed	[241]
Se, As, Sb, Te, Cd, Cu, Zn, Ni, Cr, Fe, V	Groundwater, wastewater, soil extracts	ICP-MS	Interfaces described	[242]
Sn	Sediments, natural waters, biota	ICP-MS, MS, fluorimetric detector	Isotope dilution	[243]
Hg, As, Sb Cr, V, Pb, Se, Sn	Natural waters, plankton, plants	MS, ICP-MS, AAS, X-ray methods	Sample preparation related to speciation discussed	[244]

Pb, Cd, As, Hg, Cu, Sn	Environmental samples	ICP-MS, MS, LC-MS, AAS	Capillary coating materials and target analytes discussed. SME and SPME comparison	[245]
Se	Environmental samples	AAS, AFS	Sample preservation and treatment appropriate for speciation described	[246]
Pb, Cd, As, Hg, Al, Cr	Natural waters, sediments, soils	Voltammetry	Direct speciation, possibility for in situ monitoring discussed	[247]
Hg, Cu, Pb, Cd, U, Cr, Ni, Zn	Coastal seawaters	Voltammetry, ASV	Direct speciation. In situ measurements, Ir electrode	[143]
Hg, Cr, As, Pb, Cd, Fe, Sn, Se	Natural waters	AAS, ICP-OES, ICP-MS,	Multielement speciation	[248]
Trace elements, org. compds	Environmental samples	LC, GC	Method selection guide provided. General article	[249]
As	Environmental samples	MS, ICP-MS, AAS, AFS	Validation by CRMs, sampling, and storage discussed	[250]
Hg, Cd, Pb, Zn, etc.	Natural waters	ASV, CSV AAS	Nonchromatographic separation	[251]
Se	Natural waters	GC, LC, ICP-MS	Speciation strategy for inorganic Se in waters	[252]

Note: CRM certified reference material; ICP-MS, inductively coupled plasma mass spectrometry; OES, optical emission spectrometry; MIP, microwave-induced plasma;GD, glow discharge; ID, isotope dilution; AAS, atomic absorption spectrometry;ESI, electrospray ionization; SS, solid sampling; TIMS, thermal ionization mass spectrometry; UV, ultraviolet; VIS, visible, HG-AAS, hydride-generation atomic absorption spectrometry; GF-AAS, graphite furnace atomic absorption spectrometry; DCP, direct current plasma; NAA, Neutron activation analysis; TXRF, total reflection X-ray fluorescence; FTIR, Fourier transform infrared spectroscopy; HPLC, high-performance liquid chromatography; IC, ion chromatography; GC, gas chromatography; CV-AFS, cold vapor atomic fluorescence spectrometry; FID, flame ionization detector; LC, liquid chromatography; HR-ICP-MS, high-resolution inductively coupled plasma mass spectrometry; ASV, anodic stripping voltammetry; CSV, cathodic stripping voltammetry; SME, solvent microextraction; SPME, solid-phase microextraction; TBT, tributyltin.

Katarzyna Wrobel, Kazimierz Wrobel, Alma Rosa Corrales Escobosa and Armando Gómez Ojeda

4 Food Analysis and Speciation

4.1 Topics, elements and species of interest

Food, beverages and drinking water are primary sources of minerals for humans; therefore, determination of essential and/or potentially toxic elements is an intrinsic component of food quality control. Analytical results obtained for Fe, Zn, Cu, Cr, I, Co, Mo and Se allow for monitoring dietary intakes in order to satisfy specific metabolic requirements and also to prevent/detect the unsafe states of deficiency or overload. For such purposes, acceptable range of oral intake or dietary reference intake (DRI) needs to be observed, as defined by the World Health Organization (WHO) [1, 2]. For other elements, including those probably essential for humans (Si, Mn, Ni, B, V) and those nonessential or toxic, several parameters and limits are in use, majority of them developed from the dose–response relationships observed in toxicological studies [3]. In particular, the terms and respective values often cited in the literature include oral reference dose (RfD), acceptable daily intake (ADI), tolerable daily intake (TDI) and provisional tolerable daily or weekly intake (PTDI or PTWI) [4]. Despite conceptual similarity of the above parameters, each regulating organization uses its own definition and the values reported for a given element may vary depending on the age/sex or clinical conditions of the subjects. In Table 4.1, the established values for oral intakes of few selected elements are presented; these values were taken from the websites of US Environmental Protection Agency [5], WHO [2] and Joint Food and Agricultural Organization/WHO Expert Committee on Food Additives [4, 6].

In another approach to the assessment of food safety in terms of mineral content, international organizations and food agencies have introduced legislations that specify maximum allowable concentrations of nonessential elements in various products. Noteworthy, the European Community Commission Regulation (EC) No. 1881/2006 sets maximum levels of cadmium and lead in marine food, in different types of animal meat, in vegetables, fruits, cereals and legumes. This same directive includes also regulation for total tin in canned food, for lead in infant formula, milk and its products and for mercury in marine food. Depending on the element and food type, the maximum element levels are in the range 0.05–1 mg/kg of the sample wet mass (for Sn the range is 50–200 mg/kg wet mass) [7]. In the guidance documents of Food and Drug Administration (FDA), action levels are given for As, Cd, Cr, Pb, Ni in seafood (1.5–86 µg/kg), for methylmercury (MeHg) in fish (1.0 mgHg/kg MeHg), for Pb in candy likely to be frequently consumed by children (0.5 mgPb/kg) and for inorganic As in apple juice (10 µg As/kg) [8]. Not all elements and not all food items are included in current legislation; hence, as an all-purpose rule for regulated and not-regulated

https://doi.org/10.1515/9783110366730-005

Table 4.1: Different approaches to the regulation of oral intake of selected trace elements from food.

Element	Classification	DRI (μg/day)	RfD (mg/kg bw/day)	TDI/ADI (mg/kg bw/day)	PTWI (mg/kg bw/week)
Cu	Essential	700–900	–	0.04	3.5
Mn	Essential	34–45	0.14	0.14	–
Se	Essential	40–55	0.005	0.005	–
Ni	Possibly essential	–	0.02	0.02	–
Inorg. As	Nonessential	–	0.0003	0.0003	0.015
Cd	Toxic	–	0.001	0.001	0.007
Inorg. Hg	Toxic	–	0.0001	0.0003	0.005

Note: DRI, daily reference intake for healthy adults (WHO) [2]; RfD, oral reference dose describes daily exposure that is likely to be without an appreciable risk during a lifetime (The Integrated Risk Information System, US EPA) [5]; TDI, tolerable daily intake (WHO) equivalent to ADI; ADI, acceptable daily intake (JECFA) refers to "safe" levels of intake [4]; PTWI, provisional tolerable weekly intake, useful for nonessential elements that are not cleared rapidly from the body (JECFA) [4, 6].

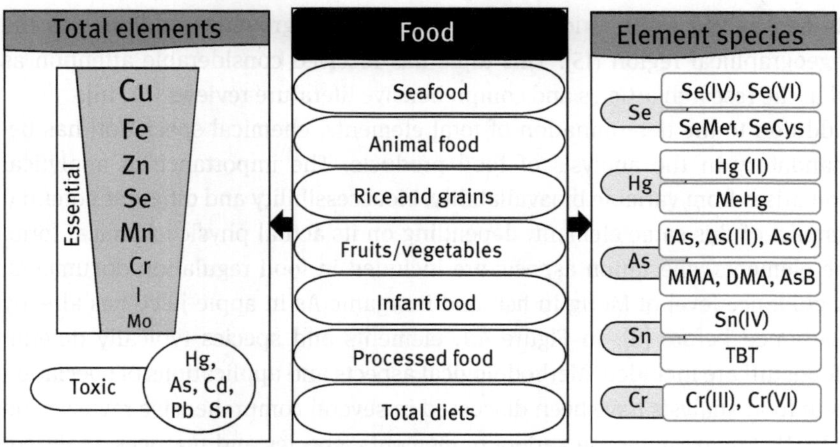

Figure 4.1: General scheme indicating products, elements and their primary species of interest in food analysis.

analytes/products, the levels of contaminants and natural toxicants should be kept as low as reasonably achievable [9].

Figure 4.1 shows a general scheme of the most analyzed food products and elements of interest. For total concentration of essential elements, practically all food types are included, whereas the determination of toxic elements is preferentially carried out in foods derived from plants, in seafoods and in mushrooms. Nonessential elements of major interest include As, Cd, Pb, Hg and Sn. At this point, the increasing demand for controlling various infant foods should be emphasized.

Globalization of food supply brings obvious advantages but it also entails new safety risks due to the potential contamination and adulteration issues. Tracking the provenance of food helps to ensure the good quality of products by controlling their origin, their processing/production process, transport and delivery. Today, different categories of foods are protected by domestic or international legislations. As an example, EC EU Protected Food Name Scheme considers three categories of traditional and speciality foods: (1) agricultural products and foodstuffs that are produced, processed and prepared in a given geographical area using recognized know-how (Protected Designation of Origin); (2) food products closely linked to the geographical area with at least one of the stages of production, processing or preparation taking place in the area (Protected Geographical Indication); and (3) products obtained by specific traditional processing and/or production methods and not associated with a specific geographical area (Traditional Specialities Guaranteed). In this regard, trace element profiling combined with statistical classification/discrimination has been widely accepted as an important tool in the assessment of food provenance [10–14]. Unlike organic markers of food origin, potentially affected by many environmental and anthropogenic factors, the concentration profile of several trace and rare earth elements has proved to be strictly associated with the growing conditions in the specific geographical region [15]. This topic has received considerable attention as reflected in the research articles and comprehensive literature reviews [14, 16].

In addition to the determination of total elements, chemical speciation has become mandatory in the analysis of food products. The importance of analytical speciation arises from variable bioavailability, bioaccessibility and different potential health impact of this same element, depending on its actual physicochemical form. For some elements, speciation aspects are included in food regulation documents; the FDA guidance level of MeHg in fish and inorganic As in apple juice has already been mentioned before [8]. In Figure 4.1, elements and species typically determined in foodstuff are included. Methodological aspects and applications of speciation schemes in food analysis have been discussed in several comprehensive reviews and books [17–21]. Among immense variety of elements, species and matrices analyzed, few specific topics have received major attention; such is the case of mercury and arsenic speciation in edible marine organisms [22–25], arsenic speciation in wheat and rice [26], selenium in dietary supplements and in biofortified yeast, vegetables or mushrooms [27–29].

In this chapter, the state of art in methodology of trace element determination and speciation in food is presented. The discussion is based on the recent advances and applications, including studies reported in the last four years. The main focus is on the specific aspects of food quality control that have been depicted in this introduction. Due to the requirement for precise quantification at low concentration levels in chemically complex food matrices, the applications of atomic spectrometry are highlighted as the preferentially selected tools. For speciation analysis, the determination of common species of mercury, arsenic, selenium, chromium and tin in different foodstuffs is reviewed.

4.2 Determination of total elements

Current regulations of mineral elements in foodstuff refer to the concentration levels in wet mass, which facilitates evaluation of the intake from edible portions [7, 8]. On the other part, quantification results are usually referred to the mass of dried or freeze-dried sample because elimination of moisture results in better stability and easier sample handling, in enhanced analytical reliability and it also allows for traceability to the certified reference materials (CRMs) [12]. To circumvent the apparent inconsistency hindering direct comparison of analytical data with the regulated values, the moisture content needs to be evaluated and it is usually done in a separate experiment taking conveniently large amount of the wet sample [30–35]. Alternatively, internal standard addition improves the precision while analyzing wet sample and such approach is especially well suited for speciation analysis [36]. Worth mentioning that for total diet studies, FDA has established specific instructions regulating the preparation and the amount of edible portion for each item [37].

Once the sample is defined, reliability of the analytical results rests on the selection of appropriate instrumental technique, suitable sample pretreatment and minimization of analytical errors by implementation of quality control schemes [38]. In the analysis of foodstuff, the majority of samples are solids or semi-solids and must be subjected to the pretreatment prior to trace element determination, unless intended to be analyzed by X-ray techniques [39]. While using optical or mass atomic spectrometry, dry-ashing or wet acid digestion is typically carried out yet without requirement for full mineralization. The main effort is centered at efficient solubilization of analytes and on the degradation of chemical matrix in a short time and in line with the principles of green chemistry. In Table 4.2, a survey of recent studies is presented indicating the pretreatment procedure applied in each case. As can be observed, microwave-assisted digestion with concentrated nitric acid has often been used for variety of matrices, including baby foods [34, 40, 41], rice and grains [8, 42], fruits [43], dairy products [44], mushrooms [45–47] and composite foods [48–51]. In many applications, hydrogen peroxide was added to enhance oxidation power; several examples are presented in Table 4.2 for the above-mentioned products [32, 39, 43, 52–65] and additionally, for vegetables [66–69], seafood [70–73], animal meat [74–77] and honey [12, 78, 79].

Conventional heating on a hot plate or in a heating block has also been reported [80–83]; however, it should be stressed that digestion carried out in open systems compromises the determination of the so-called volatile elements that can be lost at this stage. In the analysis of more difficult samples, especially those containing high lipids content (whole milk, cocoa products), mixtures of concentrated nitric, perchloric and/or sulfuric acids can be used [84–86]. For the digestion of seafood, vegetables and legumes, hydrochloric acid has sometimes been added to the nitric acid or to nitric acid/hydrogen peroxide mixtures [87–90]. Of note, hydrofluoric acid is practically not required for food analysis. Dry-ashing by application of high

Table 4.2: Determination of total elements in different food products: application contexts, chemical matrices and procedural approaches.

Food product	Purpose	Pretreatment	Elements	Technique	Validation	References
Fish, Persian Gulf	Methodological development (preconcentration)	Open-vessel acid digestion: 500 mg of dried sample + 5 mL HNO_3 + 6 mL H_2O_2, 180 °C, 1.5 h; $K_2S_2O_8$ addition, neutralization and extraction/preconcentration on Fe_3O_4 magnetic nanoparticles functionalized with triazene groups	Hg	ICP-OES	DL: 0.04 ng/mL, enrichment factor 500 CRM: 436 tuna fish (IAEA)	[83]
Fish, Central Adriatic Sea	Evaluation of contaminating elements in different species; comparison between the muscle and bone	MW digestion: 200 mg of the freeze-dried and homogenized muscle (or bone) + 7 mL HNO_3 + 1.5 mL H_2O_2 + 0.05 mL HF; diluted	As, Cd, Pb, Hg, Se, Zn, Cu	ICP-OES	DLs: 0.002–0.10 ng/g (dry mass) CRM: BCR 422 cod muscle (IRMM)	[105]
Fish, Mexico	Health risk assessment	Wet acid digestion in a capped Teflon vial: 0.25 g of the powdered sample + 5 mL conc. HNO_3, 120 °C 3 h; diluted	Hg	CV-AAS	DL: 0.012 µg /g dry mass CRMs: DORM-3 fish muscle, DOLT-4 fish liver (NRC)	[59]
Fish, seafood from FAO fishing area, Atlantic Ocean	Evaluation of human exposure through diet	Edible tissues homogenized and MW digested with HNO_3 and H_2O_2; Rh as IS	Cd, Pb	ICP-MS	DLs: 0.01–0.2 µg/kg CRMs: BCR 422 cod muscle, BCR 668 mussel tissue (IRMM); DORM-4 fish protein and DOLT-3 dogfish (NRC)	[82]
		Cold Hg vapor formed in a flow system (0.2% $NaBH_4$, 3% HCl)	Hg	FI-CV-AAS		

Fresh, canned and frozen fish, shellfish, Spain	Evaluation of health risk	MW digestion: 0.7 g of edible fish/shellfish sample + 6 mL HNO_3:H_2O (4:2) + 0.5 mL HCl, 20 min heating program; dilution MeHg determined separately For As: 1 mL of the digest + 1 mL conc. HCl + 1 mL 5% KI/ascorbic acid (room temp. 45 min); dilution and HG in a flow system using 0.2% $NaBH_4$ For Hg: 1 mL of the digest + 0.5 mL 5% $KMnO_4$ + 9 mL (1.5% HNO_3 + 1.5% H_2SO_4); CV generation in a flow system using 3% $NaBH_4$ Analyte modifiers: Pd/Mg for Cd; $NH_4H_2PO_4$ for Pb and Sn MeHg determined separately	Cd, Pb, Sn, As Hg	ETAAAS Zeeman BC HG-AAS CV-AAS	DLs: 10.7 ng/kg Hg; 0.158 µg/kg Cd; 4.35 µg/kg Pb; 0.159 µg/kg As, 1.36 µg/kg Sn CRM: BB 422 fish muscle (ERM) Recoveries evaluated after standard addition in the range 96–105%	[87]
Fish, shellfish, Ionian Sea	Evaluation of daily intake	Edible portion homogenized MW digestion: 0.5 g sample + 6 mL HNO_3 + 2 mL H_2O_2; dilution Prior to cold vapor formation, 1 mL 5% $KMnO_4$ added followed by 1 mL 1.5% $NH_2OH \times HCl$, reduction	Hg Se	CV-AAS ICP-MS	DLs: 0.006 mg/kg Hg; 0.004 mg/kg Se (wet mass) CRM: 1946 Lake Superior fish (NIST)	[70]
Shrimp, worldwide	Geographical origin, authentication (discrimination analysis)	MW digestion of the dried (60 °C, 72 h) and pulverized white muscle samples: 0.25 g + 4 mL HNO_3+ 4 mL H_2O_2; dilution For IR MS sample defatted, homogenized and pulverized	As, Cd, Pb P, S C, N	ICP-MS ICP-OES IR MS	CRM: TORT-2 lobster hepatopancreas (NRC).	[71]

(*Continued*)

Table 4.2: (Continued)

Food product	Purpose	Pretreatment	Elements	Technique	Validation	References
Mussels, Adriatic coastal area	Health risk assessment	MW digestion: 0.5 g of pooled, homogenized and freeze-dried material + 7 mL HNO_3 + 2 mL H_2O_2; dilution	Fe, Mn, Cu, Zn Co, Ni, Pb, Cd Hg	FAAS ETAAAS CV-AAS	CRM: 2976 mussel homogenate (NIST)	[72]
Red king crab, Barents Sea Norwegian coast	Food safety, geographical and gender differences	Claw and leg materials pooled separately, cooked, weighed and freeze-dried. A portion of the powdered material MW digested with HNO_3/H_2O_2 and diluted Rh used as IS and Au added to stabilize Hg	As, Cd, Hg, Pb	ICP-MS	QLs: 0.005 mg/kg for Hg and Cd, 0.01 mg/kg for As and 0.03 mg/kg for Pb (dry mass) CRM: TORT-2 lobster hepatopancreas (NRC) Analytical method previously tested in several proficiency tests	[73]
Seaweed, Spain market	Methodological development	Four methods for acid digestion of dry mass (0.5 g + 6 mL HNO_3): (1) closed stainless steel Teflon ® PTFE bombs (temp., p) (2) closed Teflon® PFA vessels (temp., p) (3) open polypropylene tubes with reflux caps in a heating block (temp., oxidizing agents) (4) closed TFM™PTFE vessels with microwave heating (temp. p)	Hg	CV-AFS DMA	DL: 10 ng/L DL: 0.5 ng/g CRM: BCR-279 sea lettuce (IRMM)	[94]
Seaweed, Chile	Classification according to the type of seaweed and the extraction zone	Wet digestion of the washed, dried and milled biomass: 1 g + 8 mL HNO_3(heating block) + 1 mL HCl (160 °C, 5 min), dilution For P, after evaporation and redissolution, vanadate-molybdate reagent added	Na, K, Ca, Mg, Fe, Cu, Zn and Mn P	FAAS Spectrophotometry	CRM: BCR-279 sea lettuce (IRMM)	[88]

Lamb muscle, USA	Estimation of daily intake as compared to TDA, RfD, PTWI	Direct thermal decomposition Closed-vessel MW digestion: 0.5–1 g sample + 15 mL HNO_3 + 3 mL H_2O_2; dilution	Hg As, Cd, Ca, Cr, Co, Cu, Fe, Pb, Mg, Mn, Ni, P, K, Se, Na, Sn, V, Zn	DMA ICP-OES	Ds: 0.4–2.9 µg/kg (0.02 µg/kg for Hg) CRM: 1643e trace elements in natural water (NIST); DORM-2 dogfish muscle and TORT-2 lobster hepatopancreas (NRC)	[74]
Beef meet, Poland	Comparative evaluation of element profiles in different beef breeds fed identically and fattened intensively	Closed-vessel MW digestion: 0.6 g of defrost meat + 5.0 mL HNO_3 + 0.5 mL H_2O_2, 35 min heating program, dilution 0.5 mg/L Y as IS, spectral interferences corrected by software algorithm	Ca, P, Mg, Na, K, Fe, Zn, Cu, Mn, Se, Co, Cr, Ni, Sr, Ba, Cd, Pb	ICP-OES	DLs: 0.03–4.0 µg/L CRM: Seronorm™ trace elements whole blood (SERO AS, Norway).	[75]
Animal food in the vicinity of gold mines, Ghana	Evaluation of daily intakes and health risk	Closed-vessel MW digestion: 0.5 g of dried material (liver, kidney, muscle) + 5 mL HNO_3 + 1 mL H_2O_2(20 min heating program); dilution Direct thermal decomposition	Cr, Mn, Fe, Co, Ni, Cu, Zn, As, Cd, Pb Hg	ICP-MS DMA	DLs: 0.001–0.226 µg/L CRMs: DORM-3 fish protein and DOLT-4 dogfish liver (NRC)	[76]
Canned meat, Romania	Evaluation of metal migration from containers	Parameters studied: storage temperature, time and humidity and varnish type, width and porosity of the container MW digestion: 1 g of the sample + 6 mL HNO_3 + 2 mL H_2O_2(30 min heating program); dilution	Cd, Pb, Cu, Fe, Zn, Sn	AAS ICP-OES	Not reported	[77]

(*Continued*)

Table 4.2: (Continued)

Food product	Purpose	Pretreatment	Elements	Technique	Validation	References
Meat patties, New Zealand market	Evaluation of meta content depending on the cooking procedure	Procedures tested: 3 min plate/ohmic heating using different plate materials and variable frequencies; conventional plate grill Dry ashing: cooked patties dried (80 °C until stable mass), then heated in a muffle furnace (500 °C, 8 h) and white ash dissolved in diluted HNO_3	Fe, Si, Mn, Cr, Ni, Mo, Nb, Cu, Co	ICP-MS	Not reported	[102]
Southern Spanish cheeses	Classification of products	Moisture content determined; 10 g of the chopped sample dried and dry ashed at 460 °C for 16 h; 2 mL of 2 N HNO_3 added drying/incineration repeated (1 h). Residual ash dissolved in 5 mL 2 N HNO_3 and diluted	Ca, Mg, P, Zn, Fe, Mn, Cu, K, Na	FAAS	DLs: 0.011–4.30 mg/L CRM: 1549 nonfat milk powder (NIST).	[33]
Turkish cheeses	Effect of packing material (plastic, tin cans)	Dry ashing: 1.0 g of the dried sample (450–500 °C, 16 h) + 1 mL conc. HNO_3 (450–500 °C, 6 h), dissolution Hot plate wet digestion: 1.0 g of sample + 12 mL HNO_3 + 4 mL H_2O_2 (130 °C, 4 h) MW digestion: 1.0 g sample + 6 mL HNO_3 + 2 mL H_2O_2 (30 min program) Ionization buffer $LaCl_3$ used for Ca	Cd, Co, Cr, Cu, Mn, Ni, Pb, Se, Zn	ICP-OES	DLs: 0.010–0.21 mg/L CRM: GBW 07605 tea (China)	[91]
Bovine and soybean milk and yogurt, Spain	Evaluation of products with respect to ADI and RDA values	MW digestion: 4 g of product + 6 mL HNO_3, dilution and IS addition (In)	Ag, Al, As, Ba, Be, Cd, Co, Cr, Cu, Fe, Hg, Mn, Mo, Ni, Pb, Sb, Sn, Tl, V, Zn	ICP-MS	DLs: 0.5–80 ng/g CRMs: BCR-151 skim milk powder, BCR-063R skim milk powder (IRMM)	[44]

Yogurt, milk, South Korea	Evaluation of food safety	Open-system digestion in a heating block (30 cm pyrex glass tubes): 1.0–2.0 g of the sample + 25 mL HNO_3 + 2 mL H_2O_2, initially at 50 °C and then temperature increased to 150–160 °C, 10–12 h; dilution	As, Pb, Cd, Cr, Cu, Ni, Zn, Tl, U, Rb, Cs, Li, Be, Ba, Sr, Bi, Cs, Ga, In, V, Co, Se	ICP-MS	DLs: 0.002–1.6 ng/g CRM: 1549 nonfat milk powder (NIST) Recoveries evaluated after two-point standard addition: 91–107%	[60]
Yogurt, Tenerife	Evaluation of dietary intake of metals	Sample homogenized and dried (70 °C, 72 h); 20 g dry ashed at 450 °C, 8 + 48 h, dissolved in 1.5% HNO_3 and diluted	Na, K, Ca, Mg, Al, B, Ba, Cd, Co, Cr, Cu, Fe, Mn, Mo, Li, Pb, Zn, Ni, Sr, V	ICP-OES	DLs: 0.001–3.655 mg/L CRMs: 1549 nonfat milk powder, 1515 apple and 1573a tomato leaves (NIST)	[92]
Milk and yogurt, Turkey	Methodological development (preconcentration)	Open-system digestion: 10 mL of animal milk (or 10 g of yogurt) + 5 mL HNO_3+ 5 mL H_2O_2 (100 °C, 15 min) + 2 mL $HClO_4$ (150 °C, 1 h). SPE: 60 mL of the diluted digest + 30 mg of the prepared aminothioazole-containing resin + 10 mL buffer (pH 3.5); after centrifugation the sorbed ions stripped with 1 M HNO_3	Pb Ni	STAT-AAS FAAS	DLs: 0.15 ng/mL Pb; 0.75 ng/mL Ni CRM: BCR 151 skim milk powder (IRMM)	[84]
Milk, Brazil	Methodological development	Sample dilution diluted 1 + 9 (v/v) in 1% HNO_3. Modifiers: W + Ru for group 1; W + Pd for group 2 Air addition during pyrolysis	Group 1: Sb, As, Bi, Pb, Se Group 2: Co, Cu, Cr, Fe, Mn	HR-CS-ETAAAS Zeeman BC	DLs: 10–53 µg/L CRMSs: 11549 nonfat milk powder and 8435 whole milk powder (NIST).	[107]

(Continued)

Table 4.2: (Continued)

Food product	Purpose	Pretreatment	Elements	Technique	Validation	References
Milk powders, Brazil	Methodological development (digestion)	MW-induced combustion: sample pellet (3 g) together with wet filter paper (50 µL NH_4NO_3 6M) collocated on the quartz holder and introduced to the MW vessel with the condenser. Combustion carried out under continuous flow of oxygen; afterward sample heated with 2 M HNO_3 and diluted	Ba, Ca, Co, Cr, Cu, Fe, Mg, Mn, Sr, V, Zn	ICP-OES	Comparison with conventional MW digestion. QLs: 0.03–0.1 µg/g CRM: 8435 whole milk powder and 1549 nonfat milk powder (NIST)	[106]
Breast milk, infant formulas and foods, Sweden	Comparative evaluation of dietary intakes	The formulas and foods prepared in deionized water as indicated and MW digested: 1 g sample + 2 mL HNO_3 (65%) + 3 mL H_2O; dilution	Mg, Ca, Mn, Fe, Cu, Zn, Mo, As, Cd Se Sb, Pb, U	ICP-MS (CRC, He) (CRC, H_2) (standard mode)	CRMs: 1849 infant/adult nutritional formula (NIST); Seronorm™ trace elements whole blood L-2 (201605) (SERO AS, Norway).	[41]
Infant formulas, drinks, Poland	Comparative evaluation of dietary intakes	Powdered sample of formula or tea (10 mg) + 5 mL of double-distilled H_2O Liquid sample (1 mL of juice or nectar) diluted in 5 mL of double-distilled H_2O	Ca, Mg, Fe, Zn	FAAS	Not reported	[108]
Infant milk formula, Chinese CRM	Methodological development (digestion)	(1) Open-system wet digestion: 0.5 g of the sample + 6 mL HNO_3 (room temp. 12 h) + HNO_3/H_2O_2 (6:2), 200 °C, 4 h on a hot plate + 10 mL HNO_3/H_2O_2, evaporated and diluted (2) Dry ashing: 0.5 g of the sample (450–500 °C, 8 h) + 1 mL conc. HNO_3 (450–500 °C, 4 h); dissolved in 2% HNO_3, filtered and diluted (3) Closed-system MW digestion: 0.5 g of the sample + 7 mL HNO_3 + 2 mL H_2O_2, 20 min heating program, evaporated on a hot plate (150 °C) and diluted	Ca, Cu, Fe, K, Mg, Mn, Na, P, Zn	ICP-OES	DLs: (1) 5–14 µg/kg; (2) 4.7–16 µg/kg; (3) 4.1–14 µg/kg; (4) 3.8–13 µg/kg CRM: GBW 08509 infant formula (China).	[95]

		(4) Slurry dispersion: 0.5 g of the sample + 20 mL 2% HNO_3, sonicated (50 °C, 2–5 min)				
Baby food EU	Comparative evaluation of daily/weekly intakes of the 0- to 9-month-old nonbreast-fed infants	Individual products and pooled basket samples analyzed, moisture content evaluated/compared. MW digestion: 0.3 g of freeze-dried sample + 8 mL HNO_3; dilution	Ca, Cu, Fe, Mn, Ni, Se, Zn Pb, Hg, Cd	ICP-OES ICP-MS	CRMs: 184 infant/adult nutritional formula (NIST); Seronorm^{TM} trace elements whole blood L-1 and L-2 (201505, 201605) (SERO AS, Norway)	[34]
Commercial Baby food, Spain	Evaluation of dietary intake of essential elements and of exposure risk to toxic minerals	Freeze-dried purée samples or dried powdered products were MW digested: 1 g + 8 mL HNO_3 + 2 mL H_2O_2(additional ultrasonication); Sc as IS	Al, As, Ba, Be, Bi, B, Ca, Cd, Co, Cu, Fe, Pb, Li, Mg, Mn, Mo, Ni, K, Se, Na, Sr, Tl, Ti, V, Zn	ICP-OES	DLs: 0.003–3 mg/kg CRMs: 1568a rice flour (NIST); TORT-2 lobster hepatopancreas (NRC); IAEA-359 cabbage	[32]
Infant cereals, Spain	Estimation of dairy intake	Closed-vessel MW digestion: 0.5 g of cereal + 7 mL HNO_3+ 2 mL H_2O_2 (55 min program + ventilation) Hg: Ar bubbling for NO_x elimination As: prereduction with HCl/KI/ascorbic acid	As, Hg	HG-AAS CV-AAS	DLs: 0.27 µg/kg Hg and 2.30 µg/kg As CRMs: 73008 rice and 73009 wheat (NCS, China)	[52]
Ready-to-eat baby vegetables, Portugal	Evaluation as a dietary source of macro- and micro-mineral nutrients	MW digestion optimized: 0.15 g of the freeze-dried food + 9 mL HNO_3:H_2O (6:3), 15 min predigestion at room temperature followed by optimized heating program; dilution 1% $CsCl_2$ as ionization buffer	K, Na, Ca, Mg Fe, Mn, Zn Cu	HR-CS-FAAS ETAAAS	DLs: 0.004–0.438 µg/g CRM: BCR-679 white cabbage (IRMM) Recoveries after two-point standard addition: 91–110%	[40]

(Continued)

Table 4.2: (Continued)

Food product	Purpose	Pretreatment	Elements	Technique	Validation	References
Rice, wheat, corn, soybean, potatoes, Brazil	Food safety in highly fertilized (P) area	Harvested products dried (65 °C), ground and sieved (<0.38 mm) MW digestion: 0.5 g of the sample + 10 mL HNO_3, 10 min heating program; dilution Soil also analyzed	As, Cd, Pb	ETAAAS	DLs: 7 µg/kg Cd, 15 µg/kg As, 40 µg/kg Pb CRMs: BCR-482 lichen, BCR-142R sandy soil (IRMM), SRM 2710 Montana soil (NIST)	[42]
Cereals, legumes, tubers, vegetables, Nigeria	Comparative evaluation of dry and wet digestion Estimation of adult daily intake of metals	Dry ashing: 1 g of the dried (100 °C) and homogenized edible part of each product heated in a muffle furnace (500 °C, 12 h) and dissolved in diluted HNO_3 Wet digestion: 1 g of the sample + 20 mL HNO_3:HCl (3:1) heated at 150 °C during 2.5 h (heating block) and diluted	Mn, Fe, Cu, Zn, Cr, Cd, Pb, Ni	FAAS	Recovery experiments carried out on spiked samples: 81.2–97.8%.	[89, 90]
Rice	Methodological development (direct sample introduction)	ICP-OES: 4 mg of pulverized sample (50–200 µm) directly weighted to the graphite boat, CCl_2F_2 addition, ETV conditions optimized, Ar line for internal standardization For ICP-MS: closed-vessel acid digestion: 0.2 g + 2.5 mL HNO_3 + 0.5 mL H_2O_2, dilution	Al, As, Co, Cu, Fe, Mg, Pb, Se, Zn	ETV-ICP-OES ICP-MS (CRC: H_2)	DLs: 0.01–6 ng/g CRMs: 1568a rice flour, 8433 corn bran and 8437 hard red spring wheat flour (NIST) Good agreement between two methods	[53]

Rice	Methodological development (digestion)	Acid extraction: 500 mg of the ground sample + 250 mg L-cysteine + 5 mL H_2O + 2 mL HNO_3, sonication (1 h), centrifugation and collection of the supernatant. After second extraction, dilution to about 1% L-Cys MW digestion: sample with L-Cys, 3 mL H_2O and 3 mL HNO_3, left overnight and then digested Au as IS	Hg	ICP-MS DMA	DL: 0.015 µg/L Memory effects eliminated by addition of 1% L-Cys to standards and to the samples Good agreement between different pretreatments CRMs: 1568a rice flour, 1570a, spinach leaves (NIST)	[8]
Imported rice and rice products, Qatar	Health risk evaluation	Open-system digestion: 0.5 g of ground rice predigested with 5 mL HNO_3 (12 h), then heated (60 °C, 2 h), 3 mL H_2O_2 added (130 °C, 3–4 h) and finally diluted.	As, Zn, Se	ICP-MS	Ds: 0.058 µg/L As, 0.34 µg/L Zn, 0.30 µg/L Se. CRM: NCS ZC 73027, China.	[61]
Vegetable oils, Taiwan	Methodological development (pretreatment)	Emulsification: 5% vegetable oil + 1.5% Triton X-100+ 50 µg/mL ascorbic acid (antifoam Y-30) ETV: Pd nanoparticles prepared and used as modifier ICP-MS: dynamic reaction cell pressurized with 1.8 mL/min H_2 for elimination of polyatomic interferences on As	Cu, As, Hg, Pb	ETV-ID-ICP-MS	DLs: 0.4–1.1 ng/g Comparison with nitric acid MW digestion Control samples: Mineral Oil Standard S23-100Z, ORG AS8-2Z, and ORG HG8-2Z (SPEX CertiPrep, Edison, NJ, USA)	[96]
Vegetable oils and solid fats, Ukraine	Methodological development (direct sample introduction)	Sample aliquot (5.0–25 g) heated (60 °C), dissolved/diluted in hexane to 50 mL; 20 µL injected to the graphite FF atomizer using Pd–Mg chemical modifier.	As, Cd, Cu, Mn, Pb, Zn	FF-ETA-AAS	DLs: 0.1–1.2 µg/kg Reference methods used for validation	[97]

(*Continued*)

Table 4.2: (Continued)

Food product	Purpose	Pretreatment	Elements	Technique	Validation	References
Tomatoes, Brazil	Evaluation of element profiles at different maturation times	Closed-vessel MW digestion: 300 mg of dried sample + 1 mL H_2O_2 + 7 mL HNO_3 (4.0; 7.1; 14.2 M tested), 18 min gradient program (25–90–180 °C); dilution	Al, Ba, Cu, Fe, Mn, Sn, Sr, Zn Cd, Co, Cr, Hg, Ni, Se, V	ICP-OES ICP-MS (CRC: 7%H_2 in He)	ICP-OES DLs: 6–98 µg/g ICP-MS DLs: 8–129 µg/g CRM: 1573a tomato (NIST).	[66]
Tomato paste soups, Turkey	Estimation of uncertainties of the method	MW digestion: 1–1.5 g of sample + 7 mL HNO_3 + 1 mL H_2O_2.	Cd, Cu, Fe, Pb, Sn and Zn	HR ICP-MS	DLs: 0.16–2.75 µg/mL CRM: 573a tomato leaves (NIST).	[67]
Fresh and canned tomatoes, Poland	Methodological development; comparative evaluation of Sn concentration depending on the storage in the open container	Open system acid digestion: 3 g of (wet) tomato sample + 10 mL HNO_3; overnight predigested and then heated on a hot plate until NO_x fumes; evaporation, redissolution with L-Cys addition (1%) HG in a flow system: 1% $NaBH_4$ + 1 M HNO_3 + 1% L-Cys	Sn	HG-ICP-OES	DL: 1.2 ng/mL Recovery evaluated in two-point standard addition: 92.9–109%.	[62]
Canned tomatoes, Greece	Study of metals migration from the container	Parameters tested: temperature, time of storage, open vs closed can, pH MW digestion: 0.5 g of the homogenized wet sample + 5 mL HNO_3 + 1 mL H_2O_2 (10 min heating program), dilution ETAAAS temperature cycle optimized for Pb–Cd; Cu–As; Cr–Ni; Fe–Mn simultaneous determination	Pb, Cd, Cu, As, Cr, Ni, Fe, Mn	ETAAAS (Zeeman BC)	DLs: 2–64 ng/g CRM: 1573a tomato leaves (NIST)	[68]

Various vegetables, Turkey	Methodological development (preconcentration)	MW digestion: 0.1–1.0 g of the freeze-dried material + HNO_3–H_2SO_4 (9:1), 20 min program; for Se(VI) reduction, 2 mL 5 M HCl added and MW heating applied. CPE: sample aliquot + complexing agent (pyrinine B) + surfactant (Ponpe 7.5/SDS) heated at 40 °C, 20 min and after centrifugation, surfactant-rich phase diluted with methanol + 1 M HNO_3	Se	FAAS	DL: 3.81 µg/L CRMs: 1547 peach leaves, 1567a wheat flour, 1568a rice flour, 1571 orchard leaves (NIST)	[86]
Fruit and vegetables, Algeria	Evaluation of the daily intake	Digestion: 3 g of the dried (105 °C, 24 h) homogenized edible part of each product dry-ashed (450 °C, 3 h), then digested with conc. HNO_3, filtered and diluted	Pb, Zn, Cu, Cr	FAAS	Not reported	[93]
Potatoes, carrots, peas, Poland	Methodological development (preconcentration)	MW digestion: 1.0 g of the sample + 8 mL HNO_3 + 2 mL H_2O_2; after cooling, diluted. Preconcentration: 100 mL of the diluted digest + 2 mg La(III) + 2 mL 5% 8-hydroxyquinoline (pH 8 with borate buffer), heated 20 min at 50 °C, centrifuged and precipitate digested in 5 mL 2 M HCl; dilution	Cd, Cu, Ni, Pb, Zn	FAAS ICP-OES	DLs: 0.31–2.9 µg/L for ICP-OES; 0.63–3.2 µg/L for FAAS (preconc. factor 60) CRM: NCS ZC85006 tomato (China) Recovery after standard addition: > 94%	[69]
Apples, pears, Poland	Methodological development (preconcentration)	Closed-vessel digestion of the dried (50 °C, 48 h) and ground fruits: 0.5 g + 6 mL HNO_3. Diluted sample (100 mL) stirred with activated carbon and filtered. Sorbent together with filter MW digested (concentration factor 60)	Cd, Co, Cu, Ni, Pb, Zn	ICP-OES	DLs: 0.17–2.60 µg/L CRM: NCS ZC85006 tomato (China)	[63]

(Continued)

Table 4.2: (Continued)

Food product	Purpose	Pretreatment	Elements	Technique	Validation	References
Apples, Croatia	Evaluation of edible portion against peel and total fruit	Open-system MW digestion: 1 g of fresh apple + 5 mL conc. HNO_3; heating program up to 110 °C, 15 min. Closed-vessel MW digestion: 0.2 g of dried apple + 5 mL HNO3 (1:1); heating program up to 200 °C, 45 min	Ag, Al, Ba, Ca, Cd, Co, Cr, Cu, Fe, K, Mg, Mn, Na, Ni, Pb, Sr, Zn	ICP-OES	Two digestion systems compared. DLs: 0.1–295 µg/L CRMs: 1643e trace elements in natural water and 1640 trace elements in water (NIST)	[43]
Reference materials, tea leaves, freeze-dried strawberry	Methodological development (sample introduction)	Slurry preparation: 8–50 mg of dried, pulverized material (10 µm) + 2 mL 5% HNO_3; homogenization in ultrasonic disintegrator directly in autosampler vials	Cr	ETAAAS (Ir/Nb permanent modifier)	DL: 86.6 ng/g CRMs: LGC 7162 Strawberry leaf, INCT tea leaves and mixed Polish herbs, IAEA 359 cabbage	[98]
Edible mushrooms, Poland	Evaluation of metal accumulation from soil, Hg loss upon cooking	Closed-vessel MW digestion: 0.5 g of dried and pulverized fruit body + 7 mL HNO_3 predigested at room temp. (24 h) and, after vessel closing 20 min heating program applied. Dilution to 10 mL For Hg: direct thermal sample decomposition in Hg analyzer	Ag, Al, Ba, Ca, Cd, Co, Cr, Cu, Fe, K, Mg, Mn, Na, Ni, Pb, P, Rb, Sr, Zn Hg	ICP-OES CV-AAS	CRMs: CS-M-1 dried fruiting bodies of mushroom Cow Bolete, INCT- tea leaves-1 and CTA- oriental tobacco leaves (ICHTJ) DL for Hg: 0.003 mg/kg dry mass	[45] [47][46]
Mushrooms *Agaricus bisporus*, Australia	Comparative evaluation between two farms, daily intake with respect to the RDI	Open-system digestion on a heating block: 1 g of the freeze-dried material + 5 mL HNO_3 +2 mL H_2O_2 (105 °C, 4–5 h), dilution	Na, K, Mg, Ca, Mn, Fe, Cu, Zn, Se, Mo, Co, Cr, Pb, Cd, Si, B, V	ICP-MS	DLs: 1–500 µg/L CRMs: 1640a and 1643e (NIST)	[103]

Edible mushrooms, Spain	Evaluation of dietary intake as compared to PTWI	MW digestion: 0.2 g of the freeze-dried sample + 3 mL HNO_3 + 0.5 mL H_2O_2 + 3 droplets of $KMnO_4$ (5%); dilution CV generation: 0.75% $NaBH_4$ + 3% HCl in a flow system.	Hg	CV-AAS	DL: 0.01 mg/kg dry mass CRM: mushroom *Cantharellus tubaeformis* (Livsmedelsverket, Sweden)	[57]
Edible mushrooms, Turkey	Evaluation of metals daily intake and antioxidant activity	Digestion: 0.25 g of dried (105 °C, 24 h) and powdered biomass + 9 mL HNO_3+ 1 mL H_2O_2 (2 × 7 min heating program); dilution	Pb, Cd, Zn, Fe, Mn, Cu, Cr, Ni, Co	ICP-OES	Not reported	[58]
Brazil honey	Comparative evaluation of two analytical procedures	(1) Dilution with acidified aqueous solution (6% HNO_3, 4% H_2O_2), 0.6% $NaBH_4$ + 6M HCl (2) No sample treatment	Hg	CV-AAS DMA	(1) DL: 30 ng/g (2) DL: 10 ng/g CRMs: 8415 whole egg powder (NIST); IAEA-336 trace elements in lichens; GBW 08301 river sediment (China)	[101]
Brazil honey	Methodological development (sample introduction)	Slurry: 70.0 g/L honey, 0.1 M HNO_3 and 3% H_2O_2; sonication 10 min (Pd–Mg modifier used for Cd)	Cd, Cr, Pb	ETAAAS	DLs: 2.0 ng/g Cd; 5.4 ng/g Pb; 9.4 ng/g Cr	[99]
Brazil honey	Multi-element fingerprints for quality control and determination of origin and authenticity	1.0 g of honey diluted, homogenized, incubated with 6M HNO_3 and again diluted ICP-MS: Rh as IS, correction equation for Zn applied	Ultra-trace: As, Cr, Sb, Be, Pb, Cd, Mn, Se, Rb, Co, Mo, U, Ni, V, Pt, La, Dy, Pr, Sm, Nd, Gd, Te, Ho, Pd, Sm, Th, Tl, Er, Ba, Bi, Ce, Yb, Tb, Hf, Eu Trace: Fe, P, Cu, Zn, Al, Mg	ICP-MS	DLs: 0.003–43.8 ng/g. The procedure compared against MW digestion in a closed system for accuracy testing	[100]

(*Continued*)

Table 4.2: (Continued)

Food product	Purpose	Pretreatment	Elements	Technique	Validation	References
Spanish honey	Quality control; classification of honey according to the botanical and geographical origin	Wet, open-vessel digestion with HNO_3 (65%) and H_2O_2 (30%), temperature 100–110 °C, 1 h, dilution	Ca, Mg, Fe, Cu, Zn Na, K	FAAS FAES	Not reported	[81]
Honey from Western Balkan area	Ionic profile as indicators of botanical origin and authenticity	Dilution, filtration (0.45 µm)	Na^+, Ca^{2+}, Mg^{2+}, NH_4^+ and several inorganic anions	Ion chromatogr. with conductivity detection	Not reported	[247]
Turkey honey	Classification according to botanical and geographical origin	MW digestion: 0.5 g sample + 9 mL HNO_3(65%) + 1 mL H_2O_2; dilution	Al, B, Ba, Ca, Cd, Co, Cr, Cu, Fe, K, Mg, Mn, Na, Ni, Pb, Sr, Zn	ICP-OES	Recoveries evaluated after standard addition: 96.64–117.97%	[78]
Hungarian honey	Association between trace elements and the floral origin; possible health risk	Open-system acid digestion in a heating block: 3 g of honey + 10 mL conc. HNO_3, predigestion at room temp., 12 h; then 60 °C, 30 min; 3 mL H_2O_2 added and 120 °C, 1.5 h, dilution	Al, Ca, Cu, Fe, K, Mg, Mn, P, S, Zn As, Cd, Cr, Mo, Pb, Se	ICP-OES ICP-MS	DLs: 0.001–0.524 mg/kg DLs: 0.003–0.048 µg/kg Recovery evaluated in spiking exp.: 86.5–109%	[80]
Croatian honey	Comparison with other European honeys	MW digestion: 0.5 g sample + 4 mL HNO_3 (65%) + 2 mL H_2O_2; dilution Hg determined directly	As, Cd, Cu, Pb Zn, Fe, Ca, Mg, Na, K Hg	ETAAAS FAAS DMA	DLs: 0.1–10 µg/kg Recovery evaluated in spiking exp.: 81–99%	[79]
Argentinian honey	Identification of the geographical origin as a function of the production area	Dry ashing at 600 °C, samples packed in Al capsules and irradiated with thermal electrons flux	Br, Ce, Co, Cr, Cs, Eu, Fe, La, Sb, Sc, Sm, Rb, Th, Zn	Instrumental neutron activation analysis	CRM: V-10 hay powder (IAEA)	[11]

Polish honey	Discrimination according to botanical and geographical origin	MW digestion: 1 g honey + 6 mL of HNO_3 + 2 mL of H_2O_2 ICP-MS: Sc, Y, Tb as IS; correction equations for Ba, Mn, Cr, Ni	Al, B, Ba, Ca, Cd, Cr, Cu, K, Mg, Mn, Na, Ni, Pb, Sr, Zn	ICP-MS	DLs: 0.004–7.9 mg/kg Recovery evaluated in spiking exp.: 91.0–99.8%	[12]
Cocoa products	Estimation of oral intakes as compared to DRI and PTDI	Open-system wet digestion (vessels with 30 cm air condenser): 0.5 g sample + 5 mL HNO_3 (70 °C, 1 h) + 1.5 mL $HClO_4$(240°C until white fumes), dilution	Pb, Cd, Ni, Fe, Cu, Zn, Mn	ETAAAS FAAS	CRMs: 1567 wheat flour (NIST), IAEA-A-11 milk powder	[85]
Latvian vegetables, cheese, eggs, honey	Association of food contamination with environmental conditions	Vegetables, cottage cheese, homogenized eggs: wet acid digestion in thermoblock, 0.5 g + 10 mL HNO_3 + 5 mL H_2O_2(160 °C), dilution Honey: hot water dissolution and acidification (0.2 mL conc. HNO_3)	Ag, As, Ba, Cd, Co, Cr, Cu, Mn, Ni, Pb, Rb, Se, Sr, V, Zn	AAS ICP-MS TXRF (eggs)	CRMs: Promochem IC-CS-CR-2 carrot root powder, ZC73017 apple powder, BCR-063R skim milk powder	[64]
Infant total Diet France	Method validation	Closed-vessel MW digestion: 0.3–0.5 g of dry food (powdered milk, freeze-dried food) or 0.5–2.0 g of fresh food (meals, yogurt, etc.) + 3 mL HNO_3; after cooling, IS (Sc + Y + In + Re + Bi) and Au added and diluted (final conc. 2 µg/L of each IS and 20 µg/L Au)	31 elements	ICP-MS	DLs: 0.006–60 µg/L CRMs: 1548a typical diet, 1566b oyster tissue (NIST); TORT-2 lobster pancreas (NRC); GBW 10014 cabbage (China); 155 whey powder (IAEA)	[48]

(*Continued*)

Table 4.2: (Continued)

Food product	Purpose	Pretreatment	Elements	Technique	Validation	References
		ICP-MS: octopol CRC, 4.3 mL/min He; 500 channels; dwell time 100 µs; total acquisition time 219 s			External quality control: participation in proficiency testing programs	
Total diet in France	Estimation of dietary exposures to trace elements	Food prepared as consumed, homogenized and dried. MW digestion: 0.2–0.6 g + 3 mL H_2O + 3 mL HNO_3 (67%), Sc + Y + In + Re + Bi (IS) addition and dilution	Li, Cr, Mn, Co, Ni, Cu, Zn, Se, Mo Sr, Ag, Sn, Fe, Te, Ga, Ge, Ba, V	ICP-MS	Internal quality control scheme CRMs: BCR 278R mussel tissue (IRMM); IAEA 407 fish tissue; INCT-MPH2 mixed Polish herbs	[49, 50]
Total diet in China	Estimation of dietary intake of essential elements	Wet acid digestion: 1.0 g for solid and 2.0 mL for liquid sample + 5 mL HNO_3 (80 °C, 1 h; 140 °C, 3 h; 160 °C 2 h); evaporation and dissolution	Na, K, Mg, Ca, Fe, Mn, Zn, Cu	ICP-OES	CRMs: from the Institute of Geophysical and Geochemical Exploration (China)	[65]
Different types of the processed food	Methodological development	Closed-vessel MW digestion: 250 mg of the freeze-dried sample + 5 mL HNO_3, predigestion at room temp. 2 h followed by 38 min heating program Online HG in a Flow Blurring®multiple nebulizer system proposed with direct aspiration of the sample + 2.2% $NaBH_4$ + 3 M HCl + 5% thiourea + IS (100 µg/L Ge). Matrix interferences (Ca, Mg, K) controlled	As, Se	ICP-OES	DLs: 2.7 µg/L for As I (228.812); 5.8 µg/L for Se I (196.026) Spiked tea samples analyzed with recoveries in the range 102–114% for As and 77–107% for Se	[51]

Different diet types, Spain	Methodological development (chemometric calibration/ prediction)	Edible portion weighed, mixed, dried, freeze-dried and homogenized For ICP-OES closed-vessel MW digestion: 1 g + 8 mL HNO_3 + 2 mL H_2O_2 (180 °C, 15 min) IR, XRF directly in the freeze-dried samples	Ca, K, Fe, Mg, Ni, Zn	IR spectrosc. XRF ICP-OES	DLs: 0.005–0.26 mg/kg CRMs: 1568a rice flour (NIST), TORT-2 lobster hepatopancreas (NRC), IAEA-359 cabbage	[39]
French Polynesian diet	Methodological development and evaluation of dietary intake	Freeze-dried composite food sample (0.1–0.5 g) extracted with 1 mL 25% TMAH (90 °C, 3 h), centrifuged and diluted For starch digestion, the samples were incubated with 1.5% α-amylase (37 °C, 12 h) prior to TMAH treatment In as IS and the integrated sample introduction system optimized	I	ICP-MS	DL: 14 µg/kg CRMs: 1568a rice flour and 1548a typical diet (NIST); GBW 10014 cabbage and GBW 10015 spinach (NIM); BC402a potato powder (ERM); BCR-422 cod muscle (IRMM)	[104]
Oral hospital diet, Brazil	Evaluation of dietary intake	Entire meals homogenated Fe: 2 g of the sample + 5 mL conc. HCl, agitation, dilution and filtering Zn, Cu, Mn: 5 g of the sample dry ashed (450 °C) and dissolved in 5% HCl Se: 2 g of the sample + 25 mL HNO_3 + 7 mL H_2O_2 (180 °C, heating block to NO_x) and dilution in 5% HCl	Fe, Zn, Cu, Mn, Se	ICP-OES	CRM: 1548a typical diet (NIST)	[54]

(*Continued*)

Table 4.2: (Continued)

Food product	Purpose	Pretreatment	Elements	Technique	Validation	References
Gluten-free food, Italy	Evaluation of daily intake and related health risk index	Closed-vessel MW digestion: 0.2 g of milled and homogenized sample (pasta, biscuits, flours, bread) + 2 mL HNO_3 + 1 m H_2O_2, 18 min heating program; dilution ICP-MS: Y, Re as IS; CRC pressurized with 4.0/0.5 mL/min He_2/H_2	As, Ba, Cd, Co, Cr, Cu, Fe, Hg, Mn, Mo, Ni, Pb, Sb, Se, Sn, V, Zn	ICP-MS	DLs: 0.1–7.0 µg/kg Replicate analyses, blank control	[55]
Food reference materials	Methodological development	Closed-vessel MW digestion: 0.3–0.5 g of sample + ^{111}Cd spike + 6 mL HNO_3 + 2 mL H_2O_2 After sample dilution, NH_4OH added and, for elimination of interferences (Mo, Zr, Sn), analyte separated by SPE chelating column	Cd	ID-ICP-MS	CRMs: 7505-a tea leaves; 7403-a swordfish tissue; 7502-white rice; 7405-a seaweed (NMIJ)	[56]

Note: ICP-OES, inductively coupled plasma–optical emission spectrometry; CRM, certified reference materials; IAEA, International Atomic Energy Agency; MW, microwave; CV-AAS, cold vapor atomic absorption spectroscopy; DORM-3, DOLT-4; NRC, National Research Council; FAO, Food and Agricultural Organization; IS, internal standard; FI, flow injection; HG, hydride generation; CV, cold vapor; ETAAAS, electrothermal atomization atomic absorption spectrometry; MS, mass spectrometry; NIST, National Institute of Standards and Technology; IR, infrared; XRF, TXRF, TORT-2, CPE, cloud point extraction; FAAS, flame atomic absorption spectrometry; AFS, atomic fluorescence spectrometry; DMA, direct mercury analyzer; RfD, oral reference dose; PTWI, provisional tolerable weekly intake; PTDI, provisional tolerable daily intake; ADI, acceptable daily intake; STAT, slotted tube atom trapping; ETV, electrothermal vaporization; ID, isotope dilution; FF, filter furnace; SPE, solid-phase extraction; NMIJ, National Metrology Institute of Japan; GBW, National Research Centre for Certified Reference Materials, China; DL, detection limit; BCR, Community Bureau of Reference; IRMM, Institute for Reference Materials and Measurements; ERM, European Reference Materials; TDA, total daily intake; QL, quantification limit; RDA, Recommended Daily Allowance; INCT, CTA, ICHTJ, Institut Chemii i Techniki Jadrowej; RDI, reference daily intake; TMAH, tetramethylammonium hydroxide.

temperature (450–650 °C) in muffle furnace during at least 8 h has been used in the analysis of dairy products [33, 91, 92], cereals and legumes [89, 90], fruits and vegetables [93], honey [11] and total diet [54]; this approach is especially convenient when large sample series need to be processed. For comparative purposes or as a part of validation scheme, two or more digestion procedures have been used for this same analysis [8, 43, 89–91, 94, 95]. Practically all authors highlight that the reagents of high purity must be used and, all vessels and laboratory material should be carefully acid-washed in order to avoid sample contamination and high blank values. Final dilution of the digest is required to assure low acid load to the analytical cell (typically, < 2% in the final solution).

Despite being the most common practice, the digestion procedures are relatively lengthy and they involve a risk of analyte(s) loss and/or sample contamination. In the analysis of foodstuff, direct sample introduction or simplified sample pretreatment has attracted research interest and few recent studies are included in Table 4.2. For the determination of nine trace elements, pulverized rice was directly analyzed by electrothermal vaporization–inductively coupled plasma–optical emission spectrometry (ETV-ICP-OES) [53]. ETV was also applied for the introduction of vegetable oil emulsion to ICP-mass spectrometry (MS); palladium nanoparticles were used as modifier/stabilizer for the determination of As, Hg, Pb and Cu [96]. In another study, vegetable oils and solid fats diluted in hexane were directly introduced to the graphite filter furnace atomizer for the determination of As, Cd, Cu, Mn, Pb and Zn by filter furnace electrothermal atomization atomic absorption spectrometry (FF-ETAAAS) [97]. The feasibility of slurry introduction to electrothermal atomic absorption spectrometry (ETAAS) was demonstrated by analyzing chromium in the freeze-dried plant tissues and fruits [98] and also by determining Cd, Cr, Pb in honey [99]. Hot water dilution and acidification of honey has also been proposed [64, 100].

The determination of total mercury in foodstuff is not only highly demanded but also challenging due to element volatility, its sorption on the parts of the introduction system and due to the related memory effects observed in various instrumental techniques. In this regard, few commercial apparatus are available for Hg determination in solid samples without any pretreatment. In a direct mercury analyzer (DMA), solid food sample is decomposed thermally in a small boat, Hg is then trapped by amalgamation with gold and upon heating, elemental Hg is released and determined by cold vapor (CV)-AAS. As shown in Table 4.2, DMA has been applied for Hg determination in seafood [94], animal meat [74, 76], rice [8] and honey [79, 101]. Alternatively, after sample digestion CV has often been chemically generated in a flow system for enhanced performance of Hg determination by AAS or atomic fluorescence spectrometry (AFS) in seafood [59, 70, 72, 82, 87, 94], baby food [52], mushrooms [45–47, 57] and honey [101].

Practically, all elements forming volatile covalent hydrides are of interest in food analysis. Hydride generation (HG) is a convenient technique because it reduces substantially matrix-based interferences; it also allows for preconcentration and offers high introduction efficiency. Few studies found in the recent literature focused on As determination in seafood [87], processed foods [51] and infant cereals [52] by HG-AAS; Sn was determined by HG-ICP-OES in canned tomatoes [62].

As to the instrumental techniques, ICP-MS is certainly the most powerful tool due to its exceptional detection power, wide dynamic range, multielemental and isotopic capabilities; on the other part, however, spectral interferences need to be efficiently eliminated and the overall cost of analysis is elevated. Therefore, in food analysis, ICP-MS is primarily used for the determination of ultra-trace contaminating/toxic elements [8, 12, 34, 44, 48–50, 53, 55, 56, 60, 61, 64, 66, 67, 70, 71, 73, 76, 80, 82, 100, 102–104]. ICP-OES is the second most popular technique in food analysis. Even though inappropriate for quantification at ultra-trace levels and despite complex emission spectra, ICP-OES is well suited for simultaneous determination of metals, metalloids and some nonmetals in foodstuff, using less sophisticated and cheaper instrumentation as compared to ICP-MS [32, 34, 39, 43, 45–47, 54, 58, 65, 66, 71, 74, 75, 77, 78, 80, 91, 92, 95, 105, 106]. In addition to the direct aspiration of digested sample to ICP-OES, extractive preconcentration procedures can be found in the recent literature and some representative examples are shortly described in Table 4.2 [63, 83]. ETAAAS is a technique of choice for trace and ultra-trace level quantification of elements in chemically complex matrices; in food analysis ETAAAS has been used for the determination of As, Cd, Pb, Sn, Ni, Co, Cu, Cr, Fe, Mn [40, 42, 72, 79, 84, 85, 87, 97–99]. Traditional limitation of AAS in terms of unielement capability has been overcome in high-resolution continuum source instrument (HR-CS-AAS), which has also proved its utility in trace element analysis of milk [107] and baby food [40]. Finally, applications of flame atomization AAS for the determination of essential elements in seafood [72, 88], dairy products [33, 84] infant formula [40, 108], vegetables and cereals [86, 89, 90, 93], cocoa products [85] and honey [79, 81], should also be mentioned.

When the goals of determination of total elements are surveyed, it is perceived that seafood, dietary products, baby/infant food and total diets from different geographical regions and specific countries are studied principally to estimate dietary intake of essential elements and to evaluate potential health risk related to the presence of contaminating/toxic elements (Table 4.2). Migration of metals from food containers and changes of element content during food processing have also been addressed. For plant-derived products and mushrooms, potential impact of environmental contamination on the concentration of elements in food is additionally considered. Such is the case of honey, which is not a rich source of essential elements (20 g of portion provides less than 1% of the recommended daily allowance for Cr, Mn, Se, S, B, K, Ca or F [80]), but the determination of contaminating elements in this food product is often reported. In this specific case, trace elements profile has been used for classification/discrimination of honey from different botanical and/or geographical regions with an aid of principal component analysis, cluster analysis, linear discriminant analysis or neural network algorithms [12, 78, 81, 100]. Finally, it should be stressed that many studies are centered at methodological development, establishing novel procedures of sample pretreatment, analyte preconcentration, sample introduction and reliable quantification.

Practically all studies considered in this section included some elements of quality control/assurance. In the great majority of works, blank control, detection/quantification limits, precision evaluated in replicate analyses and percentage recoveries based on the standard addition have been reported. Furthermore, accuracy testing by analyzing appropriate food-related CRMs has become a common practice. Worth to emphasize that in addition to the well-established CRMs form National Institute of Standard and Technology, National Research Council, International Atomic Energy Agency, Institute for Reference Materials and Measurements, there has been an increasing variety of food-related certified materials produced by individual countries, mainly from China, Poland and also Sweden (Table 4.2). Another often reported approach for accuracy checking rests on the comparative evaluation of the results obtained by two different analytical procedures. Finally, external validation by participation in different proficiency testing programs has been reported in few studies.

4.3 Speciation schemes and analysis of selected food matrices

The term "speciation analysis" is well established in chemical nomenclature and it refers to the detection, identification and/or measurement of the quantities of one or more individual species of a given element in a sample, being "chemical specie" a specific and unique molecular, electronic or nuclear structure of this element [109, 110]. Some elements are of key importance in speciation analysis, due to the diversity of their forms/species and because these species may present drastically different biological behavior. Notably, there are no general rules linking chemical form of an element with its actual effect in a living organism. As an example, lower oxidation states of arsenic and of inorganic selenium are more toxic as compared to their higher oxidation states [26, 111], whereas for chromium, its hexavalent form (Cr(VI)) is more harmful with respect to trivalent (Cr(III)), which at low concentrations is in fact beneficial for glucose metabolism [112]. Furthermore, depending on the element, the organic compounds can be more or less harmful than inorganic forms as occurs for mercury and arsenic, respectively [113].

The variety and the amount of chemical species in food depend on their natural content in the raw materials and are unavoidably affected by the composition of specific environment, its contamination and application of the agrochemical products. On the other part, the initial element status can be modified during food elaboration and storage, due to the application of food additives and also as a result of food fortification with micronutrients. For essential elements, speciation results help to evaluate the compatibility of their content in food with specific metabolic requirements. In this regard, selenium is one of the most studied elements due to its species-dependent essentiality which is observed in a relatively narrow concentration band [114–116].

On the other part, speciation results may alert on the health risks associated with the intake of potentially toxic element species, being inorganic arsenic (iAs) and MeHg the most important examples already mentioned in the previous sections. For the assessment of toxic species in foods, reliable analytical methodologies are required, which are also needed for the promotion of more specific food regulations. In addition to the determination of well-characterized element species, food analysis involves the detection/identification of less common compounds that can be formed as a result of degradation or biotransformation of the original species. Such is the case of arsenic speciation in marine organisms [117–119] or in meat from chicken treated with roxarsone [120–123]. Studies on selenium speciation in biofortified food products should also be mentioned here [30, 124–132].

Speciation analysis in food is challenging because it is necessary to identify and quantify target species in a complex chemical matrix and at the concentration levels usually far below the total element content. Additional difficulties include chemical lability of species, their sometimes uncertain identity, the lack of suitable standards and still insufficient number of appropriate CRMs [110, 133, 134]. In general scheme, speciation analysis starts with the pretreatment, which is necessary to convert the original sample to a form suitable for further processing; this step is followed by separation of element species/forms and their quantification. In principle, the first two operations should be enabled for selective and efficient extraction and preconcentration of target element forms present in the sample, without affecting their identity and quantity [20, 26, 113, 135]. In some applications, for example while using selective extraction or chemical vapor generation, these two phases can be combined into a single step [136–140]; otherwise, species separation is carried out in the sample extract typically by means of chromatographic [19, 26] or electrophoretic [121, 141–143] techniques. Atomic spectrometry techniques are the most convenient detectors in speciation schemes; however, they can be used only for the analysis of known species whose authentic standards are available [20, 135, 144]. Confirmation and identification of new or unexpected compounds is achieved by molecular mass spectrometry or other structure characterizing techniques, such as nuclear magnetic resonance, X-ray absorption and fluorescence spectrometry [119, 124, 129, 145–148].

In speciation analysis, special care is needed to preserve natural composition and distribution of species in the sample during the entire procedure and to avoid sample contamination. Accordingly, the procedures employing soft chemical conditions and minimized sample manipulation are preferred, and a clear trend exists toward using the so-called hyphenated techniques that combine separation and detection steps into one operating online system [149]. The important advantages of hyphenated techniques are minimized contamination risk, enhanced precision, short time of analysis and relatively easy data handling [113, 135]. The number and variety of combinations between separation and detection techniques have been continuously increasing in search of simplicity and feasibility of implementation in routine analysis, yet without sacrificing the required selectivity and detection power.

For nonvolatile species, liquid chromatography (LC) has been typically used because of extended range of separation mechanisms available. Nonetheless, optimum separation conditions not always can be used, since chemical lability of element species, their interactions with the stationary phase and/or the components of mobile phase (buffers, organic modifiers) may cause disruption of natural speciation existing in the original sample. In particular, ion-exchange and reverse-phase columns are the most common choice in food analysis, specifically for inorganic and organometallic species of As [26, 117, 122, 150], Se [30, 126–128], Hg [151–153] or Cr [154, 155]. ICP-MS has been a prevailing element-specific detection system due to its easy online coupling with different chromatographic modes and its capability to monitor effluents for their elemental/isotopic composition with high sensitivity, to determine target element with high selectivity over other potentially coeluting elements and to compensate for incomplete chromatographic resolution from complex matrices. Typical flow rates for standard-sized analytical columns (0.5–1.2 mL/min) are compatible with the conventional pneumatic nebulizers (concentric Meinhard, Babington or cross-flow), so the column outlet can be connected directly to the liquid inlet of the nebulizer, using short polymeric or stainless steel tubing. It should be mentioned however that certain components present in column effluent may deteriorate analytical performance of ICP-MS. Such is the case of excessive concentrations of organic solvents that cause lower plasma stability and formation of carbon deposits on the torch and the sampling cone; high salt concentration is also undesired because of nebulizer and cone clogging, plasma destabilization and a potential increase of polyatomic interferences [156–158]. The most convenient chromatographic conditions imply isocratic elution with diluted aqueous mobile phases and with low percentage of organic modifier. In addition to the careful selection of separation conditions, elimination of polyatomic interferences by collision/reaction cell (CRC) technology has often been used in food speciation analysis [31, 122, 124, 127, 159, 160]. Other atomic spectrometry techniques with continuous sample introduction, such as ICP-OES and flame atomic absorption spectrometry, have not been used as LC detectors due to insufficient detection power for food speciation purposes and due to troublesome interferences. For hydride-forming elements (mainly As, Se, Sb) and for Hg, an alternative detection system quickly gaining popularity is AFS; this technique offers relatively low detection limits and represents lower cost of both, instrument and operation, as compared to ICP-MS [120, 161–164]. Finally, the applications of LC-molecular mass spectrometry for structural characterization of target compounds in speciation studies should be highlighted. A soft electrospray ionization (ESI) is particularly well suited for ionic species, labile organometallic compounds and biomolecules [165]; its combination with HR-MS and tandem MS/MS devices (quadrupole with time of flight, quadrupole-quadrupole time of flight or Orbitrap) provides accurate mass, isotopic pattern and fragmentation spectra required for ultimate species identification [123, 124, 129, 147, 166, 167]. The instrumental setup allowing for simultaneous ICP-MS and ESI-MS detection has been used in speciation analysis [168], including food matrices [123].

Despite obvious advantages of gas chromatography (GC) in terms of high resolving power and, which is particularly important in speciation analysis, the use of nonreactive carrier gas and efficient transport of the column effluent to the element-specific detector, the application of this separation technique is limited to volatile and thermally stable solutes [169]. In food products, only very few element species are volatile (e.g., As-containing hydrocarbons [167]), so a precolumn derivatization step is required [170]. During this additional procedural step, the analytes can be efficiently separated from the sample with many-fold preconcentration, offering additional benefits of enhanced detection limits and high matrix tolerance. Typical derivatization procedures prior to GC separation include conversion of inorganic and small organometallic ions into volatile covalent compounds in aqueous media (hydrides, fully ethylated, propylated or phenylated species), conversion of larger alkylated cations with Grignard reagents to saturated nonpolar species and, conversion of ionic species to volatile chelates (dithizone, methyl chloroformate, among others) [153, 171–175]. The remarkable progress in sample pretreatment prior to GC analysis is reflected in several recent reviews, specifically focused on derivatization reactions [170, 176], different modes of immersed or headspace solid-phase microextraction (SPME) [177, 178] and liquid–liquid microextractions [179, 180]. Application of ICP-MS as the detection system has often been reported, mainly due to its high detection power and capability for quantification by isotope dilution methods [132, 167, 169, 170, 181, 182]. Molecular mass spectrometry with electron ionization has also been used in food speciation analysis [132, 153, 172–174]. Depending on the instrumental design, enhanced analytical performance for quantification of target compounds can be achieved by isotope dilution [132, 153], by selective ion monitoring [132, 153, 173] or multiple reaction monitoring (MRM) [172, 174]. For MRM, GC-MS instruments equipped with triple quadrupole provide fragmentation of selected ions in tandem MS^n operation mode; these features make modern GC-MS an attractive tool for structural characterization and determination of low analyte concentrations immersed in chemically complex matrices [183, 184]. Among other detection systems used for GC in food speciation analysis, microwave-induced plasma optical emission (atomic-emission detection, depending on the configuration) [185–187], AFS [171] and also electron capture detector (ECD) [175] should be mentioned.

Capillary electrophoresis (CE) has proved its strength in speciation analysis, based on its capability for separation of species presenting different physicochemical characteristics, no need for stationary phase, simple composition of running buffer, mild separation conditions and high resolution [188–190]. On the other part, this technique requires highly sensitive detection system with somehow problematic connection between the capillary outlet and ICP-MS introduction system; most importantly, however, separation performance in CE is strongly affected by chemical matrix of the sample. Despite several applications in food speciation analysis reported before 2010 [143], in the recent literature only few studies have used CE separation [121, 141, 142].

Element speciation in food products can also be carried out by nonchromatographic procedures and this topic has been comprehensively discussed in the recent review articles [191–193]. Nonchromatographic procedures are based on different fractionation schemes and are usually devised for separation of element oxidation states, for discrimination between organic and inorganic forms or for selective extraction of target species with or without previous derivatization. In the recent literature, applications of miniaturized liquid- and solid-phase extraction (SPE) [130, 137, 138, 140, 194–197], field-flow fractionation [198, 199], derivatization and chemical vapor generation [137, 195, 196, 200–202] can be found. For element quantification in the obtained fractions, AAS and AFS have been preferentially used [130, 137, 138, 140, 194–197, 201]. Since analyte preconcentration can be achieved during fractionation, application of highly sensitive and expensive ICP-MS is generally not required. Finally, worth to note that integration of fractionation or derivatization and detection steps into a flow system is still an attractive approach facilitating miniaturization, possible automation, high precision and environmentally friendly character [136, 139, 197, 203].

In Table 4.3, a survey over the recent literature on food speciation analysis is presented with details of sample pretreatment and analytical speciation procedure used in each study. In the following sections, a brief description is presented with an emphasis on the most commonly analyzed elements, their species and food products.

4.3.1 Mercury

Generally, foodstuff contains trace concentrations of mercury with a low health risk for consumers, and determination of total concentration is usually adequate for food quality control purposes (Table 4.2). Nonetheless, food derived from aquatic organisms may contain higher levels of total mercury with considerable contribution of extremely toxic MeHg. The great majority of speciation studies have been focused on the determination of MeHg and inorganic mercury in seafood (Table 4.3) with several comprehensive reviews available [22, 134, 204–206]. In addition to the fresh tissues of marine organisms, processed and canned products have also been analyzed [35, 153, 202]. Efficient extraction of all species without perturbation of their natural distribution is the most critical procedural step due to the chemical lability and different properties of organic versus inorganic mercury (volatility, polarity, electrical charge). Different extraction procedures have been in use [24, 206] and those reported in the recent studies include alkaline (KOH, tetramethylammonium hydroxide (TMAH)) [139, 175] or acid (HCl, HNO_3) [161, 163, 202] extractions, addition of L-cysteine [151–153], organic solvents (toluene) [138] and also enzymatic treatment with protease XIV and lipase [153]. Microwave heating or ultrasonication during extraction has often been recommended [151, 163]. In the applications of LC, separation has been achieved on reversed-phase columns with mobile phases containing L-cysteine [153, 161] or ammonium pyrrolidinedithiocarbamate (APDC) [163] as a complexing agent, pH range 2–6, and acetonitrile or methanol as organic modifier [161, 163].

Table 4.3: Food speciation analysis: representative examples from the recent literature.

Food matrix	Species	Sample treatment	Analytical speciation procedure	Additional comments	References
Fresh and canned fish, Canada	Hg(II), MeHg	Extraction: 0.2–0.7 g of the homogenized material + 20 mL 0.75% L-Cys·HCl incubated at 60 °C 2 h , centrifuged 30 min at 4 °C, supernatant filtered Enzymatic digestion: 1 g of the homogenized sample + 1 mL H_2O + $Me^{202}Hg$ 1 µg Hg/mL + 10 mL protease XIV/lipase mixture incubated in the capped tube at 37 °C, 12 h Derivatization: 500 µL of glacial CH_3COOH and 1 mL of the 2% NaBPh4 added SPME: a home-made fiber assembly inserted; sample stirred at 80 °C, 2 h and phenylated MeHg reextracted to isooctane	HPLC-ICP-MS: RP18 (150 × 2.0 mm, 4 µm) column with a guard column; 0.1% L-Cys/ 0.1% L-Cys·HCl mobile phase; flow rate 0.4 mL/min; injection volume 25 µL; chromatographic run 4 min ID-GC-MS: capillary column Zebron ZB-5 MS (30 m × 0.25 mm × 0.25 µm); He as a carrier gas; injector temperature 220 °C; column temperature gradient 50–285 °C, chromatographic run 12.5 min. EI, SIM mode for molecular ions of 5 Hg isotopes	HPLC-ICP-MS DLs: 2.0 ng Hg/g for Hg(II); 5.0 ng Hg/g for MeHg Recoveries in MeHg-spiked samples: 93% ID-GC-MS DL: 3.4 ng Hg/g for MeHg Recoveries in MeHg-spiked samples: 91% CRMs: BCR 463 tuna fish; DORM-2 (NRC) Good agreement between two methods MeHg range in fresh fish: 18–1674 ng Hg/g MeHg range in canned tune: 25.8–358 ng Hg/g	[153]
Seafood	Hg(II), MeHg	MW-assisted extraction: 0.1–1.0 g of the sample + 4 M HCl, 12 min heating program (100 °C); filtration For total Hg, MW digestion with HNO_3 + H_2O_2	HPLC-UV-CV-AFS: RP-C18 column (250 × 4.6 mm, 5 µm); isocratic elution with 20% APDC and CH_3COONH_4solution and 80% of MeOH; chromatographic run 10 min Column effluent passed by photo-oxidation unit	DLs: 1 µg/kg for tHg, 3 µg Hg/kg for MeHg and 0.4 µg Hg/kg for Hg(II) CRMs: DOLT-4 dogfish, TORT-2 lobster hepatopancreas (NRC); BCR-463 tuna fish (IRMM)	[163]

			(12 m × 0.5 mm PTFE tube coiled around UV lamp); Hg^0 generated online by mixing with 1.5% $SnCl_2$ and 4% HCl Vapor phase transported via gas/liquid separator to AFS Total Hg determined by ICP-MS	In 24 seafood samples from different geographical origins, the concentrations for tHg, MeHg and Hg(II) were in the ranges 0.07–2.33, 0.003–2.23, 0.006–0.085 mg Hg/kg, respectively	
Market seafood, Spain	MeHg	Extraction: 0.2 g of the dried sample + 0.5 mL H_2O + 5 mL HBr + 10 mL toluene (room temp.); back extraction to 6 mL 0.7% L-Cys and 0.5 mL taken for analysis	CV-AAS: the dried extract decomposed in oxygen flow, the combustion product (Hg^0) trapped by amalgamation with gold and thermally desorbed for AAS determination	DL: 0.018 µg Hg/g dry mass Recoveries in the spiked food samples: 83–105% CRMs: TORT-2 lobster hepatopancreas, DORM-3 dogfish liver (NRC); CRM: 1633c coal fly ash (NIST) MeHg concentration found in 19 fish types: 0.031–1.24 µg Hg/g	[138]
Marine and freshwater fish, Ghana	MeHg	Extraction: 0.2 g of the homogenized edible tissue + 10 mL KOH 1 M in ethanol, 1 h in water bath at 100°C and, after neutralization, 0.01% dithizone in toluene added for MeHg complexation/extraction. Prior to	GC-ECD: Glass column packed with Hg-20 A on Uniport HP AW-DMCS (3 m × 3mm, 60–80 mesh) equipped with the 2–3 cm length injection section packed with NaCl for conversion of MeHg–dithizone complex to chloride; N_2 carrier gas	DL: 0.5 µg Hg/kg CRM: DORM-2 dogfish (NRC) Recovery experiments in the real-world samples performed MeHg concentrations found in the analyzed fish tissues:	[175]

(Continued)

Table 4.3: (Continued)

Food matrix	Species	Sample treatment	Analytical speciation procedure	Additional comments	References
Blue shark, Brazil	Hg(II), organic Hg	analysis, the HgMe extract extensively cleaned-up in several liquid–liquid extraction steps Extraction: 100 mg of the freeze-dried and homogenized sample + 4 mL 25% TMAH; dilution	at 40 mL/min; column oven at 150 °C; injector port and electron capture detector oven at 200 °C FI-CV-AFS: for Hg(II), CV formed by addition of 0.1% L-Cys and 1% $SnCl_2$ in a flow system For tHg: 0.1% $KMnO_4$ added at the sample zone followed by 0.5% $SnCl_2$ Organic Hg obtained as a difference	0.009–0.107 mg Hg/kg wet mass CRM: DORM-2 dogfish (NRC) For 26 samples analyzed, tHg found in the range 0.46–2.40 mg/kg About 40% of samples had Hg concentrations >1.0 mg/kg Organic Hg accounted for 95–98% of tHg	[139]
Dried shark and shark soup, American restaurants	MeHg	Extraction: 0.05–0.3 g of the freeze-dried fish or homogenized soup + 4.75 M HNO_3	CV-AFS: Hg^0 generation in a flow system by reacting with $NaBH_4$ (details not given)	CRMs: TORT-2 lobster hepatopancreas; DORM-3 fish protein (NRC) MeHg concentrations found in fish: 9–1720 ng Hg/g dry mass; in soups: 0.01–34 ng Hg/mL MeHg in the analyzed food accounted for 62 ± 7% of tHg	[202]
DORM-2 dogfish	Hg(II), MeHg, EtHg, PhHg	Extraction: an aliquot of CRM + 5 M HCl sonicated during 30 min and centrifuged. The supernatant neutralized with NaOH and injected to the flow system Magnetic Fe_3O_4 nanoparticles obtained and characterized by	HPLC-CV-AFS: MP C_{18} column (150 × 4.6 mm, 5 μm); ACN-aqueous 0.12% L-Cys (10:90) mobile phase at pH 6.8; chromatographic run 10 min Column effluent mixed with acidified H_2O_2(pH 2) and transferred to	DLs: 0.7–1.1 mg Hg/L for Hg(II), MeHg, EtHg and PhHg CRM: DORM-2 dogfish (NRC); MeHg was the only species detected and its concentration was in agreement with the certified value	[161]

		SEM; possible mechanism of organic mercury oxidation by H_2O_2 in acidic medium in the presence of NPs presented	a minicolumn (25 × 5 mm) filled with 0.5 g Fe_3O_4 magnetic nanoparticles for organic species oxidation; afterward the sample directly transported to CV-AFS system		
Fish	Hg(II), MeHg, EtHg, PhHg	Extraction: the homogenized fish muscle + 0.1% HCl + 0.05% L-Cys sonicated during 15 min at 40 °C and centrifuged. The supernatant filtered (0.45 μm) and diluted For tHg, MW-assisted acid digestion	HPLC-ICP-MS: Two strong cation-exchange guard columns (Zorbax SCX, 2.5 × 4.6 mm, 5 μm) connected in series used as the analytical column; isocratic elution with 20 mM L-Cys or thiourea (pH 2.0); flow rate 1.5 mL/min; chromatographic run 2.5 min. For ICP-MS detection, *m*/*z* 202 for Hg and 209 for Bi (continuously introduced for internal standardization) monitored	DLs: 1.9–3.1 μg Hg/kg CRM: GBW 10029 fish tissue (China)	[151]
Fish	Hg(II), MeHg	Different drying conditions tested by combining thawing, freeze-drying and drying in an air circulation oven at temperatures from 50 to 175 °C and time periods 24, 48 h. The dried samples ground in a cryogenic mill.	HPLC-ICP-MS: Si-C18 column (250 × 4 mm, 5 μm); 0.1% L-Cys mobile phase at pH 4.0; flow rate 1 mL/min; chromatographic run 6 min For postcolumn CV generation, 0.2% $NaBH_4$ and 1.0 M HCl added using an X-type connector; after gas–liquid-phase separation, detection carried out by ICP-MS at *m*/*z* 202	DLs: 1.7 and 2.3 ng Hg/g for Hg(II) and MeHg, respectively CRMs: DORM-2 dogfish, DOLT-3 dogfish liver (NRC) Freeze-drying and oven drying below 100 °C proposed as reliable	[152]

(*Continued*)

Table 4.3: (Continued)

Food matrix	Species	Sample treatment	Analytical speciation procedure	Additional comments	References
		Extraction: 0.6% L-Cys at 20 °C during 6–24 h; after centrifugation, the supernatant collected and filtered (0.45 µm) For tHg, MW-assisted acid digestion or MW-induced combustion	Total Hg determined by ICP-MS after cold vapor generation	procedures for Hg speciation analysis	
Fish after culinary treatments	Hg(II), MeHg	Culinary treatments examined: boiling, frying, roasting with or without addition of spices, salt and lemon juice After cooking, the samples were freeze-dried and cryogenic milled. Extraction conditions as in ref. [152]	HPLC-CV-ICP-MS as in ref. [152]	CRMs: DORM-2 dogfish, DOLT-3 dogfish liver (NRC) Hg loss during frying: up to 33% No significant conversion between Hg(II) and MeHg occurred during culinary treatments	[35]
Fish	Hg(II), MeHg	Matrix solid-phase dispersion procedure: the freeze-dried and homogenized sample blended with SiO_2 to obtain a homogeneous mixture and the extraction solution (0.5 M NaCl and 4.2 M HCl) added; vortex during 1 min and centrifuged Ethylation reaction carried out in the supernatant (1% $NaBEt_4$, pH 5), the derivatized species extracted to hexane tHg determined after MW nitric acid digestion	GC-MS: Rtx-5MS column (30 m × 0.25 mm × 0.25 µm); He as a carrier gas; temperature gradient 130–270 °C; ion source and interface maintained at 230 °C and 250 °C, respectively. MS operated in SIM mode For tHg: FI-CV-ICP-MS	Extraction conditions optimized by factorial design DLs: 0.06 µg Hg/g for MeHg and 0.12 µg Hg/g for Hg(II). CRMs: DOLT-3 dogfish liver and DORM-2 dogfish muscle (NRC) Recoveries of the speciation procedure evaluated by comparing total Hg in acid-digested samples with the sum of species determined by GC-MS were close to 100%	[173]

Marine organisms, China	TBT, DBT, MBT, TPhT, DPhT, MPhT	Extraction: 5.0 g of the homogenized wet sample (1.0 g dry sample) + 500 µL TeBT (IS: tetrabutyl tin) + 10 mL tetrahydrofuran solution containing 0.6 M HCl; 5 min shaking; 20 mL 0.01% tropolone–hexane added and the mixture shaken 40 min; organic phase collected and additional extraction with 10 mL hexane done. The combined extract evaporated to ~ 2 mL, 1 mL *n*-propylmagnesium bromide added and, after sonication, the excess of Grignard reagent eliminated by the addition of 5 mL 0.5 M H_2SO_4. The centrifuged sample cleaned up on activated silica SPE column, followed by fluorosil column and evaporated to ~1 mL	GC-MS/MS: DB-1701 (30 m × 0.25 mm × 0.25 µm) column; 1.5 mL/min He carrier gas; the injector temperature 290 °C; temperature gradient 40–280 °C; injection volume 1 µL; chromatographic run 12 min Detection/quantification: *m/z* range 50–450; MRM mode with two parent–daughter ion transmission in triple quadrupole	DLs: 0.05–0.1 µg Sn/kg wet mass for six species CRM: CE477 *Mytilus edulis* (ERM) Recoveries after standard addition: 70.5–105.3% (except for MPhT: 82.2–133.5%) Sn species found in the sea products: 1.36–20.54 µg Sn/kg wet mass	[172]
Mussels (CRM and samples from the proficiency study)	TBT, DBT, MBT	Extraction: 0.2 g of freeze-dried tissue + 10 mL 10% NaOH in MeOH, ultrasonic bath 2 h; centrifuged and the supernatant (1.5 mL) transferred into a HS vial containing 7 mL of acetate buffer; IS (MBT-d9, TBT-d27) and	HS-SPME-GC-MS/MS: The PDMS fiber placed in the HS for 40 min (45 °C, agitation), then desorbed in the GC injection port (250 °C, 2 min) and introduced to GC system; DB-5MS (25 × 0.2 mm × 0.33 µm) column;	DL: 3 ng Sn/g dry mass for all species	[174]

(*Continued*)

Table 4.3: (Continued)

Food matrix	Species	Sample treatment	Analytical speciation procedure	Additional comments	References
		ethylation reagent (0.5 mL, 2% $NaBEt_4$) added; the vial tapped with a cap provided with a PTFE-lined silicone septum for sample incubation at 45 °C, 10 min 1 mL/min He carrier gas; injector temperature 250 °C; chromatographic run 12 min EI, MRM quantification	CRMs: 477 mussel tissue (ERM); two mussel tissue samples from the proficiency study Inter-laboratory reproducibility 12.6%		
Edible anemone: raw and fried	As(III), As(V), MMA(V), DMA(V), AsC, AsB, TMAO, TETRA, DMAS, GPAsC	Crude anemones battered in wheat flour and fried in olive oil; crude and processed material freeze dried for As extraction with CH_3OH-H_2O using ultrasonic bath, ultrasonic probe or focused MW (the latter finally selected)	HPLC-ICP-MS: for As(III), As(V), MMA(V), DMA(V), anion-exchange RP-X100 column; 50 mM $(NH_4)_2CO_3$ mobile phase at pH 8.5; flow rate 0.8 mL/min For AsB, TETRA, TMAO, cation-exchange LC-SCX column (250 × 4.6 mm, 5 µm); 20 mM pyridine mobile phase at pH 2.5; flow rate 1 mL/min ICP-MS detection for quantification of known and chromatographically resolved species Anion-exchange separation coupled with triple quadrupole MS in MRM mode for confirmation of DMAS and glyceryl phosphorylarsenocholine (GPAsC) coeluting with AsB, AsC	In the raw samples of anemone, the decreasing order of species was DMA(V) > AsB > DMAS > AsC > As(V) > TETRA > GPAsC > MMA(V) > TMAO After cooking, 64% of tAs was lost; about 90% was released to the cooking oil, but the order of species was the same (excluding GPAsC and MMA(V) that were not detected after cooking)	[117]

Marine organisms, Australia	Thio-methylated As species, As-sugars	Tissues freeze-dried and homogenized with three portions of CH_3OH-H_2O (1:1) either using microwave heating at 70 °C or rotational mixing at 25 °C; afterward the samples centrifuged, evaporated to dryness and redissolved in water	HPLC-ICP-MS: 14 thio-methylated As species separated on Atlantis C18 column (150 × 4.6 mm, 5 µm); 20 mM $NH_4H_2PO_4$ mobile phase at pH 3; flow rate 1 mL/min In ICP-MS detection, potential interference from $^{40}Ar^{35}Cl^+$ eliminated based on $^{35}Cl^+$, $^{35}Cl^{16}O^+$, $^{35}Cl^{17}O^+$ and $^{40}Ar^{37}Cl^+$ signals; selenium monitored at m/z 82 as a cross-check on $^{40}Ar^{35}Cl^+$ ESI-MS for characterization/ confirmation of As-sugars	Thio-methylated As species found in marine organism were thio-SO_3-riboside, thio-PO_4-riboside, thio-OSO_3-riboside, thio-Gly-riboside-sugars	[118]
Marine periwinkle, Canada	As(III), As(V), MMA(V), DMA(V), TMAO, TETRA, AsB, AsC, $As(GSH)_3$, As-sugars	Periwinkles pooled per sampling site, freeze-dried and homogenized Sequential extraction: (1) MeOH/H_2O; (2) 2% HNO_3; both extracts analyzed for tAs and for its species Extraction residue digested with HNO_3/H_2O_2	HPLC-ICP-MS: (1) anion-exchange PRP-X100 analytical column; gradient elution with (A) 4 mM NH_4NO_3 and (B) 60 mM NH_4NO_3 mobile phases at pH 8.65; flow rate 1 mL/min, In as IS (2) Cation-exchange PRP-X200 analytical column; 20 mM pyridine with HCOOH mobile phase at pH 2.7; flow rate 1 mL/min, Rh as IS For external calibration, matrix-matched standard solutions prepared X-ray absorption spectrometry: for XANES, homogenized defrost composite sample used. Frozen standard solutions were run and used for linear combination fitting tAs determined by ICP-OES.	CRMs: DORM-2 dogfish muscle, DORM-3 fish protein (NRC) Recoveries of tAs and individual As species in the spiked samples: 92–114% and 83–125%, respectively Average column recoveries: 80–100% tAs concentrations: 56–840 mg/kg dry mass As species determined by chromatographic systems: iAs, MMA(V), As-sugar, AsB, TETRA. The highest iAs concentration found: 600 mg As/kg Species detected by XANES: As(III)-O; As(III)-S, As-sugar, AsB, As(V)-O	[119]

(Continued)

Table 4.3: (Continued)

Food matrix	Species	Sample treatment	Analytical speciation procedure	Additional comments	References
Bivalve mollusks, Brazil	As(III), As(V), MMA(V), DMA(V), AsB, *p*-ASA	Pooled frozen tissues were thawed, homogenized, freeze-dried, ground and sieved (100 μm) MW-assisted extraction: 0.2 g of the sample + 6 mL $MeOH/H_2O$ (1:1, 2:1 and 3:1); 50 min heating program at 80 °C ; dilution Ultrasonic extraction: sample prepared as above sonicated during 6 min For tAs, MW digestion with conc. HNO_3	HPLC-ICP-MS: PRP-X100 (250 × 4.1 mm) column; gradient elution with (A) $(NH_4)_2HPO_4$ at pH 6.0 and (B) $(NH_4)_2CO_3$ at pH 8.5, mobile phases; flow rate 1.25 mL/min; injection volume 200 μL; chromatographic run 15 min tAs determined by ICP-OES	QLs: 1–3 ng As/g for individual species CRMs: DORM-2 dogfish muscle (NRC); BCR-627 tuna fish (IRMM); 1566b oyster tissue (NIST) Extraction efficiency in CRM: 89–101% for MW-assisted and 72–93% for ultrasonic extraction Recovery of As in speciation procedure: 90–104% AsB found as the main species; As(V) and MMA(V) < QL; two unidentified species detected	[219]
Commercial fish oils	AsHC	Extraction: 1.0 g of oil + 7.5 mL hexane + 2 × 3.75 mL $H_2O_2/MeOH$ (9:1); water–methanol phase collected, evaporated and redissolved in 1 mL $MeOH/CHCl_3/H_2O$ (60:30:8) SPE: The extract applied to a preconditioned weak anion-exchange SPE column and eluted with 12 mL $MeOH/CHCl_3/H_2O$. After evaporation, the column fractions reconstituted in 1 mL *n*-hexane and filtered (0.45 μm) For tAs, oil samples MW digested with HNO_3/H_2O_2.	GC-ICP-MS: HP-5 column (30 m × 0.32 mm × 0.25 μm); 2 mL/min He carrier gas; injection volume 1 μL; injector temperature 280 °C; chromatographic run 40 min; triphenylarsine was used for external calibration and for internal control of the retention behavior of the chromatographic system GC-MS/MS: Rtx®-5SIL MS column (30 m × 0.25 mm × 0.25 μm): 0.8 mL/min He carrier gas, injection volume 1 μL; injector temperature 280 °C; chromatographic run 43 min; detection by selected reaction monitoring mode	tAs in several commercial products: 5.9–8.7 mg/kg Three major As species detected by GC-ICP-MS These species were identified by GC-MS/MS and HPLC-QTOF-MS as dimethyl-arsinoyl hydrocarbons: $C_{17}H_{38}AsO$, $C_{19}H_{42}AsO$ and $C_{23}H_{38}AsO$ AsHC accounted for 13–35% of the tAs in the fish oils After oil decontamination by activated carbon and steam deodorization, As content decreased	[167]

			HPLC-QTOF-MS: EC-C8 column (100 × 2.1 mm, 2.7 µm); gradient elution with (A) 5 mM NH_4COOH containing 0.1% HCOOH and (B) 0.1% HCOOH in ACN mobile phases; flow rate 0.3 mL/min; injection volume 1 µL; ESI(+), *m/z* range 100–1,000		
Edible oils, canned fish	Se(IV), Se(VI), organic Se	(1) Nonpolar, polar and ionic iSe fractionated on hydrophilic polymer HLB SPE cartridge using three eluting solvents (hexane, ethanol, 3% HNO_3, respectively). For Se(IV)/Se(VI) speciation, APDC added to the last SPE fraction and Se complexes extracted to CCl_4 (2) Ultrasonic DLLE: 10 mL of oil (80 °C) + 300 µL isopropyl alcohol containing 3% HNO_3(or ionic liquid) ultrasonicated and 30 µL of the aqueous phase taken for analysis (3) For tSe, the oil samples MW digested with $HNO_3/HCl/H_2O_2$	ETAAAS: direct determination of tSe in hexane-diluted oils, in the sample fractions and extracts, Pd used as a modifier For tSe in MW-digested samples: online hydride generation with AFS detection	ETAAAS DLs: 0.03 ng/g and 0.04 ng Se/g for tSe and iSe in oil, respectively Determination of tSe, iSe and speciation Se(IV)/Se(VI) carried out in several commercial samples; the concentration range for tSe was <4–52 ng/g, and for iSe 0.38–2.62 ng Se/g For tSe, AFS DL: 5 ng/g oil	[140]
Different types of fish, China	Se(IV), SeMet, SeCys	Extraction: 2.0 g of the dried sample + 10 mL 1.5 M KOH with 2 mL MeOH shaken 1 h at 24 °C, pH adjusted to pH 6.5, supernatant filtered (0.45 µm)	HPLC-HG-AFS: anion-exchange PRP-X100 column (250 × 4.1 mm, 10 µm); isocratic elution with 30 mM $NH_4H_2PO_4$ + MeOH (39:1) mobile phase, flow rate 1.0 mL/min	DLs: 1.66 µg Se/L for SeCys, 0.990 µg Se/L for SeMet and 1.10 µg Se/L for Se(IV) Recoveries after standard addition of each species: 87.3–103%	[162]

(*Continued*)

Table 4.3: (Continued)

Food matrix	Species	Sample treatment	Analytical speciation procedure	Additional comments	References
			Online HG obtained using 2% KBH_4 and 5% HCl	Concentration ranges found in fish tissues: 0.100–0.685 µg Se/g for SeCys; ND-0.373 µg Se/g for SeMet, Se(IV) not detected	
Freshwater fish, Argentina	Se(IV), Se(VI), SeMet, $SeCys_2$, MeSeCys	Enzymatic extraction: 0.6 g of the freeze-dried defatted liver or muscle + 8 mL H_2O containing 60 mg protease XIV, incubated with shaking at 37 °C during 24 h; the supernatant filtered (0.45 and 0.22 µm) For tSe determination, acid digestion carried out with $HNO_3/H_2SO_4/H_2O_2$ in a heating block	HG-AFS: total Se determined in the digested samples, in sediments and supernatants after enzymatic hydrolysis; 0.15 mL of V_2O_5 in concentrated H_2SO_4 was added (115 °C, 20 min), followed by Se(VI) reduction in 6 M HCl (90°C, 10 min) and HG HPLC-ICP-MS: Se species in enzymatic digest (except $SeCys_2$) separated on PRP- X100 column (250 × 4.1 mm, 5 µm); gradient elution with the mobile phases containing 3 and 10 mM citrate buffer in 2% MeOH (pH 4.8); flow rate 0.5 mL/min; injection volume 50 µL; chromatographic run 40 min. For $SeCys_2$ determination, the column was Zorbax 300-SCX (250 × 4.6 mm, 5 µm); 3 mM pyridine in 2% MeOH (pH 2.1) mobile phase	DL for tSe by HG-AFS: 10.5 ng/g. CRM: DORM-2 dogfish (NRC) tSe found in muscle: 0.66–1.61 µg/g and in liver: 4.46–73.71 µg/g dry mass SeMet was a primary species in the muscle (95% of tSe); in the liver $SeCys_2$ and SeMet were found and some unidentified species were detected	[200]
Crude and cooked mussel, Slovenia	Se(IV), Se(VI), SeMet, $SeCys_2$, MeSeCys	Water/enzymatic extraction: 0.6 g of the sample + 8 mL H_2O (or water containing 60 mg protease XIV) incubated at 37–55°C, 24 h; centrifuged and both,	HPLC-ICP-MS: as in ref. [200]. ^{210}Po analyzed by alpha-particle spectrometry	tSe in mussels: 2.62–8.19 mg/kg Se species found: $SeCys_2$, SeMet plus one undefined Low column recoveries	[234]

		supernatant and pellet collected for tSe determination Buffer extraction: 0.6 g of the sample + 8 mL of 50 mM Tris-HCl buffer containing 0.5 M sucrose (pH 7.4) frozen and thawed 3× in liquid nitrogen and homogenized; after centrifugation (15,550 *g*, 60 min, 4 °C), supernatant and pellet collected For speciation analysis, supernatant filtered (0.45 μm)		obtained in speciation procedure The best extractability of Se species obtained by enzymatic extraction at 40 °C: 70% of tSe Cooking resulted in the increase of Se solubility Results obtained for Po-210 discussed	
Chicken meat spiked with AgNPs (candidate CRM)	AgNPs	Meat homogenized in ammonium carbonate buffer (pH 9), spiked with AgNPs and subjected to enzymolysis with proteinase K (37 °C, 40 min)	AF^4 – spectrophotometric DAD-ICP-MS: a detailed description of asymmetric field fractionation given in the article; DAD wavelength range 194–949 nm; ICP-MS as the final detection system used in a conventional mode (Rh as IS, Scott double-pass spray chamber, dwell time 500 ms) and in a single particle mode (cyclonic Peltier-cooled spray chamber, dwell time 3 ms) TEM: the digested meat sample and AgNP suspension diluted and 10 μL applied to a 200 mesh Formvar/carbon-coated copper grid (at least 250 NPs per sample) for the observation of particle size and shape	Size distribution of AgNPs evaluated Recovery of Ag in meat spiked with nanoparticles ~80% A residual sample matrix affected the AF^4 elution behavior TEM and single particle ICP-MS results suggested that ionic Ag might be present in the digested meat as bound to organic constituents The method in-house validated	[198, 199]

(*Continued*)

Table 4.3: (Continued)

Food matrix	Species	Sample treatment	Analytical speciation procedure	Additional comments	References
Raw and cooked chicken breast from the US markets	iAs, MMA(V), DMA(V), Rox	About 142 raw and baked meat samples individually homogenized with addition of water and freeze-dried Extraction: 0.5 g of the sample + 10 mL of the extracting solution (20 mM malonic acid with 1% H_2O_2, pH 9.5), shaken 1 h at 50 °C, centrifuged and filtered (0.22 μm) For tAs, MW-assisted digestion with conc. HNO3 followed by dilution with water and addition of 5% MeOH and IS (Ge, In).	HPLC-ICP-MS: PRP-X100 (250 × 4.1 mm, 10 μm); temperature 40 °C; gradient elution with 20–100 mM malonic acid adjusted to pH 9.5 with aqueous ammonia; flow rate 1.5 mL/min; injection volume 50 μL; chromatographic run 10 min For ICP-MS detection, *m/z* monitored: 75 and 77 ($^{40}Ar^{37}Cl^+$); 1% CO_2 as optional gas introduced after spray chamber; CRC pressurized with He	DLs: 1 μg As/kg dry mass for tAs and As species (for Rox, 2 μg As/kg) CRM: 1568a rice flour (NIST) Recoveries of As species spiked to the meat sample: 80–120% iAs concentrations higher in conventional samples (geometric mean 1.8 μg As/kg) than in antibiotic free – (0.7 μg As/kg) or organic meat (0.6 μg As/kg). In Rox-positive samples, iAs higher than in Rox-negative samples Baking increased iAs and decreased Rox conc.	[122]
Chicken meat, herbal tea, CRMs	As(III), As(V), MMA(V), DMA(V), AsC, AsB, 4NPAA, *p*-ASA, Rox	Freeze-dried and homogenized samples four times extracted with deionized water under sonication and centrifuged	CE-ICP-MS: CE separation carried out on fused silica capillary (100 cm × 50 μm) with phosphate/borate electrolyte at pH 9.2; separation time 30 min Effluent online introduced to ICP-MS through CE-ESI-MS sprayer kit. A mixture $CH_3OH–H_2O$ (95:1) with Rh internal standard introduced as the sheath flow liquid through a built-in active 1:100 flow-splitter, providing a constant flow rate of 5 μL/min	DLs: 0.9–3.0 ng As/g for 10 species separated CRMs: TORT-2 lobster hepatopancreas, DORM-3 fish protein (NRC) In tea samples, As(III), As(V) and DMA(V) were found, whereas in chicken meat As(III) was the sole species	[121]

Liver of chicken fed or not with Rox	Rox, As(III), As(V), MMA(V), DMA(V), AsB, 3-amino,*N*-acetyl	Extraction: 0.5 g of the homogenized liver sample + 5 mL extraction solution (100 mg pepsin in 5 mL 0.5% HCl) sonicated 15 min and then incubated at 37 °C, 12 h; the obtained supernatant filtered (0.45 µm)	HPLC with simultaneous ICP-MS and ESI-MS/MS detection: PRP-X110S anion-exchange column (150 × 4.1 mm 7 µm); gradient elution with (A) 5% MeOH and (B) 60 mM NH_4HCO_3 containing 5% MeOH mobile phases at pH 8.75; flow rate 2 mL/min; injection volume 30 µL; chromatographic run 20 min; eluent split to ICP-MS (80%) and to ESI-MS (20%) ESI(+) for AsB and ESI(−) for other species; MRM detection/quantification	ICP-MS DLs: 30–90 ng As/L ESI-MS/MS DLs: 0.001–2.2 µg As/L CRM: DOLT-4 fish tissue (NRC). In Rox-fed chicken, As species found in the range: 1.1–151 µg As/kg, whereas in controls: <DL-46 µg/kg Significantly higher concentrations of six species found in Rox-fed chicken as compared to controls (except AsB and *N*-acetyl)	[123]
Porcine and chicken liver, China	*p*-ASA, NIT, Rox	Extraction: (1) 2 g of the sample + 6 g silica sand + 100 µL of standard mix + extraction solution ($MeOH/H_2O$ 3:7), in accelerated solvent extractor (three cycles, 4 min each, 80 °C, 1500 psi) (2) Ultrasound extraction with CH_3OH/H_2O (1:1) at 40 °C, 20 min. For fat separation, extracts kept at −20 °C, 30 min and the withdrawn aliquot filtered (0.45 µm)	HPLC-UV-HG-AFS: Luna 5u C18 column (250 × 4.60 mm); isocratic elution with 5:95 MeOH:50 mM KH_2PO_4 (included 0.1% TFA) mobile phase at pH 1.67; temperature 25 °C; flow rate 1.2 mL/min; injection volume 100 µL; chromatographic run 20 min. Column effluent transferred to the UV irradiation unit for species oxidation and then to HG-AFS	DLs: 0.24, 0.74 and 0.41 ng As/mL for arsanilic acid (*p*-ASA), NIT and Rox, respectively Using accelerated solvent extraction, recoveries were higher than 94% for the three species The three compounds were detected in meat purchased in the local market	[120]

(*Continued*)

Table 4.3: (Continued)

Food matrix	Species	Sample treatment	Analytical speciation procedure	Additional comments	References
Total diet, Japan	iAs, As(III), As(V), MMA(V), DMA(V), TMAO, TETRA, AsB, AsC	Composite food portion homogenized and 80 g freeze-dried (except composites of nuts, water, beverages, oils) Simulated gastric digestion: 1 g of the sample + 20 g gastric juice (0.07 M HCl, 0.01% pepsin) at 37 °C, 2 h with agitation; supernatant collected	HPLC-HG-ICP-MS: TSKgel Super IC A/C column (150 +150 × 7.8 mm) with a TSK guard column; isocratic elution with 35 mM Na_2SO_4 mobile phase at pH 3.1; injected volume 20 µL. The column effluent introduced online to the photooxidation and HG system. AB determined in a separate run, without HG Octopole CRC gas: 2.0 mL/min He	DLs in food composites: 0.1–3, 0.1–3, 0.07–2, 0.08–2, 0.01–18, 0.1–3, 0.06–1 and 0.07–2 ng As/g wet mass for As(V), As(III), MMA(V), DMA(V), AsB, TMAO, TETRA and AsC, respectively CRMs: 7503 white rice flour, 7405 hijiki powder (NMIJ); BCR-627 tuna fish; TORT-2 lobster hepatopancreas (NRC) Dietary intakes for each species estimated Two products primarily contributing to iAs intake: rice and hijiki	[223]
Terrestrial animals	Hg(II), MeHg	Red deer liver and kidney and wild boar kidney freeze-dried MW-assisted extraction: 0.2 g of sample + 2 or 5 mL TMAH for liver and kidney, respectively. Derivatization: 2 mL of the extract + 5 mL 0.1 M acetate buffer (pH 3.9) + 0.5 mL or 1.3 mL 3% $NaBEt_4$ HS-SPME: fused silica fiber coated with a 100 µm PDMS inserted for 10 min at 35 °C, followed by desorption to GC injection port at 200 °C, 0.5 min	GC-AFS: DB-5 column (30 m × 0.25 mm × 0.25 µm); He carrier gas at 3 mL/min; temperature program 40–200 °C, splitless injection; chromatographic run 6 min For ASF detection, Hg^0 generated in pyrolyzer at 800 °C	DLs for liver: 31.8 and 52.5 ng Hg/g for MeHg and Hg(II) DLs for kidney: 35.3 and 58.1 ng Hg/g, respectively CRMs: NCS-ZC 71001 beef liver (China); BCR-186 pig kidney (IRMM) The HS-SPME extraction enabled for lowering the DLs 5.5–6 times as referred to the procedure without preconcentration The two Hg species were quantified in animals from the mining district area	[171]

Milk and cake packing materials	DBT, TBT, DPhT, TPhT	Ultrasonic extraction and SPE: 2.0 g of the packing material cut into pieces + 10 mL CH_2Cl_2, 25 min sonication; extract evaporated (50 °C), redissolved in 5 mL of ACN with 0.1% HCOOH, sonicated and purified on a mixed-mode cation-exchange reversed-phase SPE cartridge. The final eluent (NaOH/MeOH) evaporated, reconstituted in a mobile phase and filtered (0.45 µm)	HPLC-ESI-MS/MS: EC-C18 column (100 × 4.6 mm, 2.7 µm); temperature 30 °C; gradient elution with (A) 0.1% HCOOH in water and (B) 0.1% HCOOH in MeOH mobile phases; flow rate 0.2 mL/min; injection volume 5 µL; chromatographic run 20 min Quantification using ESI(+) and MRM mode	DLs: 0.1–0.8 µg Sn/kg for DBT, TBT, DPhT, and TPhT Recoveries of organotins in spiked samples: 68–113% In the analysis of 10 real-world samples, TBT was found in one milk plastic packaging (5.4 µg Sn/kg) and DPhT in two cake wraps (23.45 µg Sn/kg)	[147]
Human milk	Zn-citrate	Frozen samples thawed and left at 4 °C for 16 h, yielding separation of fat; defatted milk collected and centrifuged during 10 min. LMM fraction obtained by protein elimination in micro-ultrafiltration device and used for speciation analysis tZn determined in defatted milk and in LMM fraction by FAAS or ICP-MS, after appropriate dilution	HPLC-ICP-MS: CIM DEAE-1 anion-exchange tube monolithic column (6.7 × 4.2 mm, volume 1 mL); aqueous – 1.2 M NH_4NO_3 linear gradient elution (pH 5–7); flow rate 1.0 mL/min; ICP-MS detection for Zn profiling For quantification, column fractions collected and off-line introduced to ICP-MS and FAAS	Zn-citrate complex was obtained and its solution was stable for at least 5 days if stored in HEPES, MOPS or MES buffer, pH 5–7, at 4 °C. Separation of $[Zn(Cit)]^-$ and $[Zn(Cit)_2]^{4-}$ achieved in 10 min; species identity confirmed by ESI-MS/MS analysis of respective column fractions ICP-MS DL in column fractions: 0.5 ng Zn/mL (10 ng Zn/mL for FAAS) Column recoveries: 94–102%	[159]

(*Continued*)

Table 4.3: (Continued)

Food matrix	Species	Sample treatment	Analytical speciation procedure	Additional comments	References
			Polyatomic interferences at m/z 66 ($^{32}S^{34}S^+$, $^{34}S^{16}O^{16}O^+$, $^{32}S^{16}O^{18}O^+$, $^{31}P^{35}Cl^+$, $^{54}Fe^{12}C^+$) eliminated in CRC pressurized with He For identification of Zn-binding ligand, mass spectra obtained by orthogonal Z-spray ESI(−) coupled to hybrid QTOF-MS	In two milk samples, about 23% of tZn was present in LMM fraction and the citrate complex was the main species. It was suggested that citrate addition to infant formulas might enhance element bioavailability	
Selenium enriched yogurt	Se(IV), SeMet, SeMetO, $SeCys_2$,Se-proteins	The prepared yogurt defatted by extraction with cyclohexane, then dried and freeze-dried Protein fraction obtained by precipitation with acetone and resuspension of pellet in 70 mM Tris-HCl buffer containing 7 M urea at pH 7.5; after centrifugation, supernatant taken for SEC and AF^4 analyses. Se-containing protein fractions from the above separation systems subjected to SDS-PAGE, in-gel digestion, and nano-LC-ESI-LTQ-MS analyses. Protein pellet digested with Protease XIV for Se speciation by reversed phase HPLC-ICP-MS. For tSe, freeze-dried yogurt MW digested with HNO_3/H_2O_2	HPLC-ICP-MS: (1) Biosep 2000 column (250 × 4.1 mm, 5μm); 50 mM Tris-HCl and 0.05 mM KCl mobile phase at pH 6.8; flow rate 1 mL/min; injection volume 200 μL (2) Zorbax C8 column (250 × 4.60 mm, 5 μm); isocratic elution with 2% MeOH and 0.1% HFBA mobile phase; flow rate 0.8 mL/min; injection volume 100 μL AF^4-UV-ICP-MS:Cellulose regenerated membrane (cutoff 10 kDa); mobile phase as in SEC; injection flow 0.2 mL/min; cross-flow 3 mL/min; injection volume 200 μL; UV detection 280 nm ICP-MS detection: all Se isotopes monitored, CRC pressurized with 6mL/min H_2 Nano-LC-ESI-LTQ-MS: Reprosil C18 trap column; Acclaim PepMap 100-C18 analytical column (75 μm × 15 cm, 3μm); gradient elution with (A) 1% HCOOH and (B) 1% HCOOH in ACN mobile phases; flow rate 200 nL/min, injection volume 5μL	Metallomics approach to evaluate Se biotransformation by lactic bacteria during elaboration of Se-enriched yogurt tSe in the enriched yogurt: 22.8 μg/g dry mass About 70% of tSe was bound to proteins, mainly in the molecular mass range 10–70 kDa Two proteins involved in antioxidant functions were detected: thioredoxin and glutaredoxin SeCys found as the primary species after enzymatic digestion of proteins	[124]

Infant formulas and first foods, USA	As(V), As(III), DMA(V)	MW-assisted extraction for speciation analysis: 2g of sample + 20–40 mL 1% HNO_3, heated during 50 min at 55–95 °C, centrifuged or filtered (0.45 µm) and again filtered by 10 kDa spin filter For tAs: infant formulas MW digested with HNO_3/H_2O_2; purees open-vessel digested with HNO_3 and diluted before analysis	HPLC-ICP-MS: (1) PRP-X100 column; 20 mM $NH_4H_2PO_4$ mobile phase at pH 6; flow rate 1 mL/min (2) Dionex AS16 column; TMAH mobile phase; flow rate 1 mL/min	CRM: 1568 rice flour (NIST) Analytical performance of two anion-exchange columns compared: average recovery obtained for PRP-X100 was 76% and for As15 column 80% tAs found in formulas, purees and multiple-ingredient infant foods were in the range: <1–23 ng/g, with the main fraction corresponding to iAs As elution in a void volume suggested the presence of AsB or other nonanionic As species	[248]
Grain rice, baby food, Finland	As(III), As(V), MMA(V), DMA(V), AsB	MW-assisted extraction: 2 g of the homogenized sample + 1% HNO_3 (10 mL for rice, 20 mL for rice products), 65 min heating program at 55–95 °C; centrifugation, filtration (0.2 µm)	HPLC-ICP-MS: anion-exchange PRP-X100 (250 × 4.6 mm, 5 µm) with a precolumn (25 × 2.3 mm); $(NH_4)_2CO_3$ mobile phase (10–50 mM, pH 8.9); flow rate 1 mL/min; injection volume 100 µL; *m/z* 75; dwell time 200 ms	CRMs: IMEP-107 rice flour (IRMM); 1568 rice flour (NIST) AB and MMA(V) not detected in any sample iAs determined as a sum As(III) + As(V); DL 0.06 mg As/kg iAs in rice: 0.09–0.28 mg As/kg. iAs in the rice-based baby food products: 0.02–0.29 mg As/kg	[225]

(*Continued*)

Table 4.3: (Continued)

Food matrix	Species	Sample treatment	Analytical speciation procedure	Additional comments	References
Cereal-based food and infant cereals, Spain	iAs, MMA(V), DMA(V)	MW-assisted extraction: 0.25 g of the ground sample + 10 mL of the extracting solution (0.2% HNO_3 and 1% H_2O_2) for oxidation of As(III) to As(V)	HPLC-ICP-MS: anion-exchange PRP-X100 (150 × 4.1 mm, 5 µm) with a precolumn (25 × 2.3 mm); temperature 30 °C; isocratic elution with 12 mM $NH_4H_2PO_4$ mobile phase; flow rate 1 mL/min; injection volume 250 µL; *m/z* 75 and 35; dwell time 200 ms	DLs: 0.3–0.4 µg As/kg for three species CRMs: 1568a rice flour (NIST); 7503a white rice flour (NMIJ CRM) For 29 products analyzed, tAs range was 3.7–35.6 µg/kg and for iAs the range was 3.1–26.0 µg As/kg	[150]
Rice	iAs	MW-assisted extraction: 0.25 g of milled rice + 10 mL extraction solution (0.06 M HNO_3 + 3% H_2O_2), 30 min heating program at 95 °C SPE: 2 mL of the supernatant + 2 mL 0.05 M $(NH_4)_2CO_3$ introduced to a Strata SAX cartridge conditioned with MeOH; washing with 0.1–1 M CH_3COOH for elimination of MMA(V)/DMA(V) and elution of the retained As(V) with 2 mL 0.5 M HCl; dilution	HG-AFS: For the reduction of As(V) in the obtained extract, 4% KI and 0.4% ascorbic acid added; after 50 min, the sample mixed with 1% $NaBH_4$ and KI/HCl in a flow system; arsine transported to a flame supported by hydrogen evolved from HCl-$NaBH_4$ reaction. AFS detection at As emission line 193.7 nm	DL: 1.3 ng As/g Recoveries after 100 ng As/g addition: 94.73% for As(III) and 93.75% for As(V) CRMs: 1568a rice flour (NIST) and BC211 rice flour (ERM) iAs contents of several domestic and imported rice samples from local markets: 45–235 ng As/g	[137]
Brown, polished and milled rice	As(III), As(V), MMA(V), DMA(V)	Comparative evaluation of different extracting solutions and heating conditions. Extraction in a heating block: 0.5 g of the rice flour + 2 g of extracting solution (best results for 0.05 M $HClO_4$ or 0.15 M HNO_3) heated in a capped tube at 100 °C, 2 h; AsB	HPLC-ICP-MS: C18 mg column (250 × 4.6 mm); isocratic elution with a mobile phase containing 10 mM sodium 1-butane sulfonate, 4 mM malonic acid, 4 mM TMAH and 0.05% MeOH at pH 3.0; flow rate 0.75 mL/min	DLs: 0.01 ng As/g for As(III) and As(V); 0.03 ng As/g for DMA(V) As(III), As(V), DMA(V) found in the samples To avoid As(V) reduction, Ag^+ addition recommended	[160]

		added as IS; dilution MW-assisted extraction: 1.0 g of the flour + 10 g of these same extraction solutions as above, 30 min heating at 90 °C; dilution Before analysis, centrifugation and filtration (0.45 μm)	CRC gas: He 3 mL/min	CRMs: 7531-a brown rice flour, 7503-a white brown flour (NMIJ)	
Rice	As(V)	Extraction of As(V): 0.1 g of the powdered sample + 10 mL 0.5 M HNO_3 ultrasonicated 136 min Extraction for total As: after sonication, 80 nM $KMnO_4$ added for As(III) oxidation CPE: As(V)-molybdate complex in 50 mM H_2SO_4, 0.1 M ascorbic acid and 0.008% NaCl, extracted to the surfactant-rich phase (0.06% Triton X-114) without a need for heating	ETAAAS: Ir as permanent modifier, platform atomization, temperature program optimized, Zeeman BC	DL: 10 ng As/L Recoveries after standard addition: 90.8–113.1% CRM: 804 rice flour (IRMM) Only spiked samples analyzed.	[194]
Rice	As(III), As(V), MMA(V), DMA(V)	Extraction: 0.5 g of the milled rice + 2 mL 0.15 M HNO_3 heated during 2 h at 100 °C, diluted and filtered (0.45 μm) For improving the shape of As(V) chromatographic peak, 1 mM H_3PO_4 added and AsB used as IS For tAs, the sample digested with HNO_3/H_2O_2 in a heating block; Rh as IS	HPLC-ICP-MS: Discovery HS F5 column (250 × 4.6 mm, 5 μm); isocratic elution with 0.1% HCOOH + 1% MeOH mobile phase at pH 2.55; flow rate 1 mL/min; injection volume 10 μL; chromatographic run 7 min Hexapole CRC: 6 mL/min He + 3 mL/min H_2; *m/z* monitored: 75, 77 ($^{40}Ar^{37}Cl^+$)	Different columns tested with this same mobile phase, silica-based pentafluoro-phenyl column CRMs (tAs): 7503-a white rice flour (NMIJ); 1568a rice flour (NIST) Extraction efficiency: 80–102% DLs: 0.002 mg As/kg Recovery of iAs in the spiked samples: 93–111%	[31]

(*Continued*)

Table 4.3: (Continued)

Food matrix	Species	Sample treatment	Analytical speciation procedure	Additional comments	References
Rice, Italia	iAs	Extraction: 0.2 g of rice + 10 mL 1% HNO_3 left overnight and followed by 60 min MW heating program at 55–75–98 °C; 900 µL of the supernatant mixed with 100 µL of H_2O_2 and kept 12 h at 4 °C For tAs, Cd, Cr, Zn, Se and Ni, MW-assisted digestion with concentrated HNO_3 and H_2O_2	HPLC-ICP-MS: PRP-X100 column (150 × 4.1 mm, 10 µm) with a precolumn; isocratic elution with 6.66 mM $NH_4H_2PO_4$ and 6.66 mM NH_4NO_3 mobile phase at pH 6.2 Total multielement determination by ICP-MS	DL: 0.29 mg As/kg for iAs CRM: 1568a rice flour (NIST) For 101 rice samples analyzed, the range for tAs was 0.11–0.28 mg/kg and 0.06–0.11 mg As/kg for iAs Significant geographical variations found for tAs, iAs, Se and Ni	[226]
Rice products, seafood, Denmark	iAs	Extraction: 0.5 g of homogenized and dried sample + 10 mL extracting solution (0.1 M HNO_3 + 3% H_2O_2) heated in a water bath at 90 °C during 1 h, centrifuged and pH adjusted within the range 5.0–7.5 SPE: The sample extract loaded on a strong anion-exchange SPE cartridge (Strata SAX) previously preconditioned with MeOH; the cartridge washed with 0.5 M CH_3COOH and the retained iAs eluted with 0.4 M HNO_3 For seafood, MW-assisted extraction carried out with a mixture of diluted HCl + 3% H_2O_2	HG-AAS: The SPE effluent mixed with the reduction solution (30 mM KI, 28 mM ascorbic acid, 0.1% silicone, 3 M HCl) and, after 1 h at room temperature, arsine generated with 0.5% $NaBH_4$ and a gaseous phase transported to AAS	Selectivity of iAs separation and preconcentration examined by analyzing the collected SPE fractions by HPLC-ICP-MS DLs for iAs: 0.02 mg As/kg for rice and 0.08 mg As/kg for dry seafood mass The method tested in a collaborative study In the analysis of 36 rice products, iAs concentration ranged from 0.03 to 0.60 mg As/kg, whereas in 20 marine food the concentrations up to 0.14 mg As/kg were found	[195, 196]
Rice, Australia	As(III), As(V), MMA(V), DMA(V), $As(GSH)_3$	MW-assisted extraction: 0.2 g of the freeze-dried and homogenized grains + 10 mL 2% HNO_3 heated at 95 °C during 10 min program, centrifuged and the supernatant collected	HPLC-ICP-MS: As(III), As(V), DMA(V), MMA(V) separated on PRP-X100 column (250 × 4.6 mm, 10 µm); temperature 40 °C;	ETAAAS DL: 0.03 µg As/L CRMs: BC211 rice, IMEP 107 rice (IRMM); 7503a white rice (NMIJ); 1568a rice (NIST); T07151QC rice (FAPAS); 10a, 10b, 10c rice flours (NIES).	[224]

		For tAs, MW-assisted digestion with conc. HNO_3 For XANES, standards prepared in 30% glycerol and the powdered rice samples were frozen (~12 K)	phosphate buffer at pH 4.5 as a mobile phase; flow rate 1.5 mL/min To discard cationic As species: LC-SCX column (250 × 4.6 mm, 5 µm); temperature 40 °C; pyridine buffer mobile phase at pH 2.6; flow rate 1.5 mL/min Polyatomic interferences in ICP-MS corrected by monitoring *m/z* 77, 82. XANES spectra were collected in fluorescence mode using up to 15 scans. The average spectra were background corrected and analyzed by linear combination fitting tAs determined by ETAAAS with Pd/Mg modifier	Practically complete As recovery in speciation procedure XANES results gave similar proportions of total, As(III), As(V) and DMA(V) as compared to HPLC-ICP-MS results; additionally, XANES was able to distinguish between As(III) and As-thiolate	
Rice, Thailand	As(III), As(V), MMA(V), DMA(V)	Extraction: 0.25 g of the milled rice + 10 mL H_2O (90 °C, 4 h, tapped tube); the supernatant treated with *α*-amylase (37 °C, 30 min) and filtered (0.20 µm) For tAs in the extract, a portion of supernatant digested with HNO_3 instead of enzymatic treatment For tAs in rice, MW-assisted digestion with concentrated HNO_3	HPLC-ICP-MS: C18 column (150 × 4.6 mm, 5 µm); temperature 35 °C; isocratic elution with (A) 7.5 mM TBAH and 10 mM $NH_4H_2PO_4$, pH 8.25, and (B) MeOH mobile phases (95:5); flow rate 1 mL/min; injection volume 25 µL; chromatographic run 9 min	DLs: 0.4–2.0 µg As/kg for As species Recoveries obtained in the spiked rice samples for four species: 86.7–119.9% CRMs: 1568a rice (NIST); 7503-a white rice (NMIJ) tAs found in the analysis of 185 samples: 13.89–379.39 µg/kg Higher As concentrations found in red and brown rice as compared to white	[229]

(*Continued*)

Table 4.3: (Continued)

Food matrix	Species	Sample treatment	Analytical speciation procedure	Additional comments	References
Rice and rice cereal	As(III), As(V), MMA(V), DMA(V)	The samples were ground, sieved and dried at 90 °C for 2 h. MW-assisted extraction: 0.2 g powdered sample treated with aqueous solution of α-amylase (0.05%); 2 h heating at 90 °C with agitation; centrifugation and addition of *o*-ASA as IS	CE-ICP-MS: five As species separated in polymer-coated fused silica capillary (60 cm × 100 μm × 375 μm); hydrodynamic injection at 50 mBa during 8 s; 8 mM Na_2CO_3 running buffer at pH 11; 10 min run; Mira Mist CE nebulizer and make-up solution composed of 1% HNO_3 + 10% propanol; CRC pressurized with H_2, 2 mL/min	DLs: 0.15–0.27 ng As/g for individual species. CRMs: 1568-b rice (NIST; 7503-a rice (NMIJ) *o*-ASA used as IS to compensate for signal drift and variance caused by the CE injection Recovery of the speciation procedure: 92.2% and 96.2% for two real samples analyzed	[141]
Brown and white rice, Japan	As(III), As(V), MMA(V), DMA(V)	Fractionation: (1) polishing and bran separation; (2) soaking and collection of water and polished grain; (3) cooking and, after evaporation of water excess, cooked rice freeze-dried for analysis Different times of storing conditions tested Extraction for speciation: an aliquot of solid sample + 0.15 M HNO_3 heated at 100 °C up to 2 h, centrifuged	HG-AAS: determination of tAs in each fraction. For solid samples, digestion with $HNO_3/HClO_4$ was performed; 20% KI and 10% ascorbic acid were used for HG HPLC-ICP-MS: C18 mg column (250 × 4.6 mm); isocratic elution with a mobile phase composed of 10 mM sodium 1-butane sulfonate, 4 mM TMAH, 4 mM malonic acid, 0.05% MeOH; flow rate 0.75 mL/min, injection volume 20 μL	ETAAAS DLs for tAs: 0.004 mg/kg for rice, 0.02 mg/kg for bran and 0.0004 mg/kg for washing water Recoveries were in the range 90.8–98.8% DLs for iAs, MMA(V) and DMA(V): 0.001–0.002 mg As/kg in rice, 0.02–0.03 mg As/kg in bran and 0.0002–0.0004 mg As/kg in water Recoveries: 94–104.3% CRMs: 7503-a white rice, 7531-a brown rice (NMIJ) Bran removing resulted in significantly lower As content	[201]

Rice and seaweed	iAs	Extraction of iAs: 0.1 g of ground rice + 10 mL of 1% HNO_3 and 1% H_2O_2(50–75–95 °C, 40 min) For seaweed (0.2 g), three different extraction solvents tested (10 mL H_2O; 2% HNO_3 + 3% H_2O_2; 0.07 M HCl + 3% H_2O_2; 50–75–85 °C, 20 min). Prior to analysis, all samples centrifuged For tAs: MW digestion with HNO_3/H_2O_2	HG-ICP-MS: The sample mixed with 5 M HCl and 2% $NaBH_4$ in a mixing coil, volatile compounds transported through gas–liquid separator to a nebulizer with a cyclonic spray chamber. Polyatomic interferences controlled by triple quadrupole; dry versus wet plasma tested HPLC-ICP-MS: PRP-X100 column (250 × 4.6 mm 10 μm); 25 mM $(NH_4)_2CO_3$ mobile phase at pH 8.5; flow rate 1 mL/min HPLC-HG-ICP-MS: 6.2 mM NH_4NO_3 and 6.5 mM H_3PO_4 at pH 6.0 as a mobile phase	HG conditions optimized focusing on the lowest possible DMA(V)/iAs signal ratio DL: 1.1 μg As/kg for iAs. CRMs: 1568a rice flour (NIST), TORT-2 lobster hepatopancreas (NRC). Good agreement between three procedures used. Using the proposed HG-ICP-MS procedure for iAs: five sample replicates were accomplished within 4 min	[136]
Rice, Madagascar	MeHg	Extraction: 0.5 g of the ground rice + 2 mL KOH 25% in MeOH heated during 3 h at 75 °C; 6 mL dichloromethane and 1.5 mL conc. HCl added; after 1 h shaking, the sample centrifuged for phase separation; organic solvent evaporated in the presence of water Derivatization: addition of $NaBEt_4$, purge and trap (Tenax trap) tHg determined by US EPA method	GC-CV-AFS details not given	DLs: 0.02 ng/g for tHg and 0.002 ng Hg/g for MeHg CRMs: 1568a rice, 1515 apple leaves (NIST); TORT-2 lobster tissue (NRC) Recoveries after standard addition: 130% for tHg and 100% for MeHg Concentration ranges found in Madagascar rice: 0.36–510 ng/g tHg and 0.015–63 ng Hg/g MeHg	[211]

(*Continued*)

Table 4.3: (Continued)

Food matrix	Species	Sample treatment	Analytical speciation procedure	Additional comments	References
Selenium-biofortified rice	iSe, tSe	Extraction of iSe: 1 g of rice flour + 30 mL H_2O, sonicated 30 min and centrifuged Cyclohexane (5 mL) added to the supernatant, aqueous phase collected and heated 2–3 min with 6 M HCl For tSe: acid digestion with $HNO_3/HClO_4$ in a heating block under reflux.	CPE-ETAAAS: a sample aliquot mixed with 2 mM dithizone and 1% Triton X-114, pH 1 (50 °C, 20 min). The mixture centrifuged, cooled on ice-water bath and the surfactant-rich phase containing Se complex collected; the extract diluted with 0.1 M HNO_3 in 10% ethanol and injected to ETAAAS (20 µL). Analyte modifier: Pd/Mg tSe determined also by CPE-ETAAAS Organic Se evaluated as the difference	Interferences due to the foreign metal ions were not observed up to at least 50:1 molar excess DL: 0.08 µg Se/L The enrichment factor: 82 Recovery for iSe in the spiked samples: 90.3–106.0% The concentrations found in biofortified rice: 18.2–26.8 µg Se/g iSe and 124.2–143.0 µg/g tSe	[130]
Bread flour from Se-biofortified wheat	Se(IV), Se(VI), SeMet, $SeCys_2$, MeSeCys	Enzymatic hydrolysis: 0.1 g of the flour + 3 mL Tris-HCl buffer + 0.02 mg protease XIV sonicated 2 min, centrifuged (4°C, 20 min) and filtered (0.22 µm) For tSe determination by ICP-MS, MW-assisted digestion with conc. HNO_3/H_2O_2 For tSe by CNAA, wheat-flour samples heat-sealed in polyethylene vials (1.2 mL)	HPLC-ICP-MS: (1) PRP-X100 column (250 × 4.1 mm, 10 µm); 10 mM ammonium citrate with 2% MeOH mobile phase at pH 5; flow rate 1 mL/min; injection volume 100 µL, chromatographic run 15 min (2) Zorbax Rx-C8 column (250 × 4.6 mm; 5 µm); 0.1% TFA with 2% MeOH mobile phase; flow rate 1 mL/min; injection volume 100 µL; chromatographic run 20 min CRC gas: 4.5 mL/min H_2	Anion-exchange separation was selected CRMs: 1568a rice flour, 1567a wheat flour (NIST); SELM-1 selenized yeast (NRC) Recovery of Se in speciation procedure: 80–100% For tSe good agreement between ICP-MS and cyclic neutron activation analysis For different biofortification protocols, SeMet accounted for 70–100% of the tSe, whereas the max. iSe content was below 5%	[127]

Selenium-enriched sprouts of alfalfa, soy and lentil	Se(IV), Se(VI), SeMet, $SeCys_2$	Biofortification with Se(IV) in hydroponic cultures; dried and ground shoots and roots prepared Enzymatic hydrolysis: 0.3 g of the sample + 30 mg protease XIV + 10 mL 25 mM $NH_4H_2PO_4$ at pH 7.5 (37 °C, 16 h); supernatant filtered (0.45 and 0.20 μm) For tSe: *Aqua regia* mineralization and MW digestion with HNO_3/H_2O_2	HPLC-ICP-MS: PRP-X100 column (250 × 4.1 mm, 10 μm); phosphate buffer mobile phase at pH 7.0; flow rate 1.5 mL/min; injection volume 50 μL; chromatographic run 5 min For ICP-MS detection, CRC pressurized with H_2 and Rh used as IS tSe after acid digestion determined by HG-AFS after prereduction to Se(IV) with 6 M HCl	DLs: 0.01–0.07 mg Se/kg for individual species CRMs: 023-050 and 025-050 soils (Resource Matrix CRM, USA) Enzymatic extraction efficiencies: 67–120% Se(IV) found as the main species in alfalfa and soy; for lentil it was SeMet. Oxidized SeMet was detected, but not MeSeCys	[126]
Selenium-biofortified cabbage, Spain	Se(IV), Se(VI), SeMet, $SeCys_2$	In vitro gastric (pepsin) and gastrointestinal extracts (pepsin, pancreatin, amylase) of crude and boiled cabbage analyzed after filtration For tSe in cabbage and its gastrointestinal fractions; samples MW digested with HNO_3/H_2O_2	HPLC-ICP-MS: PRP-X100 column (250 × 4.1 mm, 10 μm); $NH_4H_2PO_4$ mobile phase at pH 7; flow rate 1.5 mL/min; injection volume 50 μL; chromatographic run 10 min CRC gas: H_2; Rh as IS	DLs: 0.04–0.08 mg Se/kg dry mass in cabbage samples; in gastrointestinal supernatant: 0.04–0.09 mg Se/kg Average column recovery: 74–112% iSe lixiviated during boiling; SeMet increased in the enzymolysis supernatant of boiled versus crude cabbage	[30]
Selenium-fortified pakchoi	Se(IV), Se(VI), SeMet, SeCys, MeSeCys	Biofortification obtained by growing sprouts on sand amended with Se(VI)	HPLC-ICP-MS: Inersil C18 column (250 × 4.6 mm, 5 μm); isocratic elution with 8 mM 1-butanesulfonic acid, 4 mM TFA at pH 4.5; flow rate 1.2 mL/min; chromatographic run 13.5 min	DL: below 5 ng Se/mL for individual species Recovery of speciation analysis: 97.6%	[131]

(*Continued*)

Table 4.3: (Continued)

Food matrix	Species	Sample treatment	Analytical speciation procedure	Additional comments	References
		Extraction: 0.1 g of homogenized and dried sprouts + 10 mL of methanolic solution sonicated 10 min, incubated at 60 °C for 8 h, centrifuged and diluted with a mobile phase For tSe, MW digestion with conc. HNO_3	(different columns, pH values and ion-pairing reagents tested)	MeSeCys and several unknown species found in Se-enriched pakchoi	
Selenium food supplements	Se(IV), Se(VI), SeMet, $SeCys_2$,*y*-Glu-MeSeCys	Enzymatic digestion: the powdered sample treated with protease XIV in MW digestion system at 37 °C during 30 min and centrifuged For tSe determination, wheat flour MW digested with HNO_3/H_2O_2; tablets dissolved in HNO_3	HPLC-ICP-MS: Econosphere C18 column (250 × 4.1 mm, 5 μm); gradient elution with (A) 2 mM sodium 1-pentane-sulfonate, 5 mM citric acid in 3% MeOH (pH 2.70) and (B) 5 mM NaH_2PO_4, 5 mM citric acid in 3% MeOH (pH 2.77) mobile phases; chromatographic run 6 min For ICP-MS detection, dynamic reaction cell pressurized with 1.0 mL/min CH_4 used to eliminate polyatomic interferences at *m*/*z* 78, 80 ($^{38}Ar^{40}Ar^+$ and $^{40}Ar^{40}Ar^+$)	DLs: 0.04–0.07 μg Se/L. Recoveries of the speciation procedure evaluated by relating the sum of species to tSe in acid-digested sample: 92–104% CRM: 1567a wheat flour (NIST) The main species in wheat flour was SeMe *y*-Glutamyl-Se-methyl cysteine identified in one of the Se tablets using ESI-MS/MS	[128]
Selenized yeast	SeMet	MW-assisted digestion: 0.05 g of freeze-dried yeast + ^{13}C Met and SeMet spike + 6 mL 4 M methanesulfonic acid, 165 °C, 20 min Supernatant collected for derivatization with methyl chloroformate; dissolution in $CHCl_3$	GC-ID-MS: DB-5MS column (30 m × 0.25 mm × 0.25 μm); 1.5 mL/min He carrier gas; injector temperature 280 °C; temperature gradient program 120–260 °C; chromatographic run 20 min; transfer line 260 °C Detection/quantification in SIM mode using molecular ions	CRM: SELM-1 selenized yeast (NRC)	[132]

Selenium-biofortified edible mushrooms	Se(IV), SeMet, SeOMet	*Pleurotus florida* cultivated on Se-rich wheat straw: fruiting bodies dried (40 °C) and powdered Supernatant after in vitro gastrointestinal enzymolysis collected for Se speciation and for tSe determination.	HPLC-ICP-MS: (1) SEC fractionation on Superdex 75 10/300 GL column; CH_3COONH_4 mobile phase at pH 7.5 (2) PRP-X100 column (250 × 4.1 mm, 5 µm); gradient elution with CH_3COOH 20–200 mM + 100 mM triethylamine For ICP-MS quantification, DRC pressurized with CH_4, 0.7 mL/min	tSe found in biofortified mushrooms: 141 µg/g Approximately 75% of tSe solubilized during enzymolysis In SEC analysis, Se not detected in the fraction of molecular mass >5 kDa The main species determined by anion-exchange chromatography: SeMet (73% of all Se species); traces of SeOMet, Se(IV) also found and about 25% of Se corresponded to unidentified species	[125]
Selenium-enriched edible mushrooms	SeMet, MeSeCys, $SeCys_2$	*Agaricus bisporus* biofortified by irrigation of cultures with Se(IV); caps and stalks freeze-dried and homogenized Extraction of proteins: 0.5 g of the sample + 50 mL 30 mM Tris–HCl, pH 7.5, precipitation with 90% acetone (4 °C, 48 h) and redissolution in 2 mL of Tris-HCl	HPLC-ESI-MS: derivatized amino acids separated on Zorbax Eclipse XDB-C18 Rapid Resolution HT column (50 × 2.1 mm, 1.8 µm); temperature 30 °C; gradient elution with water and acetonitrile mobile phases, both containing 0.1% HCOOH; flow rate 0.3 mL/min; chromatographic run 9.5 min. HPLC system coupled to an ESI-triple quadrupole MS operated at positive mode with MRM	HG-AFS DL: 0.1 ng Se/g dry mass In protein digests, selenocysteine detected as $SeCys_2$, SeMet and MeSeCys were separated, identified and quantified by LC-ESI-MS. For the three species, their concentrations significantly increased in Se-biofortified mushrooms as compared to controls	[129]

(*Continued*)

Table 4.3: (Continued)

Food matrix	Species	Sample treatment	Analytical speciation procedure	Additional comments	References
		Enzymatic digestion: 1 mL of extract + 30 mg trypsin (50 °C, 24 h) + 30 mg proteinase K (50 °C, 24 h); supernatant collected and filtered. Prior to HPLC-ESI-MS, amino acids in the obtained hydrolyzate were derivatized with 6-aminoquinolyl-*N*-hydroxy succinimidyl carbamate For tSe in raw material and in protein fraction: HNO_3/H_2O_2 digestion in a heating block	tSe determined by HG-AFS after prereduction to Se(IV) with conc. HCl	$SeCys_2$ was found as the most abundant species in protein fraction. Se-amino acids were not detected in nonprotein fraction	
Edible mushrooms, Poland	As(III), As(V), DMA(V)	*Xerocomus badius* collected in forests, dried, ground and sieved Extraction: 1.0 g of the biomass + 10 mL 1 M H_3PO_4 + Triton-X100, ultrasonic bath 30 min; the samples filtered, centrifuged and diluted adjusting pH to 6–6.5	HPLC-HG-AAS: (1) LC-SAX1 (250 × 4.6 mm) column; 50 mM Na_2HPO_4 and 5 mM KH_2PO_4 mobile phase; flow rate 3 mL/min; injection volume 200 μL; chromatographic run 3.5 min. (2) Zorbax SAX column (250 mm × 4.6 mm); 100 mM Na_2HPO_4 and 10 mM KH_2PO_4 mobile phase; flow rate 2.5 mL/min, injection volume 200 μL; chromatographic run 5 min Column effluent online transferred to a home-made HG unit (3% $NaBH_4$, 10 M HCl); hydrides atomized in a quartz tube (900 °C) AAS	DL: 0.04 mg As/kg for the three species Recoveries evaluated after standard addition: 93.7–104.4% Good agreement between two chromatographic systems The highest concentrations found: 27.1, 40.5, 88.3 mg As/kg for As(III), As(V) and DMA(V), respectively In noncontaminated forests, As species 0.5 mg As/kg	[231]

Tomato, mushroom, garlic	Sb(III), Sb(V)	Freeze-dried and pulverized samples extracted with water (ultrasonic bath, 30 min), centrifuged and adjusted to pH 8.0 Sb(III)–DDTC complex formed and preconcentrated on nano γ-alumina that was synthesized and characterized in this study	FIA-SPE-FAAS: in a flow system, sample extract mixed with diethyl dithiocarbamate; the Sb(III)–DDTC complex retained on microcolumn (8 × 3 mm) packed with nano γ-alumina; at this stage 1 M HNO_3 was introduced to FAAS equipped with a STAT In the next step, the complex eluted from the column with 1 M HNO_3 containing MeOH (15 s) and transported to FAAS. tSb determined in this same FIA system, after prereduction with acidified (HCl) thiourea solution Sb(V) determined as a difference	STAT-FAAS enabled for 40-fold higher sensitivity as compared to the conventional FAAS DLs evaluated for Sb(III) and Sb(V) standards: 6.0 and 8.2 ng Sb/L, respectively Recoveries obtained after standard addition to the food samples: 96–107%	[197]
Chromium nutritional supplement (yeast)	Cr(III), Cr(VI), CrPic	Ultrasonic extraction: 0.1 g of pulverized yeast tablet + 4 mL MeOH-H_2O (1:3), 20 min sonicated and filtered (0.22 µm) Cr(III) chelation: 250 µL of the sample + 250 µL 0.4 M DTCA, 60 °C, 45 min; dilution with a running buffer	CE-ICP-MS: Cr(III)-DTCA, Cr(VI), CrPic separated in a fused silica capillary (50 cm × 0.75 µm × 375 µm); electrokinetic injection; running buffer composed of 25 mM $NH_4H_2PO_4$, 6.25 mM $Na_2B_4O_7$ and 0.6 mM CTAB at pH 7.80; 8 min run; microconcentric nebulizer (100–400 µL/min); DRC gas 2.0 mL/min H_2	DLs: 0.10–0.20 ng Cr/mL for three species Cr(III) picolinate recoveries based on the product label: 93–103% CrPic was the only species detected in yeast tablets	[142]

(Continued)

Table 4.3: (Continued)

Food matrix	Species	Sample treatment	Analytical speciation procedure	Additional comments	References
Bread, tea	Cr(III), Cr(VI)	Tea infusions prepared with boiling water doubly spiked with 10 μg/L ^{50}Cr(VI) and 10 μg/L ^{53}Cr(III), left to cool down, filtered and immediately analyzed; in parallel, alkaline extraction with Na_2CO_3 also performed Spiked bread samples extracted with 10 mM NaOH with agitation (17 h); the supernatants collected and filtered	HPLC-ICP-MS: strong anion-exchange Mono Q HR 5/5 column (50 × 5 mm, 10 μm); linear gradient elution from 100% water to 100% 0.7 M NaCl; flow rate 1.5 mL/min; chromatographic run 10 min. Ions monitored for ICP-MS detection: *m*/*z* 50, 52 , 53 Species interconversion examined by isotope ratios measurements. Polyatomic interferences controlled by high-energy collision mode CRC	Cr(VI) was not detected in any of the samples analyzed; oxidation of Cr(III) in the tea and bread samples was discarded. Almost all added ^{50}Cr(VI) was reduced in tea infusions to ^{50}Cr(III), due to the presence of antioxidants. Partial reduction of ^{50}Cr(VI) was observed even in highly alkaline bread extracts (pH 12). It was concluded that the antioxidants and organic matter present in food matrices analyzed in this work prevent from Cr(VI) formation	[154]
Food products of plant and animal origin, France	Cr(VI)	Dairy products, cereals, chocolate, beverages, vegetables, fruits, eggs, meat and seafood purchased at the local market. Solid products freeze-dried. Milk whey obtained by centrifugation. Extraction: 500 μL of liquid or 0.5 g of freeze-dried sample + 10 mL NH_4OH at pH 11.5, 1 h in ultrasonic bath. For solid samples, the supernatant obtained by centrifugation and ultrafiltered (cutoff 10 kDa)	SEC-ICP-MS: for molecular mass fractionation in the entire milk, Superdex-200 HR 10/30 column (300 × 10 mm) used; 30 mM Tris-HCl mobile phase at pH 7.5; flow rate 0.7 mL/min; injection volume 100 μL; chromatographic run 45 min; *m*/*z* 52 and 53 monitored HPLC-ICP-MS: food extracts analyzed on ion-exchange CS5A Dionex column (250 × 4 mm); isocratic elution with 0.085 M HNO_3 followed by 5-min cleaning with 0.35 M HNO_3; flow rate 1 mL/min; injection volume 100 μL;	SEC-ICP-MS: spectral interference due to high C load observed HPLC-ICP-MS DLs: 1 μg Cr/L for milk and 10 μg Cr/kg for other products Recoveries obtained after Cr(VI) addition to the foodstuff: 94–99% Cr(VI) not detected in any sample Stability of Cr(VI) studied by standard addition to a semiskimmed milk; recovery significantly decreased with	[155]

		tCr determined after digestion with HNO_3/H_2O_2	chromatographic run 10 min; m/z 52, 53, 13 (^{13}C) monitored	time and temperature of storage	
Cu-rich foods	Cu(I), Cu(II)	Roasted cashews, cocoa powder, sesame seeds, sunflower seeds, lamb liver and lobster flesh analyzed For XANES: snap-frozen samples were placed in polymethyl methacrylate cuvettes, secured using one ply of kapton tape and kept in liquid nitrogen For tCu, MW-assisted digestion with HNO_3/H_2O_2	XANES: spectra were recorded in fluorescence mode. The energy ranges used were as follows: preedge region, 8779–8959 eV (10 eV steps); XANES region, 8959–9029 eV (0.25 eV steps); postedge region, 9029–9167 eV Sample spectra were fitted with those obtained for model compounds (Cu(I) acetate; Cu(II) acetate; Cu(I)-glutathione; Cu(I)-cysteine; and, Cu(II)-histidine) tCu determined by ICP-MS	Based on Cu K-edge XANES spectra and principal component analysis, up to five species of Cu were suggested, with Cu(I) dominating in all samples excluding cocoa For tCu by ICP-MS, the following CRMs were used: 1568 arice flour, 1570a spinach leaves, 1547 peach leaves and 1545 apple leaves (NIST)	[148]

Note: ICP-OES, inductively coupled plasma–optical emission spectrometry; CRM, certified reference materials; MW, microwave; CV-AAS, cold vapor atomic absorption spectroscopy; DORM-3, DOLT-4; NRC, National Research Council; FAO, Food and Agricultural Organization; FI, flow injection; HG, hydride generation; CV, cold vapor; ETAAAS, electrothermal atomization atomic absorption spectrometry; MS, mass spectrometry; NIST, National Institute of Standards and Technology; TORT-2, CPE, cloud point extraction; FAAS, flame atomic absorption spectrometry; AFS, atomic fluorescence spectrometry; DMA(V), dimethylarsinic acid; PTWI, provisional tolerable weekly intake; PTDI, provisional tolerable daily intake; STAT, slotted tube atom trapping; ID, isotope dilution; SPE, solid-phase extraction; NMIJ, National Metrology Institute of Japan; NIT, nitarsone; Rox, roxarsone, HPLC, high-performance liquid chromatography; EI, electron ionization; SIM, selective ion monitoring; UV, ultraviolet; GC-ECD, gas chromatography-electron capture detector; HS-SPME, headspace solid-phase microextraction; MRM, multiple reaction monitoring; TETRA, tetramethylarsonium ion; TMAO, trimethylarsine oxide; DMAS, dimethyl-arsinothioic acid; ESI, electrospray ionization; XANES, X-ray absorption near-edge structure; MMA(V), methylarsonous acid; QTOF, quadrupole with time of flight; CRC, collision/reaction cell; CE, capillary electrophoresis; DBT, dibutyl tin; TBT, tributyl tin; DPhT, diphenyl tin; TPhT, triphenyl tin; LMM, low molecular mass; HFBA, heptafluorobutyric acid; EPA, Environmental Protection Agency; HEPES, 4-(2-hydroxyethyl)-1-piperazineethanesulfonic acid; MOPS, 3-(N-morpholino)propanesulfonic acid; MES, 2-(N-morpholino)ethanesulfonic acid; DL, SEM, scanning electron microscopy; APDC, ammonium pyrrolidinedithiocarbamate; PTFE, Polytetrafluorethylen; NPs, nanoparticles; ACN, Acetonitrile; SDS-PAGE, sodium dodecyl sulfate polyacrylamide gel electrophoresis; LTQ, Linear Trap Quadrupole (mass spectrometer); BCR, Community Bureau of Reference; IRMM, IS, ERM, QL, GBW, TMAH, tetramethylammonium hydroxide; CNAA, cyclic neutron activation analysis; FIA, flow injection analysis; DDTC, diethyldithiocarbamate; CTAB, cetyltrimethyl ammonium bromide; DRC, dynamic reaction cell; SEC, size exclusion chromatography; AF, DAD, diode array detector; PDMS, Polydimethylsiloxane. HLB, hydrophilic-lipophilic-balanced; DLLE, dispersive liquid liquid extraction; TEM, transmission electron microscopy; TFA, trifluoroacetic acid; CIM, DEAE, diethyl-aminoethyl; TBAH, tetrabutylammonium hydroxide; DTCA, 1,2-diaminocyclohexanetetraacetic acid, AF^4, asymmetric flow field-flow fractionation.

Alternatively, two strong cation-exchange guard columns connected in series allowed for the separation of Hg(II), MeHg, EtHg and PhHg with the mobile phase containing L-cysteine or thiourea at pH 2 [151]. For sensitive, element-specific detection, column effluent has been introduced directly to ICP-MS [151–153], or to CV-AFS after online oxidation of MeHg [161, 163].

GC has a long history in mercury speciation; in the original approach, multistage extraction/purification was needed and ECD was used; this system presented several drawbacks due to instability of chloride derivatives and relatively low sensitivity [207]. Recently, dithizone was examined as a complexing agent for MeHg during sample treatment/extraction and afterward, the complex was converted to chloride for ECD detection [175]. On the other hand, alkylation of Hg species and application of atomic and molecular mass spectrometry techniques enabled for improved analytical performance [208, 209]. Specifically, when tetraethylborate ($NaBEt_4$) is used as a derivatizing agent, the reaction is carried in aqueous sample extract, and Hg(II) and MeHg are converted to $HgEt_2$ and MeHgEt, respectively [173]. Other alkylating agents include tetrapropylborate ($NaBPr_4$) and tetraphenylborate ($NaBPh_4$), both of them offer an advantage of higher hydrophobicity of derivatized species as compared to $NaBEt_4$ and also enable to distinguish between Hg(II) and EtHg, if the latter is present in the sample [153, 176]. Typically, the volatile derivatives are collected/preconcentrated by purge-and-trap or SPME techniques for their introduction to GC system [153, 206].

Nonchromatographic procedures for the determination of MeHg in foodstuff are of interest due to their relative simplicity and preconcentration capability. In the recent applications, selective extraction of MeHg or Hg(II) has been proposed prior to CV generation and AAS or AFS detection [138, 139, 202].

In addition to the seafood, a concern has arisen in relation to possible contamination of rice with MeHg [206, 210]. In a recent study, the extraction was initially carried out with potassium hydroxide-methanol, followed by addition of concentrated HCl and dichloromethane; after solvent evaporation the extract was analyzed by GC-AFS [211]. Unlike seafood, where MeHg is usually the main mercury species, rice contained relatively lower percentages of organic mercury.

Finally, due to the often reported species interconversion, a special care is needed to avoid analytical errors. In many studies, recoveries of individual species after standard addition to the real sample have been reported [35, 153, 175]; however, the most reliable approach is isotope dilution analysis while using mass spectrometry detection [22, 153].

4.3.2 Arsenic

This element has been extensively studied in food products with more than 50 different compounds detected and identified; however, for the assessment of health risk associated with food consumption, only few species are usually considered

[19, 23, 26, 212]. In particular, arsenite (As(III)), arsenate (As(V)), methylarsonous acid (MMA(V)), dimethylarsinic acid (DMA(V)), arsenobetaine (AsB) and arsenocholine (AsC) are important in the analysis of seafood with nontoxic AsB being the main species in the majority of fish types. In plant-derived foods, As(III) and As(V) are prevailing, yet DMA(V) and MMA(V) have also been determined. Noteworthy, the determination of iAs (iAs = As(III) + As(V)) has been adopted as indicator of a health risk associated with food consumption, because the two oxidation states are readily interconverted during gastrointestinal digestion and also in a course of analytical procedure [26, 133, 213] Additionally, As-based feed additives and their metabolites have been included while analyzing animal meat (specifically chicken and porcine) [120, 122, 123, 214].

Depending on the food type and species of interest, different pretreatment procedures have been used [19, 23, 26, 160, 204, 212, 215, 216]; preanalytical operations such as grounding, drying, freeze-drying, homogenization and storage have also been explored as possible factors affecting native distribution of arsenic species [26, 133, 217, 218]. For fresh and processed seafood, water–methanol extraction has been used in the recent studies [117–119, 219] and extractions with diluted nitric acid or with hot water can be found in earlier studies [121, 220–222]. For detection/identification of As-containing hydrocarbons in fish oils, hexane–water–methanol mixture was used [119], whereas for animal meat, malonic acid with H_2O_2 at pH 9.5, methanol–water or water extraction have lately been reported [120–122]. For chicken liver, extraction with simulated gastric juice was applied [123, 223]. Rice and wheat are difficult chemical matrices and both incomplete extraction and species interconversion have been observed [26, 133]. As reviewed in Table 4.3, a consensus seems to exist that microwave-assisted extraction with diluted nitric acid (hydrogen peroxide added for iAs determination) is an appropriate procedure for rice and wheat products [136, 137, 150, 194–196, 201, 217, 224–226].

As presented in numerous reviews and also depicted in Table 4.3, ion-exchange or ion-pair reversed-phase LC with ICP-MS detection are the primary tools for As speciation in food [19, 26, 227]. In particular, anion-exchange separations of As(III), As(V), MMA(V), DMA(V) and AsB have been reported in the range of pH 4.5–9.5, according to pK_{a1} values of these species (9.3, 2.3, 2.6, 6.2, 2.2, respectively) [228]. In recent studies, a clear preference can be noted for using polymeric anion-exchange PRP-X100 column and mobile phases containing carbonate, phosphate, citrate or malonate salts at millimolar concentrations (pH 4.5–9.5) [117, 119, 122, 123, 127, 150, 219, 224, 226]. For reversed-phase columns, recently reported mobile phases were composed of 1-butanesulfonate, malonic acid, TMAH, MeOH at pH 3 [160], formic acids and MeOH at pH 2.55 [31] or contained quaternary amines at pH > 8 [201, 229]. In addition to the four species mentioned earlier, reversed-phase separation of thio-organic As species in marine organisms was reported with phosphate-based mobile phase at pH 3 [118]. Due to the considerable variety of organic As species present in marine organisms, trimethylarsine oxide (TMAO), tetramethylarsonium ion (TETRA), As-sugars, thio-methylated species and As-containing hydrocarbons have been included in speciation analysis [117–119, 167]. Since separation of all these species is not possible in a single chromatographic run, a combination of anion- and cation-exchange columns was reported in the cited studies. Typically, cation-exchange separations have been used for AsB, TETRA, TMAO with pyridine-containing mobile phases at relatively low pH (pH < 4) [19, 26, 117, 227]. As an alternative to LC, CE has recently been used for separation of nine As species on chicken meat, herbal tea and seafood CRMs [121]. Even though ICP-MS detection has been used in many studies, it should be emphasized that monoisotopic arsenic suffers from $^{40}Ar^{35}Cl^{+}$ interference that needs to be eliminated. For quadrupole instruments, CRCs have been pressurized either with He as neutral gas or with reactive gases (O_2, CH_4); otherwise, chromatographic separation of chloride was obtained [26, 117, 218, 224, 229, 230]. Furthermore, postcolumn HG has been reported not only for ICP-MS but also for AFS and AAS detection [120, 223, 231]; however, only As(III), As(V), MMA(V) and DMA(V) can be converted to volatile arsines when reduced with $NaBH_4$ in acidic media whereas for other organic species such as AsB, AsC, TMAO or TETRA, their oxidation is required before HG. As to other detection systems, recent applications of molecular mass spectrometry for confirmation/identification of less common As species are included in Table 4.3 [117, 123, 167].

GC has rarely been used in As speciation in foodstuff; as shown in Table 4.3 the recent application was for the analysis of As-containing hydrocarbons in commercial fish oils [167].

As mentioned before, a special interest exists in the determination of iAs and several nonchromatographic procedures have been proposed for this purpose. In particular, after oxidative extraction and preconcentration, arsenate was reduced with KI/ascorbic acid and then analyzed by HG with AFS, AAS or ICP-MS detection [136, 137, 194–196, 201].

Finally, applications of X-ray absorption near-edge structure (XANES) for direct As speciation in marine periwinkle and rice is noteworthy; however, as stated by the authors, the weak points of this technique were high spectra noise, low sensitivity and possible overestimation of reduced As due to “in-beam” As photoreduction [119, 224].

4.3.3 Selenium

Dietary sources of selenium include seafood, animal liver and kidney, brazil nuts, milk, eggs and grains; however, the daily recommended intake of this element (40–55 µg/day) is not always assured, and supplementation has been a common practice [232, 233]. For this purpose, biofortification of yeast, vegetables or mushrooms has been extensively studied and the obtained products are the main targets of speciation analysis. A variety of selenium compounds have been found in food products; among them inorganic oxoanions selenite Se(IV) and selenate Se(VI), low molecular organic species such as selenomethionine (SeMet), selenocysteine (SeCys), selenocystine

($SeCys_2$), Se-methylselenocysteine (MeSeCys), Se-methylselenomethionine (MeSeMet), γ-glutamyl-Se-methylselenocysteine (γ-Glu-MeSeCys), and high molecular mass compounds such as Se-containing proteins, selenoenzymes or selenosugars [28, 29, 114, 204]. According to their abundance in food and importance to a good health, inorganic Se, SeMet, SeCys/$SeCys_2$ and MeSeCys are the most commonly determined compounds and the related recent studies carried out in edible oils, seafood and enriched food are presented in Table 4.3 [30, 124–132, 140, 162, 200, 234].

Since the important part of organic Se is protein-bound, enzymatic or acid hydrolysis is often required, whereas for nonprotein-bound species water extraction has usually been used. In recent studies on Se speciation in seafood, KOH/MeOH or enzymatic treatment with protease XIV has been reported [162, 200, 234]. For enriched rice, yeast, wheat flour, sprouts and vegetables, pretreatment procedures consisted of water, water/methanol, methanesulfonic acid, protease XIV treatment or simulated gastrointestinal digestion [30, 126, 127, 130–132]. In explorative studies, centered at Se biotransformation by lactic bacteria during elaboration of selenized yogurt [124] or selenized edible mushrooms [129], proteins were extracted, fractionated, enzymatically digested and the obtained extracts submitted to speciation analysis. More detailed description of sample pretreatment for Se speciation analysis in food-related products can be found in several reviews and book chapters [23, 28, 29, 114, 116, 204, 215].

The most explored hyphenated technique in Se speciation has been LC with ICP-MS detection. Specifically, Se(IV), Se(VI), SeMet, $SeCys_2$, MeSeCys have been separated on anion-exchange columns with mobile phases at pH 5–7 [126, 127, 162, 200, 234]. Furthermore, different perfluorinated and alkylsulfonic acids have been extensively used as ion-pair reagents for separation of Se compounds on reversed-phase columns; among them, heptafluorobutyric acid, 1-butanesulfonic acid and 1-pentanesulfonic acid have been recently applied [124, 127–129, 131]. As to ICP-MS detection, ionization efficiency of Se in Ar plasma is relatively low (about 30%), the most abundant ^{80}Se isotope suffers from polyatomic interference caused by argon dimer ($^{40}Ar^{40}Ar^{+}$) and mobile phase composition may have deteriorating influence on the detection sensitivity. Due to these limitations, the use of CRC pressurized with hydrogen is strongly recommended as well as the addition of methane gas for increased sensitivity and careful selection of chromatographic conditions giving preference to isocratic elution with diluted mobile phases and low concentration of organic modifier [29, 113, 116, 235]. In some applications, postcolumn HG and AFS detection has been used; however, all Se compounds eluting from the column have to be converted to Se(VI) and then reduced to Se(IV) to assure efficient H_2Se generation [162, 164, 236]. Similarly as for other elements, confirmation/identification and also quantification of Se species can be achieved by mass spectrometry detection coupled with different separation schemes [124, 129, 132].

Nonchromatographic procedures should also be mentioned. Interesting fractionation scheme based on SPE and APDC complexation with off-line ETAAAS was described for the determination of Se(IV), Se(VI) and tSe in edible oils and in canned fish [140]. ETAAS was also used after cloud point extraction of inorganic Se from biofortified rice [130]. Asymmetric flow field fractionation was applied for size fractionation of proteins from Se-enriched yogurt; the results were compared with those obtained by size-exclusion chromatography, both with ICP-MS detection [124].

4.3.4 Other elements

Due to excessive use of tin-containing biocides in the past, speciation analysis had often been carried out in environmental samples and in marine or freshwater organisms. Analytical procedures focusing on inorganic tin and its alkylated compounds are described in the comprehensive reviews [170, 237, 238]. In the recent literature, butyl- and phenyltin compounds have been determined in seafood after solvent extraction, derivatization with Grignard reagent of $NaBEt_4$ and followed by GC-MS [172, 174]. Another aspect of tin speciation in food is related to its leaching from packing materials. Whereas less harmful inorganic forms can be expected in canned food [239], polymeric wraps contain organic highly toxic forms as plasticizers, and in a recent study, diphenyl tin was detected and quantified in the leachates of two cake packing materials [147].

Determination of Cr(VI)/Cr(III) in environmental samples and in drinking water has received much attention due to the well-known toxicity of hexavalent form [240]. There have also been reports on chromium speciation in food products [241, 242]; however, a consensus seems to be reached now that natural content of antioxidants is sufficient to prevent Cr(VI) formation in food [154, 155]. On the other part, in the frame of quality control of the dietary supplements, CE-ICP-MS procedure has been proposed enabling separation of Cr(III), Cr(VI) and CrPic in 8 min. The results obtained in the analysis of commercial yeast supplements indicated that CrPic was the only species of chromium in these products [142].

Metallic nanoparticles constitute relatively new addition to the metal-based species of interest in food analysis. Several types of nanomaterials are used today as food additives, based on their antimicrobial activity, in order to improve food texture and taste and also to increase the shelf-life (elemental forms or oxides of Ti, Si, Zn, Cu, Ag, Au, among others) [243]. Fresh foods are often exposed to colloidal silver nanostructured antimicrobial salts, oxides or elemental colloids (Ag) used for disinfection. Migration from packing materials is another important source on nanomaterials containing Ti, Si, Cu, Zn, Sn, Ag that are added as plasticizers or preservation agents [244]. Environmental contamination with nanodimensional materials is another source of element forms/species in food [245]. Analytical data of potential interest for the assessment of bioaccessibility/bioavailability of nanoparticles and their possible transformations in food or during food digestion can be obtained by determination of total element content derived from nanoparticles, their size distribution, morphology, solubilization and possible chemical transformations. Typical analytical schemes involve fractionation/separation by centrifugation, size-exclusion chromatography or flow field fractionation with ICP-MS detection. Alternatively to the conventional conditions, ICP-MS can be operated in a single particle mode with the intent to simplify or avoid separation step; direct imaging of nanoparticles in food samples by powerful microscopic techniques has also been reported [246]. As shown in Table 4.3, silver nanoparticles in chicken meat were analyzed for size distribution by asymmetric flow field fractionation with ICP-MS (either in conventional or single particle mode) and compared with the results obtained by transmission electron microscopy [198, 199].

Among other topics in food analysis, speciation of oxidation states of antimony is of interest due to higher toxicity of Sb(III) versus Sb(V); as an example the recent non-chromatographic procedure based on selective SPE of Sb(III)–diethyldithiocarbamate (DDTC) complex is included in Table 4.3 [197]. Direct speciation of copper oxidation states in nuts, seeds, cocoa powder, tissues of marine and sea organisms has been undertaken by XANES using Cu(I)-acetate, Cu(II)-acetate, Cu(I)-glutathione, Cu(I)-cysteine and Cu(II)-histidine as the model compounds and it was concluded that Cu(I) was a predominated oxidation state of copper in the majority of products analyzed [148]. With the aim of improving bioavailability of zinc in infant formulas, speciation analysis has been carried out in defatted low molecular mass fraction of human milk by high-performance liquid chromatography-ICP-MS and by off-line analysis of the collected fractions by ESI-MS [159]. Zn-citrate complex was identified as the main species and the authors suggested citrate addition to the infant formula for improved Zn bioavailability.

4.4 Future perspectives in food analysis and speciation

Metals and other trace elements are naturally present in food, but there are also numerous external sources that may affect original nutritional value or cause undesired health effects related to consumption of contaminated food. Therefore, the determination of total element concentrations is the essential part of food quality control. The state of art in analytical instrumentation and methodology enables for reliable determination of target elements at low concentration levels in compliance with the existing legislation. On the other hand, not all elements and foods are included in the current regulations, which is the driving force for further development in this field. In particular, there is a need of simple and robust procedures, operating in line with the principles of green chemistry, capable of providing reliable results and feasible to be implemented in the routine laboratories. Accordingly, the main effort is centered at the sample pretreatment so the efficient solubilization/extraction of analyte(s) and degradation of chemical matrix can be obtained in a short time with minimum amount of harmless reagents and with an aid of microwave energy or ultrasonication. For the determination of ultra-trace concentrations, ICP-MS is today a technique of choice, but other, less expensive techniques such as ICP-OES, AAS and AFS are also in use after suitable preconcentration step. In this regard, a clear trend can be observed for using miniaturized extraction techniques such as SPE, dispersive liquid–liquid extraction, single drop-liquid extraction or cloud point extraction (Table 4.2).

In parallel to the determination of total elements, the interest in food speciation analysis has been steadily increasing, and today the most toxic compounds (iAs, MeHg, tributyl tin, Cr(VI)) are included in official regulation of food and drinking water. The emerging variety of elemental forms that are used in agricultural practice, in food processing and that are introduced as contaminants by human activity, together with the expanding knowledge of the specific role of element species in human nutrition and toxicology, are the main factors stimulating progress in analytical methodology. For the analysis of known species, hyphenated techniques have gained undisputed priority with ICP-MS being the most powerful detection tool. Other atomic spectrometry techniques are preferred in nonchromatographic speciation schemes or, when postcolumn chemical vapor generation is integrated (AFS). Full characterization of element species is sometimes required in food analysis, especially in the analysis of dietary supplements, highly processed foods and foods elaborated from biofortified raw material. For such purposes, the identification/confirmation of new or unexpected species relies on accurate mass measurements and structural information furnished by molecular mass spectrometry. Similarly as for total element determination, in food speciation an incessant effort can be noted in the development of simple and robust procedures enabling for precise and accurate quantification of element form/species that impose a health risk or provide health benefits.

It can be anticipated that in future, an effort will continue toward reinforcement of reliability in food analysis and speciation. Further methodological development will be centered at establishing simple and robust procedures, appropriate for being implemented in routine food control. Since standardization of the speciation methods is necessary to promote more specific food regulations, the emphasis on quality control schemes and the production of suitable CRMs is to be expected. In particular, today there are relatively few CRMs available for speciation analysis; these include food-related materials with certified values for MeHg and tHg (BCR 463, DOLT-3, DORM-2, International Atomic Energy Agency [IAEA]-350, IAEA-140, ERM-CE), those with certificate for As species and tAs (BCR 627, IMEP-107, ERM-BC21, DORM-2, National Metrology Institute of Japan-7503a) and for SeMet and tSe (SELM-1).

Finally, new directions in the determination/speciation of trace element are dictated by production of functional and transgenic foods and by increasing applications of nanodimensional metal/metalloid forms in food processing, preservation and packing.

Acknowledgments

The financial support from the National Council of Science and Technology, Mexico (CONACYT), project 178553, is gratefully acknowledged.

References

[1] Principles and Methods for the Assessment of Risk from Essential Trace Elements. Environmental Health Criteria 228. World Health Organization, Geneva, Switzerland, 2002.

[2] Dietary Reference Intakes: Vitamin A, Vitamin K, Arsenic, Boron, Chromium, Copper, Iodine, Iron, Manganese, Molybdenum, Nickel, Silicon, Vanadium, and Zinc. Institute of Medicine, National Academy Press, Washington, DC, 2001.

[3] Aggett P, Nordberg GF, Nordberg M. Essential metals: Assessing risks from deficiency and toxicity. In: Nordberg GF, Fowler BA, Nordberg M, eds. Handbook on the Toxicity of Metals. Elsevier, Amsterdam, 2015, 283–96.

[4] Herrman JL, Younes M. Background to the ADI/TDI/PTWI. Reg Toxicol Pharmacol 1999, 30, S109–S113.

[5] Diaz SC, Encinar JR, Sanz-Medel A, Alonso JI. Towards compound-independent calibration for organic compounds using online isotope dilution mass spectrometry. Anal Bioanal Chem 2012, 402, 91–97.

[6] FAO/WHO. Summary of Evaluations Performed by the Joint FAO/WHO Expert Committee of Food Additives (JECFA 1956-2003). ILSI Press International Life Science Institute, ftp://ftp.fao.org/es/esn/jecfa/jecfa61sc.pdr, 2004.

[7] Commission Regulation (EC) N° 1881/2006 of 19 December 2006 setting maximum levels for certain contaminants in foodstuff, http://eur-lex.europa.eu/LexUriServ/LexUriServ.do?uri=OJ:L:2006:364:0005:0024:EN:PDF, 2006.

[8] Drennan-Harris LR, Wongwilawan S, Tyson JF. Trace determination of total mercury in rice by conventional inductively coupled plasma mass spectrometry. J Anal At Spectrom 2013, 28, 259–65.

[9] Australia New Zealand Food Standards Code – Standard 1.4.1 – Contaminants and Natural Toxicants, 2011.

[10] Herrero Latorre C, Barciela Garcia J, Garcia Martin S, Pena Crecente RM. Chemometric classification of potatoes with protected designation of origin according to their producing area and variety. J Agric Food Chem 2013, 61, 8444–51.

[11] Pellerano RG, Uñates MA, Cantarelli MA, Camiña JM, Marchevsky EJ. Analysis of trace elements in multifloral Argentine honeys and their classification according to provenance. Food Chem 2012, 134, 578–82.

[12] Chudzinska M, Baralkiewicz D. Application of ICP-MS method of determination of 15 elements in honey with chemometric approach for the verification of their authenticity. Food Chem Toxicol 2011, 49, 2741–49.

[13] Luykx DMAM, van Ruth SM. An overview of analytical methods for determining the geographical origin of food products. Food Chem 2008, 107, 897–911.

[14] Hondrogiannis EM, Ehrlinger E, Poplaski A, Lisle M. Use of laser ablation-inductively coupled plasma-time of flight-mass spectrometry to identify the elemental composition of vanilla and determine the geographic origin by discriminant function analysis. J Agric Food Chem 2013, 61, 11332–37.

[15] Drivelos SA, Georgiou CA. Multi-element and multi-isotope ratio analysis to determine the geographical origin of foods in the European Union. Trend Anal Chem 2012, 40, 38–51.

[16] Kelly S, Heaton K, Hoogewerff J. Tracing the geographical origin of food: The application of multi-element and multi-isotope analysis. J Trends Food Sci Tech 2005, 16, 555–67.

[17] de la Guardia M, Garrigues S, eds. Handbook of Mineral Elements in Food. Wiley-Blackwell, Chichester, UK, 2015.

[18] Capar C, Szefer P. Determination and speciation of trace elements in foods. In: Otles S, ed. Methods of Analysis of Food Components and Additives. CRC Press, Boca Raton, FL, 2012, 165–210.

[19] Sadee B, Foulkes ME, Hill JS. Coupled techniques for arsenic speciation in food and drinking water: A review. J Anal At Spectrom 2015, 30, 102–18.

[20] Ruzik L. Speciation of challenging elements in food by atomic spectrometry. Talanta 2012, 93, 18–31.

[21] Caroli S, ed. The Determination of Chemical Elements in Food: Applications for Atomic and Mass Spectrometry. Wiley & Sons, Hoboken, NJ, 2007.

[22] Clémens S, Monperrus M, Donard OFX, Amouroux D, Guérin T. Mercury speciation in seafood using isotope dilution analysis: A review. Talanta 2012, 89, 12–20.

[23] Carey AM, Lombi E, Donner E, et al. A review of recent developments in the speciation and location of arsenic and selenium in rice grain. Anal Bioanal Chem 2012, 402, 3275–86.

[24] Krystek P, Rob R. Determination of Hg species in seafood. In: Caroli S, ed. The Determination of Chemical Elements in Food. John Wiley & Sons, Hoboken, NJ, 2007, 707–26.

[25] Nakazato T. Analytical methods for speciation of organic arsenicals. Biomed Res Trace Elem 2007, 18, 63–72.

[26] Tyson J. The determination of arsenic compounds: A critical review. ISRN Anal Chem 2013, Article ID 835371.

[27] Thiry C, Ruttens A, De Temmerman L, Schneider YJ, Pussemier L. Current knowledge in species-related bioavailability of selenium in food. Food Chem 2012, 130, 767–84.

[28] Pyrzynska K. Selenium speciation in enriched vegetables. Food Chem 2009, 114, 1183–91.

[29] Wrobel K, Wrobel K, Caruso JA. Application of ICP-MS for the evaluation of Se species in food related products and dietary supplements. In: Caroli S, ed. The Determination of Chemical Elements in Food. Applications for Atomic and Mass Spectrometry. John Wiley & Sons, Hoboken, NJ, 2007, 671–706.

[30] Funes-Collado V, Rubio R, López-Sánchez JF. Does boiling affect the bioaccessibility of selenium from cabbage? Food Chem 2015, 181, 304–9.

[31] Baba K, Arao T, Yamaguchi N, Watanabe E, Eun H, Ishizaka M. Chromatographic separation of arsenic species with pentafluorophenyl column and application to rice. J Chromatogr A 2014, 1354, 109–16.

[32] Mir-Marqués A, González-Masó A, Cervera ML, de la Guardia M. Mineral profile of Spanish commercial baby food. Food Chem 2015, 172, 238–44.

[33] Moreno-Rojas R, Sánchez-Segarra PJ, Cámara-Martos F, Amaro-López MA. Multivariate analysis techniques as tools for categorization of Southern Spanish cheeses: Nutritional composition and mineral content. Eur Food Res Technol 2010, 231, 841–51.

[34] Pandelova M, Levy Lopez W, Michalke B, Schramm KW. Ca, Cd, Cu, Fe, Hg, Mn, Ni, Pb, Se, and Zn contents in baby foods from the EU market: Comparison of assessed infant intakes with the present safety limits for minerals and trace elements. J Food Comp Anal 2012, 27, 120–27.

[35] Schmidt L, Bizzi CA, Duarte FA, Muller EI, Krupp E, Feldmann J, Flores EMM. Evaluation of Hg species after culinary treatments of fish. Food Control 2015, 47, 413–19.

[36] Maldonado Santoyo M, Landero Figueroa JA, Wrobel K, Wrobel K. Analytical speciation of mercury in fish tissues by reversed phase liquid chromatography – inductively coupled plasma mass spectrometry with Bi^{3+} as internal standard. Talanta 2009, 79, 706–11.

[37] Elemental Analysis Manual (EAM) for Food and Related Products. US Food and Drug Administration. http://www.fda.gov/Food/FoodScienceResearch/LaboratoryMethods/ucm2006954.htm., 2014.

[38] Murphy CJ, MacNeil JD, Capar SG. Best practices for single-laboratory validation of chemical methods for trace elements in foods. Part I – Background and General Considerations. J AOAC Int 2013, 96, 190–203.

[39] Mir-Marqués A, Garrigues S, Cervera ML, de la Guardia M. Direct determination of minerals in human diets by infrared spectroscopy and X-ray fluorescence. Microchem J 2014, 117, 156–63.

[40] Santos J, Oliva-Teles MT, Delerue-Matos C, Oliveira MBPP. Multi-elemental analysis of ready-to-eat "baby leaf" vegetables using microwave digestion and high-resolution continuum source atomic absorption spectrometry. Food Chem 2014, 151, 311–16.

[41] Ljung K, Palm B, Grandér M, Vahter M. High concentrations of essential and toxic elements in infant formula and infant foods – A matter of concern. Food Chem 2011, 127, 943–51.

[42] Branco Corguinha AP, de Souza GA, Cardoso Goncalves V, et al. Assessing arsenic, cadmium, and lead contents in major crops in Brazil for food safety purposes. J Food Comp Anal 2015, 37, 143–50.

[43] Juranovic Cindric I, Krizman I, Zeiner M, Kampic S, Medunic G, Stingeder G. ICP-AES determination of minor- and major elements in apples after microwave assisted digestion. Food Chem 2012, 135, 2675–80.

[44] Llorent-Martínez EJ, Fernández de Córdova ML, Ruiz-Medina A, Ortega-Barrales P. Analysis of 20 trace and minor elements in soy and dairy yogurts by ICP-MS. Microchem J 2012, 102, 23–7.

[45] Drewnowska M, Falandysz J. Investigation on mineral composition and accumulation by popular edible mushroom common chanterelle (*Cantharellus cibarius*). Ecotox Environ Safe 2015, 113, 9–17.

[46] Dryzaowska A, Falandysz J. Bioconcentration of mercury by mushroom *Xerocomus chrysenteron* from the spatially distinct locations: Levels, possible intake and safety. Ecotox Environ Safe 2014, 107, 97–102.

[47] Falandysz J. Distribution of mercury in Gypsy *Cortinarius caperatus* mushrooms from several populations: An efficient accumulator species and estimated intake of element. Ecotox Environ Safe 2014, 110, 68–72.

[48] Chevallier E, Chekri R, Zinck J, Guerin T, Noel L. Simultaneous determination of 31 elements in foodstuffs by ICP-MS after closed-vessel microwave digestion: Method validation based on the accuracy profile. J Food Comp Anal 2015, 41, 35–41.

[49] Noël L, Chekri R, Millour S, et al. Li, Cr, Mn, Co, Ni, Cu, Zn, Se and Mo levels in foodstuffs from the Second French TDS. Food Chem 2012, 132, 1502–13.

[50] Millour S, Noël L, Chekri R, et al. Strontium, silver, tin, iron, tellurium, gallium, germanium, barium and vanadium levels in foodstuffs from the Second French Total Diet Study. J Food Comp Anal 2012, 25, 108–29.

[51] Pereira CD, Aguirre MA, Nóbrega JA, Hidalgo M, Canals A. Aerosol generation of As and Se hydrides using a new Flow Blurring®multiple nebulizer for sample introduction in inductively coupled plasma optical emission spectrometry. Microchem J 2014, 112, 82–6.

[52] Hernández-Martínez R, Navarro-Blasco I. Survey of total mercury and arsenic content in infant cereals marketed in Spain and estimated dietary intake. Food Control 2013, 30, 423–32.

[53] Sadiq N, Beauchemin D. Optimization of the operating conditions of solid sampling electrothermal vaporization coupled to inductively coupled plasma optical emission spectrometry for the sensitive direct analysis of powdered rice. Anal Chim Acta 2014, 851, 23–9.

[54] Moreira DCF, de Sá JSM, Cerqueira IB, Oliveira APF, Morgano MA, Quintaes KD. Evaluation of iron, zinc, copper, manganese and selenium in oral hospital diets. Clin Nutr 2014, 33, 808–14.

[55] Orecchio S, Amorello D, Raso M, Barreca S, Lino C, Di Gaudio F. Determination of trace elements in gluten-free food for celiac people by ICP-MS. Microchem J 2014, 116, 163–72.

[56] Zhu Y, Chiba K. Determination of cadmium in food samples by ID-ICP-MS with solid phase extraction for eliminating spectral-interferences. Talanta 2012, 90, 57–62.

[57] Ostos C, Perez-Rodriguez F, Moreno Arroyo B, Moreno-Rojas R. Study of mercury content in wild edible mushrooms and its contribution to the Provisional Tolerable Weekly Intake in Spain. J Food Comp Anal 2015, 37, 136–42.

[58] Sarikurkcu C, Tepe B, Kocak MS, Uren MC. Metal concentration and antioxidant activity of edible mushrooms from Turkey. Food Chem 2015, 175, 549–55.

[59] Spanopoulos-Zarco P, Ruelas-Inzunza J, Meza-Montenegro M, Osuna-Sanchez K, Amezcua-Martınez F. Health risk assessment from mercury levels in bycatch fish species from the Coasts of Guerrero, Mexico (Eastern Pacific). Bull Environ Contam Toxicol 2014, 93, 334–8.
[60] Khan N, Jeong IS, Hwang IM, et al. Analysis of minor and trace elements in milk and yogurts by inductively coupled plasma-mass spectrometry (ICP-MS). Food Chem 2014, 147, 220–4.
[61] Rowell C, Kuiper N, Al-Saad K, Nriagu J, Shomar B. A market basket survey of As, Zn and Se in rice imports in Qatar: Health implications. Food Chem Toxicol 2014, 70, 33–9.
[62] Welna M, Szymczycha-Madeja A. Evaluation of optimal conditions for determination of low tin content in fresh and canned tomato samples using hydride generation Inductively Coupled Plasma Optical Emission Spectrometry. Food Anal Methods 2014, 7, 127–34.
[63] Feist B, Milkula B. Preconcentration of heavy metals on activated carbon and their determination in fruits by inductively coupled plasma optical emission spectrometry. Food Chem 2014, 147, 302–6.
[64] Vincevica-Gaile Z, Klavins M, Rudovica V, Viksna A. Research review trends of food analysis in Latvia: Major and trace element content. Environ Geochem Health 2013, 35, 693–703.
[65] Jiang J, Lu S, Zhang H, et al. Dietary intake of human essential elements from a Total Diet Study in Shenzhen, Guangdong Province, China. J Food Comp Anal 2015, 39, 1–7.
[66] Bressy FC, Brito GB, Barbosa IS, Teixeira LSG, Korn mGA. Determination of trace element concentrations in tomato samples at different stages of maturation by ICP OES and ICP-MS following microwave-assisted digestion. Microchem J 2013, 109, 145–9.
[67] Yenisoy-Karakas S. Estimation of uncertainties of the method to determine the concentrations of Cd, Cu, Fe, Pb, Sn and Zn in tomato paste samples analysed by high resolution ICP-MS. Food Chem 2012, 132, 1555–61.
[68] Raptopoulou KG, Pasias IN, Thomaidis NS, Proestos C. Study of the migration phenomena of specific metals in canned tomato paste before and after opening. Validation of a new quality indicator for opened cans. Food Chem Toxicol 2014, 69, 25–31.
[69] Feist B, Mikula B. Preconcentration of some metal ions with lanthanum-8-hydroxyquinoline co-precipitation system. Food Chem 2014, 147, 225–9.
[70] Copat C, Vinceti M, D'Agati MG, et al. Mercury and selenium intake by seafood from the Ionian Sea: A risk evaluation. Ecotox Environ Safe 2014, 100, 87–92.
[71] Ortea I, Gallardo JM. Investigation of production method, geographical origin and species authentication in commercially relevant shrimps using stable isotope ratio and/or multi-element analyses combined with chemometrics: An exploratory analysis. Food Chem 2015, 170, 145–53.
[72] Jovic M, Stankovic S. Human exposure to trace metals and possible public health risks via consumption of mussels Mytilus galloprovincialis from the Adriatic coastal area. Food Chem Toxicol 2014, 70, 241–51.
[73] Julshamn K, Valdersnes S, Duinker A, Nedreaas K, Sundet JH, Maage A. Heavy metals and POPs in red king crab from the Barents Sea. Food Chem 2015, 167, 409–17.
[74] Ikem A, Shanks B, Caldwell J, Garth J, Ahuja S. Estimating the daily intake of essential and nonessential elements from lamb m. longissimus thoracis et lumborum consumed by the population in Missouri (United States). J Food Comp Anal 2015, 40, 126–35.
[75] Pilarczyk R. Concentrations of toxic and nutritional essential elements in meat from different beef breeds reared under intensive production systems. Biol Trace Elem Res 2014, 158, 36–44.
[76] Bortey-Sam N, Nakayama SMM, Ikenaka Y, et al. Human health risks from metals and metalloid via consumption of food animals near gold mines in Tarkwa, Ghana: Estimation of the daily intakes and target hazard quotients (THQs). Ecotox Environ Safe 2015, 111, 160–7.
[77] Buculei A, Amariei S, Oroian M, Gutt G, Gaceu L, Birca A. Metals migration between product and metallic package in canned meat. LWT – Food Sci Technol 2014, 58, 364–74.

[78] Yücel Y, Sultanoglu P. Characterization of Hatay honeys according to their multi-element analysis using ICP-OES combined with chemometrics. Food Chem 2013, 140, 231–7.

[79] Bilandzic N, Gacic M, Dokic M, Sedak M, Ivanec Sipusic D, Koncurat A, Tlak Gajger I. Major and trace elements levels in multifloral and unifloral honeys in Croatia. J Food Comp Anal 2014, 33, 132–8.

[80] Czipa N, Andrási D, Kovács B. Determination of essential and toxic elements in Hungarian honeys. Food Chem 2015, 175, 536–42.

[81] de Alda-Garcilope C, Gallego-Picó A, Bravo-Yagüe JC, Garcinuño-Martínez RM, Fernández-Hernando F. Characterization of Spanish honeys with protected designation of origin "Miel de Granada" according to their mineral content. Food Chem 2012, 135, 1785–8.

[82] Zaza S, Balogh K, Palmery M, Pastorelli A, Stacchini P. Human exposure in Italy to lead, cadmium and mercury through fish and seafood product consumption from Eastern Central Atlantic Fishing Area. J Food Comp Anal 2015, 40, 148–53.

[83] Rofouei MK, Rezaei A, Masteri-Farahani M, Khani H. Selective extraction and preconcentration of ultra-trace level of mercury ions in water and fish samples using Fe_3O_4-magnetite-nanoparticles functionalized by triazene compound prior to its determination by inductively coupled plasma-optical emission spectrometry. Anal Methods 2012, 4, 959–66.

[84] Er C, Senkal BF, Yaman M. Determination of lead in milk and yoghurt samples by solid phase extraction using a novel aminothioazole-polymeric resin. Food Chem 2013, 137, 55–61.

[85] Rehman S, Husnain SM. Assessment of trace metal contents in chocolate samples by Atomic Absorption Spectrometry. J Trace Elem Anal 2012, 1, 1–11.

[86] Ulusoy HI, Yılmaz O, Gürkan R. A micellar improved method for trace levels selenium quantification in food samples, alcoholic and nonalcoholic beverages through CPE/FAAS. Food Chem 2013, 139, 1008–14.

[87] Olmedo P, Pla A, Hernández AF, Barbier F, Ayouni L, Gil F. Determination of toxic elements (mercury, cadmium, lead, tin and arsenic) in fish and shellfish samples. Risk assessment for the consumers. Environ Int 2013, 59, 63–72.

[88] Astorga-España MS, Rodrıguez Galdon B, Rodrıguez Rodrıguez EM, Díaz Romero C. Mineral and trace element concentrations in seaweeds from the sub-Antarctic ecoregion of Magallanes (Chile). J Food Comp Anal 2015, 39, 69–76.

[89] Akinyele IO, Shokunbi OS. Concentrations of Mn, Fe, Cu, Zn, Cr, Cd, Pb, Ni in selected Nigerian tubers, legumes and cereals and estimates of the adult daily intakes. Food Chem 2015, 173, 702–8.

[90] Akinyele IO, Shokunbi OS. Comparative analysis of dry ashing and wet digestion methods for the determination of trace and heavy metals in food samples. Food Chem 2015, 173, 682–4.

[91] Bakircioglu D, Bakircioglu Kurtulus Y, Ucar G. Determination of some traces metal levels in cheese samples packaged in plastic and tin containers by ICP-OES after dry, wet and microwave digestion. Food Chem Toxicol 2011, 49, 202–7.

[92] Luis G, Rubio C, Revert C, Espinosa A, Gonzalez-Weller D, Gutierrez AJ, Hardisson A. Dietary intake of metals from yogurts analyzed by inductively coupled plasma optical emission spectrometry (ICP-OES). J Food Comp Anal 2015, 39, 48–54.

[93] Cherfi A, Abdoun S, Gaci O. Food survey: Levels and potential health risks of chromium, lead, zinc and copper content in fruits and vegetables consumed in Algeria. Food Chem Toxicol 2014, 70, 48–53.

[94] Fernandez-Martınez R, Rucandio I, Gomez-Pinilla I, Borlaf F, Garcıa F, Larrea MT. Evaluation of different digestion systems for determination of trace mercury in seaweeds by cold vapour atomic fluorescence spectrometry. J Food Comp Anal 2015, 38, 7–12.

[95] Tang J, Ying Y, Pan XD, Jiang W, Wu PG. Elements analysis of infant milk formula by ICP-OES: A comparison of pretreatment methods. Accred Qual Assur 2014, 19, 99–103.

[96] Hsu WH, Jiang SJ, Sahayam AC. Determination of Cu, As, Hg and Pb in vegetable oils by electrothermal vaporization inductively coupled plasma mass spectrometry with palladium nanoparticles as modifier. Talanta 2013, 117, 268–72.

[97] Zhuravlev Z, Zacharia A, Gucer S, Chebotarev A, Arabadji M, Dobrynin A. Direct atomic absorption spectrometry determination of arsenic, cadmium, copper, manganese, lead and zinc in vegetable oil and fat samples with graphite filter furnace atomizer. J Food Comp Anal 2015, 38, 62–8.

[98] Dobrowolski R, Pawlowska-Kapusta I, Dobrzynska J. Chromium determination in food by slurry sampling graphite furnace atomic absorption spectrometry using classical and permanent modifiers. Food Chem 2012, 132, 597–602.

[99] Kulek de Andrade C, Egéa dos Anjos V, Felsner ML, Reyes Torres Y, Pércio Quináia S. Direct determination of Cd, Pb and Cr in honey by slurry sampling electrothermal atomic absorption spectrometry. Food Chem 2014, 146, 166–73.

[100] Batista BL, da Silva LRS, Rocha BA, et al. Multi-element determination in Brazilian honey samples by inductively coupled plasma mass spectrometry and estimation of geographic origin with data mining techniques. Food Res Int 2012, 49, 209–15.

[101] Pereira Vieira H, Nascentes CC, Carvalhinho Windmoller C. Development and comparison of two analytical methods to quantify the mercury content in honey. J Food Comp Anal 2014, 34, 1–6.

[102] Wang R, Farid MM. Corrosion and health aspects in ohmic cooking of beef meat patties. J Food Eng 2015, 146, 17–22.

[103] Rao Koyyalamudi S, Jeong SC, Manavalan S, Vysetti B, Pang G. Micronutrient mineral content of the fruiting bodies of Australian cultivated Agaricus bisporus white button mushrooms. J Food Comp Anal 2013, 31, 109–14.

[104] Leufroy A, Noël L, Bouisset P, et al. Determination of total iodine in French Polynesian foods: Method validation and occurrence data. Food Chem 2015, 169, 134–40.

[105] Perugini M, Visciano P, Manera M, Zaccaroni A, Olivieri V, Amorena M. Heavy metal (As, Cd, Hg, Pb, Cu, Zn, Se) concentrations in muscle and bone of four commercial fish caught in the central Adriatic Sea, Italy. Environ Monit Assess 2014, 186, 2205–13.

[106] Pereira JSF, Pereira LSF, Schmidt L, Moreira CM, Barin JS, Flores EMM. Metals determination in milk powder samples for adult and infant nutrition after focused-microwave induced combustion. Microchem J 2013, 109, 29–35.

[107] Freschi GPG, Fortunato FM, Freschi CD, Gomes Neto JA. Simultaneous and direct determination of As, Bi, Pb, Sb, and Se and Co, Cr, Cu, Fe, and Mn in milk by electrothermal atomic absorption spectrometry. Food Anal Methods 2012, 5, 861–6.

[108] Molska A, Gutowska I, Baranowska-Bosiacka I, Noceń I, Chlubek D. The content of elements in infant formulas and drinks against mineral requirements of children. Biol Trace Elem Res 2014, 158, 422–7.

[109] Michalke B. Element speciation definitions, analytical methodology, and some examples. Ecotoxicol Environ Safe 2003, 56, 122–39.

[110] Caruso JA, Klaue B, Michalke B, Rocke DM. Group assessment: Element speciation. Ecotoxicol Environ Safe 2003, 56, 32–44.

[111] Nagy G, Benko I, Kiraly G, et al. Cellular and nephrotoxicity of selenium species. J Trace Elem Med Biol 2015, 30, 160–70.

[112] Wiernsperger N, Rapin J. Trace elements in glucometabolic disorders: An update. Diabetol Metab Syndr 2011, 2, 70.

[113] Wrobel K, Wrobel K, Caruso JA. Speciation small molecules. In: Beauchemin D, Matthews D, eds. Elemental and Isotope Ratio Mass Spectrometry Vol. 5. Elsevier, Oxford, UK, 2010, 222–33.

[114] Dumont E, Vanhaecke F, Cornelis R. Selenium speciation from food source to metabolites: A critical review. Anal Bioanal Chem 2006, 385, 1304–22.

[115] Rayman MP, Infante HG, Sargent M. Food-chain selenium and human health: Spotlight on speciation. Br J Nutr 2008, 100, 238–53.

[116] Pedrero Z, Madrid Y. Novel approaches for selenium speciation in foodstuffs and biological specimens: A review. Anal Chim Acta 2009, 634, 135–52.

[117] Contreras-Acuña M, García-Barrera T, García-Sevillano MA, Gómez-Ariza JL. Speciation of arsenic in marine food (*Anemonia sulcata*) by liquid chromatography coupled to inductively coupled plasma mass spectrometry and organic mass spectrometry. J Chromatogr A 2013, 1282, 133–41.

[118] Maher WA, Foster S, Krikowa F, Duncan E, St John A, Hug K, Moreau JW. Thio arsenic species measurements in marine organisms and geothermal waters. Microchem J 2013, 111, 82–90.

[119] Whaley-Martin KJ, Koch I, Reimer KJ. Determination of arsenic species in edible periwinkles (*Littorina littorea*) by HPLC–ICPMS and XAS along a contamination gradient. Sci Tot Environ 2013, 456–457, 148–53.

[120] Cui J, Xiao Y, Dai L, Zhao X, Wang Y. Speciation of organoarsenic species in food of animal origin using accelerated solvent extraction (ASE) with determination by HPLC-Hydride Generation-Atomic Fluorescence Spectrometry (HG-AFS). Food Anal Methods 2013, 6, 370–9.

[121] Liu L, He B, Yun Z, Sun J, Jiang G. Speciation analysis of arsenic compounds by capillary electrophoresis on-line coupled with inductively coupled plasma mass spectrometry using a novel interface. J Chromatogr A 2013, 1304, 227–33.

[122] Nachman KE, Baron PA, Raber G, Francesconi KA, Navas-Acien A, Love DC. Roxarsone, Inorganic Arsenic, and Other Arsenic Species in Chicken: A U.S.-Based Market Basket Sample. Environ Health Perspect 2013, DOI:10.1289/ehp.1206245.

[123] Peng H, Hua B, Liu Q, et al. Liquid chromatography combined with atomic and molecular mass spectrometry for speciation of arsenic in chicken liver. J Chromatogr A 2014, 1370, 40–9.

[124] Palomo M, Gutiérrez AM, Pérez-Conde MC, Cámara C, Madrid Y. Se metallomics during lactic fermentation of Se-enriched yogurt. Food Chem 2014, 164, 371–9.

[125] Bhatia P, Aureli F, D'Amato M, Prakash R, Cameotra SS, Nagaraja TP, Cubadda F. Selenium bioaccessibility and speciation in biofortified Pleurotus mushrooms grown on selenium-rich agricultural residues. Food Chem 2013, 140, 225–30.

[126] Funes-Collado V, Morell-Garcia A, Rubio R, López-Sánchez JF. Study of selenocompounds from selenium-enriched culture of edible sprouts. Food Chem 2013, 141, 3738–43.

[127] Galinha C, Sánchez-Martínez M, Pacheco AMG, et al. Characterization of selenium-enriched wheat by agronomic biofortification. J Food Sci Technol 2014, DOI 10.1007/s13197-014-1503-7.

[128] Hsieh YJ, Jiang SJ. Determination of selenium compounds in food supplements using reversed-phase liquid chromatography–inductively coupled plasma mass spectrometry. Microchem J 2013, 110, 1–7.

[129] Maseko T, Callahan DL, Dunshea FR, Doronila A, Kolev SD, Ng K. Chemical characterisation and speciation of organic selenium in cultivated selenium-enriched *Agaricus bisporus*. Food Chem 2013, 141, 3681–7.

[130] Sun M, Liu G, Wu Q. Speciation of organic and inorganic selenium in selenium-enriched rice by graphite furnace atomic absorption spectrometry after cloud point extraction. Food Chem 2013, 141, 66–71.

[131] Thosaikham W, Jitmanee K, Sittipout R, Maneetong S, Chantiratikul A, Chantiratikul P. Evaluation of selenium species in selenium-enriched pakchoi (*Brassica chinensis Jusl* var parachinensis (Bailey) Tsen & Lee) using mixed ion-pair reversed phase HPLC–ICP-MS. Food Chem 2014, 145, 736–42.

[132] Yang L, Maxwell P, Mester Z. Microwave-assisted acid digestion protocol for the determination of methionine and selenomethionine in selenium-enriched yeast by species specific isotope dilution GC-MS. Anal Methods 2013, 5, 525–29.
[133] de la Calle MB, Baer I, Robouch P, et al. Is it possible to agree on a value for inorganic arsenic in food? The outcome of IMEP-112. Anal Bioanal Chem 2012, 404, 2475–88.
[134] Ibanez-Palomino C, Lopez-Sanchez JF, Sahuquillo A. Certified reference materials for analytical mercury speciation in biological and environmental matrices: Do they meet user needs?; a review. Anal Chim Acta 2012, 720, 9–15.
[135] Caruso JA, Montes-Bayon M. Elemental speciation studies-new directions for trace metal analysis. Ecotoxicol Environ Safe 2003, 56, 148–63.
[136] Musil S, Petursdottir AH, Raab A, Gunnlaugsdottir H, Krupp E, Feldmann J. Speciation without chromatography using selective hydride generation: Inorganic arsenic in rice and samples of marine origin. Anal Chem 2014, 86, 993–9.
[137] Chen G, Chen T. SPE speciation of inorganic arsenic in rice followed by hydride-generation atomic fluorescence spectrometric quantification. Talanta 2014, 119, 202–6.
[138] Ruiz-de-Cenzano M, Rochina-Marco A, Cervera ML, de la Guardia M. Speciation of methylmercury in market seafood by thermal degradation, amalgamation and atomic absorption spectroscopy. Ecotox Environ Safe 2014, 107, 90–6.
[139] Gustinelli Arantes de Carvalho G, Alves Manoel Degaspari I, Branco B, Canário J, Ferreira de Amorim A, Kennedy VH, Ferreira JR. Assessment of total and organic mercury levels in blue sharks (*Prionace glauca*) from the South and Southeastern Brazilian coast. Biol Trace Elem Res 2014, 159, 128–34.
[140] Lopez-García I, Vicente-Martínez Y, Hernández-Córdoba M. Nonchromatographic speciation of selenium in edible oils using dispersive liquid-liquid microextraction and electrothermal atomic absorption spectrometry. J Agric Food Chem 2013, 61, 9356–61.
[141] Qu H, Mudalige TK, Linder SW. Arsenic speciation in rice by capillary electrophoresis/inductively-coupled plasma mass spectrometry: Enzyme-assisted water phase microwave digestion. J Agric Food Chem 2015, 63, 3153–60.
[142] Chen YQ, Chen JF, Xi Z, Yang GD, Wu Z, Li JR, Fu FF. Simultaneous analysis of Cr(III), Cr(VI) and chromium picolinate in foods using capillary electrophoresis-inductively coupled plasma mass spectrometry. Electrophoresis 2015, doi: 10.1002/elps.201500015.
[143] Timerbaev AR. Element speciation analysis using capillary electrophoresis: Twenty years of development and applications. Chem Rev 2013, 113, 778–812.
[144] Wrobel K, Wrobel K, Caruso JA. Metal Analysis (ICP-OES and ICP-MS). In: Scott RE, Lukehart CM, eds. Applications of Physical Methods to Inorganic and Bioinorganic Chemistry. Encyclopedia of Inorganic Chemistry (vol 11). John Wiley & Sons, Chichester, 2007, 205–25.
[145] Wu B, Becker JS. Imaging techniques for elements and element species in plant science. Metallomics 2012, 4, 403–16.
[146] Schaumloffel D, Tholey A. Recent directions of electrospray mass spectrometry for elemental speciation analysis. Anal Bioanal Chem 2011, 400, 1645–52.
[147] Li T, She Y, Wang M, et al. Simultaneous determination of four organotins in food packaging by high-performance liquid chromatography–tandem mass spectrometry. Food Chem 2015, 181, 347–53.
[148] Ceko MJ, Aitken JB, Harris HH. Speciation of copper in a range of food types by X-ray absorption spectroscopy. Food Chem 2014, 164, 50–4.
[149] Heumann KG. Hyphenated techniques – the most commonly used method for trace elemental speciation analysis. Anal Bioanal Chem 2002, 373, 323–4.

[150] Llorente-Mirandes T, Calderon J, Centrich F, Rubio R, Lopez-Sanchez JF. A need for determination of arsenic species at low levels in cereal-based food and infant cereals. Validation of a method by IC-ICPMS. Food Chem 2014, 147, 377–85.

[151] Chen X, Han C, Cheng H, Wang Y, Liu J, Xu Z, Hu L. Rapid speciation analysis of mercury in seawater and marine fish by cation exchange chromatography hyphenated with inductively coupled plasma mass spectrometry. J Chromatogr A 2013, 1314, 86–93.

[152] Schmidt L, Bizzi CA, Duarte FA, Dressler VL, Flores EMM. Evaluation of drying conditions of fish tissues for inorganic mercury and methylmercury speciation analysis. Microchem J 2013, 108, 53–9.

[153] Wang Z, Forsyth D, Belisle S, Beraldin F, Sparling M, Trudelle RA, Lapointe P, Bellon-Gagnon P. Comparison of two analytical methods for the analysis of methylmercury in fish. Food Anal Methods 2013, 6, 157–63.

[154] Novotnik B, Zuliani T, Scancar J, Milacic R. Chromate in food samples: An artefact of wrongly applied analytical methodology? J Anal At Spectrom 2013, 28, 558–66.

[155] Vacchina V, de la Calle i, Séby F. Cr(VI) speciation in foods by HPLC-ICP-MS: Investigation of Cr(VI)/food interactions by size exclusion and Cr(VI) determination and stability by ion-exchange on-line separations. Anal Bioanal Chem 2015, DOI 10.1007/s00216-015-8616-3.

[156] Wrobel K, Wrobel K, Caruso JA. Liquid chromatography with inductively coupled plasma mass spectrometric detection for element speciation: Clinical and toxicological applications. In: Winecker RE, Bertholf R, eds. Chromatographic Methods in Clinical Chemistry and Toxicology. John Wiley & Sons, Chichester, 2007, 217–75.

[157] Montes-Bayón M, DeNicola K, Caruso JA. Liquid chromatography – inductively coupled plasma mass spectrometry. J Chromatogr A 2002, 1000, 457–76.

[158] Agatemor C, Beauchemin D. Matrix effects in inductively coupled plasma mass spectrometry: A review. Anal Chim Acta 2011, 706, 66–86.

[159] Milacic R, Ajlec D, Zuliani T, Zigon D, Scancar J. Determination of Zn-citrate in human milk by CIM monolithic chromatography with atomic and mass spectrometry detection. Talanta 2012, 101, 203–10.

[160] Narukawa T, Suzuki T, Inagaki K, Hioki A. Extraction techniques for arsenic species in rice flour and their speciation by HPLC-ICP-MS. Talanta 2014, 130, 213–20.

[161] Ai X, Wang Y, Hou X, Yang L, Zheng C, Wu L. Advanced oxidation using Fe_3O_4 magnetic nanoparticles and its application in mercury speciation analysis by high performance liquid chromatography-cold vapor generation atomic fluorescence spectrometry. Analyst 2013, 138, 3494–501.

[162] Xie X, Feng C, Ye M, Wang C. Speciation determination of selenium in seafood by high-performance ion-exchange chromatography-hydride generation-atomic fluorescence spectrometry. Food Anal Methods 2014, doi: 10.1007/s12161-014-0055-9.

[163] Zmozinski AV, Carneado S, Ibanez-Palomino C, Sahuquillo A, Lopez-Sánchez JF, da Silva MM. Method development for the simultaneous determination of methylmercury and inorganic mercury in seafood. Food Control 2014, 46, 351–9.

[164] Sanchez-Rodas D, Corns WT, Chen B, Stockwell PB. Atomic fluorescence spectrometry: A suitable detection technique in speciation studies for arsenic, selenium, antimony and mercury. J Anal At Spectrom 2010, 25, 933–46.

[165] Jirásko R, Holcapek M. Structural analysis of organometallic compounds with soft ionization mass spectrometry. Mass Spectrom Rev 2011, 30, 1013–36.

[166] Rosen AL, Hieftje GM. Inductively coupled plasma mass spectrometry and electrospray mass spectrometry for speciation analysis: Applications and instrumentation. Spectrochim Acta Part B 2004, 59, 135–46.

[167] Sele V, Amlund H, Berntssen MHG, Berntsen JA, Skov K, Sloth JJ. Detection of arsenic-containing hydrocarbons in a range of commercial fish oils by GC-ICPMS analysis. Anal Bioanal Chem 2013, 405, 5179–90.
[168] Sola-Vázquez A, Costa-Fernández JM, Pereiro R, Sanz-Medel A. Plasma-based mass spectrometry for simultaneous acquisition of elemental and molecular information. The Analyst 2011, 136, 246–56.
[169] Easter RN, Caruso JA, Vonderheide AP. Recent developments and novel applications in GC-ICPMS. J Anal At Spectrom 2010, 25, 493–502.
[170] Zachariadis GA. In situ derivatization of metals and organometallics using borate reagents in gas chromatographic speciation studies. J Chromatogr A 2013, 1296, 47–69.
[171] Berzas Nevado JJ, Rodríguez Martín-Doimeadios RC, Guzmán Bernardo FJ, Rodríguez Fariñas N, Patiño Ropero MJ. Mercury speciation analysis in terrestrial animal tissues. Talanta 2012, 99, 859–64.
[172] Cui Z, Qian Y, Ge N, Zhang J, Liu Y, Cao Y. Determination of butyltin and phenyltin analogues in sea products by Grignard derivatization and gas chromatography-triple quadrupole tandem mass spectrometry. Anal Methods 2014, 6, 9333–9.
[173] Duarte FA, Soares BM, Vieira AA, Pereira ER, Maciel JV, Caldas SS, Primel EG. Assessment of modified mtrix solid-phase dispersion as sample preparation for the determination of CH_3Hg^+ and Hg^{2+} in fish. Anal Chem 2013, 85, 5015-22.
[174] Noventa S, Barbaro I, Formalewicz M, et al. A fast and effective routine method based on HS-SPME–GC–MS/MS for the analysis of organotin compounds in biota samples. Anal Chim Acta 2015, 858, 66–73.
[175] Voegborlo RB, Matsuyama A, Adimado AA, Akagi H. Determination of methylmercury in marine and freshwater fish in Ghana using a combined technique of dithizone extraction and gas–liquid chromatography with electron capture detection. Food Chem 2013, 124, 1244–8.
[176] Wu P, He L, Zheng C, Hou X, Sturgeon RE. Applications of chemical vapor generation in non-tetrahydroborate media to analytical atomic spectrometry. J Anal At Spectrom 2010, 25, 1217–46.
[177] Vas G, Vékey K. Solid-phase microextraction: A powerful sample preparation tool prior to mass spectrometric analysis. J Mass Spectrom 2004, 39, 233–54.
[178] Lord HL, Zhan W, Pawliszyn J. Fundamentals and applications of needle trap devices: A critical review. Anal Chim Acta 2010, 677, 3–18.
[179] Sarafraz-Yazdi A, Amiri A. Liquid-phase microextraction. Trend Anal Chem 2010, 29, 1–14.
[180] Viñas P, Campillo N, López-Gacía I, Hernández-Córdoba M. Dispersive liquid-liquid microextraction in food analysis. A critical review. Anal Bioanal Chem 2014, 406, 2067–99.
[181] Leach AM, Heisterkamp FC, Adams FC, Hieftje GM. Gas chromatography-inductively coupled plasma time of flight mass spectrometry for the speciation analysis of organometallic compounds. J Anal At Spectrom 2000, 15, 151–5.
[182] Profrock D, Prange A. Inductively coupled plasma-mass spectrometry (ICP-MS) for quantitative analysis in environmental and life sciences: A review of challenges, solutions, and trends. Appl Spectrosc 2012, 66, 843–68.
[183] Paterson AC, MsAlister GC, Quarmby ST, Griep-Raming J, Coon JJ. Development and characterization of GC-enabled QLT-Orbitrap for high-resolution and high-mass accuracy GC/MS Anal Chem 2010, 82, 8618–28.
[184] Kolachova K, Cajka T, Sandy C, Hajslova J, Pulkrabova J. High throughput sample preparation in combination with gas chromatography coupled to triple quadrupole tandem mass spectrometry (GC-MS/MS): A smart procedure for (ultra)trace analysis of brominated flame retardants in fish. Talanta 2013, 105, 109–16.

[185] Pereiro IR, Diaz AC. Speciation of mercury, tin, and lead compounds by gas chromatography with microwave-induced plasma and atomic-emission detection (GC-MIP-AED). Anal Bioanal Chem 2002, 372, 74–90.

[186] Dimitrakakis E, Haberhauer-Troyer C, Abe Y, Ochsenkuhn-Petropoulou M, Rosenberg E. Solid-phase microextraction-capillary gas chromatography combined with microwave-induced plasma atomic-emission spectrometry for selenite determination. Anal Bioanal Chem 2004, 379, 842–8.

[187] da Silva CA, Tessier E, Tavares Kutter V, Wasserman JC, Donard OFX, Silva-Filho EV. Mercury speciation in fish of the Cabo Frio upwelling region, SE-Brazil. Braz J Oceanogr 2011, 59, 259–66.

[188] Liu JM, Cheng JK. Elemental speciation analysis in capillary electrophoresis. Electrophoresis 2003, 24, 1993–2012.

[189] Michalke B. Capillary electrophoresis-inductively coupled plasma-mass spectrometry: A report on technical principles and problem solutions, potential, and limitations of this technology as well as on examples of application. Electrophoresis 2005, 26, 1584–97.

[190] Kannamkumarath SS, Wrobel K, Wrobel, K, B'Hymer C, Caruso JA. Capillary electrophoresis-inductively coupled plasma-mass spectrometry: An attractive complementary technique for elemental speciation analysis. J Chromatogr A 2002, 975, 245–66.

[191] Vieira MA, Grinberg P, Bobeda CRR, Reyes MNM, Campos RC. Non-chromatographic atomic spectrometric methods in speciation analysis: A review. J Hazard Mater 2009, 64, 459–76.

[192] Gonzalvez A, Armenta S, Cervera Ml, de la Guardia M. Non-chromatographic speciation. Trend Anal Chem 2010, 29, 260–68.

[193] Rezende HC, Almeida ILS, Coelho LM, Coelho NMM, Marques TL. Non-chromatographic methods focused on speciation of arsenic and selenium in food and environmental samples. Sample Prep 2014, 2, 31–48.

[194] Costa BE, Coelho NM, Coelho LM. Determination of arsenic species in rice samples using CPE and ETAAS. Food Chem 2015, 178, 89–95.

[195] Rasmussen RR, Hedegaard RV, Larsen EH, Sloth JJ. Development and validation of an SPE HG-AAS method for determination of inorganic arsenic in samples of marine origin. Anal Bioanal Chem 2012, 403, 2825–34.

[196] Rasmussen RR, Qian Y, Sloth JJ. SPE HG-AAS method for the determination of inorganic arsenic in rice – results from method validation studies and a survey on rice products. Anal Bioanal Chem 2013, 405, 7851–7.

[197] Tian J, Huang T, Lu J. Speciation of inorganic antimony in food and water samples by flow injection on-line nano γ-alumina micro-column solid-phase extraction coupled with slotted tube atom trapping flame atomic absorption spectrometry. Food Anal Methods 2013, 6, 1080–9.

[198] Loeschner K, Navratilova J, Grombe R, Linsinger TPJ, Kōbler C, Mōlhave K, Larsen EH. In-house validation of a method for determination of silver nanoparticles in chicken meat based on asymmetric flow field-flow fractionation and inductively coupled plasma mass spectrometric detection. Food Chem 2015, 181, 78–84.

[199] Loeschner K, Navratilova J, Kobler K, Molhave K, Wagner S, von der Kammer F, Larsen EH. Detection and characterization of silver nanoparticles in chicken meat by asymmetric flow field flow fractionation with detection by conventional or single particle ICP-MS. Anal Bioanal Chem 2013, 405, 8185–95.

[200] Kristan U, Arribére MA, Stibilj V. Selenium species and their distribution in freshwater fish from Argentina. Biol Trace Elem Res 2013, 151, 240–46.

[201] Naito S, Matsumoto E, Shindoh K, Nish T. Effects of polishing, cooking, and storing on total arsenic and arsenic species concentrations in rice cultivated in Japan. Food Chem 2015, 168, 294–301.

[202] Nalluri D, Baumann Z, Abercrombie DL, Chapman DD, Hammerschmidt CR, Fisher NS. Methylmercury in dried shark fins and shark fin soup from American restaurants. Sci Tot Environ 2014, 496, 644–8.
[203] Bendicho C, Lavilla I, Pena-Pereira F, Romero V. Green chemistry in analytical atomic spectrometry: A review. J Anal At Spectrom 2012, 27, 1831–57.
[204] Yin YG, Liu JF, Jiang GB. Recent advances in speciation analysis of mercury, arsenic and selenium. Chin Sci Bull 2013, 58, 150–61.
[205] Carro AM, Mejuto MC. Application of chromatographic and electrophoretic methodology to the speciation of organomercury compounds in food analysis. J Chromatogr A 2000, 882, 283–307.
[206] Ferreira SLC, Lemos VA, Silva LOB, et al. Analytical strategies of sample preparation for determination of mercury in food matrices – A review. Microchem J 2015, doi: 10.1016/j.microc. 2015.02.012.
[207] Westöö G. Determination of methylmercury compounds in foodstuffs. I. Methylmercury compounds in fish, identification and determination. Acta Chem Scand 1966, 20, 2131–7.
[208] Berzas Nevado JJ, Rodríguez Martín-Doimeadios RC, Krupp EM, et al. Comparison of gas chromatographic hyphenated techniques for mercury speciation analysis. J Chromatogr A 2011, 1218, 4545–51.
[209] Gao Y, Shi Z, Long Z, Wu P, Zheng C, Hou X. Determination and speciation of mercury in environmental and biological samples by analytical atomic spectrometry. Microchem J 2012, 103, 1–14.
[210] Rothenberg SE, Windham-Myers L, Creswell JE. Rice methylmercury exposure and mitigation: A comprehensive review. Environ Res 2014, 133, 407–23.
[211] Rothenberg SE, Mgutshini NL, Bizimis M, Johnson-Beebout SE, Ramanantsoanirina A. Retrospective study of methylmercury and other metal(loid)s in Madagascar unpolished rice (Oryza sativa L.). Environ Pollut 2015, 196, 125–33.
[212] Amaral CDB, Nóbrega JA, Nogueira ARA. Sample preparation for arsenic speciation in terrestrial plants – A review. Talanta 2013, 115, 291–9.
[213] Feldmann J, Krupp EM. Critical review or scientific opinion paper: Arsenosugars – a class of benign arsenic species or justification for developing partly speciated arsenic fractionation in foodstuffs? Anal Bioanal Chem 2011, 399, 1735–41.
[214] Conklin SD, Shockey N, Kubachka K, Howard KD, Carson MC. Development of an ion chromatography-inductively coupled plasma-mass spectrometry method to determine inorganic arsenic in liver from chicken treated with roxarsone. J Agric Food Chem 2012, 60, 9394–404.
[215] Wrobel K, Wrobel, K, Caruso JA. Pretreatment procedures for characterization of arsenic and selenium species in complex samples utilizing coupled techniques with mass spectrometric detection. Anal Bioanal Chem 2005, 381, 317–31.
[216] Francesconi KA, Kuehnelt D. Determination of arsenic species: A critical review of methods and applications, 2000-2003. Analyst 2004, 129, 373–95.
[217] D'Amato M, Aureli F, Ciardullo S, Raggi A, Cubadda F. Arsenic speciation in wheat and wheat products using ultrasound- and microwave-assisted extraction and anion exchange chromatography-inductively coupled plasma mass spectrometry. J Anal At Spectrom 2011, 26, 207–13.
[218] Raber G, Stock N, Hanel P, Murko M, Navratilova J, Francesconi KA. An improved HPLC–ICPMS method for determining inorganic arsenic in food: Application to rice, wheat and tuna fish. Food Chem 2012, 134, 524–32.
[219] Santos CMM, Nunes MAG, Barbosa IS, et al. Evaluation of microwave and ultrasound extraction procedures for arsenic speciation in bivalve mollusks by liquid

chromatography–inductively coupled plasma-mass spectrometry. Spectrochim Acta Part B 2013, 86, 108–14.

[220] Garcia-Salgado S, Quijano MA, Bonilla MM. Arsenic speciation in edible alga samples by microwave-assisted extraction and high performance liquid chromatography coupled to atomic fluorescence spectrometry. Anal Chim Acta 2012, 714, 38–46.

[221] Jesus JP, Suárez CA, Ferreira JR, Giné MF. Sequential injection analysis implementing multiple standard additions for As speciation by liquid chromatography and atomic fluorescence spectrometry (SIA-HPLC-AFS). Talanta 2011, 85, 1364–8.

[222] Petursdottir AH, Gunnlaugsdottir H, Jorundsdottir H, Mestrot A, Krupp EM, Feldmann J. HPLC-HG-ICP-MS: A sensitive and selective method for inorganic arsenic in seafood. Anal Bioanal Chem 2012, 404, 2185–91.

[223] Oguri T, Yoshinaga J, Tao H, Nakazato T. Inorganic arsenic in the Japanese diet: Daily intake and source. Arch Environ Contam Toxicol 2014, 66, 100–12.

[224] Maher W, Foster S, Krikowa F, Donner E, Lombi E. Measurement of inorganic arsenic species in rice after nitric acid extraction by HPLC-ICPMS: Verification using XANES. Environ Sci Technol 2013, 47, 5821–7.

[225] Rintala EM, Ekholm P, Koivisto P, Peltonen K, Venalainen ER. The intake of inorganic arsenic from long grain rice and rice-based baby food in Finland – low safety margin warrants follow up. Food Chem 2014, 150, 199–205.

[226] Sommella A, Deacon C, Norton G, Pigna M, Violante A, Meharg AA. Total arsenic, inorganic arsenic, and other elements concentrations in Italian rice grain varies with origin and type. Environ Pollut 2013, 181, 38–43.

[227] Leermakers M, Bayens W, De Gieter M, Smedts B, Meert C, De Bisschop HC, Morabito R, Quevauviller P. Toxic arsenic compounds in environmental samples: Speciation and validation. Trend Anal Chem 2006, 25, 1–10.

[228] Bissen M, Frimmel FH. Arsenic – A review. Part I: Occurrence, toxicity, speciation, mobility. Acta Hydrochim Hydrobiol 2003, 31, 9–18.

[229] Nookabkaew S, Rangkadilok N, Mahidol C, Promsuk G, Satayavivad J. Determination of arsenic species in rice from Thailand and other Asian countries using simple extraction and HPLC-ICP-MS analysis. J Agric Food Chem 2013, 61, 6991–8.

[230] Hsieh YJ, Jiang SJ. Application of HPLC-ICP-MS and HPLC-ESI-MS procedures for arsenic speciation in seaweeds. J Agric Food Chem 2012, 60, 2083-89.

[231] Niedzielski P, Mleczek M, Magdziak Z, Siwulski M, Kozak L. Selected arsenic species: As(III), As(V) and dimethylarsenic acid (DMAA) in *Xerocomus badius* fruiting bodies. Food Chem 2013, 141, 3571–77.

[232] Navarro-Alarcon M, Cabrera-Vique C. Selenium in food and in the human body: A review. Sci Tot Environ 2008, 400, 115–41.

[233] Kieliszek M, Blazejak S. Selenium: Significance, and outlook for supplementation. Nutrition 2013, 29, 713–18.

[234] Kristan U, Planinšek P, Benedik L, Falnoga I, Stibilj V. Polonium-210 and selenium in tissues and tissue extracts of the mussel Mytilus galloprovincialis (Gulf of Trieste). Chemosphere 2014, 119, 231–41.

[235] Wang W, Chen ZL, Davey DE, Naidu R. Speciation of selenium in biological samples by ion chromatography with inductively coupled plasma mass spectrometry. J Liq Sep Rel Technol 2010, 33, 1151–73.

[236] Chen YW, Belzile N. High performance liquid chromatography coupled to atomic fluorescence spectrometry for the speciation of the hydride and chemical vapour-forming elements As, Se, Sb and Hg: A critical review. Anal Chim Acta 2010, 671, 9–26.

[237] Dietz C, Sanz J, Sanz E, Munoz-Olivas R, Camara C. Current perspectives in analyte extraction strategies for tin and arsenic speciation. J Chromatogr A 2007, 1153, 114–29.
[238] Abalos M, Bayona JM, Compano R, Granados M, Leal C, Prat MD. Analytical procedures for the determination of organotin compounds in sediment and biota: A critical review. J Chromatogr A 1997, 788, 1–49.
[239] Blunden S, Wallace T. Tin in canned food: A review and understanding of occurrence and effect. Food Chem Toxicol 2003, 41, 1651–62.
[240] Gomez V, Callao MP. Chromium determination and speciation since 2000. Trend Anal Chem 2006, 25, 1006–15.
[241] Soares ME, Vieira E, Bastos Mde L. Chromium speciation analysis in bread samples. J Agric Food Chem 2010, 58, 1366–70.
[242] Metze D, Jakubowski N, Klockow D. Speciation of chromium. In: Cornelis R, Caruso JA, Crews H, Heumann K, eds. Handbook of Elemental Speciation II – Species in Environment, Food, Medicine and Occupational Health. John Wiley & Sons, Chichester, UK, 2005.
[243] Cushen M, Kerry J, Morris M, Cruz-Romero M, Cummins E. Nanotechnologies in the food industry – recent developments, risks and regulation. Trends Food Sci Technol 2012, 24, 30–46.
[244] Llorens A, Lloret E, Picouet PA, Trbojevich R, Fernandez A. Metallic-based micro and nanocomposites in food contact materials and active food packaging. Trends Food Sci Technol 2012, 24, 19–29.
[245] Klaine SI, Alvarez PJJ, Batley GE, et al. Nanomaterials in the environment: Behavior, fate, bioavailability, and effects. Environ Toxicol Chem 2008, 27, 1825–51.
[246] Mbundi L, Gallar-Ayala H, Rizwan Khan M, Barber JL, Losada S, Busquets R. Chapter Two – Advances in the analysis of challenging food contaminants: Nanoparticles, bisphenols, mycotoxins, and brominated flame retardants. Adv Mol Toxicol 2014, 8, 35–105.
[247] Fermoa P, Beretta G, Maffei Facino R, Gelmini F, Piazzalunga A. Ionic profile of honey as a potential indicator of botanical origin and global environmental pollution. Environ Pollut 2013, 178, 173–81.
[248] Brian Jackson VT, Punshon T, Cottingham K. Arsenic concentration and speciation in infant formulas and first foods. Pure Appl Chem 2012, 84, 215–23.

Anna Ruszczyńska and Eliza Kurek

5 Trace Element and Speciation Analysis of Biological Samples

5.1 Trace element analysis of biological samples

The role of elements in living organisms has been studied since centuries. Most of the elements are essential for living organisms but only at trace levels; a part of elements present at trace levels is toxic. Therefore, the researchers focused on the total content of elements in various biological samples first. Most of the researches refer to human health and processes in living organisms. The term biological sample has a very broad range and concerns such materials as biological fluids (e.g. blood, serum, urine and breast milk), plant and animal tissues, food (not processed and processed) as well as other materials whose analysis can bring knowledge about their biological functions. Diversity of studied materials in determined analytes as well as diversity in their physical and chemical form requires special and individual procedures using all possible tools and methods of analytical chemistry. Because of the problematic matrix of biological materials and necessity of determination at trace levels of elements, biological samples demand precise and accurate analytical methods.

The history of trace element analysis has begun along with atomic absorption spectrometry (AAS) development. For most elements and samples, AAS with flame atomization (FAAS) is specific, free of interferences and in its concentration range of parts per million (ppm) a problem of sample contamination is not afflictive. AAS with graphite furnace (GF) atomizer gave the opportunity for direct determination of much lower concentration of elements than in flame atomizers. It is worth to mention that electrothermal atomizers pose some problems with interferences, especially in the case of biological materials with substantial matrix compared with the analyte. Even though a great sensitivity made this method very popular, FAAS or GF AAS system is still most frequently used in trace element analysis in biological, medical and clinical areas [1]. Essential elements such as Fe, Zn and Cu, and toxic elements like Pb and Cd present in ppm levels in biological samples are determined directly with the use of FAAS. But in many cases, elements of interest are present at concentration levels in the parts per billion (ppb) range or below. In that situation, widely available GF AAS or another sufficient analytical methods like mass spectrometry (MS) and neutron activation analysis are used. Nevertheless, most of them are not free from troublesome interferences and a problem of contamination at this range becomes significant. In case of single element analysis, GF AAS method is effective but for fast multi-element determination at the trace and ultra-trace concentration levels for liquid samples, ICP MS and ICP optical emission spectrometry (OES) are the most frequently used methods [1, 2]. Additionally an ICP MS method allows a precise and accurate isotope analysis of

https://doi.org/10.1515/9783110366730-006

chemical elements and this kind of analysis – isotope dilution analysis – provides the possibility of absolute quantification for elements with two and more isotopes in any sample material to obtain accurate trace element concentration. Another advantage of ICP MS method is the possibility of association with different sample introduction systems suitable for liquid samples, dry aerosols and vapours, for example, traditional pneumatic nebulization, electrothermal vaporization, direct injection nebulization, ultrasonic nebulization and laser ablation (LA). An introduction system can improve the limit of detection and decrease some characteristic interferences of the ICP MS method. LA sampling is very attractive due to the very little or no sample preparation requirements as well as reduction of spectral interferences. Depending on the laser used for sampling, this method can offer high spatial resolution (10–100 μm for nanosecond laser, below 1 μm for femtosecond laser) with low sample uptake. The attraction of application of LA as a sample introduction system for ICP MS in elemental analysis of biological materials started being reported at the end of 20th century [3, 4]. A problem with fully quantitative analysis was always stressed at that time and it still persists due to the matrix-dependent non-stoichiometric effects occurring during sampling, aerosol fractionation during transport to the plasma and processes in the plasma: vaporization, atomization and ionization [5]. Despite this, the LA ICP MS has been increasingly exploited for the analysis of various matrices over the last decades [6–8] and distribution analysis carried out with LA ICP MS for bioimaging animal and plant tissues demonstrates the usefulness and importance in biomedical research, bioaccumulation and bioavailability studies in humans, animals and plants [9].

Accuracy of an analytical determination should be established for example by analysing the same samples by two independent methods or by analysing reference materials whose matrix is identical to that of the samples. Nowadays for some laboratories might be much easier to find and buy suitable biological reference materials with useful range of analyte concentrations and matrix than to have an access to second analytical method. Table 5.1 presents some examples of methods used for trace elemental analysis in biological samples.

5.1.1 Sampling, sample storage and sampling treatment

Despite the analytical technique suitable for determining elements in biological samples, there are several stages prior to analysis which are important and can be a source of analytical error in trace element analysis of biological samples: sampling, sampling pretreatment, sample storage and preparation.

In case of collecting blood samples with disposable stainless steel needles, it is analysed that nickel, chromium or manganese may cause contamination problems as these elements are major components of stainless steel. The same source of contamination originates from containers used to store the samples but analyte loss by adsorption is also possible. Usually polypropylene or polyethylene materials are considered as the best storage containers but in some cases when long time freezing is applied permeability to frozen water vapour can be observed together with gradual

Table 5.1: Diversity of biological samples analysed by various methods for elemental content.

Sample	Analyte	Technique	LOD (µg/L)	RSD (%)	References
Hair, nails, liver	Zn(II)	FAAS	2.2	<1.2	[10]
Food	Cr(III), Cu(II), Fe(III), Pb(II), Pd(II), Zn(II)	FAAS	0.1–5.3	<5	[11]
Fruit (apple, grape, nectarine, green beans, cantaloupe)	Cu(II)	FAAS	0.65	3	[12]
Fruit juices, seawater	Cu	FAAS	0,15	2.7–6	[13]
Juices, alcoholic beers	Sn(IV)	FAAS	0.33	2.1–6.2	[14]
Vegetable oil, fat samples	Mn, As, Pb, Cu, Cd, Zn	GF AAS	4.0–0.8	3–8	[15]
Wine, beer, milk, kefir, yoghurt, juice, lemonade	Pb, As, Cd	GF AAS	0.1–2.0	3–6	[16]
Vinegar	Cd	GF AAS	0.004	1.07–2.33	[17]
Body fluids	Cr, Ni, Al, V	ICP AES	1–2	1	[18]
Body fluids	Al, Be, Cd, Co, Cr, Hg, Mn, Ni, Pb, V	ICP MS	0.001–0.05	1–5	[19]
Animal tissue, plant leaves	Cu, Ag, Se, Pb, As, Cd, V, Cr, Mn, Ni, Sn, Sb, Te, Bi, Co	ETV ICP MS	0.0013–1.4 µg/g	–	[20]
Mouse kidney	Pt, Zn, Cu	LA ICP MS	50 fg	–	[21]
Leaves	Cu, K, Mg, Mn, P, S, B	LA ICP MS	–	–	[22]
Grounded vanilla	Mn, Mg, P, Ca, Fe, Cu, Zn, Rb, Sr, Ba	LA ICP MS	35 µg/kg	<10	[23]

Note: FAAS, atomic absorption spectrometry with flame atomization; GF AAS, graphite furnace atomic absorption spectrometry; ICP, inductively coupled plasma; AES, atomic emission spectroscopy; MS, mass spectrometry; ETV, electrothermal vaporization; LA, laser ablation.

moisture loss. Sometimes freeze-drying before storage might be the best solution but due to time and money-consuming procedure only in case of benefit of dry sample during analytical process. Because a great part of biological samples constitute biological fluids it is necessary to remember that they may segregate during freezing which might be a problem if only a part of the entire sample is used for analysis.

The same problem with contamination or loss of analyte coming from sampling equipment and storage containers happens during the sample preparation procedure because of the use of reagents or even airborne particulates. Therefore, in some cases clean rooms of laminar flow hoods are essential. Analyte adsorption on surfaces, incomplete solubilization or dissolution or extraction processes may lead to analyte loss.

Not to be underestimated is the analyst's work style. Plant samples acquired from natural conditions, fields, meadows, forests or gardens require removal of dust or soot cover on their surfaces. This could be achieved by rinsing but it has to be remembered that in case of complete elemental composition studies potassium ions will be lost during this process.

Most of the biological samples are composed of rich matrix which may cause suppression or enhancement of the analyte signal. In that case, analysis of analyte recovery and standard addition method might not be successful. In case of all natural materials there is also a risk of heterogeneity of a sample to be analysed. To obtain a representative material homogenization is needed and this could be achieved by milling techniques. Then in case of trace element analysis the contamination coming from parts of mill might be significant.

Fluids are a great part of biological samples and are possible to analyse just after water dilution. This saves a lot of time and seems to be much more economical but on the other hand the deposition of organic components from the body may cause clogging of nebulizers, torch injectors and cones during ICP MS analysis. Too high dilution factors applied to these samples bring closer the level of analyte's concentration to the detection limits.

Because most of the analytical methods used for trace element determination demand liquid samples and also organic residues usually cause analytical biases, it is common to mineralize biological materials prior to analysis. Depending on the sample type, the element as well as the number of elements to be measured, there are different digestion methods to be used for digestion of organic matter. The most commonly employed are wet digestion and dry ashing. The last method is especially problematic in case of volatile elements such as Hg, As and Se which could be lost during the process but also because of hygroscopic feature of ash and its adsorbing tendency to the porcelain or quartz surfaces of cups. Therefore, when discussing trace element determination it is worth to mention closed systems for dry ashing, among them a microwave-induced combustion allowing the digestion of organic samples in oxygen-pressurized closed vessels.

There are many wet digestion procedures described in the literature but definitely the use of closed vessels with microwave heating minimizes the risk of analyte losses and improves the efficiency of oxidation of the matrix [24, 25]. Especially in biological samples there is a possibility that an analyte might be present in some unknown volatile organic species and lost even in lower temperatures than expected. Some plant samples are digested easily in typical conditions of pressure and temperature used in microwave digestion ovens (40–80 bar and 220–250 °C) and with the use of concentrated HNO_3. But in cases of plant samples containing high concentrations of not easily dissolved elements such as Si, Al, Cr, special digestion methods are required, for example, for rice straw an addition of HF is needed. If high concentrations of HF are needed, a suitable caution for non-HF-resistant elements of instrumentation must be held [26].

5.2 Element speciation of biological samples

However, when knowledge of biological functions of trace elements and their metabolism increased, there is a need not only to determine their total content but also to identify and quantitatively determine their different chemical forms. Therefore and together with recent development of trace and ultra-trace analysis, element speciation analysis received a global attention. The aim of speciation analysis is to identify and determine the concentration of various forms of elements in different materials. There were a lot of speciation definitions [27, 28]. The latest presented at the *IUPAC Gold Book* says that “speciation in chemistry is distribution of an element amongst defined chemical species in a system” [29] and speciation analysis concerns all analytical operations leading to the identification and/or quantification of one or more chemical forms of a given element in the sample [30]. Whereas the process of classification of an analyte or a group of analytes from a certain sample according to physical or chemical properties is called fractionation [31].

In a complex biological matrix, trace elements might be present as different species (free anions, chelates or bound to simple metabolites) and speciation analysis identifies and quantifies these biologically active compounds which are bound to trace elements. All analytical measurements have to be accurate and precise and require a systematic check for the trace element and the biocompound. Speciation analysis in biological materials is an extremely important task allowing the research in many branches of science and life: toxicology, environment, biology or human health. Nevertheless, it is a very difficult task. The modern speciation analysis identifies and quantifies one or more than one chemical species in a sample and determines biological parameters.

A fundamental tool for the speciation analysis is the combination of a selective separation technique, which ensures that the analyte compound leaves the column unaccompanied by other species of the analyte element, with a sensitive and specific detection method permitting for a quantitative and qualitative determination of the target element. The low concentration of the trace elements present in biological matrices usually at the level of ppm or ppb and complicated matrix constitutes main problems of speciation analysis. Properly chosen hyphenated technique is a solution for best results in speciation analysis of trace elements in biological samples. The most popular separation techniques are chromatography, electrochromatography and gel electrophoresis combined with specific detection, very often ICP MS or ICP atomic emission spectroscopy. The last component of the analytical strategy in speciation analysis is identification and characterization of discovered species. Recently, ICP MS offering low detection limits is frequently used as the detection method. One of the advantages of this method is the ease of its coupling with high-pressure liquid chromatography (HPLC). It is worth to remember though that coupled system of HPLC–ICP MS allowing for identification of compounds has a serious limitation due to the lack of standards for many existing in real sample species. At the same time it is difficult to assume that existing in real sample complexing agents can change times of retention of an analyte. In this case, a method of detection specific not for an element but for a whole compound should be used. Usefulness of tandem mass spectrometry (MS/MS) with soft ionization methods (e.g. electrospray, ESI) in identification and structural characterization of molecules at trace levels in biological matrices was demonstrated at the end of 20th century for the first time and its popularity is still growing [32]. Metabolomic studies based on low-resolution instrumentation (triple quadrupole, ion trap) [33, 34] at the beginning but with time had been replaced with high-resolution systems (time-of-flight, TOF) and combined in MS/MS tandems.

5.2.1 Sampling, sample storage and sample treatment

Speciation analysis is carried out in three stages: 1 – preparation of a sample, 2 – separation of chemical species, 3 – their identification and possibly quantification of identified species. There is a wide range as well of matrix as chemical compounds identified among biological samples and therefore a lot of different sample preparation methods exist for speciation analysis purposes. The analytical procedure starts of course with sampling but except samples' representativeness a special attention has to be paid to keep a constant speciation of an analyte in the investigated materials during collection procedures. Even more troublesome is storage of collected samples and then washing, drying, fragmenting and homogenizing.

In the case of plant materials and possible dust-borne deposits on its surfaces, which can interfere the results of speciation analysis, its decontamination is highly recommended prior to analysis. This can be achieved with rinsing plant tissues with deionized water with the addition of detergent. The more complicated problem is with coated soil plant roots. Decontamination damages root cells during removal of soil particles and therefore plants cultivated in hydroponics are more appropriate for speciation analysis in plant roots.

The main purpose of sampling treatment in speciation analysis is **extraction** of interested compounds from a sample without any changes in its speciation. Soft extraction techniques meet these requirements. Low efficiency of this process is not a problem as the clue is stable speciation. For elements bound to peptides, water (with eventual acid addition) or organic solvents extraction might be a solution. Increase of it causes peptide denaturation and destroys cell membrane increasing the efficiency of elemental extraction. Other solvents used for extractions, for example, from plant tissues might be highly complex solvents containing buffers, electrolytes, reducing agents, antibacterial agents and enzyme inhibitors. Enzymatic extraction is commonly used due to the many advantages: enzymes do not demand higher temperatures (usually 35–50 °C) and work in biological range of pH. It is time-consuming though. There are a lot of parameters that have an impact on the efficiency of this process, for example, time, temperature, pH, stirring, shaking, using ultrasound baths, sample mass to extractant volume ratio. Ultrasound bath shortens the time of efficient extraction significantly. But very often species cannot be efficiently extracted by the use of only one extractant. In this case, sequential and parallel extraction techniques have been developed. Sequential extraction increases the risk of primary speciation disorders but at the same time it gives total element content and individual species in solvent fractions. During this process a series of different extractants in specified order (water, dilute salt solutions, enzymes, organic solvents, acids and bases) are applied to the same sample in order to subdivide the total element content and individual species in solvent fractions.

5.3 Elements

Even though there are multi-elemental methods of detection available, nowadays speciation analysis is considered usually for only one element. The choice of the method depends on the researched material and element of interest.

Selenium is an essential trace element with a very narrow range from deficient, essential and toxic doses in living organisms. Therefore, an extensive effort has been done in recent years to develop effective analytical techniques of accurate determination of this element as well as identification of its species. Due to low concentrations in most biological materials, selenium speciation analysis is a challenging task. The hyphenation of HPLC, gas chromatography (GC) and capillary electrophoresis with ICP MS detection enabled speciation of low concentrated selenium in real samples. But except excellent sensitive ICP MS detection efficient extraction, improvement of chromatographic resolution, development of complementary molecular MS techniques for standard-less peak identification was needed.

Selenium compounds in biological samples such as human urine and serum are determined and identified just after centrifugation and dilution with the use of separation techniques coupled to sensitive determination methods: ion chromatography (IC) coupled with ICP MS method [35], HPLC coupled with ICP MS [36, 37], size exclusion chromatography (SEC) coupled with ICP MS, liquid chromatography (LC) coupled with ESI MS/MS [38]. In most biological materials, different extraction procedures of selenium compounds are applied using various extractants, predominantly water [35, 39–44] and enzymes [45–48]. In case of volatile selenocompounds, GC is used as a separation method, but because most of selenocompounds are not volatile, LC is preferred and HPLC as a separation method is one of the most frequently applied. In case of homogenous and representative samples, it is also possible to use X-ray absorption spectra technique and X-ray absorption near-edge structure (XANES) method [49]. Selenium is not considered an essential plant nutrient but more selenium species has been identified in plants than in animals, for example, inorganic selenate (Se(VI)) and selenite (Se(IV)), organic forms such as selenomethionine, selenocysteine, Se-methyl-selenocysteine, γ-glutamyl Se-methyl-selenocysteine, selenoproteins. Some plants also volatilize selenium as dimethyl-selenide or dimethyl-diselenide.

Aluminium is the third most abundant element in the Earth's crust and the most abundant metal. After a sharp reduction in fish population and also the complete disappearance of some species in the lakes of the United States and Europe, some studies concentrated on the effects of aluminium from acidified soils on the viability of fish and other aquatic organisms. It has been proved that aluminium forms both organic and inorganic monomeric complexes and the polymerization produces high molecular compounds migrating to the natural water. The most reactive and toxic forms of aluminium in an aqueous medium are Al^{3+}, $Al(OH)^{2+}$, $Al(OH)_{2^+}$ and very labile complexes of aluminium. All these forms are included in the so-called inorganic monomeric aluminium. It's been also indicated that aluminium in body fluids can be toxic to humans according to its species; however, the mechanism is still not clear. And because of very low levels of aluminium concentration in biological materials a sensitive method of detection such as ICP MS coupled to HPLC is frequently used in speciation analysis of this element [50–52]. Fast protein LC with ICP OES and electrospray MS/MS detection have been used for Al speciation in plant tissues and for separation of positively charged monomeric aqua- and hydroxyl-aluminium species, positively charged low molecular mass like aluminium organic complexes such as Al-citrate and Al-malate [53].

Cadmium is relatively poorly distributed in the Earth's crust but on the other hand it has been wildly distributed over the environment by human activities. Additionally, cadmium has a dangerous impact on the health of humans and animals and is easily absorbed by plants from soils. The mechanism of cadmium uptake by plants is proportional to its concentration in the environment. The ability to accumulate cadmium in plants depends not only on its quantity in the soil but also on its species. Therefore, speciation of cadmium complexes (phytochelatins) in plant material is frequently the subject of intensive studies [54–56]. Except excellent methods used for cadmium speciation such as SEC–ICP MS, ESI–TOF MS, HPLC–ESI–MS, a μ-XANES method was developed for cadmium speciation in plant tissues and allowed determination of $CdSO_4$, CdS, $CdCl_2$, $CdNO_3$, Cd-acetate and CdO [57, 58]. The localization of cadmium in the roots and leaves of a plant *Arabidopsis thaliana* was investigated using scanning electron microscopy coupled with energy-dispersive X-ray analysis and synchrotron-based micro-X-ray fluorescence [59]. The species of cadmium (Cd-cysteine, Cd-glutathione, Cd-pectine and CdS) were established by Cd_{III}-edge μ-XANES, and the cadmium concentrations were determined using ICP MS method. Speciation of cadmium in bean plant extracts performed with the use of ESI–TOF–MS showed that only a small fraction of the xylem sap Cd (less than 10%) may bind to certain thiol compounds, while the large fraction may be free of Cd^{2+} [60].

All *vanadium* compounds should be considered toxic. Vanadium plays a very limited role in human biology and is much more important in marine environments than terrestrial ones. Again an application of LC–ICP MS or HPLC–ICP MS methods allows for easy and sensitive speciation of V(III), V(IV) and V(V) in biological materials [61, 62]. SEC–ICP MS and ESI–MS/MS characterize the unknown vanadium metabolites [63].

There is evidence for the essentiality of *arsenic* as a trace mineral in birds and in mammals. However, the biological mechanism for its essential function is not known. Arsenic is mostly associated with its toxicity in living organisms. The progress of arsenic speciation studies has linked geochemistry with plant uptake, translocation and storage as well as metabolism and toxicity. Plants are organisms responsible for the transfer of arsenic between terrestrial environment and animals and due to the health risks posed to humans a lot of interest is given to the biological and metabolic pathways of arsenic in plants. Both, inorganic and organic, forms of arsenic are toxic and its poisonous properties have been known since ancient times. If there are no problems with detection of low arsenic content usually performed with ICP MS, the separation step is much more problematic. Moreover, extraction as well as storage can change the native speciation in biological samples. HPLC technique is the most successful separation technique applied in arsenic speciation analysis. IC–ICP MS method has frequently been used for As speciation in various biological materials [64, 65]. Arsenobetaine (AsB), inorganic arsenite (As^{III}), dimethylarsinic acid (DMA), monomethylarsonic acid (MMA), inorganic arsenate (As^{V}), 3-nitro-4-hydroxyphenylarsonic acid (Roxarsone) and *N*-acetyl-4-hydroxy-*m*-arsanilic acid were identified in breast muscle tissues [66]. HPLC–ICP MS and hydride generation atomic fluorescence spectroscopy were used as the most highly recommended techniques for speciation and total arsenic determination of As(III), As(V), MMA and DMA, AsB [67, 68]. HPLC ICP MS and LC ICP MS methods are also successfully applied in *lead* [54, 69], *thallium* [70–73], *nickel* [74, 75], *manganese* [76–78] and *chromium* [79–81] speciation. Chromium speciation can also be performed using nanoparticles magnetic solid-phase extraction and FAAS detection [82].

Not only speciation of elements considered as toxic is under investigation. *Copper* is an element very well known for its positive and important functions in organisms and therefore its speciation is widely studied in various biological materials [83]. XANES method allowed to determine Cu(I)-acetate, Cu (II)-acetate, Cu (I)-glutathione, Cu (I)-cysteine and Cu(II)-histidine in food and marine samples [84, 85]. Analysis showed that copper species in this material could be divided into groups of the Cu-S bond species, the Cu-O or the Cu-N bond species. Application of SEC–ICP-MS allowed for speciation analysis of copper proteins: transcuprein, ceruloplasmin and albumin in body fluids [86, 87]. Chromatographic analysis showed that copper is mostly bound to ceruloplasmin present in human serum, and small part of this element exists bound to albumin and approximately 1% is ionic or bound to low-molecular-weight fractions. Another essential element for living organisms is *zinc*. Sample extraction for zinc speciation was performed mainly in water [88, 89] with SEC–ICP MS but in human serum speciation was performed without this stage with the use of anion-exchange HPLC coupled with ICP MS [87].

There is no question that total element analysis is demanded for clinical assessment and control (deficiencies of essential elements, presence of toxic elements or toxic overloading, pharmacological levels) but that information is not sufficient. The role of some trace and ultra-trace elements in living organisms is rich and varied. This role as well as metabolic behaviour, bioavailability and toxicity of an element in biological materials is determined by its actual chemical form. Most of analytical procedures used for analytical speciation involve a lot of stages, including extraction, preconcentration, cleaning, derivatization, separation and detection. In case of biological materials except low concentration of elements, there is a great problem as complex matrix. But without any excuses all analytical measurements have to be accurate and precise and require a systematic check for both the trace element and the biocompound.

References

[1] Taylor A, Day MP, Hill S, Marshall J, Patriarca M, White M. Atomic spectrometry update: review of advances in the analysis of clinical and biological materials, foods and beverages. J Anal At Spectrom 2015, 30, 542–79.

[2] Su CK, Sun YCh. Consideration of inductively coupled plasma mass spectrometry techniques for characterizing the dissolution of metal based nanomaterials in biological tissues. J Anal At Spectrom 2015, 30, 1689–705.

[3] Ward NI, Durrant SF, Gray AL. Analysis of biological standard reference materials by laser ablation inductively coupled plasma mass spectrometry. J Anal At Spectrom 1992, 7, 1139–46.

[4] Durrant SF, Ward NI. Laser-ablation inductively coupled plasma mass spectrometry for the multielemental analysis of biological materials – a feasibility study. Food Chem 1994, 43(3), 317–23.

[5] Limbeck A, Galler P, Bonta M, Bauer G, Nischkauer W, Vanhaecke F. Recent advances in quantitative LA-ICP-MS analysis: Challenges and solutions in the life sciences and environmental chemistry. Anal Bioanal Chem 2015, 407, 6593–617.

[6] Durrant SF, Ward NI. Recet biological and environmental applications of laser ablation inductively coupled plasma mass spectrometry (LA ICP MS). J Anal At Spectrom 2005, 20, 821–9.

[7] Pozebon D, Scheffler GL, Dressler VL, Nunes MAG. Review of the application of laser ablation inductively coupled plasma mass spectrometry (LA-ICP-MS) to the analysis of biological samples. J Anal At Spectrom, 2014, 29, 2204–28.

[8] Wang LM, Becker JS, Wu Q, Oliveira MF, Bozza FA, Schwager AL, Hoffman JM, Morton KA. Bioimaging of copper alterations in the aging mouse brain by autoradiography laser ablation inductively coupled plasma mass spectrometry and immunohistochemistry. Metallomics 2010, 2, 348–53.

[9] Becker JS, Matusch A, Wu V. Bioimaging mass spectrometry of trace elements – recent advance and applications of LA-ICP-MS: A review. Anal Chim Acta 2014, 835, 1–18.

[10] Yilmaz S, Tokaliglu S, Sahan S, Ulgen A, Sahan A, Soykan C. On-line preconcentration/determination of zinc from water, biological and food samples using synthesized chelating resin and flame atomic absorption spectrometry. J Trace Elem Med Biol 2013, 27, 85–90.

[11] Ozdemir C, Sacmaci S, Kartal S. A coprecipitation procedure for the determination of some metals in food and environmental samples by flame atomic absorption spectroscopy. Anal Methods 2013, 5, 3977–983.

[12] Rezvani M, Ebrahimzadeh H, Aliakbari A, Khalilzadeh A, Kasaeian M, Amini M. Novel modified carbon nanotubes as a selective sorbent for preconcentration and determination of trace copper ion in fruit samples. J Sep Sci 2014, 37, 2559–65.
[13] Ferreira SLC, Lemos VA, Moreira BC, Spinola Costa AC, Santelli RE. An on-line continuous flow system for copper enrichment and determination by flame atomic absorption spectroscopy. Anal Chim Acta 2000, 403, 259–64.
[14] Gurkan R, Altunay N. Determination of total Sn in some canned beverages by FAAS after separation and preconcentration. Food Chem 2015, 177, 102–10.
[15] Zhuravlev A, Zacharia A, Gucer S, Chebotarev A, Arabadji M, Dobrynin A. Direct atomic absorption spectrometry determination of arsenic, cadmium, copper, manganese, lead and zinc in vegetable oil and fat samples with graphite filter furnace atomizer. J Food Compos Anal 2015, 38, 62–8.
[16] Qadir MA, Ahmed M, Shahzad S. Determination of aluminum by electrothermal atomization atomic absorption spectrometry in serum to characterize hemodialysis toxicity. Anal Lett 2015, 48, 147–53.
[17] Junior MMS, Silva LOB, Leao DJ, Ferreira SLC. Analytical strategies for determination of cadmium in Brazilian vinegar samples using ETAAS. Food Chem 2014, 160, 209–213.
[18] Ohtsu N, Ashino T, Kimura H, Takada K, Hanawa T. Investigation for analytical procedure for determination of trace metallic ions in simulated body fluids by inductively coupled plasma atomic emission spectrometry. J Mater Sci Mater Med 2007, 18, 429–33.
[19] Bocca B, Alimonti A, Petrucci F, Violante N, Sancesario G, Forte G, Senofonte O. Quantification of trace elements by sector filed inductively coupled plasma mass spectrometry in urine, serum, blood and cerebrospinal fluid of patients with Parkinson's disease. Spectrochim. Acta, Part B 2004, 59, 559–66.
[20] Pozebon D, Dressler VL, Curtius AJ. Determination of trace elements in biological materials by ETV-ICP-MS after dissolution or slurry formation with tetramethylammonium hydroxide. J Anal At Spectrom 1998, 13, 1101–5.
[21] Moreno-Gordaliza E, Giesen Ch., Lazaro A, Esteban-Fernandez D, Humanes B, Canas B, Panne U, Tejedor A, Jakubowski N, Milagros Gomez-Gomez M. Elemental bioimaging in kidney by LA-ICP-MS as a tool to study nephrotoxicity and renal protective strategies in cisplatin therapies. Anal Chem 2011, 83, 7933–40.
[22] Wu B, Chen Y, Becker JS. Study of essential element accumulation in the leaves of a Cu-tolerant plant Elsholtzia splendens after Cu treatment by imaging laser ablation inductively coupled plasma mass spectrometry (LA-ICP-MS). Anal Chim Acta 2009, 633, 165–72.
[23] Hodrogiannis EM, Ehrlinger E, Popalski A, Lisle M. Use of laser ablation inductively coupled plasma mass spectrometry time of flight to identify the element composition of vanilla and determine the geographic origin by discriminant function. J Agric Food Chem 2013, 61, 11332–7.
[24] Muller ALH, Muller EI, Barin JS, Flores EMM. Microwave-assisted digestion using diluted acids for toxic element determination in medicinal plants by ICP-MS in compliance with United States pharmacopeia requirements. Anal Methods 2015, 7, 5218–24.
[25] Bocca B, Alimonti A, Forte G, Petrucci F, Pirola C, Senofonte O, Violante N. High-throughput microwave-digestion procedures to monitor neurotoxic elements in body fluids by means of inductively coupled plasma mass spectrometry. Anal Bioanal Chem 2003, 377, 65–70.
[26] Fridlund S, Littlefield S, Rivers J. The use of modified microwave digestion dissolution for the quantitative-determination of aluminum, silicon, and iron in biological-materials by inductively-coupled plasma spectrometry. Commun Soil Sci Plant Anal 1994, 25, 7–8, 933–41.
[27] Cornelis R, De Kimpe J. Elemental speciation in biological fluids. J Anal At Spectrom 1994, 9, 945–50.
[28] Szpunar J, Lobinski R, Prange A. Hyphenated techniques for elemental speciation in biological systems. Appl Spectrosc 2003, 57(3), 102a–12a.

[29] IUPAC. Compendium of Chemical Terminology, 2nd ed. (The Gold Book). Compiled by A.D. McNaught and A. Wilkinson. Blackwell Scientific Publications, Oxford (1997). XML on-line corrected version: http://goldbook.iupac.org (2006-) created by M. Nic, J. Jirat, B. Kosata; updates compiled by A. Jenkins. ISBN 0-9678550-9-8. doi:10.1351/goldbook, Last update: 2014-02-24; version: 2.3.3., DOI of this term: doi:10.1351/goldbook.ST06861.Original PDF version: http://www.iupac.org/goldbook/ST06861.pdf

[30] Templeton DM, Ariese F, Cornelius R, Danielsson LG, Muntau H, Van Leeuwen HP, Łobiński R. Guidelines for terms related to chemical speciation and fractionation of elements. Definitions, structural aspects, and methodological approaches (IUPAC Recommendations 2000). Pure Appl Chem 2000, 72(8), 1453–70.

[31] Hulanicki A. Współczesna chemia analityczna. Wybrane zagadnienia, Warszawa 2001, Wydawnictwo Naukowe PWN SA.

[32] Casiot C, Vacchina V, Chassaigne H, Szpunar J, Potin-Gautier M, Lobinski R. An approach to the identification of selenium species in yeast extracts using pneumatically-assisted electrospray tandem mass spectrometry. Anal Commun 1999, 36, 77–80.

[33] Plumb RS, Stumpf CL, Gorenstein MV, Castro-Perez JM, Dear GJ, Anthony M, Sweatman BC, Connor SC, Haselden JN. Metabonomics: The use of electrospray mass spectrometry coupled to reversed-phase liquid chromatography shows potential for the screening of rat urine in drug development. Rapid Commun Mass Spectrom 2002, 16(20), 1991–6.

[34] Lafaye A, Junot C, Ramounet-Le GB, Fritsch P, Tabet JC, Ezan E. Metabolite profiling in rat urine by liquid chromatography/electrospray ion trap mass spectrometry. Application to the study of heavy metal toxicity. Rapid Commun Mass Spectrom 2003, 17(22), 2541–9.

[35] Wang R-Y, Hsu Y-L, Chang L-F, Jiang S-J. Speciation analysis of arsenic and selenium compounds in environmental and biological samples by ion chromatography–inductively coupled plasma dynamic reaction cell mass spectrometer. Anal Chim Acta 2007, 590, 239–44.

[36] Hongwei Y, Chunying C, Yuxi G, Bai L, Zhifang C. Selenium speciation in biological samples using a hyphenated technique of high-performance liquid chromatography and inductively coupled plasma mass spectrometry. Chin J Anal Chem 2006, 34(6), 749–53.

[37] González de Vega R, Fernández-Sánchez ML, Iglesias HG, Prados MC, Sanz-Medel A. Quantitative selenium speciation by HPLC-ICP-MS (IDA) and simultaneous activity measurements in human vitreous humor. Anal Bioanal Chem 2015, 407, 2405–13.

[38] Lu Y, Rumpler A, Francesconi KA, Pergantis SA. Quantitative selenium speciation in human urine by using liquid chromatography–electrospray tandem mass spectrometry. Anal Chim Acta 2012, 731, 49–59.

[39] Michalska-Kacymirow M, Kurek E, Smolis A, Wierzbicka M, Bulska E. Biological and chemical investigation of Allium cepa L. response to selenium inorganic compounds. Anal Bioanal Chem 2014, 406, 3717–22.

[40] Larsen EH, Lobinski R, Burger-Meÿer K, Hansen M, Ruzik R, Mazurowska L, Rasmussen PH, Sloth JJ, Scholten O, Kik C. Uptake and speciation of selenium in garlic cultivated in soil amended with symbiotic fungi (mycorrhiza) and selenite. Anal Bioanal Chem 2006, 385, 1098–1108.

[41] Ogra Y, Kitaguchi T, Ishiwata K, Suzuki N, Iwashita Y, Suzuki KT. Identification of selenohomolanthionine in selenium-enriched Japanese pungent radish. J Anal At Spectrom 2007, 22, 1390–6.

[42] Slekovec M, Goessler W. Accumulation of selenium in natural plants and selenium supplemented vegetable and selenium speciation by HPLC-ICPMS. Chem Speciation Bioavailability 2005, 17(2), 63–73.

[43] Tie M, Li B, Liu Y, Han J, Sun T, Li H. HPLC–ICP-MS analysis of selenium speciation in selenium-enriched Cordyceps militaris. RSC Adv 2014, 4, 62071–5.

[44] Mazej D, Falnoga I, Veber M, Stibilj V. Determination of selenium species in plant leaves by HPLC–UV–HG-AFS. Talanta 2006, 68, 558–68.

[45] Kápolna E, Fodor P. Speciation analysis of selenium enriched green onions (Allium fistulosum) by HPLC-ICP-MS. Microchem J 2006, 84, 56–62.
[46] Pedrero Z, Elvira D, Cámara C, Madrid Y. Selenium transformation studies during Broccoli (Brassica oleracea) growing process by liquid chromatography–inductively coupled plasma mass spectrometry (LC–ICP MS). Anal Chim Acta 2007, 596, 251–6.
[47] Cankur O, Yathavakilla SKV, Caruso JA. Selenium speciation in dill (Anethum graveolens L.) by ion pairing reversed phase and cation exchange HPLC with ICP-MS detection. Talanta 2006, 70, 784–90.
[48] Funes-Collado V, Morell-Garcia A, Rubio R, López-Sánchez JF. Selenium uptake by edible plants from enriched peat. Sci Hortic 2013, 164, 428–33.
[49] Eiche E, Bardelli F, Nothstein AK, Charlet L, Göttlicher J, Steininger R, Dhillon KS, Sadana US. Selenium distribution and speciation in plant parts of wheat (Triticum aestivum) and Indian mustard (Brassica juncea) from a seleniferous area of Punjab, India. Sci Total Environ 2015, 505, 952–61.
[50] Chen B, Zeng Y, Hu B. Study on speciation of aluminum in human serum using zwitterionic bile acid derivative dynamically coated C18 column HPLC separation with UV and on-line ICP-MS detection Talanta 2010, 81, 180–6.
[51] Shang DR, Zhao YF, Zhai YX, Ning JS, Ding HY, Sheng XF. Development of a new method for analyzing free aluminum ions (Al3+) in seafood using HPLC-ICP-MS. Chinese Science Bulletin 2013, 58(35), 4437–42.
[52] Street R, Drábek O, Szákova J, Mládková L. Total content and speciation of aluminium in tea leaves and tea infusions Food Chem 2007, 104, 1662–9.
[53] Bantan-Polak T, Mitrović B, Milacic R. The use of fast protein liquid chromatography with ICP-OES and ES–MS–MS detection for the determination of various forms of aluminium in the roots of Chinese cabbage. Anal Chim Acta 2005, 540, 83–9.
[54] Barałkiewicz D, Kózka M, Piechalak A, Tomaszewska B, Sobczak P. Determination of cadmium and lead species and phytochelatins in pea (Pisum sativum) by HPLC–ICP-MS and HPLC–ESI-MS^n. Talanta 2009, 79, 493–8.
[55] Persson P, Hansen TH, Holm PE, Schjoerring JK, Hansen JChB, Nielsen J, Cakmak I, Husted S. Multi-elemental speciation analysis of barley genotypes differing in tolerance to cadmium toxicity using SEC-ICP-MS and ESI-TOF-MS. J Anal At Spectrom 2006, 21, 996–1005.
[56] Połeć-Pawlak K, Ruzik R, Abramski K, Ciurzyńska M, Gawrońska H. Cadmium speciation in Arabidopsis thaliana as a strategy to study metal accumulation system in plants. Anal Chim Acta 2005, 540, 61–70.
[57] Fu X, Dou C, Chen Y, Chen X, Shi J, Yu M, Xu J. Subcellular distribution and chemical forms of cadmium in Phytolacca americana L. J Hazard Mater 2011, 186, 103–107.
[58] Akhter MF, Omelon ChR, Gordon RA, Moser D, Macfie SM. Localization and chemical speciation of cadmium in the roots of barley and lettuce. Environ Exp Bot 2014, 100, 10–19.
[59] Isaure M-P, Fayard B, Sarret G, Pairis S, Bourguignon J. Localization and chemical forms of cadmium in plant samples by combining analytical electron microscopy and X-ray spectromicroscopy. Spectrochim Acta, Part B 2006, 61, 1242–52.
[60] Hazama K, Nagata S, Fujimori T, Yanagisawa S, Yoneyama T. Concentrations of metals and potential metal-binding compounds and speciation of Cd, Zn and Cu in phloem and xylem saps from castor bean plants (Ricinus communis) treated with four levels of cadmium. Physiol Plant 2015, 154, 243–55.
[61] Colina M, Gardiner PHE, Rivas Z, Troncone F. Determination of vanadium species in sediment, mussel and fish muscle tissue samples by liquid chromatography–inductively coupled plasma-mass spectrometry. Anal Chim Acta 2005, 538, 107–115.

[62] Sherry Li X, Chris Le X. Speciation of vanadium in oilsand coke and bacterial culture by high performance liquid chromatography inductively coupled plasma mass spectrometry, Anal Chim Acta 2007, 602, 17–22.
[63] Nischwitz V, Davies JT, Marshall D, González M, Gómez Ariza JL, Goenaga-Infante H. Speciation studies of vanadium in human liver (HepG2) cells after in vitro exposure to bis(maltolato)oxovanadium(IV) using HPLC online with elemental and molecular mass spectrometry. Metallomics 2013, 5, 1685–97.
[64] Mandal BK, Ogra Y, Anzai K, Suzuki KT. Speciation of arsenic in biological samples. Toxicol Appl Pharmacol 2004, 198, 307–18.
[65] Pizarro I, Gómez M, Cámara C, Palacios MA. Arsenic speciation in environmental and biological samples. Extraction and stability studies. Anal Chim Acta 2003, 495, 85–98.
[66] Liu Q, Peng H, Lu X, Le XCh. Enzyme-assisted extraction and liquid chromatography mass spectrometry for the determination of arsenic species in chicken meat. Anal Chim Acta 2015, 888, 1–9.
[67] Mandal BK, Ogra Y, Suzuki KS. Speciation of arsenic in human nail and hair from arsenic-affected area by HPLC-inductively coupled argon plasma mass spectrometry. Toxicol Appl Pharmacol 2003, 189, 73–83.
[68] Zhao D, Li HB, Xu JY, Luo J, Ma LQ. Arsenic extraction and speciation in plants: Method comparison and development. Sci Total Environ 2015, 523, 138–45.
[69] Chen Y, Huang L, Wu W, Ruan Y, Wu Z, Xue Z, Fu FF. Speciation analysis of lead in marine animals by using capillary electrophoresis couple online with inductively coupled plasma mass spectrometry. Electrophoresis 2014, 35, 1346–52.
[70] Nolan A, Schaumloffel D, Lombi E, Ouerdane L, Łobiński R, McLaughlina M. Determination of Tl(I) and Tl(III) by IC-ICP-MS and application to Tl speciation analysis in the Tl hyperaccumulator plant Iberis intermedia. J Anal At Spectrom 2004, 19, 757–61.
[71] Krasnodębska-Ostręga B, Asztemborska M, Golimowski J, Strusińska K. Determination of thallium forms in plant extracts by anion exchange chromatography with inductively coupled plasma mass spectrometry detection (IC-ICP-MS). J Anal At Spectrom 2008, 23, 1632–5.
[72] Pałdyna J, Beata Krasnodębska-Ostręga B, Sadowska M, Gołębiewska J. Indirect speciation analysis of thallium in plant extracts by anodic stripping voltammetry. Electroanalysis 2013, 25(8), 1926–32.
[73] Krasnodębska-Ostręga B, Sadowska M, Ostrowska S. Thallium speciation in plant tissues—Tl(III) found in Sinapis alba L. grown in soil polluted with tailing sediment containing thallium minerals. Talanta 2012, 93, 326–9.
[74] Ouerdane L, Mari S, Czernic P, Lebrun M, Lobinski R. Speciation of non-covalent nickel species in plant tissue extracts by electrospray Q-TOFMS/MS after their isolation by 2D size exclusion hydrophilic interaction LC (SEC HILIC) monitored by ICP-MS. J Anal At Spectrom 2006, 21, 676–83.
[75] Vacchina V, Mari S, Czernic P, Marquee's L, Pianelli K, Schaumlolffel D, Lebrun M, Lobinski R. Speciation of nickel in a hyperaccumulating plant by high-performance liquid chromatography-inductively coupled plasma mass spectrometry and electrospray MS/MS assisted by cloning using yeast complementation. Anal Chem 2013, 75(11), 2740–5.
[76] Quintana M, Klouda AD, Ochsenkuhn-Petropoulou M, Michalke B. Size characterization of manganese species from liver extracts using size exclusion chromatography inductively coupled plasma mass spectrometry. Anal Chim Acta 2005, 554, 130–5.
[77] Diederich J, Michalke B. Enhanced extract preparation of native manganese and iron species from brain and liver tissue. Anal Bioanal Chem 2011, 399,1799–806.
[78] Michalke B, Lucio M, Berthele A, Kanawati B. Manganese speciation in paired serum and CSF samples using SEC-DRC-ICP-MS and CE-ICP-DRC-MS. Anal Bioanal Chem 2013, 405, 2301–9.

[79] Vacchina V, de la Calle I, Séby F. Cr(VI) speciation in foods by HPLC-ICP-MS: investigation of Cr(VI)/food interactions by size exclusion and Cr(VI) determination and stability by ion-exchange on-line separations. Anal Bioanal Chem 2015, 407, 3831–9.

[80] Novotnik B, Zuliani T, Scancar J, Milacic R. Content of trace elements and chromium speciation in neem powder and tea infusions. J Trace Elem Med Biol 2015, 31, 98–106.

[81] Novotnik B, Zuliani T, Scancar J, Milacic R. Chromate in food samples: An artefact of wrongly applied analytical methodology? J Anal At Spectrom 2013, 28, 558.

[82] Wu Y, Zhang J, Liu J, Chena L, Denga Z, Hana M, Weia X, Yua A, Zhanga H. Fe3O4@ZrO2 nanoparticles magnetic solid phase extraction coupled with flame atomic absorption spectrometry for chromium(III) speciation in environmental and biological samples. Appl Surf Sci 2012, 258, 6772–6.

[83] Abollino O, Aceto M, Bruzzoniti MC, Mentasti E, Sarzanini C. Speciation of copper and manganese in milk by solid-phase extraction/inductively coupled plasma-atomic emission spectrometry. Anal Chim Acta 1998, 375, 299–306.

[84] Cekoa MJ, Aitkenb JB, Harris HH. Speciation of copper in a range of food types by X-ray absorption spectroscopy. Food Chem 2014, 164, 50–54.

[85] Tan QG, Wang Y, Wang WX. Speciation of Cu and Zn in two colored oyster species determined by Xray absorption spectroscopy. Environ Sci Technol 2015, 49, 6919–25.

[86] El Balkhi S, Poupon J, Trocello JM, Massicot F, Woimant F, Laprévote O. Human plasma copper proteins speciation by size exclusion chromatography coupled to inductively coupled plasma mass spectrometry. solutions for columns calibration by sulfur detection. Anal Chem 2010, 82, 6904–10.

[87] Malavolta M, Piacenza F, Basso A, Giacconi R, Costarelli L, Pierpaoli S, Mocchegiani E. Speciation of trace elements in human serum by micro anion exchange chromatography coupled with inductively coupled plasma mass spectrometry. Anal Biochem 2015, 421, 16–25.

[88] Ovca A, Johannes T. van Elteren, Falnoga I, Šelih VS. Speciation of zinc in pumpkin seeds (Cucurbita pepo) and degradation of its species in the human digestive tract. Food Chem 2011, 128, 839–46.

[89] Weber G, Konieczyński P. Speciation of Mg, Mn and Zn in extracts of medicinal plants. Anal Bioanal Chem 2003, 375, 1067–73.

Wiktor Dmitruk and Zuzanna Brożek-Mucha

6 Forensic Analysis of Microtraces

6.1 Characterisation of forensic microtraces

People commit a crime or undergo accidents through the medium of things and leave traces at the crime scene. Criminal investigation concerns people and objects. Although the trace in forensic sciences is being defined as every change in the environment that happened at the site and moment of a crime, the subject of physical and chemical analysis for forensic purposes is the physical evidence. Various materials may become traces of a crime. Sometimes they occur in very small quantities, and so they are called microtraces. They are unique for the case and their small amount ought to be sufficient for carrying out all the necessary examinations. The analytical results to be obtained are expected to meet forensic and legal requirements.

The main aim of the studies on the chemical composition and properties of microtraces is their identification, that is establishing the class of the material or an object they might originate from (*classification problem*) as well as their comparison with a reference material, whenever accessible for investigations (*discrimination problem*). Also more advanced studies are being performed, for example to determine mutual relationships between the physical and chemical properties of microtraces and their source as well as to obtain the knowledge on the processes of their formation and *the primary and secondary transfer* from one substrate to another one.

The material in the form of microtraces is being characterised by its minor amount measured in micro-, nano- and even picograms and also by the difficulty in its separation from the substrate. Moreover, a microtrace being evidence in a forensic case should not be completely consumed or damaged in the course of chemical analysis. Thus, the analytical methods of the primary choice for forensic analysis are "non-destructive" ones, such as optical and electron microscopy, microspectrometric methods, for example in the UV-VIS range and the infrared, as well as the X-ray spectrometry and X-ray diffraction. They enable one to establish the chemical composition of the examined material and the crystalline phase of the particular components in solid state.

6.2 Analytical methodology

In order to identify the examined materials, it is necessary to determine their chemical composition – mostly qualitative – and to establish some of their physical and chemical properties. Methods of elemental analysis have a special place. Three kinds of measuring techniques that utilise X-rays are most often used in the analysis of traces: X-ray fluorescence spectrometry (XRF), micro-XRF (μ-XRF) and scanning electron microscopy coupled with energy-dispersive X-ray spectrometry (SEM/EDX). X-rays are

https://doi.org/10.1515/9783110366730-007

applied in trace examination in order to observe the morphology of the trace and determine its elemental composition. They also provide information, which can help to establish if the suspected person was involved in a crime. X-ray diffraction is another method occasionally used by forensic experts which analyses the crystalline structure of the material rather than its elemental content and has the ability to provide definite identification of its components.

Among the above-mentioned techniques, SEM/EDX is particularly useful. It provides multiple investigative possibilities in case of very small samples of chemical substances as well as complex materials in solid phase; it provides the possibility of simultaneous examination of the chemical composition as well as the morphology. Moreover, an additional benefit of usage of the technique is no need of destruction of the sample, neither during the preparation, nor in the course of the analysis.

SEM/EDX can be applied for the identification and comparative studies of evidence materials such as fragments of glass and paints, metals, photocopy toners, chosen inorganic poisons, skeletons of diatoms, morphology of textile fibres, endings of filaments of car light bulbs, as well as post-blast and gunshot residues (GSRs).

The XRF method, however, is irreplaceable in investigations of gunshot holes occurring on various bases. A fragment of fabric cut from clothing or a fragment of skin or bones together with the gunshot wound can be placed directly in the measuring dish of the X-ray spectrometer and analysed for the presence of metals originating from the primer, the bullet and the cartridge. Their disclosure confirms the use of a firearm and contributes to inferences about estimations of the shooting distance.

The observation of the morphology of single fibres using SEM enables identification of the kind of natural fibres. The form of the surface is very characteristic for cotton (ribbon), wool (scale on surface), silk (with characteristic node) and jute fibres. On the other side, the chemical fibres have very similar form. The observation of cross section of fibres helps in discrimination of chemical fibres. All the information obtained provides confirmation of a contact between two people (e.g. the victim and the suspect) or a person and tool used in the investigated act. Moreover, examining damaged fibre contributes to determination of the mechanism of its separation from the textile (i.e. whether it was cut or torn, what kind of tool was used). The observation of endings of fibres is useful for this purpose. It is possible to find out what was the reason for damage of the textile (biochemical, fire and explosion).

Observation of filament of lighting bulb originated from car involved in a road accident provides information whether during the accident the lights were on or off (droplet of metal/fragile fracture).

When comparing soil or geological samples, the observation and identification of traces such as minerals as well as pollen and spore provide more characteristic information.

When analysing biological material originated from stomach or lungs of a victim, the presence of small silica skeletons of diatoms points to the cause of death by drowning.

6.3 Evidence materials

6.3.1 GSR – inorganic gunshot residue

GSRs belong to the forensic traces of the greatest evidential value due to their rare chemical and physical characteristics as well as to their low prevalence in the environment. GSRs are usually defined as gaseous, liquid and solid products of subsequent detonation of the primer and combustion of the propellant that are formed, when a firearm is discharged. Their examination plays an important role in establishing some circumstances of a crime with the use of a firearm. When revealed around the gunshot hole and on the clothing and body of the shooter, they provide information for concluding about the shooting distance and the kind of ammunition used and, most importantly, they serve to link the suspect to the shooting [1–4].

Inorganic component of GSR contains metallic particles originating from the ammunition primer revealing characteristic properties and therefore providing a high evidential value of using a firearm [5]. Only few particles, of total mass number greater than 100 pg, can be accepted as the evidence relating an individual involved in a shooting incident. The fundamental criterion of the identification of metallic gunshot particles is their chemical composition, for example, rare set of lead, antimony and barium related to traditional ammunition types, and their morphology reflecting the kinetics of the explosive processes during a gunshot, embracing rapid cooling of droplets of molten metal such as presented in Figures 6.1–6.4.

The most successful and up-to-date technique for the analysis of GSR particles is SEM/EDX. The combination of information on the elemental composition as well as the morphologic features of an analysed object contributes to differentiation of any metallic trace from a GSR.

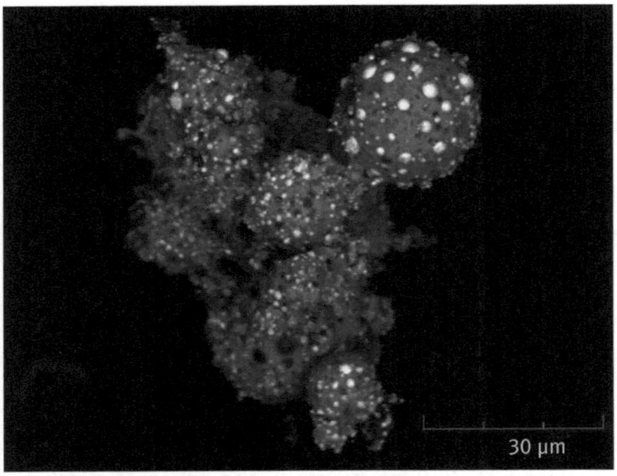

Figure 6.1: The consistent GSR particle composed of lead, calcium, barium and silicone (PbCaBaSi).

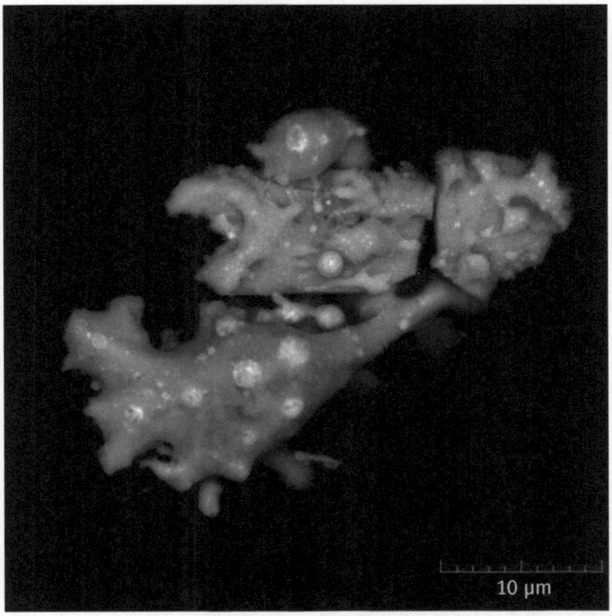

Figure 6.2: The characteristic GSR particle composed of lead, barium and antimony (PbSbBa).

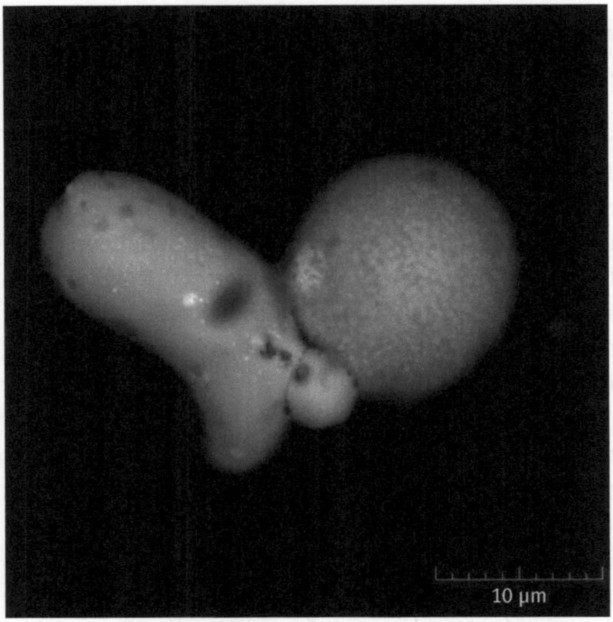

Figure 6.3: The characteristic airborne GSR particle composed of lead, barium and antimony (PbSbBa).

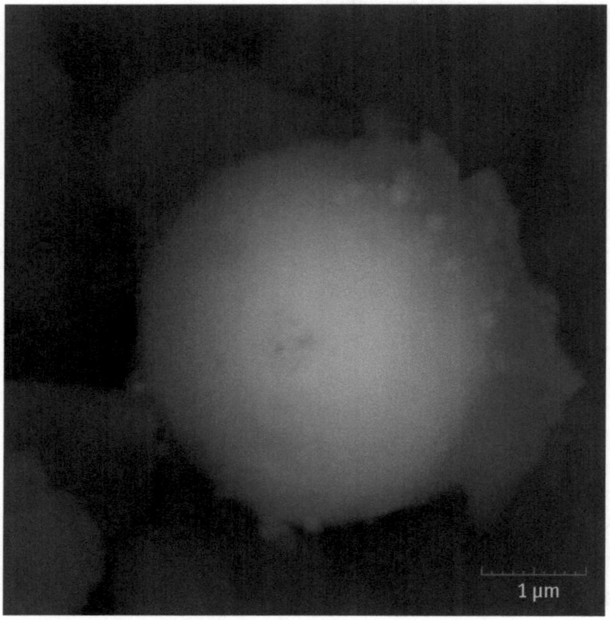

Figure 6.4: The consistent airborne GSR particle composed of sulphur, tin and antimony (SSnSb).

Usually, half-inch diameter aluminium stubs with an adhesive layer of double-sided tape are used for sampling, that is for collecting the GSR from hands, face, hair, clothing (Figure 6.5).

When the sample has been secured on the stub, it is searched for the presence of spherical metallic particles of defined diameter and chemical composition – manually or using suitable software for automatic search. Imaging with the use of backscattered electrons enables to focus on heavy metal particles keeping light element debris such as fibres, hair and epidermis below the limit of detection. Detection of this signal indicates differences of the chemical contents of phases that can be perceived as different shades of grey. Observation of the images of single or grouped particles reflects their complex nature and the extreme conditions they undergo at various stages of their formation.

Chemical composition and properties of GSR depend directly on the kinds of materials used in the production of the ammunition. J. S. Wallace classified the most common primer mixtures into eight classes, among which only one includes all modern solutions that relay mainly on the substitution of heavy element toxic compounds with materials of more abundant composition [2]. Thus, in the majority of cases the evaluation of analytical findings can be based upon a formal classification scheme for gunshot particles being worked out and applied by the experts in Western Europe and the USA since about 1980 taking into account the experiences with traditional ammunition, see for example [4]. Particles containing lead, antimony and barium

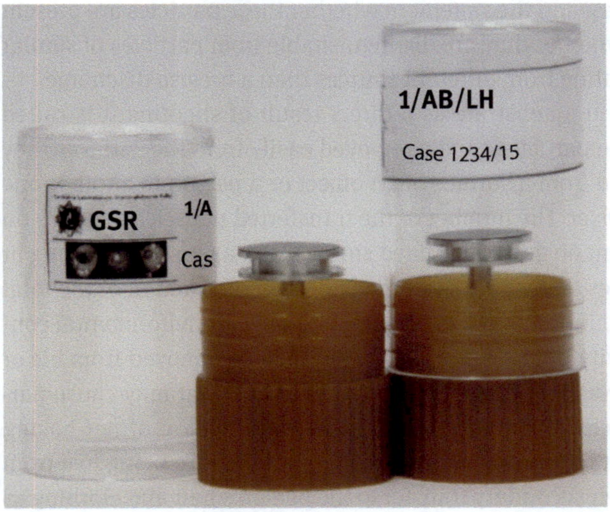

Figure 6.5: The standard half-inch microscopic stubs used for collecting microtraces.

(Pb–Sb–Ba) are nowadays classified into *characteristic* class allowing to state with a very high probability that they originated from a firearm discharge. Particles containing two- or one-component combinations of these elements (Pb–Sb, Pb–Ba, Sb–Ba, Pb, Sb and Ba) as well as the ones containing barium, calcium and silicon (Ba–Ca–Si) were classified into the class *consistent with GSR* that allow to infer on their origin from a firearm discharge with a high probability [6, 7]. Torre et al. [8] demonstrated that a single particle containing lead, antimony and barium of an oval shape resembling the morphology of GSR can be formed in car brake pads. One of the authors revealed primer residues that occurred not as a result of a firearm discharge but of a detonation of a primer component in the area of an ammunition factory. For these very rare cases, the three-component particles cannot be called *unique*; nevertheless, especially when prolific in a sample, they provide a high evidential value of a gunshot. Establishing the presence of GSR in the sample collected from hands or clothing of a person allows relating the person with a high or very high probability, either with shooting, or being in the nearest vicinity of a firing gun, or getting in contact with an object highly contaminated with GSR.

Due to a variety of chemical composition of primer mixtures applied in production of ammunition worldwide, the formal scheme of classification of GSR and so the system of the evaluation of the evidence sometimes might be inadequate for preparing reliable expert's reports. Romolo and Margot [9] proposed another individual attempt to the assessment of the evidential value of metallic particles based on the mutual consistency of particles found in the items of certain case, called *case-to-case* approach, rather than comparing them with the arbitrary classification scheme. The majority of works published earlier concerned only the identification of lead, antimony and barium particles, and the statement whether these particles are present in the studied material and whether they are distinguishable from particles of similar chemical contents but originating from different sources than a firearm discharge.

When GSR is deposited on a substrate as a direct result of shooting it is called *primary transfer*. However, the particles can be removed easily from the surfaces they are deposited, and transferred from a surface of an object or a person to another one as a result of a *secondary transfer*. The number of the transferred particles depends on the number of particles present on a contaminated surface (e.g. a person's clothing or hands) as well as its properties and would most likely be a small fraction of the total number of particles deposited initially [10–15]. Depending on the environmental conditions and the physical activity of the shooter, particles may be removed from his or her hands, for example, by washing hands soon after shooting that may cause partial or complete loss of particles. It has been found that even in case of not having washed hands within the first four hours after shooting, particles are being lost from the shooter's hands due to their secondary transfer to the person's hair and clothing as well as to the environment. The chemical composition of particles does not influence the process of their loss, contrary to their shape. It was observed that particles remaining for longer at certain surfaces reveal sizes not only within the submicron–micron range but also greater ones, however, usually of irregular shape and more developed surface [13].

The loss of particles from one substrate and their transfer to another one, for example, to persons and objects that were not present at the site, when shooting occurred, are of great importance for forensic science examiners. This corresponds to the contamination with GSR of certain occupational environments at a level that depends on how often the firearm has been used. Forensic laboratories gather their own knowledge and experience on the persistence and prevalence of GSR in an environment to be able to assess the risk of contamination of pollution in a given case. This is perceived as a major issue and reflected in recent publications [14, 15] concerning the interpretation of the presence of GSR in certain circumstances and the primary and secondary transfer of GSR particles. However, not many of them attempted a study of physical and chemical mechanisms ruling these processes because they are complex and the examination of GSR is time-consuming.

A comprehensive information on the prevalence of GSR in various populations as a result of the primary and secondary transfers as well as a measure of factors that influence these processes and states is presented in the publications of one of the authors [15]. The examinations were necessary for the forensic science laboratory in dealing with the risk of contamination not only as the measure of internal controls but also as a background of the evaluation of the evidence within the forensic expertise concerning shooting cases, since particular experiences of laboratories in other countries could not have been adopted due to different social conditions they were obtained.

The subject of these studies involved 273 specimens collected from people of various occupations, both users and non-users of firearms. As a result, information on the primary and secondary transfer of GSR particles as well as their prevalence in certain environments was obtained. Frequent users of firearms create a class of the individuals highly contaminated with GSR. Occasional users of firearms such as police officers or foresters can be a source of contamination after using a gun. For this group, a criterion of low-risk contamination was established as 5 h since recent shooting. In a group of 100 individuals declaring no use of firearms, a single spherical Pb–Sb–Ba particle was found on the hands of only one person.

These studies provided a contribution to a possible assessment of contamination risks and to formulating recommendations for minimising them, while dealing with GSR samples being transported from the crime scene to the forensic laboratory. The study also advanced understanding of the mechanisms of transfer and persistence of GSR and further improved the quality of the expert opinion in relating an individual to the fact of using a gun. Whereas a single particle provides a limited evidential value, then finding more numerous groups of GSR particles makes the evidence of greater confidence; however, its evidential value can be assessed in relation to interdisciplinary examinations of the case.

In order to fulfil the rule of versatile utilisation of the information obtained from GSR examinations to elucidate a crime, it is also recommended to strengthen the communication between crime scene investigators and the experts working in the forensic laboratories prior to collection of evidence, if possible.

From the literature review and research experiences with the application of SEM/EDX, we have found that this analytical method is capable not only to identify GSR but also to reveal the relationships between its chemical and morphological properties and its source as well as the mechanisms of their formation and dissipation in the environment. This kind of examinations became possible with a research methodology focusing primarily on a precise and repeatable procedure of detection and identification of particles in the automatic manner, and second, on the representation of the obtained analytical data in the form of frequencies of occurrence of particles of a certain chemical class (or a class of certain range of effective diameters) within the whole population of particles revealed in a specimen.

On the basis of this methodology some relationships were established between the chemical and morphological properties of populations of particles and factors, such as the type of ammunition, the distance from the gun muzzle to the target, the type of a substrate, on which the particles are deposited and the time that had lapsed between shooting and collecting the specimens.

Significant differences in the chemical composition of GSR originating from various types of ammunition depending mainly on the composition of primer mixture were established. This gave rise to working out the method of group identification of ammunition used in a shooting incident, when the cartridge case is not available for examinations [16–18]. Chemical and morphological examinations of particles collected from the shooter's hands and from the interior of physically the same cartridge revealed either similarities or discrepancies depending on the chemical composition of the primer but also of other parts of the cartridge. This ought to be taken into account in the expert opinion in order to avoid false elimination of the possible involvement of the suspect in a shooting incident [19].

The first publications on the expansion of GSR and the possible range of distance they may reach starting from the gun muzzle presented a visual assessment of the vapour cloud photographed by means of a high-speed camera [1, 3]. Later publications focused solely on the numbers of metallic particles and neglected the chemical aspect of their dispersion [16]. Only recently it has been established that the distribution of particles in the surroundings of a shooting gun is not uniform [20, 21]. The numbers of particles found in the specimen, their chemical composition and morphological features depend on the distance from the muzzle of the shooting gun and the type of the substrate the particles are deposited. For example, the ratio of antimony to lead particles originating from some ammunition lots manufactured by Mesko Metal Works, Skarżysko-Kamienna, Poland, increases with the distance from the muzzle in the direction of shooting and decreases in the opposite direction. On average the equivalent circle diameter of particles increases with the distance from the muzzle.

Results of this study gave rise to elaboration of the method for determination of a shooting distance from the physical and chemical examinations of GSR pattern around the gunshot wounds and damages – extending the possibilities of shooting distance estimation with range of about 50–100 cm, in addition to the three categories commonly used until now: (i) contact or a nearest vicinity shot (0–about 1 cm), (ii) close distance shot (about 1–50 cm) and (iii) a distant shot.

A consistency of the elemental composition of the residue remaining inside the cartridge case with that of the primer was observed, whereas the elemental composition of the airborne particles (e.g. collected from the shooter's hands) differed. Most likely, it resulted from the interactions of products of reaction of a primer and propellant with other ammunition parts (i.e. materials of cartridge case, the projectile core and jacket) as well as the interior surface of the gun barrel. Thus, it was established that GSR occurring in the subsequent stages of the explosives reaction of the primer and the propellant propagates inside the cartridge case and the gun barrel only in one direction – towards the muzzle, and they do not return or mix with each other. The fact of variability of the chemical composition of particles depending on the distance from the shooting gun, both in the direction of shooting and in the opposite direction could be utilised for establishing the positions of the part-takers in the shooting incident in relation to the firearm, and hence, to the direction of shooting.

Each of the worked out aspects of chemical analysis of metallic particles may significantly contribute to the reconstruction of the course of a shooting incidence. In spite of the fact that usually the physical and chemical examinations lead to a group identification and are complementary to other kind of information included in case files, in specific cases they may be decisive in establishing the circumstances of the incident.

Due to the rapid development of modern types of ammunition, forensic examiners face challenges related to the identification and properties of resulting new type of GSR. Therefore, examinations nowadays focus on the GSR originating from lead-free ammunition embracing inorganic and organic residue, broader range of the particle sizes as well as their internal structure [21–23].

In a broad meaning, GSRs may embrace the mutual transfer of a projectile to target and the material the target is made of onto the surface of a projectile. Moreover, quite often in casework, the recovered projectiles are damaged or disintegrated in such a way that the comparative physical examination of marks originating from the barrel bore is impossible to achieve that would otherwise lead to linking the projectile to the firearm it had been fired from. Thus, chemical analysis of lead isotopic compositions in projectiles and primers in shooting incident investigations may bring in some information relating to a firearm, fired ammunition and the gunshot entry in a particular shooting.

Zeichner et al. [24] have demonstrated in 2006 that analysis of lead isotopic composition may provide valuable evidence in investigating specific scenarios of shooting incidents. For instance, in a shoot-out, where several firearms and ammunition brands are involved, it may be feasible to point out which ammunition and/or firearm caused a particular gunshot entry if the ammunition brands involved (projectiles and primers) differ considerably in their lead isotopic composition. Multiple-collector inductively coupled plasma mass spectrometry (MC-ICP-MS) was used to analyse lead isotopic compositions in projectiles, cartridge cases, firearm discharge residues in barrels of firearms and in the gunshot entries. Plain lead .22 calibre and plated ammunition and 9 mm Luger full metal jacket ammunition were employed in shooting experiments using semiautomatic pistols. Cotton cloth was the target material and two firing distances were tested: 1 cm (near contact) and 2 m distances. It was observed that various mechanical or chemical means of cleaning do not completely remove lead deposits from barrels of firearms. Nonetheless, it was shown that, when the ammunition brands used in consecutive firings differed considerably, it was possible to discriminate between targets shot by the different ammunition brands. Similarly it was possible to relate the lead deposits in the barrel to one of the two ammunition brands. Thus, analysis of lead isotopic compositions may be valuable in specific scenarios of shooting incident investigations.

6.3.2 Airbag deployment residues

For safety reasons, airbags that are installed in front of the driver and passenger seats of vehicles also play an important role from the point of view of forensic examiners. When deployed, airbags come into contact with the body and clothing of the occupants of the vehicle, with a certain impact, causing a mutual transfer of solid particles, hair, fibres and imprints of the airbag on the body and the clothing of the occupants. In this manner, it may be possible to determine who was the driver or the passenger. The most successful methods for determination of the driver seem to be analysis of DNA transferred into the surface of the airbag during an accident. However, forensic experience has shown that cases, where the investigator requested to compare inorganic microtraces collected from suspect clothes with debris from the airbag, are not so exceptional.

In former times, talc and starch were commonly used for reducing coherence of neoprene inner linings and these powders were responsible for an airbag cloud of powder during the impact. Nowadays, inorganic microtraces have their source of origin in chemical components, which are used in the gas generators. Considering the way propellants are produced, gas generators can be divided into two categories: pyrotechnic and hybrid models. Hybrid models contain high-pressure containers with a nontoxic gas with the membrane that is opened pyrotechnically and, for instance, nitrogen flows out and inflates the airbag. In the pyrotechnical system, gas is produced as the effect of consumption of solid propellant, the most popular being sodium azide till mid-1990s [25]. The sodium azide reacts with oxidising agents such as alkaline earth nitrates (e.g. $Sr(NO_3)_2$, $Ba(NO_3)_2$), metal oxides (e.g. CuO, Fe_2O_3), sulphides (e.g. MoS_2) and sulphur. The slag-forming agents such as silicon dioxide, aluminosilicates, alkaline earth metal or transition metal oxides, as well as other compounds could also be added. The inflating gas is being cooled on filters and the majority of solid products are retained.

The common way of dealing with investigation of airbag deployment residues is utilisation of software dedicated for standard GSR detection. The first step involves collection of residues from the interior side of the airbag for assessing their elemental composition, which is the base for creation of a new class of interest for automatic runs. Robert E. Berk [26] detected large population of "zirconium-rich" particles in passenger-side airbag from a 1999 Honda Accord. These particles contained zircon as a main element, but the majority of them had traces of potassium and chloride or titanium as well. In the same study, the authors performed the analysis of a driver-side airbag from a 2005 Nissan Altima and discovered particles with strontium-, aluminium- and iron-rich particles.

The particles containing strontium as a main component with the presence of Na, Mg, K, Si on minor level and with traces of Cu and Fe were discovered among microtraces collected from the inner side of the driver's airbag of Suzuki Swift by one of the authors (WD). The surface morphology of these "strontium-rich" particles is similar to GSR and is affected by pressure and high temperature during formation process (see Figure 6.6(a) and 6.6(b)).

These specific features with connection to elemental composition allow to verify the source of origin of particles discovered on the automatic stage of analysis.

6.3.3 Paint

Quite often forensic experts examine small particles of paint coats which are frequently collected in connection with such incidents as car accidents, robberies and burglaries. They occur in the form of micro-fragments of paint coat, frequently with an area of several mm^2 or less, or visible smears of paint in the form of coloured streaks on

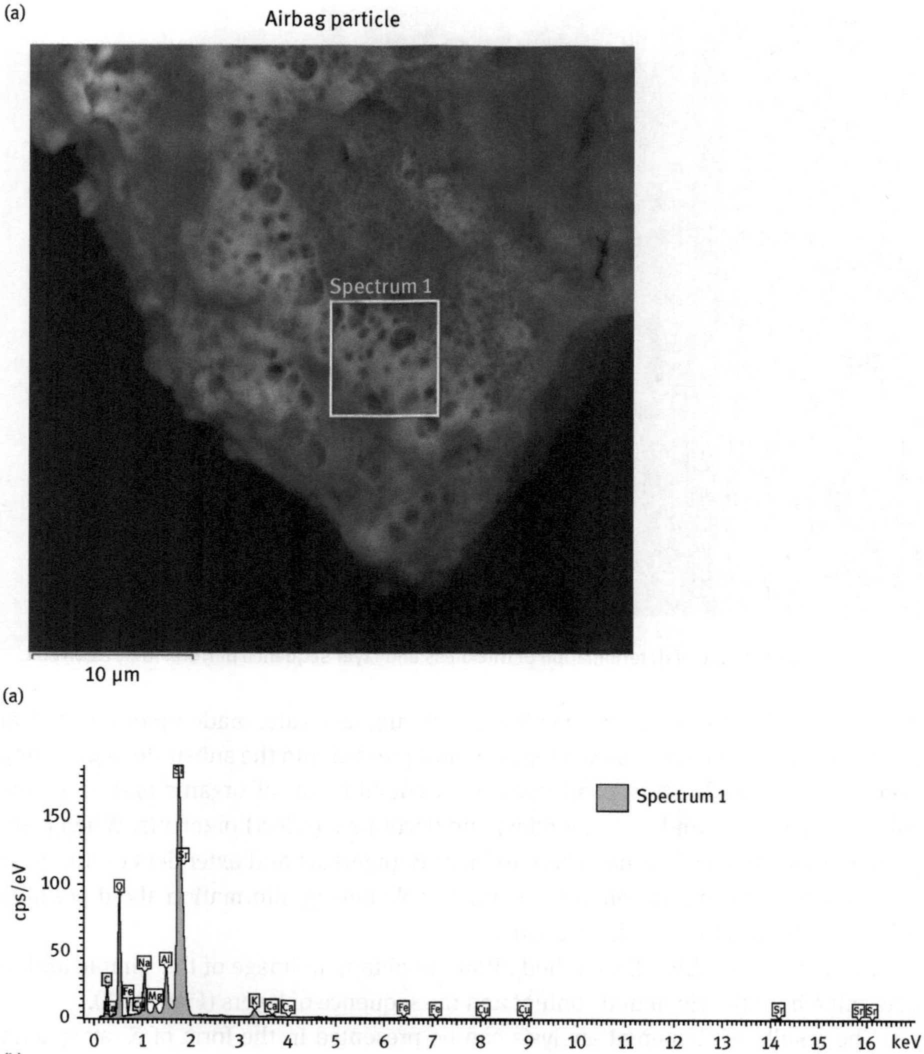

Figure 6.6: The BSE image (a) and spectrum (b) of a particle discovered inside the airbag of a Suzuki Swift car.

the clothing of persons being involved in these events or on other substrates. The aim of paint analysis is to establish the degree of similarity between the questioned paint trace and the sample originating from the suspect (from his or her vehicle, tools used, etc.). The identification analysis is also carried out, which leads to the determination of the type of paint, its application, manufacturer and the year of production.

Typically, fragments of paint reveal a multilayer structure. Each layer (about 10–50 cm thick) is made up of painting material and is a mixture of many chemical

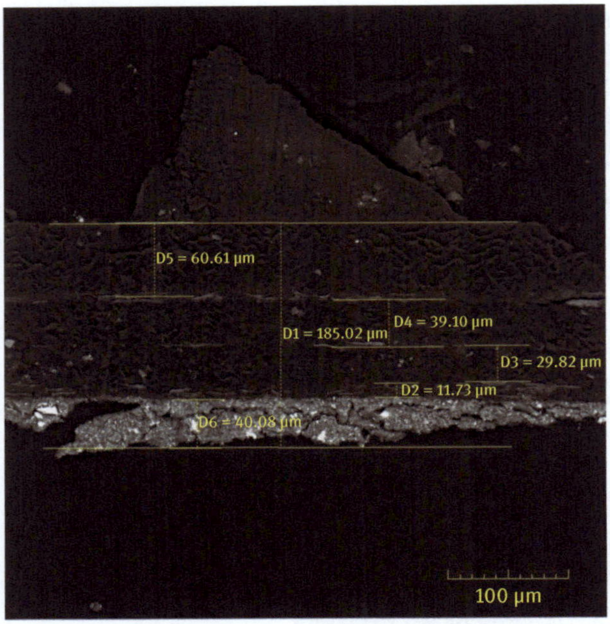

Figure 6.7: The example of determination of thickness and layer sequence performed by SEM/EDX.

compounds. Paint smears, on the other hand, are, as a rule, made up of one or two layers of painting material mixed together and pressed into the substrate (e.g. among fibres of the fabric). Every paint contains a combination of organic and inorganic colouring pigments and also extenders and decorative (effect) pigments. When comparing paint samples it is necessary to identify pigments and extenders on the base of the elemental composition of the paint sample having information about possible pigment sets used in the paint industry.

Application of SEM/EDX method allows to obtain an image of the sample and to determine both the elemental content and the sequence of layers (Figure 6.7).

The results of elemental analysis can be presented in the form of X-ray spectra and in the form of mapping of the sample, which shows the distribution of detected elements. However, elemental maps are generally not quantitative and may lack the sensitivity to demonstrate minor differences between samples. SEM/EDX is also helpful in revealing the paint traces in smears on tools used by a perpetrator, clothing of victim or on bullet that penetrates the painted surface.

6.3.4 Glass

6.3.4.1 Introduction

The environment we are living in is full of glass objects that are used as tools, containers or window panes and could be shattered during the criminal activity such as fights, burglaries, robberies or traffic accidents. Glass fragments are known to transfer onto the body and clothing of a suspect and a whiteness person. The amount and size of glass samples collected at the scene is highly dependent on many factors such as glass thickness and its physical size, the distance from the glass object, the type of a suspect's garments and finally it is highly affected by time which lapsed between the incident and the collection of evidence.

The fundamental issue is preservation of the evidence material when transported to the laboratory for examinations. Proper packing containers or wrappings should be chosen to reduce the risk of interference and the subsequent contamination. Glass containers should be avoided in favour of paper packages, sticky foils, small plastic bags or envelopes. Most often the substrate the microtraces are recovered from, such as garment, is intact, naturally dried if required and packed separately. From the perspective of requirements of the chain of custody each individual, who handled the evidence, is identified in protocols and the package should be marked and sealed in such a way that one can easily verify its integrity in a laboratory's reception.

6.3.4.2 Preliminary examinations

Fragments of glass may have various sizes. Fragments found at the crime scene are rather big, whereas the ones revealed on clothing, hair and body of persons or found on the surface of shoe sole are tiny – quite often with linear dimension less than 1 mm (Figures 6.8 and 6.9).

Figure 6.8: The women shoe – medium of glass evidence.

Figure 6.9: The shoe soil with a glass fragment.

There are numerous techniques applied in forensic laboratories for recovering glass from garments. Bigger fragments can be removed from clothing by picking up with tweezers. The smaller ones are usually found in the debris collected as a result of shaking or brushing the clothing over the sheet of clean paper or glass-free metal cone. Adhesive tapes are also used by some examiners as well-vacuuming techniques particularly inside the vehicles or over big surfaces such floors, carpets and so on.

The first stage of routine examination of selected transparent materials from shake off is confirmation that one deals with glass fragments. Glass particles are differentiated, for instance, from grains of sand by characteristic conchoidal fracturing, sharp edges, reaction to a hotpoint, hardness and finally by its dark shade in all orientations under the polarised light.

Subsequently, glass fragments are characterised by physical features such as thickness, colour, fluorescence, surface features and the refractive index (RI).

When both smooth surfaces are available, the thickness of glass fragments is determined with the use of calliper and according to Standard Operating Procedure (SOP) introduced in a laboratory – usually at least three independent measurements are required.

Determination of the colour of glass fragment involves a certain degree of subjectivity and is performed by the expert against the white background in natural light. If the size of colourless fragment is big enough the shade of the colourless glass is determined as well. The numerical colour scale coding is introduced in Central Forensic Laboratory of the Police (Warsaw, Poland) (CFLP) as presented in Table 6.1.

Some glass fragments, if exposed to shortwave (254 nm) or long wave (350 nm) UV light, produce characteristic illumination, which is caused by specific additives such as uranium oxides or lead oxides. Glass made by the float process is also characterised by luminescence caused by tin contamination introduced by pouring liquid glass onto the molten tin bed. These observations are available only if one of the original surfaces is present and glass fragment has the appropriate size. The presence of the original surface gives more opportunities for identifying the source of glass.

Table 6.1: The numerical scale used for describing colour of glass evidence.

Code	Colour	Colourless glass shade
0	Colourless	Light green
1	White	Green
2	Yellow	Light blue
3	Orange	Blue
4	Red	
5	Pink	
6	Blue	
7	Green	
8	Grey	
9	Brown	
10	Black	

Characteristic scratches, or presence of car paint droplets, traces of black foil may indicate that this specific glass fragment originates from a car pane.

The last but most common routine analytical technique used in forensic laboratories in glass analysis is RI measurement, serving for either comparative or classification purposes. The method has been known since 19th century and relays on shifts in light velocity in transparent medium in comparison with light velocity in void. RI can be measured in many ways, but thermal immersion method is predominant in forensic laboratories for low instrumentation costs and its high performance in routine work [27, 28]. Glass fragments with sharp edges are immersed in the oil of known RI characteristics and heated. The temperature dependence of oil is known, and at certain temperature the edge of glass sample disappears under the contrast phase microscope which indicates the same value of RI of the oil and glass.

Although thermal immersion method has its advantages, there are several factors like sample size requirements, small changes in RI among the same type of glass which make the analyst search for other analytical tools with more discriminative power.

6.3.4.3 The elemental analysis of glass fragments

Physical dimensions of seized glass determine further analytical procedure. If the initial techniques are insufficient, then establishing the elemental composition of bulk glass samples is required. The elemental data, especially for trace components, can be used by a forensic scientist for comparison of samples of glass and also for ascertaining the kind of object they could have originated from. The composition of the same type of glass due to worldwide standardisation and restrictive quality controls of raw materials becomes more uniform. The main elements such as Na, Ca, Si and Al are present in the same glass categories at nearly the same level.

Table 6.2: Raw materials used in glass industry.

Oxides	Raw materials and common additives
SiO_2	Silica sand
Al_2O_3	Alumina sand, kaolinite ($Al_2Si_2O_5(OH)_4$)
Na_2O	Soda ash, sodium lye, sulphate (Na_2SO_4), nitratine ($NaNO_3$), Na_2CO_3
CaO	Limestone, chalk ($CaCO_3$)
K_2O	Saltpetre (KNO_3)
Li_2O	Synthetic Li_2CO_3, lepidolite ($K_2Li_4Al_2F_4Si_8O_{22}$), spodumene $LiAl(SiO_3)_2$, amblygonite ($LiAl(F,OH)PO_4$), petalite ($LiAl(Si_2O_5)_2$)
MgO	Dolomite ($CaMg(CO_3)_2$), $Mg(OH)_2$, MgO
PbO	Minium ($Pb_2^{2+}Pb^{4+}O_4$), litharge (PbO)
ZnO	Zinc white
BaO	Baryte ($BaSO_4$)

The typical soda lime glass is known as an amorphous, non-crystalline material manufactured mainly from melted sand, soda ash, limestone with various types of additives such as alumina, soda and only less than 1% is reserved for the rest of ingredients. The typical list of raw ingredients used in glass industry is presented in Table 6.2.

The differences concern other elements originating from additives, which are added to improve the utilisation properties of glass, are connected with its later application or originate from impurities in raw materials used in the production process. Their concentration is significantly lower (at trace level). The typical list of oxides used as colorants is presented in Table 6.3.

The methods that are most widely used in forensic laboratories for determination of major and minor elements in glass are μ-XRF, SEM/EDX, laser-induced breakdown spectroscopy (LIBS) and laser ablation (LA) ICP-MS. Most of all the choice of available analytical methods are a compromise between the cost and time of research and, where possible, a possible destruction of evidential material plays a significant role as well. It implies that radiochemical methods like neutron activation analysis, atomic spectroscopy methods such as atomic absorption or atomic emission are rarely used on everyday basis.

6.3.4.4 Scanning electron microscopy coupled with energy-dispersive X-ray spectrometry

If physical size of glass sample recovered from suspect's clothing might be the issue, then most often SEM/EDX method is applied in the routine examination and evaluation of glass evidence. Scanning electron microscopy combined with energy-dispersive detectors (EDX) is a well-known technique, which is capable to work with small fragments and provides information about major and minor elements.

Table 6.3: Typical colourants used in glass industry.

Colour	Colourant
Blue	CuO + COO (10:1)
Blue-green	FeO + COO (10:1)
Turquoise	CuO + Fe_2O_3 (150:12)
	CuO + Cr_2O_3 (150:7)
	CuO + COO (150:1)
Spring green	Cr_2O_3 + CuO (2:1)
Celery	Cr_2O_3 ~ 0.2%
	CdS + U_3O_8 (3:1)
Green	U_3O_8 + CuO (3:2)
Green olive	Fe_2O_3+ Cr_2O_3 (4:1)
Pink	Au ~ 0.005%
Red-violet	Mn_2O_3+ Se (10:1)
Purple-violet	Mn_2O_3+ CoO (70:1)
Yellow	CdS ~1% + S
	U_3O_8 ~ 0.6%
	$AgNO_3$
	Ce_2O_3 TiO_2
Black	FeS + Mn_2O_3
	FeS + NiO
	FeS + CoO
Colourless	CdS
	CdS + Se
	C + S

The pre-treatment of glass samples usually involves cleaning with distilled water or ethanol, crushing, embedding in a resin and then polishing the surface until it is flat, if necessary.

Examinations performed by one of the authors (ZBM) for certified reference glass samples revealed that the analytical results obtained for small fragments placed directly on the stubs with adhesive tapes are reliable and comparable with results being achieved for larger fragment with smoothed surfaces [29].

For qualitative and semiquantitative determination, a glass fragment is placed on a stub and coated with carbon. As a result of examinations characteristic signals for the major and minor elements are produced, and its qualitative and quantitative chemical content is in many cases sufficient for distinguishing the questioned glass sample from the reference material. Small colourless objects may mimic glass and therefore determination of elemental composition with SEM/EDX is sufficient to prove that they originated from the natural source. The typical spectrum and secondary electron image of a soda lime glass fragments is presented in Figure 6.10.

The milestone work by Reeve et al. [30] was the first one, where SEM/EDX method was used for discrimination of all but two of 81 glass samples that were

(a)

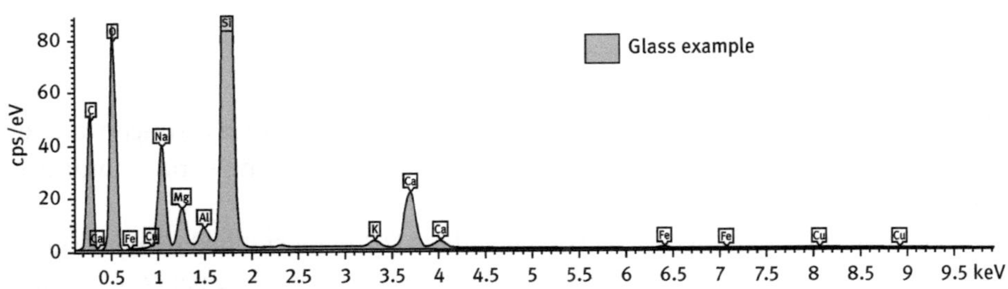

(b)

Figure 6.10: The example of SE detector image of glass fragment originating from a car window (a) and its X-ray spectrum (b).

indistinguishable by density and refracting index. Elemental compositions were reported in a form of ratios referenced to concentration of calcium. Andrasko and Maehly [31] in 1976 reported that they were able to distinguish all but two of 40 samples of window glass by means of measurement of RI, density and two chemical methods: SEM/EDX and atomic emission spectroscopy. The chemical composition was expressed as concentration ratios of main elements: Na, Al, Mg Ca and K.

The example of semiquantitative analysis of reference material (BR FS1 Breitländer Eichproben und Labormaterial GmbH) performed by SEM/EDX is presented in Table 6.4. A flat, carbon-coated surface of the glass was analysed in three points and for each point three X-ray spectra were obtained. The acquisition parameters were as follows: acceleration voltage 20 kV, magnification 500×, beam current 550 pA, approximately 10k counts/s and acquisition time 300 live seconds. The results of quantification were presented in the form of wt.% of oxides, and values were recalculated by the acquisition software. According to the reference, 15 of 29 components were successfully detected with the smallest concentration of equal to 0.1 wt.%. Heavier elements such as Y_2O_3 with concentration even higher than Fe_2O_3 remained undetected because of higher energy required for excitation of line $K_{\alpha 1}$ = 14.957 eV.

The classification system of glass evidence based on elemental composition obtained from SEM/EDX was introduced in the Institute of Forensic Research, Cracow. Examinations were performed for glass fragments originating from various utility groups (window panes, bulbs, jars, bottles, etc.) and revealed significant differences in qualitative and quantitative chemical composition. In this manner, the suitability of SEM/EDX method was established for the identification of the class of objects the fragments originated from (*classification* problem) when no reference material is available as well as when the comparative material is provided for the examinations (*discrimination* problem) [32, 33].

In contrary to some opinions, SEM/EDX proved to be an accurate and precise method for examinations of very small glass fragments without the need of time-consuming sample preparation and avoiding the loss of evidential material, however, allowing discrimination of elements in glass at concentration higher than 0.1 wt.%. With this in mind, the method was used further for the evaluation of primary and secondary transfer of glass fragments on individuals taking part in breaking a window pane that simulated a burglary [34].

In order to obtain detailed information on *primary and secondary transfer* of glass fragments as well as on the level of *contamination* of glass fragments of garment belonging to persons not taking part in breaking a glass object a number of experiments simulating burglaries were performed. A breaking-in individual was to damage double- or triple-glazed window and was accompanied by two persons playing the role of passive witnesses. In this research on the *primary transfer* the following factors were taken into account: glass type, the distance from the window, the kind of garment and time intervals between the experiment and the collection of garment for examinations. As a result of examinations of garments of individuals, who did not take part in breaking glass in recent time, it was found that about 80% of garment was free from *contamination* with glass fragments. After examining additional layer of clothes put on the garment used while breaking windows, it was established that in 70% of garments of "burglar" and in 30% of garments of individuals accompanying a "burglar" *the secondary transfer* of glass fragments took place.

Results of the performed study provided important information from the point of view of forensic expert in cases concerning the reconstruction of a crime, where

Table 6.4: Results of semiquantitative analysis of BR FS1 glass standard by scanning electron microscopy coupled with energy-dispersive X-ray spectrometry.

No.	Oxide	wt.%	wt.%/CaO	S1	S2	S3	S4	S5	S6	S7	S8	S9	S_min	S_max	S_avg	Δε	S_avg/CaO
1	B_2O_3	**2.08**	**0.73**	nd	nd	nd	nd	nd	nd	nd	nd	nd	nd	nd	nd	nd	nd
2	F	**2.90**	**1.02**	4.27	4.27	4.31	4.39	4.31	4.32	4.47	4.43	4.40	4.27	4.47	4.36	**1.46**	**1.54**
3	Na_2O	**1.20**	**0.42**	0.90	0.98	0.93	0.98	1.01	1.00	0.94	0.97	0.94	0.90	1.01	0.96	**−0.24**	**0.34**
4	MgO	**0.82**	**0.29**	0.80	0.84	0.82	0.86	0.82	0.85	0.82	0.85	0.81	0.80	0.86	0.83	**0.01**	**0.29**
5	Al_2O_3	**3.85**	**1.36**	5.29	3.85	4.56	3.77	4.19	3.82	5.01	4.24	5.18	3.77	5.29	4.45	**0.60**	**1.57**
6	SiO_2	**59.60**	**20.99**	59.62	60.74	60.40	60.89	60.59	60.82	60.02	60.57	60.00	59.62	60.89	60.38	**0.78**	**21.33**
7	K_2O	**18.40**	**6.48**	16.48	16.46	16.44	16.24	16.28	16.31	16.02	16.29	16.16	16.02	16.48	16.29	**−2.11**	**5.75**
8	CaO	**2.84**	**1.00**	2.67	2.89	2.82	2.90	2.89	2.90	2.82	2.87	2.81	2.67	2.90	2.83	**−0.01**	**1.00**
9	TiO_2	**0.04**	**0.01**	nd	nd	nd	nd	nd	nd	nd	nd	nd	nd	nd	nd	**nd**	**nd**
10	V_2O_5	**1.70**	**0.60**	1.73	1.81	1.77	1.76	1.75	1.78	1.72	1.77	1.69	1.69	1.81	1.75	**0.05**	**0.62**
11	Cr_2O_3	**0.27**	**0.10**	nd	nd	nd	nd	nd	nd	nd	nd	nd	nd	0.00	0.00	**−0.27**	**0.00**
12	Fe_2O_3	**0.07**	**0.02**	0.10	0.06	0.07	0.05	0.06	0.08	0.11	0.05	0.08	0.05	0.11	0.07	**0.00**	**0.03**
13	CoO	**0.25**	**0.09**	0.26	0.25	0.24	0.26	0.25	0.28	0.25	0.25	0.25	0.24	0.28	0.26	**0.01**	**0.09**
14	CuO	**1.80**	**0.63**	1.82	1.96	1.90	1.94	1.88	1.95	1.84	1.88	1.83	1.82	1.96	1.89	**0.09**	**0.67**
15	Ga_2O_3	**0.09**	**0.03**	nd	nd	nd	nd	nd	nd	nd	nd	nd	nd	nd	nd	**nd**	**nd**
16	Rb_2O	**0.16**	**0.06**	nd	nd	nd	nd	nd	nd	nd	nd	nd	nd	nd	nd	**nd**	**nd**
17	Y_2O_3	**0.45**	**0.16**	nd	nd	nd	nd	nd	nd	nd	nd	nd	nd	nd	nd	**nd**	**nd**
18	ZrO_2	**0.74**	**0.26**	0.28	0.30	0.29	0.33	0.30	0.33	0.30	0.28	0.33	0.28	0.33	0.30	**−0.44**	**0.11**
19	Nb_2O_5	**0.38**	**0.13**	nd	nd	nd	nd	nd	nd	nd	nd	nd	nd	nd	nd	**nd**	**nd**
20	CdO	**0.20**	**0.07**	nd	nd	nd	nd	nd	nd	nd	nd	nd	nd	nd	nd	**nd**	**nd**
21	In_2O_3	**0.26**	**0.09**	3.95	3.83	3.68	3.83	3.80	3.78	3.82	3.76	3.74	3.68	3.95	3.80	**3.54**	**1.34**
22	SnO_2	**0.20**	**0.07**	nd	nd	nd	nd	nd	nd	nd	nd	nd	nd	nd	nd	**nd**	**nd**
23	Sb_2O_3	**0.25**	**0.09**	1.35	1.31	1.28	1.32	1.41	1.34	1.38	1.36	1.37	1.28	1.41	1.35	**1.10**	**0.48**
24	Cs_2O	**0.13**	**0.05**	nd	nd	nd	nd	nd	nd	nd	nd	nd	nd	nd	nd	**nd**	**nd**
25	BaO	**0.34**	**0.12**	0.48	0.45	0.48	0.49	0.46	0.46	0.46	0.43	0.41	0.41	0.49	0.46	**0.12**	**0.16**
26	Ce_2O_3	**0.39**	**0.14**	nd	nd	nd	nd	nd	nd	nd	nd	nd	nd	nd	nd	**nd**	**nd**
27	Sm_2O_3	**0.18**	**0.06**	nd	nd	nd	nd	nd	nd	nd	nd	nd	nd	nd	nd	**nd**	**nd**
28	Ta_2O_5	**0.36**	**0.13**	nd	nd	nd	nd	nd	nd	nd	nd	nd	nd	nd	nd	**nd**	**nd**
29	PbO	**0.05**	**0.02**	nd	nd	nd	nd	nd	nd	nd	nd	nd	nd	nd	nd	**nd**	**nd**

Note: nd, not detected.

breaking of glass objects takes place as well as in determination whether suspects were involved in the crime.

It is necessary to apply statistical methods in the characterisation of glass evidence by its elemental composition, such as cluster analysis or LR [35].

6.3.4.5 Micro-X-ray fluorescence

The second most frequently used method for forensic glass analysis is μ-XRF, of which operating principle relies on detection of photons emitted as an effect of interaction of X-ray radiation with sample surface. μ-XRF offers lower detection capabilities, fair – good precision, limited sample preparation time and finally it is another non-destruction technique. The "dark side" of the technique involves the requirement of having reasonably clean and smooth sampling area, which should be at least 1 mm^2 and 0.5 mm in thickness. The surface of a glass sample can be cleaned with ultrasonic cleaner or washed with nitric acid, rinsed with demineralised water, rinsed again in ethanol and finally dried at 100 °C. According to the guidelines issued by the ENFSI [36], recommended acquisition parameters for qualitative analysis of glass should be as follows: X-ray tube voltage 40 kV, 2,500 counts/s, measuring time 100 live seconds and dead time kept below 40%. SOP should contain procedure for performing calibration with metal standard and glass sample of known composition, which contains light, medium and heavy elements that should be measured regularly to maintain the quality of results.

Similar to SEM/EDX method, the researcher compares intensities of X-ray signals or elemental ratios. Naes et al. [37] performed the analysis of 41 different automotive non-float glass fragments obtained from 14 different cars that make 820 pairs. The authors identified six elements (K, Ca, Ti, Fe, Sr and Zr) from which they utilised six intensity ratios: Ca/Fe, Sr/Zr, Ca/K, Fe/Zr, Fe/Sr and Fe/Ti for comparison. All but 14 pairs were indistinguishable according to three SD criteria that make 98.3% pairs successfully discriminated. Similar discriminative power for μ-XRF was obtained by Hicks et al. [38], who were able to discriminate 86.8% pairs of 129 pairs of glass that were indistinguishable by means of RI measurement.

6.3.4.6 Laser ablation-inductively coupled plasma-mass spectrometry

LA-ICP-MS technique links excellent sensitivity and precision with almost no requirements for sample preparation. Being fast, precise and almost non-destructive, only small ablation crater of several dozen micrometres in diameter is created. LA-ICP-MS came to forensic scientists' attention quite early. A huge discrimination power is the effect of combination of a high sensitivity and isotopic information about the questioned sample. The heterogeneity of trace elements in float glass [39] and availability of commercial certified material for glass influenced rapid increase in application of LA-ICP-MS in differentiation among glass samples as forensic evidence materials [40].

One of the early applications in forensic analysis of glass sample was the work performed by Watling et al. [41] in 1997. The author analysed 62 glass samples, where the majority constituted float and container glass types and proved that the method is capable to distinguish samples from each other.

The increasing application of LA-ICP-MS in European forensic laboratories and outcome of inter-laboratory tests for quantitative analysis of float glass were the factors contributing to issuing of set of recommendations by ENFSI [42]. According to the document, each piece of glass evidence should be broken at least into two parts; one should be preserved for RI analysis and the other one glued onto an adhesive tape and analysed. Other researchers recommend crushing, embedding in epoxy resin and polishing and finally washing in 2% nitric acid and rinsing in distilled water [43] or wiping with acetone [44].

The recommended and being most commonly used glass standards are SRM 612 and 610 manufactured by the National Institute of Standards and Technology, FGS 1 and FGS 2 manufactured by Natural Isotopes and Trace Elements in Criminalistics and Environmental Forensics that provide measurement/information on the content of major and minor elements present in glass such as ^{23}Na, ^{25}Mg, ^{27}Al, ^{29}Si, ^{39}K, ^{42}Ca, ^{49}Ti ^{57}Fe accompanied with trace elements ^{7}Li, ^{55}Mn, ^{85}Rb, ^{88}Sr, ^{90}Zr, ^{118}Sn, ^{137}Ba, ^{139}La, ^{140}Ce, ^{146}Nd, ^{178}Hf and ^{208}Pb. This can be useful in some applications, as instead of measuring the whole isotopic array, estimation of specific isotopic ratios such as $^{208}Pb/^{206}Pb$ and $^{207}Pb/^{206}Pb$ [43] is performed.

Helium as a carrier gas for ablation should be used with a typical repetition rate of the laser equal to 10 Hz, but for plasma stabilisation additional flow of argon is required. The diameter of ablation crater should correspond to the size of the glass evidence and be chosen to guarantee the repeatability of measurements. The recommended 50 s of ablation should be followed by 30 s of gas blank and each unknown glass sample should be analysed at least four to six times. Each series, containing no more than two unknown samples, is followed by either multiplied measurements of calibration standard or quality control standard. For quantification purposes, ^{29}Si from glass is treated as internal standard under the assumption that in float glass average weight ratio of silicon dioxide is close to 72 wt%.

Obtained quantitative results require further chemometric processing that is partially done by instrument software; however, the assessment of pairwise comparison requires common statistic tests such as *t*-tests, analysis of variance or Tukey's Honestly Significant Difference (HSD). In the work mentioned earlier [37], the authors found indistinguishable only 11 pairs of glass and that value referred to all pairs provides a discrimination power of 98.7%. The combination of results obtained from other techniques such as RI allows to even increase this value.

6.3.5 Evidential materials – summary

The subject of interest of forensic scientists is various materials that frequently occur in very small quantities, forming microtraces. Being unique for a crime they become an evidence and ought to be characterised by means of contemporary analytical methods possibly avoiding consumption of the material.

Forensic scientists perform the analytical activity of identifying and/or measuring the quantities of one or more individual chemical species in a sample, they study the distribution of an element among defined chemical species in a system. They search for specific forms of an element defined as to isotopic composition, electronic or oxidation state, and/or complex or molecular structure to be able to differentiate an evidence material from a common pollution or impurity. To demonstrate the specificity of a detected trace in relation to a crime a forensic analyst takes into account not only its chemical state but also its physical properties and background information on the circumstances that influenced occurrence of the trace.

Authors of publications in the field of forensic sciences very often present data on trace element speciation, but never use the term "*speciation*".

References

[1] Schwoeble AJ, Exline DL. Current Methods in Forensic Gunshot Residue Analysis. CRC Press, Boca, Raton, FL, London, New York, Washington DC, 2000.

[2] Wallace J. Chemical Analysis of Firearms, Ammunition and Gunshot Residue. CRC Press, Boca Raton, FL, London, New York, 2008.

[3] Dietrich H. Distribution of gunshot residues – the influence of weapon type. Forensic Sci Int 2012, 220, 85–90.

[4] Dalby O, Butler D, Birkett JW. Analysis of gunshot residue and associated materials – a review. J Forensic Sci 2010, 55, 924–43.

[5] Brożek-Mucha Z. Scanning electron microscopy and X-ray microanalysis for chemical and morphological characterisation of the inorganic component of gunshot residue: Selected problems. Biomed Res Int 2014, Article ID 428038, 11 pages, doi:10.1155/2014/428038.

[6] ASTM Standard E1588-10e1. Standard Guide for Gunshot Residue Analysis by Scanning Electron Microscopy/ Energy Dispersive X-ray Spectrometry. ASTM International, West Conshohocken, PA, 2010.

[7] ENFSI Guide for Gunshot Residue Analysis by Scanning Electron Microscopy/Energy-Dispersive X-Ray Spectrometry. ENFSI Firearms/GSR Working Group guidelines, 2008.

[8] Torre C, Mattutino G, Vasino V, Robino C. Brake linings: A source of non-GSR particles containing lead, barium and antimony. J Forensic Sci 2002, 47, 494–504.

[9] Romolo FS, Margot P. Identification of gunshot residue: A critical review. Forensic Sci Int 2001, 119, 195–211.

[10] Geusens S, Charles N. A study of the potential risk of gunshot residue transfer from special units of the police to arrested suspects. Forensic Sci Int 2012, 216, 78–81.

[11] French J, Morgan R, Davy J. The secondary transfer of gunshot residue: An experimental investigation carried out with SEM-EDX analysis. X-ray Spectrom 2014, 43, 56–61.

[12] French J, Morgan R. An experimental investigation of the indirect transfer and deposition of gunshot residue: Further studies carried out with SEM–EDX analysis. Forensic Sci Int 2015, 247, 14–17.

[13] Brożek-Mucha Z. Chemical and morphological study of gunshot residue persisting on the shooter by means of scanning electron microscopy and energy dispersive X-ray spectrometry. Microsc Microanal 2011, 17, 972–82.

[14] Gauriot R, Gunaratnam L, Moroni R, Reinikainen T, Corander J. Statistical challenges in the quantification of gunshot residue evidence. J Forensic Sci 2013, 58, 1149–55.
[15] Brożek-Mucha Z. On the prevalence of gunshot residue in selected populations – an empirical study performed with SEM-EDX analysis. Forensic Sci Int 2014, 237, 46–52.
[16] Brożek-Mucha Z, Jankowicz A. Evaluation of the possibility of differentiation among various types of ammunition by means of GSR examinations with SEM-EDX. Forensic Sci Int 2001, 123, 39–47.
[17] Brożek-Mucha Z, Zadora G. Grouping of ammunition types by means of frequency of occurrence of GSR. Forensic Sci Int 2003, 135, 97–104.
[18] Brożek-Mucha Z, Zadora G. SEM-EDX – a useful tool for forensic examination. Mater Chem Phys 2003, 81, 345–8.
[19] Brożek-Mucha Z. Comparison of cartridge case and airborne GSR – a study of their elemental contents and morphology by means of SEM-EDX. X-ray Spectrom 2007, 36, 398–407.
[20] Brożek-Mucha Z. Variation of the chemical contents and morphology of gunshot residue in the surroundings of the shooting pistol as a potential contribution to a shooting incidence reconstruction. Forensic Sci Int 2011, 210, 31–41.
[21] Romolo FS, Christopher ME, Donghi M, et al. Integrated Ion Beam Analysis (IBA) in Gunshot Residue (GSR) characterisation. Forensic Sci Int 2013, 231, 219–28.
[22] Christopher M, Warmenhoven J, Romolo FS, et al. A new quantitative method for gunshot residue analysis by ion beam analysis. Analyst 2013, 16, 4649–55.
[23] Melo LGA, Martiny A, Pinto AL. Nano characterization of gunshot residues from Brazilian ammunition. Forensic Sci Int 2014, 240, 69–79.
[24] Zeichner A, Ehrlich S, Shoshani E, Halicz L. Application of lead isotope analysis in shooting incident investigations. Forensic Sci Int 2006, 158, 52–64.
[25] Meyer R, Köhler J, Homburg A. Explosives, 2002, 5th ed. Wiley-VCH Verlag GmbH & Co, Weinheim.
[26] Berk RE. Automated SEM/EDS analysis of airbag residue. *I: Particle identification. J Forensic Sci 2009, 54, 60–68.
[27] Locke J, Underhill M. Automatic refractive index measurement of glass particles. Forensic Sci Int 1985, 27, 247–60.
[28] Zoro JA, Locke J, Day RS, Badmus O, Perryman AC. An investigation of refractive index anomalies at the surfaces of glass objects and windows. Forensic Sci Int 1988, 39, 127–41.
[29] Brożek-Mucha Z. X-ray microanalysis of glass for forensic purposes – research on the persistence of glass fragments on clothing. X-ray Spectrom 2009, 38, 58–67.
[30] Reeve B, Mathiesen J, Fong W. Elemental analysis by energy dispersive X-ray: A significant factor in the forensic analysis of glass. J Forensic Sci 1976, 21, 291–306.
[31] Andrasco J, Maehly AC. Discrimination between samples of windows glass by combining physical and chemical techniques. J Forensic Sci 1978, 23, 250–62.
[32] Zadora G, Brożek-Mucha Z, Parczewski Z. A classification of glass microtraces. Prob Forensic Sci 2001, XLVII, 137–43.
[33] Zadora G, Brożek-Mucha Z, Parczewski A. A differentiation of glass microtraces of the same qualitative composition. Prob Forensic Sci 2001, XLVII, 144–6.
[34] Brożek-Mucha Z, Chochół A, Trzcińska B. Complex analysis of evidence for establishing circumstance of death. Prob Forensic Sci 2001, XLVII, 300–7.
[35] Ramos D, Zadora G. Information – theoretical feature selection using data obtained by scanning electron microscopy coupled with and energy dispersive X-ray spectrometer for the classification of glass traces. Anal Chim Acta 2011, 705, 2007–17.
[36] ENFSI Best practice manual for forensic glass examination. Appendix E Micro X-ray spectrometry in comparative glass analysis, 2009.

[37] Naes BE, Umpierrez S, Ryland S, Barnett C, Almirall JR. A comparison of laser ablation inductively coupled plasma mass spectrometry, micro X-ray fluorescence spectroscopy, and laser induced breakdown spectroscopy for the discrimination of automotive glass. Spectrochim Acta, Part B 2008, 63, 1145–50.
[38] Hicks T, Monard Sermier F, Goldmann T, Brunelle A, Champod C, Margot P. The classification and discrimination of glass fragments using non destructive energy dispersive X-ray microfluorescence. Forensic Sci Int 2003, 137, 107–18.
[39] Trejos T, Almirall JR. Sampling strategies for the analysis of glass fragments by LA-ICP-MS Part I. Micro-homogeneity study of glass and its application to the interpretation of forensic evidence. Talanta 2005, 67, 388–95.
[40] Orellana FA, Gálvez CG, Roldán MT, Garcia-Ruiz C. Application of laser-ablation inductively-coupled plasma-mass spectrometry in chemical analysis of forensic evidence. Trends Anal Chem 2013, 42, 1–34.
[41] Watling RJ, Lynch BF, Herring D. Use of laser ablation inductively coupled plasma mass spectrometry for fingerprinting scene of crime evidence. J Anal At Spectrom 1997, 12, 195–203.
[42] ENFSI Best practice manual for forensic glass examination. Appendix G Quantitative forensic float glass analysis using LA-ICP-MS, 2009.
[43] Sjåstad K-E, Andersen T, Simonsen SL. Application of laser ablation inductively coupled plasma multicollector mass spectrometry in determination of lead isotope ratios in common glass for forensic purposes. Spectrochim Acta, Part B 2013, 89, 84–92.
[44] van Eltaren JT, Tennent NH, Šelih VS. Multi-element quantification of ancient/historic glasses by laser ablation inductively coupled plasma mass spectrometry using sum normalization calibration. Anal Chim Acta 2009, 644, 1–9.

Andrzej Wyciślik

7 Industrial Analysis and Speciation

7.1 Introduction

The development of metallurgical analytics is closely related to the currently used technology of steel and metal alloys production and to the increasing demand for high-quality metallurgical products. Steel remains the basic construction material and is still regarded by many as the most durable, reproducible and innovative alloy. What best illustrates this statement is the increase in world steel production from 200 million tonnes in 1950 to 1.8 billion tonnes at present. Metallurgical analytics and other application areas in analytical chemistry, especially material and environmental analytics, benefit from general advances in research methodology and new technical and instrumental solutions. Studies of new steel grades and alloys are also an important contribution to the general development of analytical techniques and methods, including new solutions in the construction of laboratory apparatus and control and measuring equipment. This chapter presents a brief historical outline of the development of metallurgical analytics, indicating application fields for the analytical techniques in metallurgy and specifying additionally a wide range of the analysed materials. Application requirements in the field of metallurgy and materials engineering have been and are the factor that defines the direction of studies in the field of metallurgical analytics, acting at the same time as a driving force for the development of analytical methods and technical solutions.

Much attention has been paid to the issues of speciation and speciation analysis, currently covered by extensive studies in various fields of science and technology as well as medicine, toxicology and environmental protection. Speciation is the occurrence of various physical and chemical forms of a particular species in the material under examination, while speciation analysis is the tool used in the identification and quantification of these forms in the examined object. Thus, assuming, with some simplifications, that speciation is the occurrence of various "chemical forms" of the element in the sample, and speciation analysis helps to identify and quantify these forms, by analogy, with respect to metallurgical samples, equivalent phase identification and phase analysis means are used, including also chemical analysis. The principles of chemical and electrochemical phase isolation techniques have been described, stressing their potentials and limitations. The complexity of the electrochemical phase isolation process has been underlined. Chemical and electrochemical parameters influencing the anodic phase extraction process have been indicated. Chemical analysis carried out in the area of metallurgy and materials engineering on the isolated and separated phases satisfies, though not always, the requirements of quantitative speciation analysis. In this respect, particular attention deserves the determination of the chemical composition of the isolated γ' intermetallic phase.

https://doi.org/10.1515/9783110366730-008

Various methods used for the analysis of isolated phases, especially carbide phases, have been compared, with emphasis put on the determination of their chemical composition. Additional opportunities to use in this research area own design solutions and technical modifications (to mention as an example portable capillary bearing and cover with precisely drilled holes for vertical insertion of spectrometer capillary to the Teflon sampling manifold) were indicated, including practical application of the flow injection technique combined with flame atomic absorption spectrometry (FAAS) using small volumes of sample solutions. The option of using less sensitive analytical lines has also been exposed as a means for the determination of the content of many elements in a wide range of concentrations with presentation of the results of own tests. Small volumes of the available test materials used in the determination of the chemical composition of the isolates and absence of certified reference materials are a great challenge that all specialists involved in the speciation analysis of multicomponent alloys must face. A solution in this respect can be the method of one base reference standard developed by the author of this chapter. The method of one reference standard involves the use of several different weighed portions of one base (starting) reference standard to which, after or in the course of dissolution, appropriate quantities of the main matrix components are introduced to bring this content in both reference standard and test sample to a similar, with some tolerance, level, thus compensating for its effect on the element being determined. The base reference standard is selected so as to best match the sample to be analysed. It is also important that the upper and lower concentration limits of the individual elements are situated in the rectilinear part of reference graphs. Most important techniques to prepare separated phases for analysis were disclosed together with the developed analytical diagrams. Based on literature data, the possibilities and several technical variants of aluminium determination by atomic absorption spectrometry (AAS) in steels were discussed, and a division into acid-soluble aluminium and total aluminium was proposed as a typical example of the speciation analysis in metallurgy. Selected results of the research on complex Fe–Ni alloys, conducted with great skills and competence by K. Ducki, served as a background for the identification of numerous phases occurring in these materials. Unfortunately, in this particular case, quantitative determination of the content of the identified phases was not possible. Thus, as regards speciation (phase identification), the entire anticipated research cycle has been accomplished with success, but as regards the speciation analysis (quantitative evaluation of individual structural constituents), the goal has not been reached.

In this chapter, taking into account the harmful effect of welding dust on the health of welders, a sampling station for the welding dust and a universal method for the determination of total dust emission during welding process have been presented. This can certainly serve a broader purpose than only the examination of chromium speciation in test samples. The developed methodology enables comprehensive studies of dust generated during welding, considered in terms of its weight and chemical composition. This reveals numerous relationships that occur during the welding dust emission process with reference to the type of the welding process, the selected welding consumables and the technological conditions of the process. Quantitative dependencies and their in-depth analysis are the basis for technological modifications introduced to welding processes in order to reduce the total dust emission, the amount of toxic substances, especially those that are particularly harmful to the health of welders and, last but not least, to reduce the negative impact of the welding process on the natural environment. The effort to achieve all those goals clearly implies the need for studies of the environmental issues as a basis for implementation of the environmental management system in enterprises. Moreover, let us not forget that welding is one of the most dynamically developing technological processes in a wide range of industries. As an additional contribution to the general discussion, the determination of total chromium content and chromium (VI) content in the welding dust was presented as a nearly model example of the speciation and speciation analysis carried out on complex metallurgical materials, including those used in the welding practice.

The importance of speciation in the development of metallurgy and materials engineering, on the one hand, and in the development of material analytics, on the other, was emphasized. The speciation analysis of complex metallurgical materials fits perfectly this pattern.

7.2 A brief historical overview of the development of metallurgical analytics

F. Harkort, the first historically recorded metallurgical chemist, was employed in the steel industry in 1826. In the following years, the number of chemists working in the steel industry (blast furnaces, steelworks and foundries) began to grow steadily. The first chemical laboratory to study raw materials, auxiliary materials and finished products was established in 1863 in Gussstahlfabrik in Essen. In 1879, the Bessemer process was introduced in the steelworks, followed by the Siemens-Martin process. Thanks to these solutions, new types of steel started being produced, and further production plants with small metallurgical laboratories were established, developing "wet" methods of chemical analysis for specific industrial applications. There was also a marked increase in the employment of chemists in metallurgical laboratories [1–4].

It is worth pointing out that already in the last decade of the 19th century in Germany, and so in a country where iron and steel industry was at the highest technical and technological level and enjoyed the fastest development in Europe, some tendencies related to the standardization of methods used in metallurgical laboratories appeared. These tasks were entrusted to the Verein Analytischer Chemiker Commission in Berlin. In 1911, Chemikerausschuss des Vereins Deutscher Eisenhüttenleute VDEh (*Chemists Committee of the VDEh*) was formed. Its main tasks included, among others, the standardization of analytical methods, the systematization and limitation of methods introduced into laboratory work, the development of weighing methods, the development of gas analysis methods and the assurance of reliability (correctness) of results using reference and comparative tests. Furthermore, it was also argued that many of the methods used were laborious and time consuming. An improvement in this respect was achieved by the gradual introduction of instrumental methods, especially spectrophotometry. The effect of these actions was lowering the limits of detectability and determinability of the analysed elements. In the years 1920–1940, further development in methods of the analysis of steel and metal alloys (analysis of alloying additives) was noted, introducing into the analytical practice – polarographic, potentiometric, systematically developed spectrophotometric and spectrographic methods [1–5].

The next decade faced a further comprehensive development of metallurgical analytics, and its dynamics clearly gained momentum after the Second World War. Analysts have already indicated the need for gradual automation of the process of analysis. The major advances in the 1950s in the development of methods and techniques for the chemical analysis of steel and alloys were related to the application of analytical procedures resulting from the development of optical emission spectrometry (OES) and X-ray fluorescence (XRF). In modern analytical metallurgical laboratories, due to the development of post-treatment methods of steel processing, computerized systems of spectrometric analysis (OES and XRF) are already used. The determination of the chemical composition of steel and pig iron, and also of non-ferrous metal alloys, usually takes 2–4 min [5].

In the 1950s, there was also progress in X-ray techniques used in chemical analysis of ores and fluxes. Previously, the methods used in this field were mainly based on classical time- and labour-intensive weight and titration procedures, limited only to the identification of some basic elements such as iron, carbon, sulphur and phosphorus, as well as oxides, like SiO_2, CaO, MgO and Al_2O_3. Currently, full chemical analysis of iron ore by XRF technique takes 10–20 min, depending on the sample preparation method used [1–5].

As early as in the 1930s, the importance of determining the basicity of metallurgical slags, especially steelmaking slags, was widely discussed in the aspect of the technological process of steelmaking, but it took two decades of intensive studies to develop rapid methods for calculation of this parameter, since it was not until the 1950s that rapid "titration" techniques to determine the basicity of steel slags were developed. Currently, the analysis of steel and blast furnace slag done by XRF takes only 6 min [5].

Other techniques that have found a lasting place in the routine activity of analytical metallurgical laboratories included AAS since the 1960s, and since the late 1970s, inductively coupled plasma-atomic emission spectrometry (ICP-AES) has been mainly used for the determination of microadditives and harmful impurities, and later also to determine the content of the main alloying constituents. Currently, these techniques, in addition to OES and XRF, are the basic tools used in the determination of the chemical composition of raw materials, as well as semi-finished and finished products [5]. Although in terms of the speed of execution of AAS and ICP-AES analyses, these two techniques cannot compete with OES and XRF, mainly because it is necessary to convert the test materials to the state of solution, in the control and arbitration analysis of melts and also in the analysis of samples characterized by "strong" tendencies for segregation of alloying constituents, they render invaluable services to analysts [6]. This is due to the possibility of taking chip samples from different, usually strictly defined or selected, sites of the material analysed, depending on the scope of the research being carried out. Additionally, in many metallurgical laboratories, these techniques are used to determine the chemical composition of own reference and correction samples for OES and XRF. In smaller metallurgical laboratories with modest equipment, these techniques are also used to control the chemical composition of alloy samples taken in the course of the melting process. However, a necessary condition in this case is to develop a quick dissolution process for a given group of alloys.

Significant progress has been made, especially in the last half of the 20th century, in the studies and production of various certified reference materials for use in the field of metallurgy. In the last four decades of the 20th century, there was also a dynamic development in the production of very modern analysers used for rapid determination of carbon and sulphur content as well as the content of nitrogen, oxygen and hydrogen in steels and alloys.

In the 1990s, in the specialized analytical laboratories of metallurgical plants producing multicomponent steels and alloys for the most responsible parts of machinery and equipment, the combined technique of ICP and mass spectrometry (MS) started being used for the determination of microadditives and "harmful" elements forming low melting phases.

The identification of phases and inclusions, and their analysis are important for the quality of metallurgical products [6] and for the proper functioning of chemical and metallurgical laboratories. Direct phase identification is possible owing to the use of the following methods: X-ray diffraction, electron diffraction, scanning electron microscopy (SEM), X-ray microanalysis, Auger electron spectroscopy and electron backscattered pattern (EBSP) technique. Electron probe microanalysis (EPMA), electron spectroscopy for chemical analysis and chemical microanalysis (CMA) are used for the analysis of chemical compositions. AAS and ICP techniques are mainly applicable in the chemical analysis of electrolytically isolated phases [1–6]. This field of studies is part of the currently expanding research area, which includes speciation and speciation analysis. Phase identification basically fulfils the criteria for speciation. Much more difficult and complex in confrontation with phase analysis is the speciation analysis. The problems here concern quantitative requirements, that is, the determination of the mass fraction of the individual phases (precipitates) and, above all, the precise determination of their chemical composition.

In surface studies, it is the OES with glow discharge spectroscopy (GDS) that plays the leading role, while organic analysis in the metallurgy most commonly uses gas chromatography coupled with MS (GC-MS) and nuclear magnetic resonance (NMR) spectroscopy. Progress is also being made in studies of the chemical composition of metallurgical waste. Particularly intensive development has been observed in recent years in studies of dust and gas pollution during welding processes [7–9].

7.3 Fields of application of the analytical techniques in metallurgy

The application of analytical techniques to routine analysis in metallurgy covers four basic areas:
1. Control of delivered raw materials
2. Control of production process
3. Inspection of final product
4. Analysis of waste materials, by-products and other similar products

This is illustrated in Figure 7.1, further dividing the area into testing of semi-finished products, inoculants and master alloys. Depending on the specifics, profile and volume of production in a given plant, this range of studies can cover both the control area of delivered raw materials and the control area of production process. This depends on whether the materials are purchased or produced by the plant in question within the planned production cycle. More often, however, this area is located in the field of control of delivered raw materials.

7.4 Range of materials

A detailed list of materials analysed in the analytical metallurgical laboratories is shown in Figure 7.2. They have been grouped into the following five areas: raw materials and semi-finished products, auxiliaries, finished products, waste and by-products.

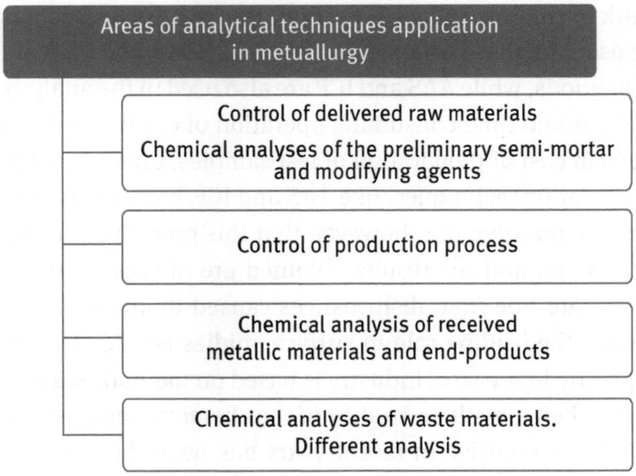

Figure 7.1: The areas of analytical techniques application in metallurgy.

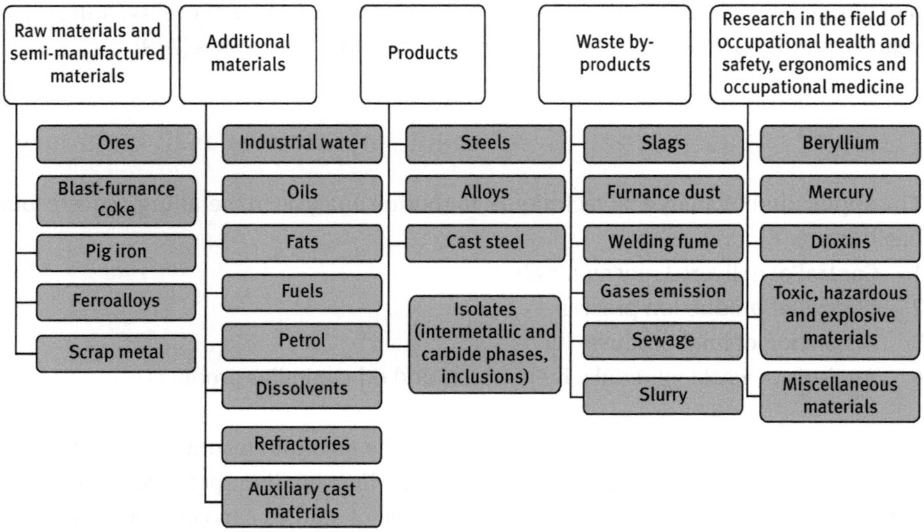

Figure 7.2: Application of chemical analysis in different metallurgical materials.

Additionally, areas of health and safety at work, ergonomics, occupational medicine and occupational disease prevention, recently gaining importance, have also been incorporated into this compendium.

7.5 Developments in metallurgical analytics

Taking into account the specific character of the steel and alloy production cycle, the techniques of chemical analysis of metallurgical samples are strongly dominated by the methods of OES with spark excitation and XRF. Further, the techniques such as AAS, ICP-AES and ICP-MS are used. In phase composition studies, EPMA and CMA are currently the most important methods, while AAS and ICP are also used in the analysis of isolates. Considering the labour and time-consuming operation of electrochemical phase isolation, and thus the high cost of preparation of test samples, direct analysis of isolates using instrumental isolation techniques, like AAS and ICP, has recently lost some of its importance. It is worth pointing out, however, that this procedure of testing isolates still has many followers, and the results obtained are of great interest, mainly due to the fact that they are free from disturbances caused by the presence of matrix. The technique playing the leading role in surface studies is GDS. The organic analysis used most frequently in the steel industry is based on methods such as GC-MS and NMR. Progress is also being made in studies of the chemical composition of metallurgical waste. Particularly intensive in recent years has been the development of studies of dust and gaseous pollutants formed in welding processes. Metal welding is seen as a three-dimensional (3D) process: dirty, dusty and dangerous. The toxicity of these pollutants, associated with undesirable physicochemical reactions that occur after ingestion into the human body, poses a serious risk and is the cause of many occupational diseases in the welders [7–9]. During welding process, under the effect of high temperature and welding arc radiation, welding fumes are generated from the base material, the secondary material, the protective coating of the base material, the shielding gas and the surrounding air. Welding fumes (biphasic condensate aerosol) are a mixture of finely dispersed particulates (welding dust) and various gases forming the dispersion phase. The welding dust produced by plasma arc acting on the base and secondary material consists of simple and complex oxides, silicates, fluorosilicates, fluorides, chromates, dichromates and metal carbonates. The diameter of the aerodynamic particles of the welding dust is from 0.01 to 1 μm [7]. The chemical particles of the welding dust range from 0.01 to 1 composition of the welding dust depending on the type of welded materials, welding methods and parameters of welding technology. Dust from the welding process using coated electrodes and metal powder wires has more complex chemical composition and structure than the dust emitted during arc welding with solid wires and shielding gas. Iron, manganese and silicon dioxide are the basic constituents of dust formed during welding of non-alloy steels with solid wires, while dust formed during welding of high-alloy steel also contains compounds of chromium, nickel, molybdenum and niobium. Welding of steel with coated electrodes and metal powder wires additionally results in the separation of sodium, potassium, calcium and magnesium. Significant achievements in this area of research have been done by the Welding Institute in Gliwice. Modern test stations for sampling of dust produced by metal inert gas (MIG)/metal active gas (MAG), tungsten inert gas (TIG) and manual metal arc welding (MMA) have been designed and executed [7–9]. The methodology of dust sampling during welding of different grades of steel was also developed, taking into account the applied welding techniques.

Market economy conditions put new challenges to the staff of chemical metallurgical laboratories in the form of skills development and work organization, as well as the extended scope of tasks and the efficient management of laboratory functions [10]. Laboratory workers should systematically broaden their professional qualifications, and technical and methodical competences. This opportunity is created by the development and implementation of knowledge system in a research facility [11]. The use of knowledge system makes the work more organized and systematic. The rational use of knowledge resources enables employees to intensify research and development, and experimentation and services. Functioning of research laboratories under conditions of market economy results in the fact that knowledge management in both the parent company and its laboratory units has become an inherent aspect of the development and gaining competitive edge. Increasing competition in the services sector poses a threat to individual chemical laboratories in terms of obtaining orders for specialized research and services. Knowledge systems in research laboratories increase the synergy of cooperation between teams and within their structure, and enable continuous control of the results of research and services and also continuous improvement of research procedures and methodologies. When the knowledge management system is

started, the first step is gaining reliable information, taking into account the specific nature of the laboratory (situational analysis subsystem), and then distributing the knowledge among employees (human resources improvement subsystem) and using it properly (implementation subsystem). Knowledge management in modern chemical laboratories becomes the basis of staff professionalism; covers a wide range of qualifications, skills and competences; and is derived from the development strategy of the research unit and/or parent unit (enterprise, university, research and development centre). However, assuming some generalizations, the following modules can be distinguished:
- Management of the laboratory unit
- Staffing
- Organization of work on the measurement bench
- Research and service works
- Archiving knowledge
- Market research and marketing

The directions in development of metallurgical analytics also determine the modernization aspects of technical, material and economic processes, taking into account environmental protection requirements. Advanced methods of sample preparation for examinations and development of modern non-destructive spectrometric methods will make a significant contribution to the development of metallurgical analytics. The process of automation, robotics and computerization throughout the analytical cycle will be complementary to providing high quality products. The development and application of laser spectroscopy in microstructure examinations and trace analysis is also envisaged.

7.6 Determination of chemical composition of metal alloys and other industrial materials

The determination of chemical composition of various grades of steel, metal alloys, metal ores, rocks and soils, cements and slags, industrial fuels, pigments and paints, and of different kinds of waste materials, using for this purpose AAS and OES with ICP has already been comprehensively described in books, monographs, and numerous publications in scientific and technical journals. Therefore, it is very difficult to choose from among the rich variety of complex reference materials the most valuable publications, since the scope of the presented studies includes the determination of major alloying elements, trace elements and harmful impurities. This applies to both basic techniques and other accompanying solutions, to mention as an example hydride generation techniques, small sample volume techniques and also coupled techniques. Applications of AAS technique to study the chemical composition of different materials are summarized in Refs. [12–20], while the use of ICP-AES is presented in Refs. [21–23, 89].

7.7 The concept of speciation in relation to metallurgy and materials engineering

Thus, assuming, with some simplifications, that speciation is the presence of various "chemical forms" of the element in the sample, and speciation analysis serves to identify those forms and quantify them, by analogy, with respect to metallurgical samples, equivalent phase identification and phase analysis means are used, including also chemical analysis. Moreover, it should be emphasized that both the concepts and the specific phase studies in the field of steel and metal alloys had been used much earlier, that is still before the term *speciation analysis* was introduced in analytical chemistry in the mid-1970s. They have been and still are inherent in the development of technology for the manufacture of various types of metal alloys. Moreover, owing to these studies, the scope of their industrial applications was systematically extended. This has ultimately led to a situation where identification of phases occurring in a given material has become a very important element of metallography, or, more accurately, of materials science. Knowing the chemical composition of the test material as well as its "technological history", the phase equilibrium diagram often enables us to make phase identification even when the microstructure deviates from the state of equilibrium. Phase identification basically fulfils the criteria for speciation. Much more difficult and complex in confrontation with phase analysis is the speciation analysis. The problems here concern quantitative requirements, that is the determination of the mass fraction of the individual phases (precipitates) and, above all, the precise determination of their chemical composition.

It is also worthwhile to introduce here the relationship between concentrations expressed in mass percentages, usually obtained by chemical analysis of isolated phases (isolates), and concentrations expressed in atomic percentages, which determine the percent content of atoms of the individual components in the total number of atoms present in the isolate, useful for the development of quantitative formulas of specific phases:

$$A_i = \frac{W_i/M_i}{\sum_{i=1}^{n} W_i/M_i} \cdot 100\% \tag{7.1}$$

$$W_i = \frac{A_i \cdot M_i}{\sum_{i=1}^{n} A_i \cdot M_i} \cdot 100\% \tag{7.2}$$

where A_i is the atomic percentage of the ith element; M_i the atomic mass; W_i the weight percentage of the ith element.

This section outlines some of the issues that characterize the area of metallurgy and materials engineering, focusing mainly on the chemical composition of electrolytically isolated carbide phases present in tool steels and isolates of the intermetallic γ' phase determined by the technique of FAAS. Based on the literature data, as a typical example of metallurgical speciation analysis, the possibility of aluminium content determination by AAS in steels with division into acid-soluble aluminium and total aluminium was presented. The possibility of implementing own solutions and technical modifications in the application of flow injection technique using small volumes of sample solutions combined with flame atomic absorption method was also indicated. The most important techniques for the preparation of isolated test phases were presented together with the analytical schemes developed. The use of less sensitive analytical lines was exposed as a means allowing for the determination of the content of many elements within a very wide range of concentrations. Testing of isolated phases is closely related with the systematically growing requirements for high-quality metallurgical products. It is also an important element in the design of new, complex metal alloys for the most responsible components of machinery and equipment, and in the development of technological processes for the manufacture of these alloys in metallurgical enterprises. Additionally, the similarities and differences between phase analysis and speciation analysis in studies of metallurgical materials were outlined.

7.8 Chemical and electrochemical phase extraction

Chemical and electrochemical anodic extraction of phases and inclusions is used as a tool in the determination of their content, structure and chemical composition [25–30].

7.8.1 Chemical phase extraction

Chemical methods of phase extraction are rarely used nowadays, and this is due to a number of constraints and drawbacks, mainly associated with a very long duration of the dissolution process, difficulties in selecting suitable digestion mixtures and conditions for selective conversion of the sample matrix into the state of solution, ensuring at the same time the required degree of purity in the isolate obtained. These methods only apply to the separation of the most chemically resistant phases, like NbC, TaC, TiC and VC carbides as well as some M (C, N) carbonitrides [31, 32]. In some cases, this method is also useful in the identification of previously determined forms of occurrence of different elements in steels and/or non-ferrous metal alloys, for example in the separation of the metallic form of alloy component from its oxide or stable carbide counterpart. Then the analysis cycle normally consists of two steps, including the selection of suitable acids dissolving the matrix and the metallic part of the element and melting, with the use of flux, the remaining (oxide or carbide) portion of this element, followed by its conversion to the state of solution. At the same time, these two steps provide the basis for carrying out two separate series of measurements of the concentration of the individual parts of the component being analysed. It is also possible to use only the first step of sample preparation, which will allow us to determine only the content of the acid-soluble part. In contrast, to determine the total content of a given element, complex mixtures of acids are used, applying additionally specialized equipment, for example, teflon bombs or microwave mineralizers [12, 33–35]. Electrochemical methods of extraction are used to separate the complex carbides, nitrides and, above all, intermetallic phases, which play a fundamental role in complex materials research, especially in evaluating the quality of various grades of steel and multicomponent metal alloys [36–48].

7.8.2 Electrochemical phase extraction

Electrochemical phase extraction can be performed in two ways: galvanostatic, that is at constant current density and, more frequently, potentiostatic, that is with constant potential [25, 26, 28, 29]. In the latter case, electronic potentiostats, automatically supporting the potential of the electrode, are used. In the anodic phase extraction process, the following condition must be met: the dissolution potential of the primary metallic phase (solid solution) should be more negative than the solubility potential of the extracted phase. Typically, the extracted phase is precipitating on the anode in the form of fine dispersion powder. The test sample operates as an anode with appropriately selected electrolyte and voltage. These parameters are essential in the phase isolation process, which can also be regarded as a targeted and selective corrosion [6]. Schematically, with some simplifications, using the data contained in Ref. [6], the principle of electrochemical phase extraction is shown in Figures 7.3 and 7.4.

From the flat surface of the sample (Figure 7.3(a)), over time, the solid solution and possibly also less stable phases are dissolved, and the precipitate to be examined, separated from the reaction surface, becomes more prominent (Figure 7.3(b)). As a result of decreasing adhesion, individual particles of this isolate start “falling off” and may be collected (Figure 7.3(c)). The isolation process can be interrupted at any moment, when the total amount of isolate is sufficient to carry out the planned tests (Figure 7.3(d)). Once the isolation process has been completed, the isolate can be separated from the sample by, for example, ultrasounds, depending on the size and amount of particles. This method, as demonstrated by comparative studies using electron microscopy, does not lead to any damage and/or distortion of the separated isolate particles [25, 26]. The final stage leading to the formation of an isolate is centrifugation of the electrolyte. Determining, and more specifically optimizing, the conditions for the selective isolation of specific precipitates in casting metal alloys is a very complex task, and as such requires a number of ancillary studies in the field of

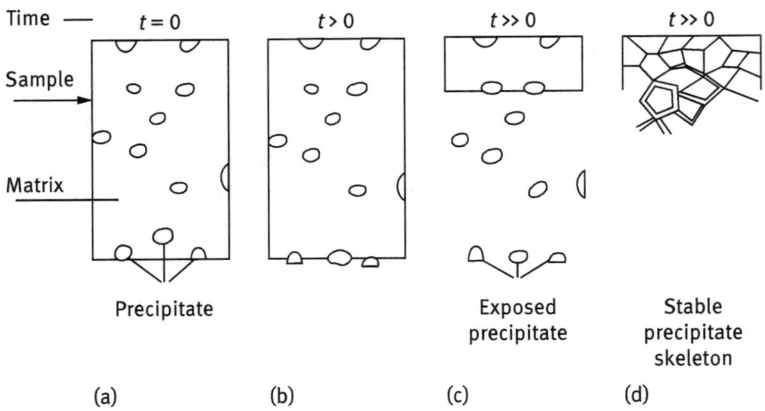

Figure 7.3: The principle of electrochemical phases extraction [26]. Explanation of a), b), c), d) is in text (page 313).

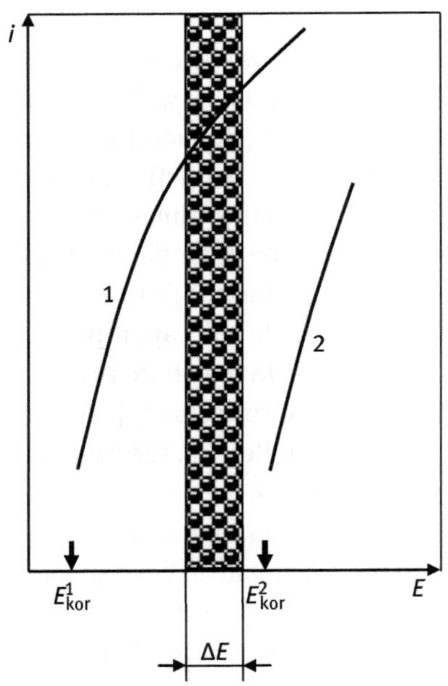

Figure 7.4: Schematic presentation of potential and current density selection in isolation process [25]. 1,2 – anode polarization curves of steel (alloy) and isolated phase; $E^1{}_{\mathrm{kor}}$, $E^2{}_{\mathrm{kor}}$ – corrosion potentials of steel (alloy) and isolated phase; ΔE – range of potentials of potentiostatic phase separation.

physical chemistry and electrochemistry, inorganic chemistry, chemistry of complex compounds and chemical thermodynamics [25, 26, 28–30, 32]. All these studies include, but are not limited to, specifying the detailed characteristics of both the test material and the electrolyte used. With respect to the test sample, the primary concern is its chemical composition (inhomogeneity), structure (crystal lattice defects) and material condition. As far as electrochemical parameters are concerned, the first ones to be mentioned here are the dissolution potential or the anode current density, while the most important parameters of the electrolyte include its chemical composition,

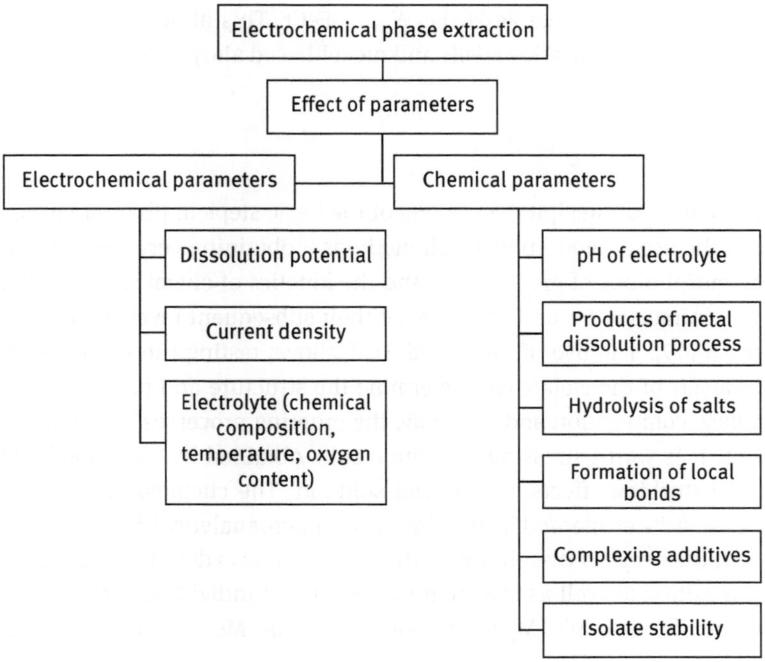

Figure 7.5: Electrochemical and chemical parameters influencing on the phases of isolation process [29].

pH and oxygen content. Process temperature is also important. Chemical and electrochemical factors that exert some control over the electrochemical phase extraction process are shown in Figure 7.5. Moreover, sample geometry, cathode material, type of diaphragm used, electrolyte preparation and dosage also deserve some attention [25, 26].

The electrolyte solution should consist of a solvent, which is a mildly dissociated liquid (e.g. H_2O, CH_3OH, C_2H_5OH) characterized by a sufficiently high dielectric permeability to ensure the ionization of the introduced substances, of the proper electrolyte capable under certain polarization conditions of undergoing adsorption on the surface of the phases which form protective layers, and also of suitable complexing additives (acids: citric, tartaric, oxalic and their salts) preventing the formation of hydrolysis products – WO_3, MoO_3, Nb_2O_5 and so on [25, 26, 29]. In the case of phase analysis of multicomponent nickel alloys using electrochemical phase extraction, in electrolytes containing halogen ions – Cl^-, Br^-, I^-, F^- – thenickel electrode activity is protected, while passive state is due to the presence of anions such as NO_3^-, SO_4^{2-}, PO_4^{3-} and ClO_4^- [25, 26, 29]. This means that in electrolytes containing ions of halogen groups, the solid solution of nickel, matrix and chemically unstable intermetallic compounds (Ni_3Al, $Ni_3(Al,Ti)$, Ni_3Ti, Ni_3Nb, $NiAl$) polarize more weakly and dissolve at a faster rate than carbides (NbC, TiC, Cr_7C_3, $Cr_{23}C_6$), borides (Cr_2B, Ni_2B_3), nitrides (CrN,

Cr_2N) and intermetallic phases like Fe_2W, Fe_7W_6, σ- FeCr. This allows quantitative separation of these phases in the alloy steels and nickel-based alloys [25, 26, 36].

7.9 Techniques of testing isolates

Electrochemical isolation of precipitates, as one of the basic steps in phase analysis, enables – through the use of modern research methods – obtaining very valuable information on the morphology of metal alloys and the kinetics of chemical reactions. The process of obtaining isolates and methods for their subsequent investigation are shown in Figure 7.6 [49]. The use of analytical SEM allows testing the isolate morphology. X-ray analysis of the isolate can determine the structure and parameters of the lattice, the phase composition and, possibly, the ordering processes taking place in the precipitates. It is worth stressing that the results obtained by these methods are free from the disturbing effects of the solid solution. The chemical analysis of the isolate made by a "wet method", or using other microanalytical instrumental techniques without the need for dissolution of this isolate, allows determining the exact chemical composition as well as the quantitative ratio of individual components present in the isolate, for example, $M_{23}C_6$ carbide – $(Cr_{16.3}Ni_{3.6}Mo_{3.1})C_6$or $(Cr_{21}Mo_2)C_6$ [26, 31, 32, 35, 36].

The methods for testing isolates shown in Figure 7.6 can be considered as baseline. Only with the contribution of other complementary methods summarized in Table 7.1, the investigation of isolates provides complete information on the chemical composition, phase composition and structure.

To avoid damage to the sample fixture, the electrochemical phase isolation process must be terminated early enough. The residual may be subjected to further studies, in particular by SEM and X-ray microanalysis to determine the type of particles and their location in the test material [25–30]. Chemical analysis of the electrolyte gives information on the composition of the solid solution of the test alloy.

The parameter most important in the quantitative evaluation of the phase isolation process is the yield W [28], which is determined as

$$W = \frac{m_i}{m_p} = \frac{m_i}{m_i + m_r} \tag{7.3}$$

where m_i is the mass of isolate, m_p the mass of sample, m_r the mass of matrix.

Sometimes, instead of the yield W, the phase ratio B is determined:

$$B = \frac{m_i}{m_r} \tag{7.4}$$

Electrochemical methods of phase extraction also have specific limits of use. On the one hand, they are controlled by electrochemical factors related to changes of concentration in the sample matrix during isolation process and by the associated processes

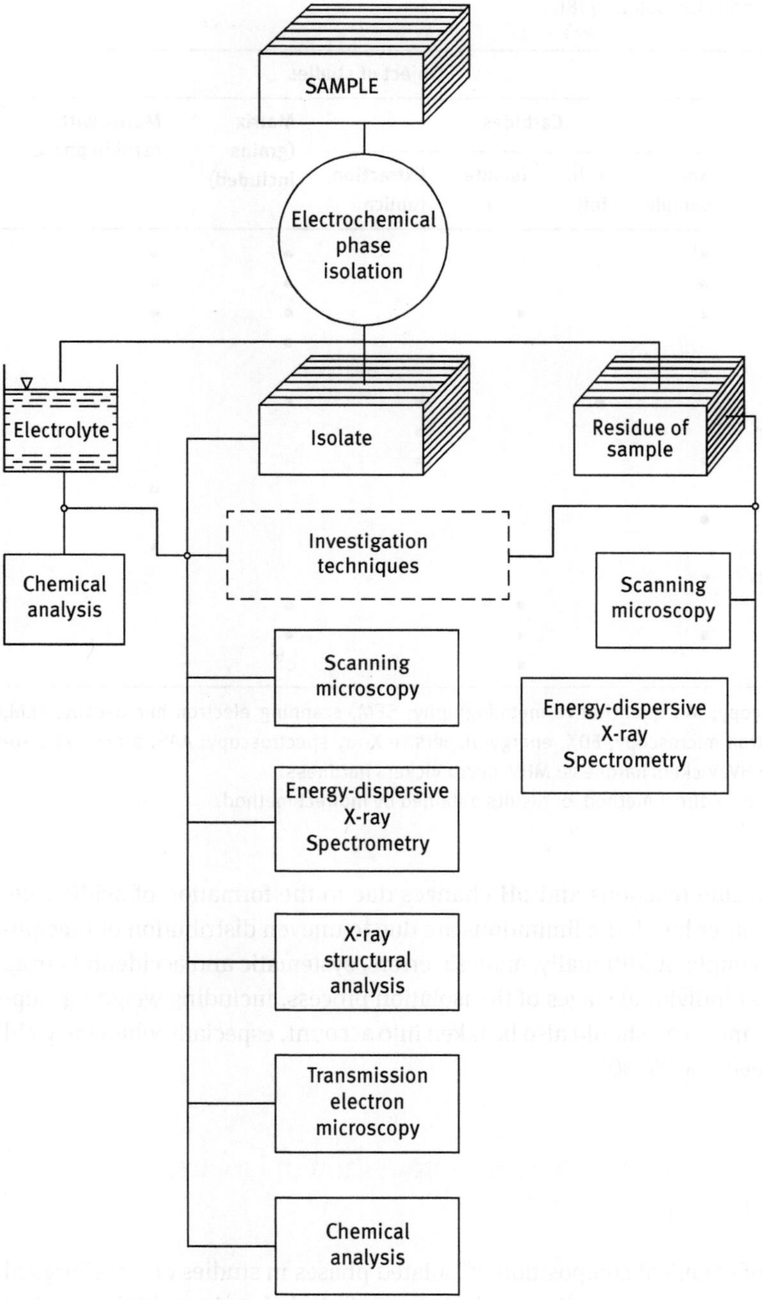

Figure 7.6: Techniques for investigating isolates.

Table 7.1: Test methods for isolates [38].

Applied methods or measure of the examined property[a]	Object of studies					
	Carbides				Matrix (grains included)	Matrix with carbide phase
	Solid sample	Thin foil	Isolate	Extraction replica		
MS	•[b]				•	•
MS-MI	•				•	•
SEM	•		•		•	•
SEM-MI					•	
MS+SEM-MI	•					
TEM		•		•	•	
Electron diffraction				•		
MS+SEM+TEM-MI	•			•		
HV						•
MHV	•					
Dilatometry						•
X-ray microanalysis	•					
X-ray diffraction			•		•	
SEM-EDX	•		•		•	
AAS			•		∘[b]	

[a] MS, light microscopy; MI, quantitative metallography; SEM, scanning electron microscopy; TEM, transmission electron microscopy; EDX, energy-dispersive X-ray spectroscopy; AAS, atomic absorption spectrometry; HV, Vickers hardness; MHV, micro Vickers hardness.
[b] •, results obtained by direct method; ∘, results obtained by indirect method.

of precipitation, side reactions and pH changes due to the formation of acidic salts [25–30]. On the other hand, the limitations are due to uneven distribution of precipitates in the test sample. Additionally, analysis errors, systematic and accidental errors that may occur at individual stages of the isolation process, including weighing, separating, drying, and so on should also be taken into account, especially when the yield W does not exceed 1% [25–30].

7.10 Determination of chemical composition of isolates – speciation analysis

Determination of chemical composition of isolated phases in studies of metallurgical materials is equivalent to speciation analysis in analytical chemistry. With chemical or electrochemical separation of structural constituents, very small amounts of isolates are obtained, ranging from several to several dozen milligrams, and sometimes even 0.5 mg. So, it is the amount of test material that presents a lot of difficulties for analysts dealing with the chemical analysis of isolates, regardless of the choice of analytical methods and techniques. In chemical analysis, in addition to spectrophotometric methods, rarely gravimetric and titrimetric methods, depending on the number of constituents determined and the amount of isolate obtained, other methods applied include X-ray spectrometry (XRF, energy-dispersive X-ray spectroscopy [EDX]), FAAS and flameless (graphite furnace) AAS, OES-ICP, ICP-MS, GDS and other techniques [6, 13, 14, 17, 20, 24, 26, 28, 29, 36, 50, 52–66].

One of the methods that has proved very useful in studies of the chemical composition of isolates is AAS [6, 12–17, 20]. That is why, in this chapter, more attention has been devoted to this method, using in particular own research and long experience of the author in this area. Among instrumental methods, AAS takes a special place. This is due to its numerous advantages, such as high sensitivity, selectivity, relatively high speed of analysis, and above all, the ability to label almost all metallic elements of the periodic table in a very wide range of concentrations. AAS is used to determine the content of metallic elements in numerous materials and in many fields of science, technology, medicine and protection of the natural environment [12, 17].

7.11 AAS in metallurgy

AAS was introduced into chemical metallurgical laboratories in the 1960s, and since the 1970s, apart from OES and XRF, it has been considered one of the leading methods for determining the chemical composition of raw materials, and semi-finished and finished products [1–4, 6, 13, 14, 18, 20, 53].

Quantitative determination of elements in steel and other metallurgical materials by the AAS method is based on plotted standard calibration curves [6, 12–18], or it uses the technique of standard additions, or single-reference standard calibration method developed by the Silesian University of Technology [6, 20, 54, 96].

The variety and complexity of the steel grades and alloys used in various industries, as well as the introduction of new materials with very complex chemical compositions designed for the most responsible machine parts require the development of numerous analytical procedures. All of them are expected to offer high accuracy and precision of the determination of the content of both major alloying elements, often varying in a very wide range of values, as well as the accompanying elements and harmful impurities. In such cases, especially when new materials are still in the testing stage, the AAS method is extremely useful. It should be noted, however, that in spite of its undoubted advantages, such as the high sensitivity and accuracy of multitask determinations, the ability to perform analyses in a fairly wide range of analyte concentrations and, what is particularly remarkable, its selectivity allowing the determination of the content of several elements from one starting sample, this method is by no means a universal tool. AAS also has its drawbacks referring mainly to the development of means and conditions for the dissolution of complex metallurgical samples and limits in the determination of some microadditives and alloy constituents, combined with the need to correct numerous interactions between elements. This is particularly true in the case of elements such as B, W, Nb, Ta, La, Ce, Zr and Hf. Additionally, the AAS method does not allow for the determination of carbon and sulphur, while the determination of the content of silicon and phosphorus can only be performed to a limited extent [6, 12–18]. The limits for the determination of selected metallic elements by the AAS method can be grouped into the following concentration ranges: 10^{-5}% – Mg; 10^{-4}% – Ag, Be, Ca, Li, Mn, Zn; 10^{-3} – 10^{-2}% – Al, As, Ba, Bi, Cd, Co, Cr, Cu, In, Mo, Ni, Pb, Pd, Rh, Ru, Sb, Sr, Te, Tl, Yb; 10^{-2} – 10^{-1}% – Ga, Ge, Hg, Se, Si, Sm, Sn, Ti, V, Y and Zr; above 0.10% – B, Ce, Gd, Hf, Ir, La, Nb, Nd, Re, Ta, U, W [12–18]. Due to the fact that in metallurgical samples the determination of the content of metallic elements covers a wide range of concentrations, it is difficult to assess in a consistent way the accuracy and precision of the obtained results. With respect to the vast majority of elements determined by the AAS method, the relative standard deviations are from 0.2% to 5%. Only in the determination of carbide-forming elements, such as niobium, tantalum, zirconium, tungsten, titanium or vanadium, the relative standard deviations can exceed these values even twice. Although the use of AAS technique in material analytics mostly relates to the low and trace content of elements, it can be added that as a result of the continuous development of atomic absorption spectrometers in both optics and electronics, it has now become possible to determine the high content of elements in steels and metal alloys with an accuracy close to classical methods [12–18].

The increasing number of atomic absorption spectrometers used in materials engineering, metallurgy, foundry and metal processing, and also in research centres of these industries, necessitates systematizing and even unifying a variety of analytical procedures for chemical and phase analysis of multicomponent metallic materials as well as further research in this extremely important field, to improve the quality not only of materials already produced but also of new alloys that are in the design or development phase of their manufacturing technology. Determination of chemical composition of isolated intermetallic and carbide phases by AAS is included in the range of speciation analysis of metallic materials. The small mass of test materials, the wide range of concentrations of the determined elements and the lack of certified reference standards create additional difficulties in determining the content of elements in isolates, regardless of the choice of analytical method and technique. Therefore, using the method of AAS in this research area, several additional technical and methodical modifications discussed later have been introduced in order to make the most of the advantages offered by this technique.

7.12 Flow injection technique combined with flame atomic absorption method

For various reasons, in some cases, in the speciation (phase) analysis of steel and alloys, the mass of the obtained isolate is insufficient for the chemical analysis carried out by conventional methods. This happens most often when studies and experiments are conducted on alloys that are still in the process of developing technologies of their production or optimization of the thermal and thermochemical treatment parameters. Then the chemical analysis of the separated phases or inclusions should be preceded by qualitative and X-ray diffraction studies, absorbing the already insignificant quantities of isolates. Sometimes the insufficient mass of isolates is associated with difficulties arising from the electrochemical phase isolation process itself [1–3, 6, 25–30]. In such cases, a very useful solution is the use of a modified version of the atomic absorption method, that is a flow injection technique, which involves the construction of special teflon sampling manifolds and special technique of analysis, allowing the introduction of small quantities of samples, that is 50–100 µl, without loss to the flame [6, 56–61, 63–77, 127], and introduction of metered portions of analysed solutions to the flame instead of continuous suction, as is the case in conventional AAS technology.

This allows conducting analyses in a very "economical" way. To achieve this purpose, however, a special sampling manifold of crater type (Figure 7.7) or funnel type (Figure 7.8) made of teflon should be constructed, since this material enables testing of solutions containing, among others, hydrofluoric acid and ammonium fluoride. Teflon sampling manifold was made of a teflon block with dimensions of 120 mm × 120 mm × 120 mm. On the top surface, there were cone-shaped recesses with an opening angle of at least 80° and a diameter of 15 mm. The recesses (craters) were located at a distance of 25 mm from each other, adopting this value as a value of the distance between two adjacent vertices of the cones. Each recess had a capacity of 0.5 ml. To carry out a series of measurements, several teflon manifolds should be made

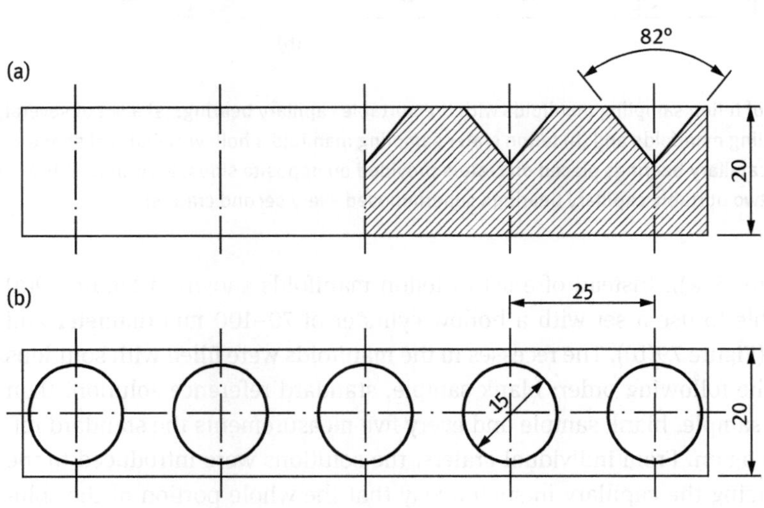

Figure 7.7: The scheme of teflon sampling crater manifold: (a) side view and (b) view from the top [66].

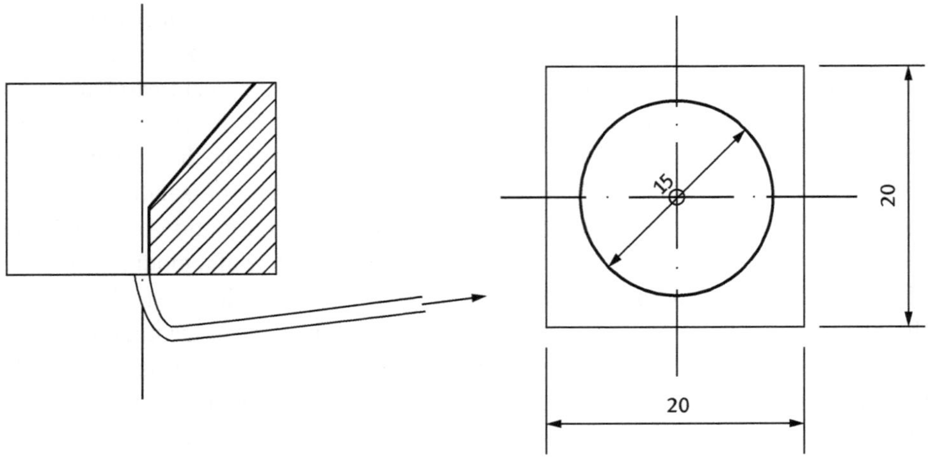

Figure 7.8: The scheme of funnel manifold [66].

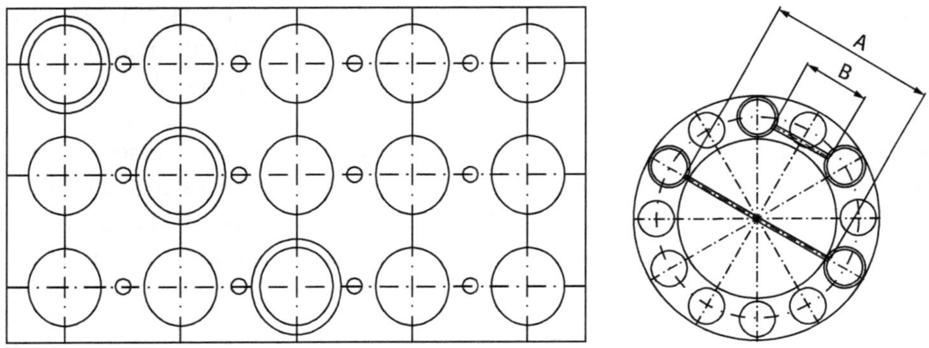

(a) (b)

Figure 7.9: Types of teflon sampling manifolds with the portable capillary bearing: (a) a set of several crater teflon sampling manifolds and (b) teflon crater sampling manifold a hollow cylindrical shape (A – two portable capillary bearings placed on craters provided on opposite sides, according to the diameter; B – the two portable capillary bearings superimposed every second craters).

available (Figure 7.9(a)). Instead of a set of teflon manifolds shown in Figure 7.9(a) it is also possible to use a set with a hollow cylinder of 70–100 mm diameter and 120 mm height (Figure 7.9(b)). The recesses in the manifolds were filled with solutions introduced in the following order: blank sample, standard reference solution, then alternately test sample, blank sample and every five measurements the standard reference solution again. From individual craters, the solutions were introduced to the flame by immersing the capillary in such a way that the whole portion of the solution was sucked in. Instead of a crater manifold, in the flow injection technique, a sampling funnel system can also be used (Figure 7.7). For this purpose, a cube with only one recess was cut off from the crater manifold, and then in the top part of thus formed crater a hole was drilled, allowing the tip of the capillary to be inserted to a depth of 1 mm. The inserted tip of the capillary was fastened with glue to the cut-off crater. The length of the capillary connecting thus prepared teflon funnel manifold with the spectrometer nebulizer should not exceed 25 mm. For mounting the funnel manifold, a small tripod or holder is sufficient, or if the height of the funnel does not exceed 14 mm, the factory-fitted washer can be used to attach it to a Perkin-Elmer (model 503) spectrometer. Using both crater and funnel sampling manifolds, 100 μl portions of the solution were measured, having established previously that with these portions of the solution, changes in the suction rate of the solution ranging from 4 to 7 ml/min would not cause measurable changes in the recorded absorbance signals for elements determined in different groups of isolates. Based on current studies it has been found that technical issues are the factors determining the accuracy, the precision and the reproducibility of the obtained results. In the case of sampling funnel manifolds, they are associated with vertical positioning of the micropipette and with the experimentally determined distance from the funnel preventing the introduced solution from splashing, which results in discontinuous suction. On the other hand, using a sampling crater manifold, which in practice is more convenient to handle, the capillary tip must be held in vertical position – perpendicular to the recess. Too vigorous insertion of the capillary can result in its bending, which will immediately manifest itself as a change in the height of the measured signal recorded in the shape of a peak. Improvements in this area have been sought in design modifications of the structure of crater manifolds and in the use of additional capillary support and stiffening systems. Of the few tested solutions, the most advantageous seems to be the construction of a plexiglass capillary bearing (Figure 7.10) [66, 67]. The use of this element has improved the accuracy and reproducibility of test results and greatly facilitated making routine analyses, especially as regards the isolates of the intermetallic γ' phase [66].

An alternative solution for the portable capillary bearing is a pillar-shaped cover (lid), made of plexiglass or polystyrene with appropriately drilled holes. Its characteristic feature is that the drilled holes are located exactly above the centres of the recesses in the teflon sampling manifold, and the side walls of the cover encompassing the teflon sampling manifold provide a fixed and stable position of the cover [6, 68]. Schematically, this is illustrated in Figure 7.11 (the shape of the cover is shown in the horizontal projection, the cover is placed on the teflon sampling manifold). Moreover, to ensure accurate insertion of the spectrometer capillary into the bottom of the cavity to suck out the entire metered amount of solution, a small rigid ring is applied on the outer wall of the capillary, providing an effective lock preventing the capillary bending.

The use of flow injection technique combined with the FAAS method in determining the chemical composition of isolates has proved to be an effective and relatively inexpensive solution.

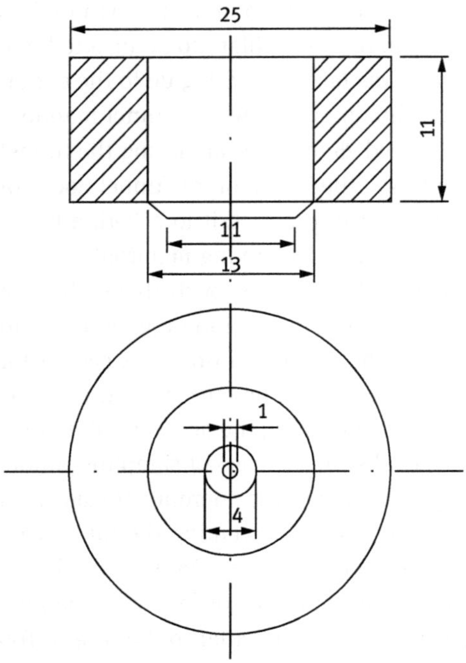

Figure 7.10: The scheme of portable capillary bearing [66, 67].

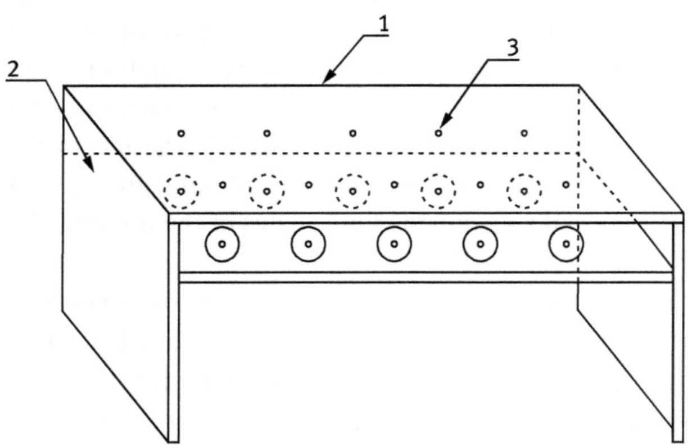

Figure 7.11: Schematic presentation of the cover (lid) with precisely drilled to a vertical input spectrometer capillary teflon sampling manifold (manifolds) 1, Cover; 2, sidewalls; 3, drilled holes for introducing the capillary into the craters of teflon sampling manifold [6, 68].

7.13 Application of alternative analytical lines

Quite often, when determining by AAS the chemical composition of metal alloys, isolated carbides and intermetallic phases, galvanic baths and industrial waste, it is necessary to determine the content of major components present in a wide range of concentrations, sometimes exceeding several and several dozen percent by weight.

If this is the case, then the use of AAS to measure the high content of elements present in the test sample will require one or more of the following operations [6, 12–18, 62, 63, 78] to be performed:

- Making appropriate dilutions of sample solutions
- Reducing (if possible) the weight of the samples tested
- Reducing the thickness of the flame-retardant layer (smaller burner head or rotation thereof)
- Using a “differential” measurement method
- Using flame zones poorer in the determined element
- Change of flame or change of flame characteristics
- Reducing the current of the lamp
- Substoichiometry (combination of classical methods with atomic absorption) [79]
- Use of less sensitive analytical lines

One of the possible effective solutions to this problem is in the use of less sensitive analytical lines [62]. Table 7.2 presents the limits of the straightness of standard reference graphs for the elements most commonly observed in a wide range of samples, taking also into account the results of own studies [6]. The use of less sensitive analytical lines enables the individual elements to be determined in a wide range of concentrations, avoiding very often the time-consuming dilution of sample solutions and reference standards. By analysing the data in Table 7.2, some practical notes and tips can be suggested.

High manganese content (15–30% by mass) can also be determined using, besides the frequently applied less sensitive 403.1 nm line, 2,450 times “weaker” 321.7 nm line, which gives straightness of standard reference graphs up to 7,500 µg/ml.

Applying the most sensitive cobalt analysis line, the measured concentration range does not exceed 5 µg/ml. Using the 352.7 nm line for cobalt determination, it is possible to analyse this element up to 200 µg/ml. Changing the acetylene–air flame, commonly applied in cobalt content determination, to acetylene–nitrous oxide flame extends the rectilinear range of the calibration graph up to 500 µg/ml, and thus with the sample weight of 0.5 g and 100 ml final volume of solution, it becomes possible to determine cobalt in nickel alloys up to 10% by mass without the need for dilution of sample solutions. Likewise, in the isolates of γ' phase (the sample weight of 20 mg and 25 ml final volume of solution), it is possible to determine cobalt content up to 10% by mass. The 372.0 nm line can be used to determine iron content in nickel and cobalt alloys, and in carbide isolates in the range of 20–60 µg/ml. Higher iron concentrations can be determined using the 305.9 nm line (23 times less sensitive than the main Fe line, i.e. 248.3 nm). In this case, the straight line in the standard reference curve will extend up to 120 µg/ml.

Table 7.2: Alternative analytical lines for the determination of Al, Mn, Mo, Fe, Ni, Co, Cu, Cr and Zn [6, 62].

Element	Analytical line (nm)	Straightness range of standard reference graphs (µg/ml)
Al	309.3	60
	394.4	120
	237.3	160
Mn	279.5	3
	403.1	25
	321.7	7,500
Mo	313.3	60
	317.0	80
	319.4	100
Fe	248.3	5
	372.0	60
	305.9	120
Ni	232.0	5
	341.5	25
	352.5	60
	351.5	120
Co	240.7	5
	352.7	200 (500 in C_2H_2–N_2O)
Cu	324.8	5
	327.4	25
	244.2	2,700
Cr	357.9	5
	359.4	15
	429.0	80
Zn	213.9	1
	307.6	5,200

The four analytical lines for nickel listed in Table 7.2 cover the range of its determination from 5 to 120 µg/ml. The compared alternative lines for copper and chromium allow the determination of the content ranging from 5 to 2,700 µg/ml (Cu) and from 5 to 80 µg/ml (Cr), respectively.

It is worth noting that the use of the 317.0 nm line for molybdenum determination prolongs the straight line in the standard reference graph from 60 (the main Mo 313.3 nm line) to 80 µg/ml. For aluminium, the 394.4 nm line doubles the detection range of this element from 60 (309.3 nm line) to 120 µg/ml. High concentrations of zinc, that is over 5 mg/ml, can be measured with a 307.6 nm line.

It is still to be noted here that for some elements, such as tungsten, characterized by a relatively low sensitivity of determination, routine analysis often uses a less sensitive analytical line of 400.9 nm rather than the main line of 255.1 nm. This is due to the fact that, using the 400.9 nm line, a better ratio of the absorbance signal to the photon noise is obtained (fluctuations lowering the precision of the determinations).

Actually, a very similar accuracy and precision of the determination of these elements using the most sensitive analytical lines and diluting the solutions and using alternative less sensitive analytical lines was obtained.

7.14 Calibration methods used in FAAS, including methods based on one reference standard

In the FAAS, different calibration methods [6, 12–18] are used, depending on the type of material to be analysed, the range of measured concentrations and the nature of interactions between elements. In preparation of standard reference solutions for the determination of the chemical composition of metallurgical samples, it is necessary to take into account both the interactions between elements and the interferences caused by acids and other substances used in sample preparation. Therefore, most often, synthetic reference solutions are prepared which, in addition to the variable amounts of the element to be measured, also contain the main constituent or constituents of the matrix. To compensate for the effect of acids used in the sample preparation process, the reference solutions should contain the same amounts of acids as the test sample solutions. The best results are achieved by complete evaporation of both test sample solution and standard reference solution, introducing next equal volumes of acids into them. Various techniques of standard additions and modifications thereof [12–18] are also used in studies of the chemical composition of metallurgical samples. Chemical metallurgical laboratories also used for calibration of natural specific standards or, so-called internal standards. Internal standards are samples of materials derived from the current production of a given plant, repeatedly analysed by various methods and analytical techniques. The author of this chapter has developed a calibration method for FAAS using one base reference standard [6, 66, 93, 94]. The method involves the use of several different weighted portions of one base (starting) reference standard to which, after or in the course of dissolution, appropriate quantities of the main matrix component are introduced to bring this content in both reference standards and test samples to a similar (with some tolerance) level, thus compensating for its effect exerted on the element being determined. The base reference standard is selected so as to best match the type of the sample to be analysed. It is also important that the upper and lower concentration limits of the individual elements are comprised in the rectilinear part of reference graphs. In order to determine the chemical composition of metal alloys and isolated phases, the starting pressures and the gas flow rates were used in accordance with the manufacturer's instructions, adjusting the flow rates in each case to obtain the maximum stable absorbance readings of the analysed elements. The optimum measuring height (burner height relative to the radiation beam) was set using the results of own studies [6, 49, 62, 64, 88]. Figures 7.12–7.16 illustrate the effect of changes in burner height on the absorbance of carbide-forming elements such as titanium, niobium, tantalum, chromium and molybdenum. For the

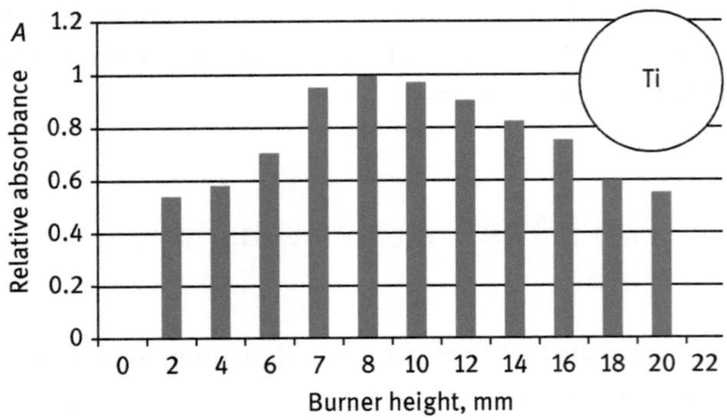

Figure 7.12: The effect of changes in burner height on the absorbance of titanium.

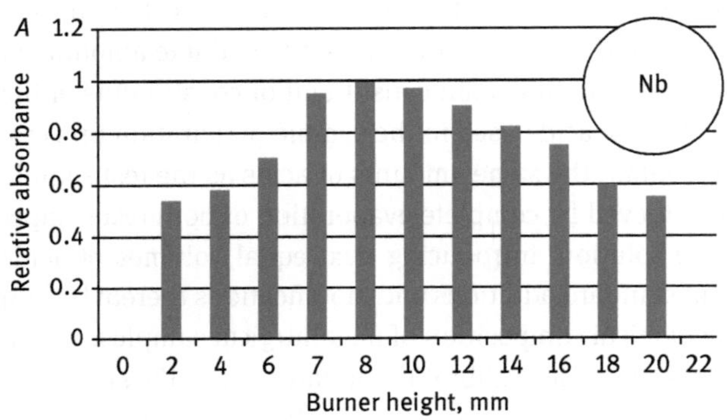

Figure 7.13: The effect of changes in burner height on the absorbance of niobium.

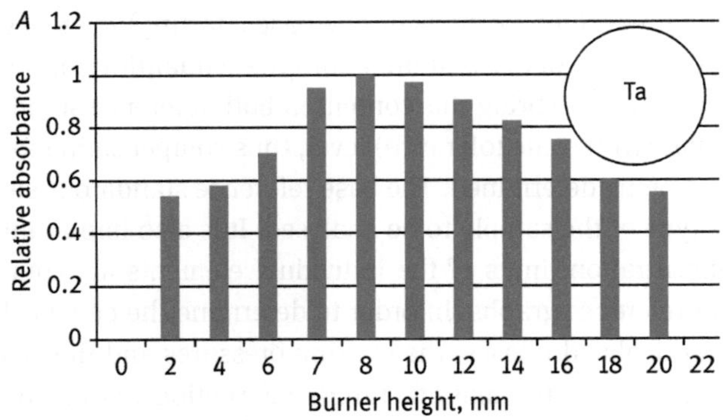

Figure 7.14: The effect of changes in burner height on the absorbance of tantalum.

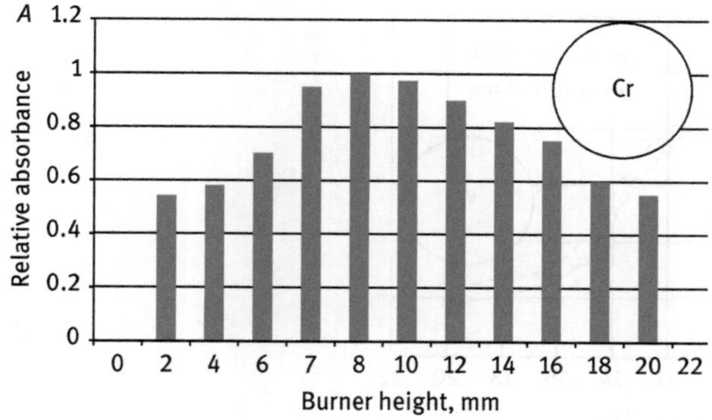

Figure 7.15: The effect of changes in burner height on the absorbance of chromium.

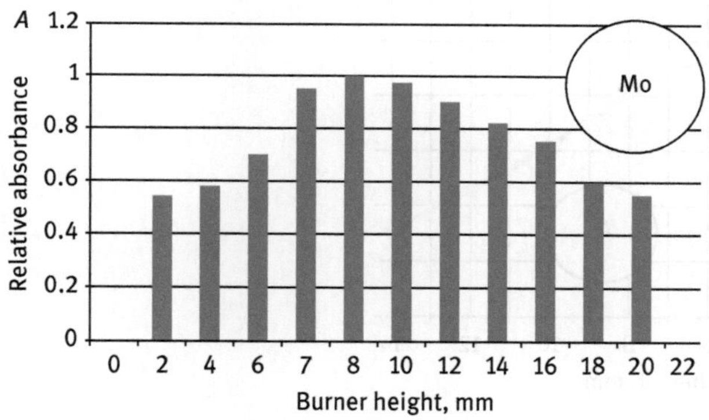

Figure 7.16: The effect of changes in burner height on the absorbance of molybdenum.

determination of tungsten content in metal alloys and in the isolates of carbides and intermetallic phases, a less sensitive 400.9 nm line is often used, and this is mainly due to its favourable signal-to-noise ratio. Therefore, for this element, the impact of changes in burner height was shown using both 255.1 and 400.9 nm analytical lines (Figure 7.17). Additionally, Figure 7.18 illustrates the results of studies of the importance of burner head positioning in aluminium determination. In all the cases, similar to the case of vanadium [49], a strong relationship was observed to exist between the vertical position of the burner head relative to the radiation beam emitted by individual hollow cathode lamps. After adjusting the optimal height of burner head, the flow of gases (acetylene and nitrous oxide) was adjusted to obtain the maximum stable absorbance readings of the analysed elements. Table 7.3 gives detailed measurement parameters for the elements most commonly determined in the tested isolates.

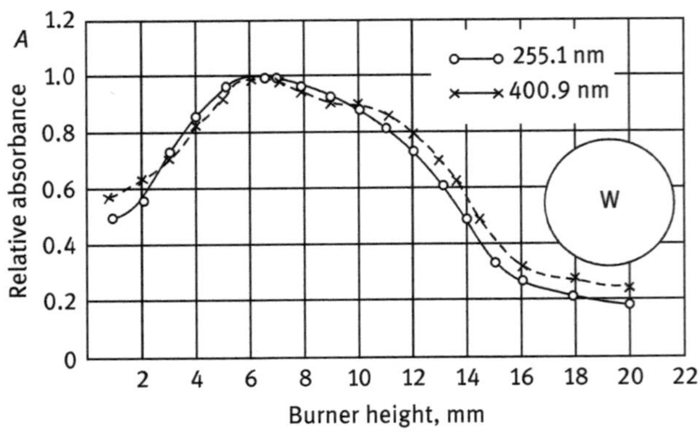

Figure 7.17: The effect of changes in burner height on the absorbance of tungsten.

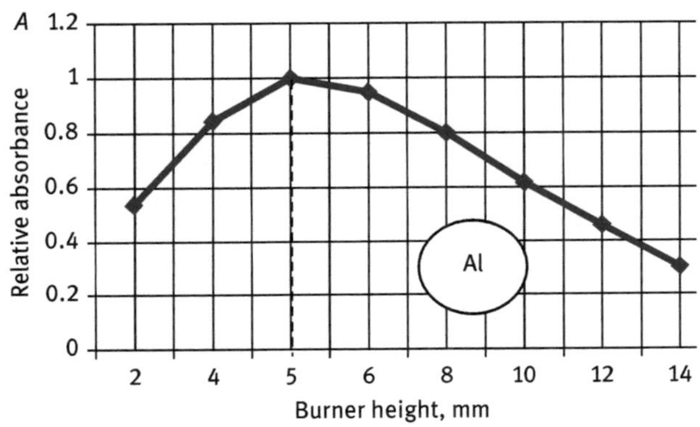

Figure 7.18: The effect of changes in burner height on the absorbance of aluminium.

7.15 Determination of aluminium in steel as a typical example of speciation and speciation analysis in metallurgy

The problem of aluminium determination in steel using AAS has already been described in detail in many papers; therefore, in this chapter only the leading aspects of this methodology are briefly outlined, citing the most commonly used analytical procedure [12, 33, 51, 80–85]. Aluminium in steel is determined in a fairly wide range of concentrations, but most commonly in the range of 0.002–0.060%. Sometimes metallurgical experts require the determination of aluminium even from 10^{-4}% [55]. On the other hand, in the analysis of high-alloy manganese–aluminium steel it is necessary to determine this element even up to about 15% [86]. Taking into account the range of concentrations so wide, aluminium is determined by both flame and flameless atomic absorption techniques [80–85]. Bosch et al. [82] have developed a rapid procedure for the determination of acid-soluble aluminium in steels in the concentration range of 0.002–0.060%. According to the described procedure, the determination

Table 7.3: Operating parameters.

Element	Measuring range (%)	Analytical line (nm)	Slit (nm)	Flame	Type of flame[a]	Burner height (cm)	Lamp current (mA)	Integration period (s)	Linear working range (μg/cm³)	Supplementary information
Cr	5–40	357.9	0.7	C_2H_2–N_2O	R	8	25	3	10	
Co	0–30	240.7	0.2	C_2H_2–airC_2H_2–N_2O	O	5	30	3	6	
		352.7					500			
Al.	0.5–7	309.3	0.7	C_2H_2–N_2O	R	6	25	3	80	
Ti	0–5	365.3	0.2	C_2H_2–N_2O	R	6	36	3	150	Add $AlCl_3$ to 1,000 μg Al/cm^3
Mo	0–12	313.3	0.7	C_2H_2–N_2O	R	7	30	3	60	
		317.0					80			
Nb	0.5–7	334.4	0.2	C_2H_2–N_2O	R	5	36	10	900	
Ta	0.5–6	271.5	0.2	C_2H_2–N_2O	R	5	36	10	1,000	Add 0.1 M NH_4F
W	0–11	255.1	0.2	C_2H_2–N_2O	R	7	36	10	1,400	
		400.9	0.7				500			
V	0–2	318.4	0.7	C_2H_2–N_2O	R	6	36	3	100	Add $AlCl_3$ to 1,000 μg Al/cm^3 and add 2% NH_4Cl
Fe	0–5	248.3	0.2	C_2H_2–air	O	6	30	3	5	
		372.0					60			
Zr	0–1	360.1	0.2	C_2H_2-N_2O	R	6	36	3	1,000	

[a] R, reducing; O, oxidizing.

of aluminium content in 20 samples takes the time of only 11 min. The standard deviation is then 0.001%. Shaw and Ottaway [83] described the determination of aluminium in steel using flameless technique. To determine the acid-soluble aluminium in steels, nitric (V) acid was used for the digestion of samples, and the resulting solutions were directly introduced into the graphite furnace. For the determination of total aluminium, the residue was fused with a mixture of sodium carbonate and sodium tetraborate and dissolved in nitric (V) acid. Persson, Frech and Cedergren [84, 85] utilized flameless technique for the determination of aluminium in various grades of steel. For the preparation of samples, a mixture of hydrochloric acid and nitric (V) acid was used. Ammonium sulphate (VI) was next added to remove the effect of hydrochloric acid. Due to the fact that iron content reduces the sensitivity of aluminium determination by about 20%, it is recommended to add appropriate iron additives to reference solutions to compensate for the content of this element in both test samples and standard reference samples. The same operation should be performed with respect to chromium, molybdenum and nickel when aluminium is determined in alloy steels. In a series of samples analysed at the Faculty of Materials Science and Metallurgy of the Silesian University of Technology, the total aluminium content was about 0.023–0.026% and the soluble aluminium content was 0.018–0.021%.

One of the most commonly used analytical procedures for the determination of aluminium content in steel is the methodology developed by König, Schmitz and Thiemann [81]. Its essence consists in preparing a 2 g weighed portion of the test sample and dissolving it in 25 ml of acid mixture comprising three parts of hydrochloric acid, one part of nitric (V) acid and two parts of water, gently heating the samples on a heating plate. After separation of silicon dioxide, acid-soluble aluminium is analysed. To determine total aluminium content, after separation, the insoluble residue is fused with 1 g of the carbonate and sodium tetraborate mixture. The resulting melt is dissolved in 5 ml of hydrochloric acid (1:1) and transferred to a 100 ml volumetric flask to make up to volume with water. The most favourable results are obtained when natural specific reference standards are used. One can also use synthesized standard reference solutions. For the determination of aluminium, the analytical line used most often is that of 309.3 nm, with the monochromator slot width of 0.7 nm. Very important for the accuracy and precision of test results is optimization of the measurement conditions (Figure 7.18).

7.16 Tool alloys – speciation and speciation analysis – determination of chemical composition of carbide phase isolates

In the Department of Materials Science of the Silesian University of Technology, a research has recently been carried out on the concept of tool alloys with trace distribution of carbide segregations and on the development and implementation of a technology for the manufacture of semi-finished products and tools from these materials. In the investigated group of non-ledeburite tool steels, by major changes introduced to the chemical composition with respect to conventional high-speed steels, the reduced level of segregations and changes in the phase composition of carbides were obtained. The essence of the solution, addressed mainly to the discussed grade of high-speed steels, consisted in the replacement of conventional primary carbides of M_6C, M_2C and MC types with MC-type carbides formed by elements characterized by higher affinity for carbon than tungsten, molybdenum and vanadium, that is titanium and niobium [40, 41]. The chemical composition of the investigated steel, designated as $SM_5Ti_3Nb_2$, is shown in Table 7.4. In this high-speed steel of non-ledeburite class, in which tungsten and part of molybdenum and vanadium were replaced by titanium and niobium, the obtained phase composition and microstructure differed from other common steel grades included in this group. The prevailing carbides were large (over 10 μm) MC-type carbides in the shape of polyhedrons, often octahedrons, evenly distributed in the matrix, additionally accompanied by M_6C and $M_{23}C_6$ carbides [41]. This was confirmed by the results of X-ray phase analysis, demonstrating the presence of MC, M_6C and $M_{23}C_6$ carbides (Figure 7.19) [30, 41]. Secondary carbides of M_6C and $M_{23}C_6$ types were present in the structure as spheroidal particles or coagulated conglomerates. Based on the results of EDX analysis, it was observed that carbides with the predominance of molybdenum and tungsten formed M_6C-type carbides, while chromium-rich $M_{23}C_6$ particles with the predominance of Ti, Nb and V formed MC carbides. In the images obtained by SEM (scanning electron imaging – SEI), the M_6C carbides were brighter than the $M_{23}C_6$ particles (MC and $M_{23}C_6$ carbides have a similar greyscale level) – Figure 7.20 [41].

Table 7.4: Chemical composition of tested $SM_5Ti_3Nb_2$ tool steel [61].

Element	Content in % by mass
C	1.98
Mn	0.46
Si	0.54
P	0.016
S	0.017
Cr	4.26
W	0.10
V	1.16
Mo	5.25
Nb	2.10
Ti	3.19
Al	0.30
Zr	0.18
Fe	Matrix

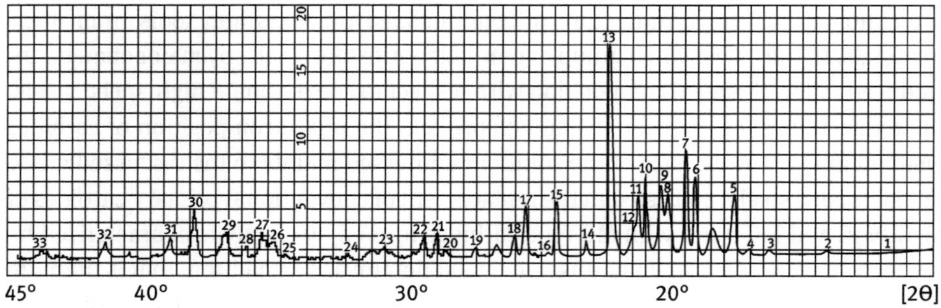

No.	Measuring data	Data (table ICDD)								
		NbC (38-1364)			M23C6(35-0783)			M6C(41-1351)		
	d [Å]	*d* [Å]	I/I_1	(hkl)	*d* [Å]	I/I_1	(hkl)	*d* [Å]	I/I_1	(hkl)
1	3.73							3.3496	<3	(331)
2	3.18							3.3496	<3	(331)
3	2.76							2.7774	22	(400)
4	2.64							2.5487	30	(331)
5	2.55	2.5808	100	(111)						
6	2.355				2.3830	23	(420)			
7	2.32							2.2677	50	(422)
8	2.24							2.2677	50	(422)
9	2.21	2.2352	80	(200)						
10	2.16				2.1761	24	(422)	2.1380	100	(511)
11	2.12							2.1380	100	(511)
12	2.11							2.1380	100	(511)
13	2.03				2.0519	100	(511)			
14	1.955							1.9639	39	(440)
15	1.87				1.8840	20	(440)			
16	1.85							1.8516	4	(442)
17	1.785				1.7766	12	(600)			
18	1.76							1.7566	4	(620)
19	1.67				1.6857	2	(620)			
20	1.61							1.6748	4	(662)
21	1.59	1.5804	35	(220)	1.6066	6	(622)			
22	1.565							1.5556	16	(551)
23	1.50							1.4463	16	(553)
24	1.44							1.4463	16	(553)
25	1.35							1.3572	20	(733)
26	1.34	1.3476	26	(311)				1.3572	20	(733)
27	1.32				1.3326	2	(800)			
28	1.30							1.3093	52	(822)
29	1.28	1.2903	9	(222)	1.2927	2	(820)	1.2828	18	(555)
30	1.24				1.2307	6	(751)			
31	1.22				1.2227	15	(662)			
32	1.16				1.1698	4	(911)			
33	1.11	1.1175	5	(400)	1.1174	1	(931)			

Figure 7.19: Results of X-ray chase analysis of carbide isolate of tested steel [41].

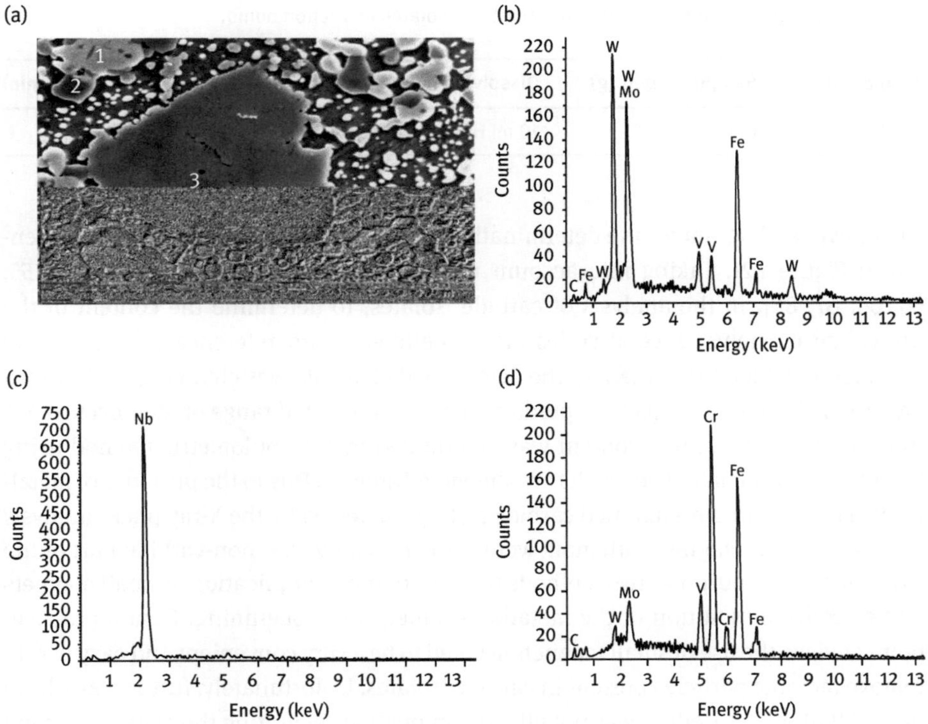

Figure 7.20: Carbide particles and conglomerates present in tested steel [24]: (a) FE SEM, SEI, ×3,500; (b) EDX spectrum of carbides nos. 1 and 4 (M_6C type) from (a); (c) EDX spectrum of carbide no. 3 (MC type) from (a); (d) EDX spectrum of carbide no. 2 ($M_{23}C_6$ type) from (a).

Knowing the average chemical composition of the carbide phase, when several types of carbides are present in the steel microstructure and form solid secondary solutions, the determination of their stoichiometric formulas is practically impossible. It is possible, however, to predict the precipitation and dissolution of carbides or their transformations during heat treatment and to indirectly determine the chemical composition of the matrix. These studies, carried out by the method of AAS using a Perkin-Elmer Model 503 absorption meter, were based on the use of carbide isolates obtained by electrochemical isolation (5% HCl, 20 °C, 20 mA/cm^2) from sections of 25 mm diameter rods of the test steel previously subjected to different cycles of heat treatment. The determination of iron and manganese content was carried out in an oxidizing acetylene–air flame, while other metallic elements were analysed in the reducing acetylene–nitrous oxide flame. Concentrations of the examined elements were measured using the following analytical lines: Cr – 357.9 nm, Mo – 317.0 nm, Ti – 364.3 nm, Nb – 334.4 nm, V – 318.4 nm, Mn – 279.5 nm, W – 400.9 nm, Zr – 360.1 nm, Fe – 372.0 nm. Samples were dissolved in a mixture of hydrochloric acid, nitric (V) acid and hydrofluoric acid using a Teflon bomb. The most effective mixture of acids is presented in Table 7.5.

Table 7.5: Conditions for the dissolution of carbide isolates in a Teflon bomb.

Temperature	Sample weight (g)	Dissolving mixture	Dissolution time (min)
150 °C	0.2	10 ml HCl + 5 ml HNO_3+ 3 ml HF	50

An analytical diagram for the determination of elements in carbide isolates is presented in Figure 7.21. Taking into account numerous inter-element effects [12, 20, 55, 63, 64, 87, 88], in the analysis of carbide isolates, to determine the content of the individual elements, several series of synthetic standard reference solutions were prepared in which the content of the determined elements was changing in appropriate proportions according to the predetermined or predicted range of their occurrence in tested isolates. Carbon content was determined by the coulometric method using Strohlein's Coulomate. The results are shown in Table 7.6. Due to the presence of metallic constituents in the examined carbide isolates, detected by the X-ray phase analysis and using small magnets, attempts were made to remove the non-carbide part of the isolates. For this purpose, two methods were used, that is application of small magnets and selective dissolution of the metallic residue. At the beginning, from a practical point of view, the use of small magnets seemed to be a very convenient and easy way to remove metallic particles present in carbide isolates. Unfortunately, further tests have proved that this procedure was not effective enough in separating the metallic residue from the matrix because, due to the high degree of the isolate refinement, the magnet – while attracting the metallic components – was also "catching" the fine particles of the carbide phase. This was fully confirmed by chemical analysis of particles separated by the magnet, which, besides iron, contained significant amounts of niobium, titanium, vanadium, zirconium and tungsten, and thus of the elements forming an integral part of the carbide phase. Moreover, the reproducibility of this method was next to none, because each time different results of the chemical analysis carried out on the magnet-separated parts of isolates were obtained. Even attempts to remove the isolate particles from the magnet by gentle shaking have proved to be insufficient. Slightly better reproducibility was obtained by spreading the isolates on a large surface before separation and removal with magnet of the non-carbide part of the isolate [73].

Selective dissolution in acids leaving carbides undissolved has proved to be a much more effective way to remove the metallic residue. Initially, attempts were made to use for this purpose hydrochloric acid in different concentrations, knowing that it does not dissolve the MC-type carbides of niobium, titanium and tungsten. However, taking into account the fact that besides the MC-type carbides, the isolated carbide phases might also contain carbides of M_6C and $M_{23}C_6$ types, hydrochloric acid was excluded from the studies as an agent partially dissolving carbides of chromium, molybdenum and vanadium [6]. Based on the results of own experiments on the dissolution of metallurgical samples and using guidelines given by the technical literature [41], a 2:1 diluted orthophosphoric (V) acid, often included in the steel sample

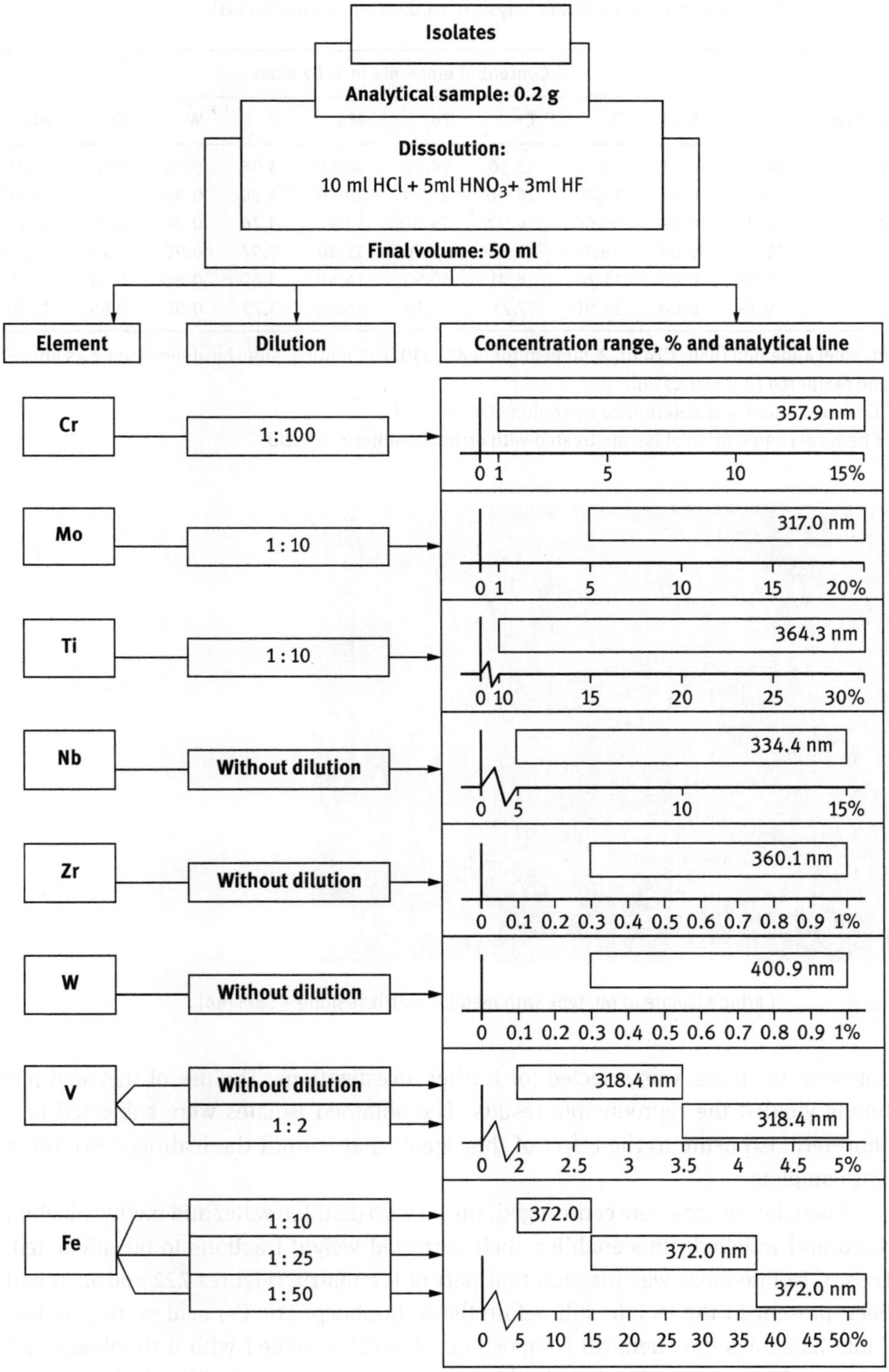

Figure 7.21: Analytical diagram for the determination of elements in carbide phase isolates [20].

Table 7.6: The results of AAS chemical analysis of carbide phase isolates [41].

	Content of elements in % by mass									
Isolate[a]	C[b]	Mn	Ti	Cr	Fe	Mo	V	W	Zr	Nb
1	10.52	0.41	12.60	13.30	19.90	18.40	3.96	0.85	0.65	9.35
1′[c]	9.54	0.41	13.70	14.00	21.50	19.70	4.30	0.86	0.62	9.40
2	8.29	0.19	16.60	4.92	25.00	12.90	3.24	0.84	0.75	12.21
2′	15.62	0.09	26.20	1.85	8.76	13.30	3.97	0.90	0.72	14.70
3	8.94	0.09	15.20	8.31	10.50	14.40	3.60	0.86	0.61	9.10
3′	9.19	0.08	18.80	7.21	8.10	16.80	3.72	0.90	0.55	11.30

[a]1, steel annealed (850 °C/4 h); 2, steel hardened (1,150 °C/7 min); 3, steel hardened (same as above) and tempered (2 × 540°C/1 h).
[b]Carbon content was determined by coulometric method.
[c]Chemical composition of isolate treated with orthophosphoric (V) acid.

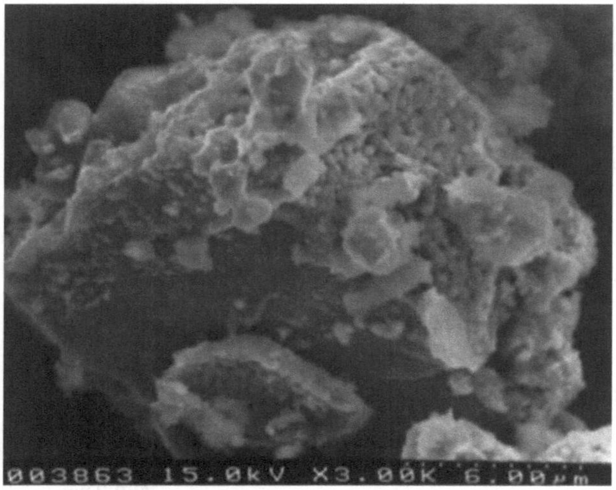

Figure 7.22: Carbide isolate of MC type with metallic matrix residue – SEM [84].

digestion mixtures, was selected for further investigations. The use of this acid has finally yielded the reproducible results. The obtained isolates were subjected to a short-term exposure to the effect of this agent, that is until the hydrogen evolution was complete.

Then the isolates were centrifuged, rinsed with distilled water and methyl alcohol, dried and weighed, thus enabling their corrected weight fractions to be calculated. Loss of isolate mass was the sign that part of the matrix (Figures 7.22 and 7.23) had been present in the isolate still before the orthophosphoric (V) acid started acting. Table 7.6 shows the chemical composition of isolates treated with orthophosphoric (V) acid. The process of isolate "etching" with orthophosphoric (V) acid has proved to be very efficient, and the weight fraction of the isolate obtained from steel in the

Figure 7.23: Treated carbide isolate of MC type after etching with orthophosphoric (V) acid – SEM [84].

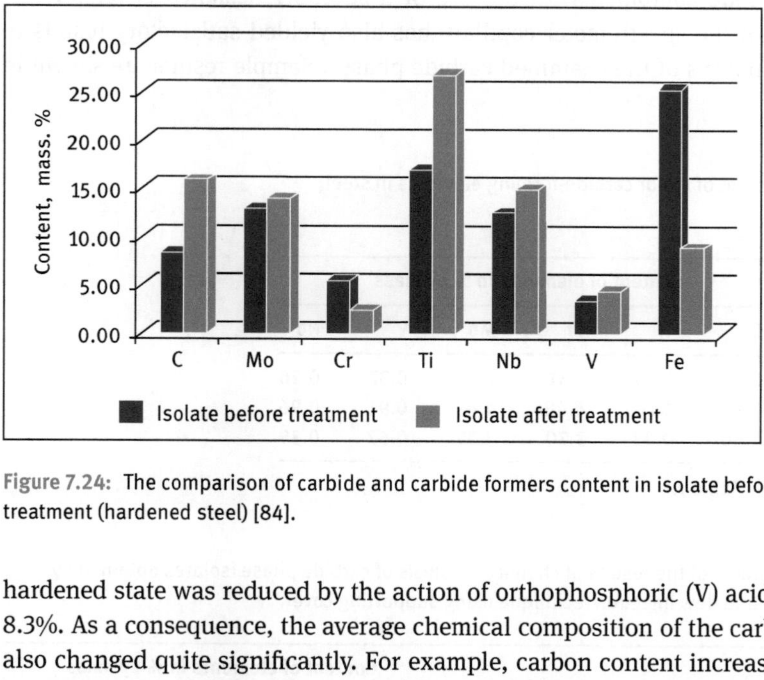

Figure 7.24: The comparison of carbide and carbide formers content in isolate before and after treatment (hardened steel) [84].

hardened state was reduced by the action of orthophosphoric (V) acid from 15.8% to 8.3%. As a consequence, the average chemical composition of the carbide phase has also changed quite significantly. For example, carbon content increased from 8.29% to 15.62%, while chromium content decreased from 4.92% to 1.85% (Figure 7.24).

From the results of the AAS analysis of the average chemical composition of carbide isolates extracted from the steel in annealed, hardened and tempered condition it follows that during austenitizing process the dissolution affects mainly carbides with the high metal-to-carbon atomic ratio, rich, moreover, in chromium, molybdenum and iron, identified as M_6C- and $M_{23}C_6$-type carbides [20, 41, 73]. Carbon-rich

MC-type carbides with high levels of titanium and niobium do not dissolve. As a result of tempering, chromium content in the isolate increases nearly three times (compared to the hardened state), mainly due to the appearance of the secondary $M_{23}C_6$-type carbides.

Some attention deserves the fact that when orthophosphoric (V) acid was reintroduced to the once acid-treated isolate, total absence of any chemical reactions was observed. This confirms the correct choice of the acid, which dissolves the metallic residue, while leaving the carbide phases untouched.

The results of the chemical analysis of the steel samples and carbide isolates, as well as the sample weight and the weight of the isolate were used in calculations of the chemical composition of the steel matrix (Table 7.7). The chemical composition of the test steel matrix in annealed condition resembled other typical grades of high-speed steels. It showed a very low content of carbon, and similar content of chromium and of total remaining carbide-forming elements. During hardening, the matrix was enriched with carbon, chromium and molybdenum. The precipitation process during tempering has impoverished the matrix mainly in chromium and molybdenum.

At the end of this section, it is worth pointing out that the application of flow injection technique combined with the use of a specially designed cover for vertical insertion of the spectrometer capillary has also yielded satisfactory results in the chemical analysis of the examined carbide phases. Sample results are shown in Table 7.8.

Table 7.7: The content of major carbide-forming elements in steel matrix [61].

	Content of elements in % by mass					
Sample no.[a]	C	Ti	Cr	Mo	V	Nb
1	0.07	0.55	1.81	1.61	0.37	0.26
2	0.74	1.11	4.48	4.49	0.91	0.96
3	0.60	0.22	3.70	3.05	0.67	0.35

[a] Designations as in Table 7.6.

Table 7.8: Comparison of the results of chemical analysis of carbide phase isolates obtained by conventional AAS and flow injection technique using supporting cover.

	Content of elements in % by mass				
Element	Mo	Cr	Ti	Nb	V
Conventional AAS method	13.7	1.85	26.2	14.7	3.97
Flow injection technique using supporting cover	13.6	1.82	26.0	14.5	3.95

Note: AAS, atomic absorption spectrometry.

As mentioned earlier, quality requirements imposed on metallurgical products determine the need for comprehensive studies, including, but not limited to, chemical composition, phase composition, structure examinations and testing of mechanical properties. This applies equally well to the steel grades already produced and to the new high alloyed grades and metal alloys for special applications. Besides full chemical analysis, very important in this field is the identification of phases that occur in a given material along with the determination of their chemical composition. Carbide phases [37, 40, 41, 61] play an important role in the structure and properties of materials. Therefore, various methods, such as X-ray microanalysis or chemical analysis of electrolytically separated carbide isolates, are used to determine their chemical composition. Knowing how important is the accurate analysis of the chemical composition of the isolated carbide fractions, modern instrumental methods and techniques, such as AAS and ICP-OES, are used for this purpose. Unfortunately, all these modern methods require conversion of the isolated and separated phases to the state of solution. The key to success in obtaining fully dissolved isolate for further analysis is proper selection of the dissolving mixture. It is all the more important that in the vast majority of cases, samples available for analysis are of a very small weight ranging from several dozen to several hundred milligrams. So, with incorrectly selected acid mixture, the digestion process cannot be repeated, and re-isolation must be done.

Taking into account the above-mentioned methodological problems, based on the results of own studies and guidelines given in the technical literature, mixtures dissolving selected MC-type carbides, namely TiC, NbC, TaC, VC, WC and ZrC, were compared. All of the aforementioned carbides belong to the group of the most chemically resistant phases, and therefore proper selection of acid mixtures for their digestion is of major importance in the process of their preparation for chemical analysis.

Using the results of own studies [6, 20, 65, 66, 90] and long experience in the analysis of alloy steels and complex nickel- and cobalt-based alloys and intermetallic and carbide phases, and basing on the literature data [37, 91], in work [87], the acid mixtures most effectively dissolving the carbides of titanium, niobium, tantalum, vanadium, tungsten and zirconium are compared. Taking into account the durability and chemical resistance of carbides of these elements, the choice of acids and/or mixtures of acids used for their digestion is rather modest. Additionally, the selection is limited by the tendency of these elements to hydrolyse [6, 25–29]. The specific nature of the analytical methods and techniques applied also imposes certain limitations of both technical and instrumental nature. Using, for example, FAAS, it is strongly recommended to allow for much weaker absorbance signals of these elements attenuated by sulphuric (VI) acid, the presence of which, combined with a relatively low sensitivity of determination of niobium, tantalum, zirconium and tungsten, significantly reduces the precision of the results obtained. Application of hydrofluoric acid, in turn, requires the use of acid-proof nebulizers. The complexity of isolated carbide phases is the reason why in some cases it may prove necessary to slightly modify the mixtures listed in work [87]. This is due to the fact that the isolated carbide phases may comprise different types of carbides that are identifiable but not quantifiable. In such cases, only average chemical composition of the isolated carbide phase is given, and the choice of dissolving mixtures and conditions of the dissolution process requires further careful analysis and practical experience. For example, it may be necessary to increase the temperature during the process of the isolate dissolution, using Teflon decomposition vessels (Teflon bombs). Sometimes, instead of the expected MC-type carbide such as TiC, other types of carbides like $(Ti_{0.8}Mo_{0.2})C$ may appear. Nitric (V) acid is the preferred agent for the dissolution of molybdenum carbides. If chromium carbides, for example, Cr_7C_3 and $Cr_{23}C_6$, are present in the isolate, then hydrochloric acid is required to dissolve them. Depending on the weight of the analysed carbide isolate, typically 5–10 ml of individual acids or mixtures thereof are used.

Due to the fact that carbide isolates contain various metallic components that, although not included in the composition of the carbide phase, are still present there, it becomes necessary to remove them, as otherwise they would interfere with the obtained results of the chemical analysis of the isolates [64–66, 73]. For this purpose, the most preferable method is selective dissolution of the metallic residue by means of suitably selected acids which do not dissolve the examined carbides. Paper [97] lists acids that do not dissolve certain compounds, and as such are applicable in the removal (digestion) of the non-carbide part of the isolate. The most versatile of these acids is the diluted orthophosphoric (V) acid. Hydrochloric acid is also applicable in various concentrations, mainly because it does not dissolve MC-type carbides of niobium, titanium and tungsten. At this point, however, it should be remembered that hydrochloric acid partially dissolves carbides of chromium, molybdenum and vanadium.

Acid mixtures given in paper [87] are not the only possible means and tools used in the preparation of carbide isolates for chemical analysis. Teflon bombs are also helpful, as they enable carrying out the digestion process at elevated temperatures of, for example, 160 °C. Carbide isolates can also be prepared for analysis by fusing the isolate with Na_2O_2 sample of a fourfold mass in a muffle furnace at 520 °C for 1 h and then dissolving the dry residue in hydrochloric acid [31, 57, 92, 93]. Studies of the chemical composition of isolated carbide phases are closely related to the systematically increasing requirements for the high quality of metallurgical products. Although the latest instrumental methods and research techniques are used for the analysis of these phases, the ongoing problem is the development of an efficient method for the dissolution of carbide isolates. The acid mixtures listed in this chapter are an important aid for professionals in chemical metallurgical laboratories specializing in phase analysis, especially when they use atomic absorption or OES as well as UV-VIS spectrophotometry. Considering the fact that some isolated carbide phases can contain different types of carbides that are qualifiable but not quantifiable, slight modifications to the acid mixtures given in paper [87] may be necessary to ensure that all types of the identified carbides are equally well dissolved. Then the average chemical composition of the isolated carbide phases is given.

A significant barrier to studies of the dissolution process of carbide isolates is the small mass of the isolated samples. Therefore it is believed that, in order to ensure full conversion of these samples to the state of solution, the use of Teflon bombs is recommended, since they allow increasing the temperature during the isolate dissolution process instead of additional experimentation with acid selection. Some suggestions concerning the choice of acids or mixtures thereof for the digestion of individual carbides may also help in the selection of mixtures that digest samples of the starting materials, that is steels or, more broadly, tool alloys [94, 95], characterized by high content of the carbide-forming elements.

Acids that do not dissolve individual carbide isolates may prove useful in the removal (dissolution) of non-carbide part of the isolate. This operation is necessary to obtain correct results of the chemical analysis carried out on individual isolates.

It is also worth noting that studies have been carried out to better explore the potentials of the microwave-assisted acid-based dissolution of carbide isolates and even at this stage of the work, the results obtained are very encouraging as regards further possible applications of this technique [94].

7.17 Speciation and speciation analysis in multicomponent nickel-based alloys

Multicomponent nickel alloys (superalloys or hyperalloys) can contain up to 18 alloying constituents, including nickel, cobalt, chromium, molybdenum, aluminium, titanium, niobium, tantalum, tungsten, vanadium, zirconium, boron, hafnium, silicon and carbon. All these alloys have a multiphase microstructure, which consists of four basic phases, i.e. γ solid solution, γ' intermetallic phase, carbides, and topologically closed packed intermetallic phases of the σ and μ type. Taking into account the complexity of the chemical composition of these alloys, the suggested breakdown into individual categories is of an illustrative nature only. Nickel is the main alloying constituent and it determines the alloy structure, phase stability, and technical and technological properties. Cobalt reduces aluminium and titanium solubility in γ solution, and also affects the amount of γ' phase and the extent of its stability. The presence of aluminium, titanium and niobium increases hardness of the nickel alloys, mainly through precipitation of the dispersed γ'-Ni_3 (Al, Ti) phase. Carbon, the content of which is up to 0.2% by mass, reacts with carbide-forming elements and forms various carbide phases of the MC, $M_{23}C_6$, M_6C and M_7C_3 types, characterized by different morphologies. Interactions of carbide-forming elements make their atoms highly interchangeable, for example (Ti, Nb)C. These carbides can also contain tungsten and molybdenum. The presence of the latter two elements weakens the MC-type carbide bonds first, to break them down completely next and result in the formation of $M_{23}C_6$

and M_6C carbides, when the temperature exceeds 815 °C. $M_{23}C_6$ carbides are mainly formed in alloys with high chromium content. Most often, in addition to the $Cr_{23}C_6$ carbides, phases of the $(Cr_{21},Mo_2)C_6$ or $(Cr_{21}Mo,W)C_6$ type also appear. On the other hand, M_6C carbides are formed in alloys containing more than 6% by mass of molybdenum and tungsten. MC carbides containing niobium and tantalum are very stable and do not dissolve during the heat treatment process carried out at 1,200–1,260 °C [37]. The creep resistance of multicomponent nickel alloys is mainly determined by the presence of γ' phase, due to a complex of its very specific properties [20, 37, 45]. The choice of the chemical composition of the creep-resistant nickel-based alloys determines the amount of the γ' phase precipitates. In many cases, the volume fraction of the γ' phase precipitates in the structure of nickel superalloys is up to 68%. Depending on the amount of alloying constituents added to nickel alloys and their final content in these alloys, and on the applied variant of alloy heat treatment, the chemical composition of the γ'- Ni_3 (Al, X) phase, where X = Ti, Ta, Nb, can vary in a very wide range of values. This is due to the fact that cobalt can replace nickel, while titanium, niobium and tantalum can occupy the position of aluminium in the ordered γ' phase lattice. Molybdenum, chromium and iron can replace both aluminium and nickel in the Ni_3Al compound [37]. Typically, in the precipitates of γ' phase, the content of five to ten metal elements is determined. Chemical composition of the γ' phase precipitates is determined during analysis of γ' phase isolates. Various reagents and extraction conditions are used to isolate and extract the γ' phase from multicomponent nickel alloys [20, 27, 66]. In studies carried out at the Silesian University of Technology [54, 56], the γ' phase was isolated by anodic dissolution in a reagent containing 20 ml $HClO_4$, 50 ml HNO_3, 1,000 ml CH_3OH, at a current density of 0.1 A/cm^2 and temperature ranging from 0 to 5 °C.

The method of determining the content of metallic elements in γ' phase isolates using AAS was described in detail in Refs. [6, 20, 66]. The 20 mg weighed portion of the isolate was placed in a 100 ml Teflon beaker, adding 10 ml of the digesting mixture (HNO_3: HF: H_2O = 1: 1: 1) and heating the whole gently on a heating plate until complete dissolution of the isolate. Then 10 ml of HCl was added and the contents of the beaker were carefully evaporated to dryness. This operation was repeated to determine cobalt content in the γ' phase isolate. The beaker walls were then rinsed with a small amount of water and 2 ml of concentrated hydrochloric acid and 1 ml of concentrated hydrofluoric acid were added. The solution was heated to dissolve the salt, boiled, cooled to room temperature and transferred quantitatively to a 25 ml graduated flask. Then 0.2 g of ammonium fluoride was added, water was added to the mark and the whole was mixed thoroughly.

Taking into account numerous inter-element effects that occur during analysis of the γ' phase isolates [12, 55, 64, 87], synthetic standard reference solutions were prepared in parallel with isolate samples by adding varying amounts of base standard reference solutions of all determinants occurring in γ' phase to cover the expected concentrations of these elements. The total mass of metals in each reference standard should be constant and equal to 20 mg. To maintain the same conditions during preparation of standard reference solutions, the order of the individual operations must be the same as when preparing the test sample solutions, that is evaporating with hydrochloric acid, adding a certain amount of hydrofluoric acid and ammonium fluoride, and supplementing with water to 25 ml. The blank samples were all reagents used in the analysis.

Due to the effect of aluminium on the measurements of titanium and vanadium concentration [12, 17, 55, 64, 87], aluminium content was the first one to be determined in the γ′ phase isolate, and then, to identify the other two elements, the additional synthetic reference standard with aluminium content equal to the content of this element in the test sample (isolate) was prepared. When more than one isolate (50–200 mg) was available for the determination of titanium and vanadium content, separate solutions of test samples and reference standards were prepared. Before adding water to the mark in volumetric flasks, aluminium was added in the form of aluminium chloride to obtain the concentration of about 1,000 μg Al/ml. The content of these two elements was also checked with standard additions. The situation was similar during determination of niobium and tantalum. Separate solutions were prepared using larger isolate weights or performing the extraction with methyl isobutyl ketone from the medium of hydrofluoric acid and hydrochloric acid [95]. Detailed analytical diagrams are presented in Figure 7.25. Given the complex chemical composition of the γ′ phase, it was decided to work on diagram, since in the isolates of this phase, usually five to even ten elements are determined within a fairly wide range of concentrations. The comprehensive analytical diagram shown in Figure 7.25 gives the concentration ranges of the analysed elements taking into account the calculated dilution of the test sample solutions. The obtained results of analyses of the γ′ phase isolates are given in Table 7.9. The precision of determinations measured with the coefficient of variation is given in Table 7.10.

In addition to the conventional analytical procedure using AAS, chemical composition of the γ′ phase isolates was also calculated by the method of one base reference standard [54]. The procedure of selecting the base reference standard for analysis of γ′ phase isolates is discussed in detail in Ref. [54]. In the case of γ′ phase isolates, the base reference standard should have the chemical composition similar to that of the isolates tested. It is also important that the upper and lower concentration limits of elements are situated in the linear working range of calibration curves. Therefore, in order to obtain a large isolate volume selected as a baseline, the extraction process must be repeated two or three times. The determination of the chemical composition of the γ′ phase using one base reference standard method was carried out in two additional variants, that is equal weighed portions (total content of components), or the same nickel content in standard solution and sample solution. The concept of the same nickel content in sample solution and standard solution should be treated with some approximation, since it is not possible to accurately reproduce this content in sample solutions.

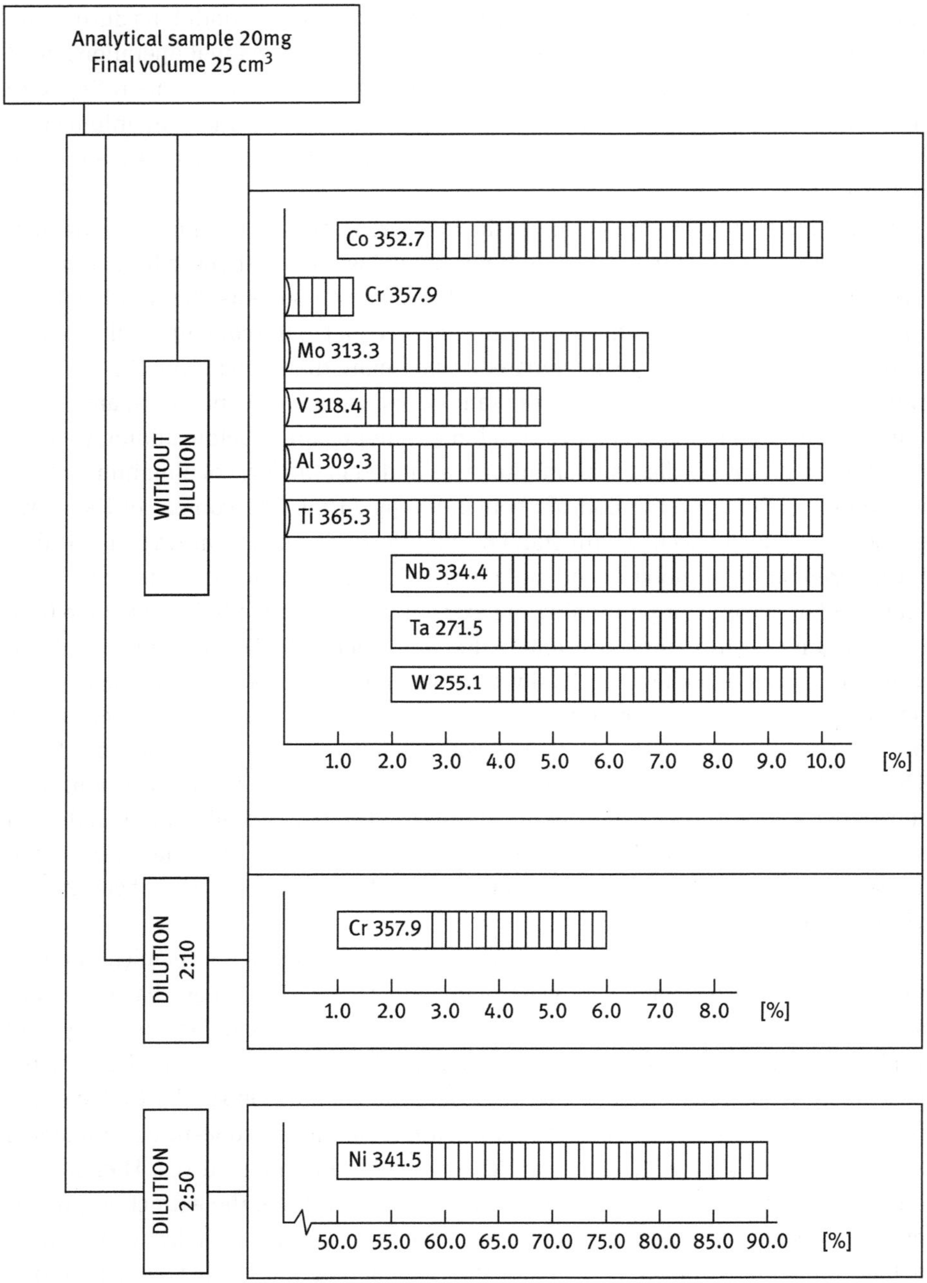

Figure 7.25: The comprehensive analytical diagram for determination of γ′ phase chemical composition by AAS method [66].

Table 7.9: Chemical composition of alloys tested and the results of chemical analysis of γ′ phase isolates [49].

Alloy grade		Element content													Concentration unit
		C	Cr	Co	Ti	Al	Ta	W	Mo	B	Zr	Fe	Nb	Ni	
Alloy 1		0.13	10.50	6.52	3.00	4.60	–	3.99	4.99	0.03	–	2.00	–	64.24	% weight
		0.61	11.45	6.27	3.55	9.66	–	1.23	2.95	0.15	–	2.03	–	62.05	% atomic
Alloy grade	**Ageing temp. (K)**	**Chemical composition γ′ phase**													**Concentration unit**
Alloy 1	1,123	–	3.02	3.08	3.71	6.10	–	4.56	1.98	–	–	–	–	R	% weight
		–	3.23	2.91	4.34	12.61	–	1.38	1.15	–	–	–	–	R	% atomic
	1,223	–	4.73	3.09	3.84	6.56	–	4.97	2.32	–	–	–	–	R	% weight
		–	5.10	2.94	4.50	13.65	–	1.51	1.35	–	–	–	–	R	% atomic
	1,323	–	3.88	3.28	5.41	6.49	–	5.22	2.62	–	–	–	–	R	% weight
		–	4.19	3.12	6.34	13.51	–	1.59	1.53	–	–	–	–	R	% atomic
		Element content													
Alloy grade		**C**	**Cr**	**Co**	**Ti**	**Al**	**Ta**	**W**	**Mo**	**B**	**Zr**	**Fe**	**Nb**	**Ni**	**Concentration unit**
Alloy 2		0.15	11.69	6.78	2.89	4.37	2.03	5.20	4.65	0.01	0.09	0.02	–	62.12	% weight
		0.73	13.04	6.67	3.50	9.41	0.65	1.64	2.81	0.08	0.06	0.02	–	61.41	% atomic
Alloy grade	**Ageing temp. (K)**	**Chemical composition γ′ phase**													**Concentration unit**
Alloy 2	1,123	–	2.70	2.81	4.87	7.24	1.16	4.49	1.79	–	–	–	–	R	% weight
		–	2.88	2.65	5.65	14.92	1.16	1.35	0.98	–	–	–	–	R	% atomic
	1,223	–	2.87	2.98	5.11	6.95	1.08	4.55	1.46	–	–	–	–	R	% weight
		–	3.07	2.81	5.93	14.32	1.08	1.37	0.84	–	–	–	–	R	% atomic
	1,323	–	3.18	2.98	5.31	7.16	1.00	4.81	1.62	–	–	–	–	R	% weight
		–	3.40	2.81	6.16	14.75	1.00	1.45	0.93	–	–	–	–	R	% atomic

(*Continued*)

Table 7.9: (Continued)

Alloy grade		Element content													Concentration unit
		C	Cr	Co	Ti	Al	Ta	W	Mo	B	Zr	Fe	Nb	Ni	
		Element content													
Alloy grade		C	Cr	Co	Ti	Al	Ta	W	Mo	B	Zr	Fe	Nb	Ni	Concentration unit
Alloy 3		0.17	13.17	–	0.88	6.69	–	–	4.44	0.01	0.07	1.00	2.32	71.25	% weight
		0.78	13.68	–	0.99	13.40	–	–	2.50	0.07	0.04	0.97	1.35	65.61	% atomic
Alloy grade	Ageing temp. (K)	Chemical composition γ' phase													Concentration unit
Alloy 3	1,123	–	4.06	–	1.12	7.62	–	–	3.06	–	–	–	3.78	R	% weight
		–	4.29	–	1.28	15.54	–	–	1.75	–	–	–	2.23	R	% atomic
	1,223	–	4.30	–	1.16	7.25	–	–	2.82	–	–	–	3.81	R	% weight
		–	4.56	–	1.33	14.82	–	–	1.62	–	–	–	2.26	R	% atomic
	1,323	–	4.52	–	1.29	7.67	–	–	2.56	–	–	–	4.51	R	% weight
		–	4.77	–	1.47	15.61	–	–	1.46	–	–	–	2.66	R	% atomic
		Element content													
Alloy grade		C	Cr	Co	Ti	Al	Ta	W	Mo	B	Zr	Fe	Nb	Ni	Concentration unit
Alloy 4		0.07	12.35	–	0.64	6.69	–	–	4.58	0.01	0.094	0.98	2.22	72.36	% weight
		0.31	12.92	–	0.73	13.50	–	–	2.60	0.06	0.056	0.96	1.30	67.11	% atomic
Alloy grade	Ageing temp. (K)	Chemical composition γ' phase													Concentration unit
Alloy 4	1,123	–	3.50	–	1.23	8.11	–	–	2.65	–	–	–	4.71	R	% weight
		–	3.69	–	1.40	16.47	–	–	1.51	–	–	–	2.77	R	% atomic
	1,223	–	3.66	–	1.30	8.04	–	–	2.51	–	–	–	4.56	R	% weight
		–	3.86	–	1.49	16.35	–	–	1.43	–	–	–	2.69	R	% atomic
	1,323	–	4.20	–	1.51	7.98	–	–	2.37	–	–	–	4.99	R	% weight
		–	4.34	–	1.73	16.23	–	–	1.35	–	–	–	2.94	R	% atomic

Note: R, rest.

Table 7.10: Precision of the determination of elements in γ′ phase isolates.

Element determined	Coefficient of variation (%)
Co	2.6
Cr	2.3
Mo	3.8
V	7.2
Al	2.8
Ti	3.1
Nb	10.9
Ta	11.2
W	7.8

The obtained results, some of which are listed in Table 7.11, fully confirm the right strategy adopted in the analytical procedure and usefulness of the calibration method based on one reference standard for the chemical analysis of γ′ phase isolates. Comparing the results of the determination of the content of selected elements obtained by the method of synthetic reference standards and by the method of one reference standard performed in two variants, a very good compatibility was observed. Slightly more accurate results were obtained using the method of one base reference standard in a variant of the same nickel content. The reason was high nickel content in the γ′ phase isolates reaching about 70%. At such concentration levels, compensating for the nickel effect on the determined elements requires no special justification. In the case of titanium content determination, the results were somewhat higher in both cases. This is probably associated with slight changes in the aluminium concentration, which strongly affects the absorbance limit of titanium. Additionally, the results obtained were statistically verified by comparing two means using Student's t-test. All of the results obtained were verified positively, which means that the differences in the determination of individual elements gave comparable results no matter which of the two variants of the method of one reference standard was applied [54]. This calibration method is very useful in metallurgical practice, for example, to optimize melting parameters or heat treatment variants selected for a given alloy. Then several γ′ phase isolates are analysed, and in the case of alloys similar in chemical composition, even several dozens of such analyses are performed. When planning such studies, that is the determination of the chemical composition of the γ′ phase isolates obtained for a specific group of nickel-based alloys, it is worthwhile to put a little more effort at the beginning and perform double extraction of the γ′ intermetallic phase selected as a baseline, to later save a lot of time and labour input in the determination of the chemical composition of the isolates without the need to prepare the whole series of synthetic standard reference solutions.

In addition to the conventional AAS method, attention was drawn to the benefits resulting from the use of flow injection technique with an originally designed cover

Table 7.11: The results of chemical analysis of γ′ phase isolate carried out by the AAS method based on one reference standard.

Isolate/type of reference standard	Content in % by mass								
	Al	Ti	Mo	Cr	Nb	Ta	W	Co	Ni
Isolate 1	7.67	1.29	2.56	4.52	4.51	–	–	–	Rest
Base reference standard (equal Ni content in reference standards)	7.65	1.30	2.55	4.51	4.49	–	–	–	Rest
Base reference standard (equal weight of reference standards)	7.68	1.32	2.54	4.49	4.48	–	–	–	Rest
Isolate 2 (synthetic)	7.73	–	–	2.31	–	8.95	9.70	5.73	Rest
Base reference standard (equal Ni content in reference standards)	7.72	–	–	2.29	–	8.90	9.66	5.72	Rest
Base reference standard (equal weight of reference standards)	7.70	–	–	2.27	–	8.88	9.65	5.72	Rest
Isolate 3 (synthetic)	6.85	3.02	–	2.75	1.42	–	12.53	7.53	Rest
Base reference standard (equal Ni content in reference standards)	6.81	2.99	–	2.77	1.38	–	12.57	7.50	Rest
Base reference standard (equal weight of reference standards)	6.82	3.00	–	2.76	1.39	–	12.57	7.51	Rest

for vertical positioning of the capillary, combined with the flame atomic absorption to suck in the metred 100 μl portions of the solution of the analysed elements. Table 7.14 gives sample results of chemical analysis of γ′ phase isolates obtained by the flow injection technique and FAAS combined with the use of supporting cover. The results obtained were comparable to the results obtained by the sole AAS method. They were also statistically verified with Student's *t*-test.

With the small volume of the available γ′ phase isolates, in the tests described in Tables 7.11 and 7.12, two "isolates" were prepared by synthesis from the standard reference solutions of individual elements, using an automatic Hamilton's Dispenser-Diluter device for diluting and dispensing the specific portions of solutions.

It would be difficult to unambiguously indicate the weights of isolates in individual groups that might be considered sufficient for the analysis carried out by the

Table 7.12: Comparison of the results of chemical analysis of γ′ phase isolates obtained by the conventional AAS and by the flow injection technique using a supporting cover.

Method of analysis	Content in % by mass								
	Al	Ti	Mo	Cr	Nb	Ta	W	Co	Ni
Conventional AAS method	7.67	1.29	2.56	4.52	4.51	–	–	–	Rest
Flow injection technique with supporting cover	7.64	1.27	2.53	4.49	4.45	–	–	–	Rest
Conventional AAS method	7.73	–	–	2.31	–	8.95	9.70	5.73	Rest
Flow injection technique with supporting cover	7.69	–	–	2.26	–	8.88	9.64	5.69	Rest
Conventional AAS method	6.85	3.02	–	2.75	1.42	–	12.53	7.53	Rest
Flow injection technique with supporting cover	6.80	2.98	–	2.78	1.37	–	12.58	7.49	Rest

Note: AAS, atomic absorption spectroscopy.

conventional AAS. This depends mainly on the number of the examined elements. In the analysis of carbide isolates and the isolates of γ′ intermetallic phase, due to the complex chemical composition, this limit is difficult to determine. For the analysis of γ′ phase isolates, 5–10 mg sample is usually required. In the case of nickel alloys free from niobium, tantalum and tungsten, about 2–5 mg of the isolate is sufficient for the analysis of this phase. Using flow injection technique for the determination of the chemical composition of the isolates, the obtained precision of the determination of each element, measured with the coefficient of variation, was as follows: Al: 3.5–4.5%, Mn 0.7–1.8%, Cr 3.0–3.5%, Mo 3.5–4.5%, Fe 1.8–2.6% and Co 2.5–3.0%. In the determination of tungsten content, the coefficient of variation was about 8.7–9.2%. In the determination of niobium and tantalum it was 11.6–12.4%.

In own studies, the focus was on the determination of the chemical composition of the isolates of the γ′ intermetallic phase, although carbide phases were also examined. The difficulties related with the electrochemical extraction of carbide phases have finally resulted in a more frequent use of the X-ray microanalysis as a tool for the chemical analysis of carbide phases. For example, in alloy no. 3 with a relatively low molybdenum content (2.5 at%), there was a tendency for the formation of MC carbide. The results of the chemical analysis of this carbide carried out by the technique of X-ray microanalysis are shown in Table 7.13. They indicate that this is an (Nb, Ti, Mo)C-type carbide containing the dissolved nickel and other elements.

Table 7.13: The results of chemical analysis of the carbide phase in alloy no. 3.

Element	Content (at.%)
C	22.2
Nb	37.5
Ti	4.15
Mo	5.80
Zr	1.44
Cr	3.80
Al	5.10
Ni	19.5
Fe	0.50

7.18 Studies of speciation in creep-resistant Fe–Ni alloys

In studies of iron–nickel alloys carried out by K. Ducki at the Silesian University of Technology [45], a modified austenitic $X_5NiCrTi_{25-15}$ grade, belonging to the group of creep-resistant Fe–Ni alloys precipitation hardened with γ' phase – Ni_3 (Al, Ti), was developed and implemented in the production of tools used by non-ferrous metal processing plants. The chemical composition of the test alloy is given in Table 7.14.

Studies of phase transformations and changes in morphology of the microstructural phase constituents, with attention focused on their effect on the strength and plastic properties of the tested Fe–Ni alloy, were conducted under conditions simulating the heat treatment process. The alloy was subjected to solution heat treatment (980 °C/2 h/water) and prolonged ageing at 715, 750 and 780 °C for 0.5 to 500 h with air cooling (Figure 7.26).

The purpose of long-term ageing was to study the precipitation process and provide quantitative description of the kinetics of the γ' phase growth; another purpose was to predict the mechanical properties and thermal stability of alloy microstructure under conditions of long-term exposure to the effect of high temperature. The basis for selecting the ageing temperature in the range of 715–780 °C and the time of up to 500 h was the possibility of simulating the alloy operation for 100,000 h in the range of 630–690 °C [45].

Table 7.14: Chemical composition of $X_5NiCrTi_{25-15}$ alloy used for testing.

Content of elements (mass.%)															
C	Si	Mn	P	S	Cr	Ni	Mo	V	W	Ti	Al	Co	B	N	Fe
0.05	0.56	1.25	0.026	0.016	14.3	24.5	1.35	0.42	0.10	1.88	0.16	0.08	0.007	0.0062	55.3

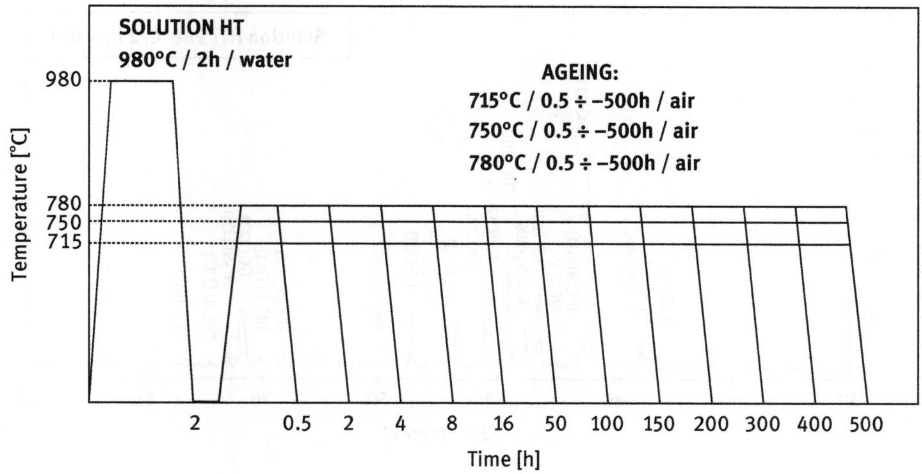

Figure 7.26: Scheme of heat treatment of the Fe–Ni alloy [45].

To accomplish the scientific goals, the structure of the Fe–Ni superalloy was examined, and technological and mechanical properties were tested. This section presents only fragmentary results of phase analysis studies performed by Ducki [45].

Isolates of precipitates present in the tested Fe–Ni alloy in both solution heat-treated and aged conditions were obtained by electrolytic etching in a solution of 14 ml HCl, 100 ml C_2H_5OH. The current density of 100 A/m^2 and process duration of 24 h were applied. The isolates were dried at 60 °C for 8 h. After the process of electrolytic separation, the isolate and the sample residue were weighed. The phase composition of the isolates was determined from the X-ray diffraction patterns generated by a Philips PW 1140 X-ray diffractometer using monochromatic Co-K_α radiation in a graphite monochromator configuration on the bent beam side. Based on positions and intensities of the interference lines found in the diffractograms and their comparison with standard data contained in the ICDD database, phase identification was performed.

In order to trace and describe the precipitation process of intermetallic and carbide phases, and to determine the precipitates morphology and perform relevant identifications, the alloy substructure was examined in the solution heat-treated condition and after long-term ageing at high temperatures. The Fe–Ni alloy substructure after 980°C/2 h/water treatment was observed to comprise the twinned austenite with low (about 0.3 wt.%) content of non-dissolved precipitates, characterized by different sizes, different morphology and high density of dislocations. Alloy microstructure also contained the precipitates of titanium compounds, such as TiC carbide, $Ti_{0.3}NO_7$ carbonitride, $TiNO_3$ nitride, $Ti_4C_2S_2$ carbosulphide and particles of Ni_2Si phase and MoB boride (Figure 7.27).

Application after solution heat treatment of single-stage ageing at 715, 750 and 780 °C carried out for the time from 0.5 to 500 h has initiated the precipitation process.

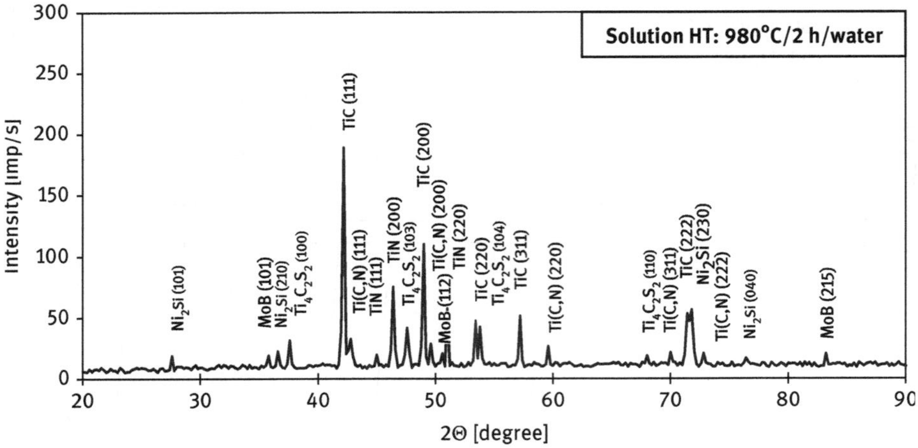

Figure 7.27: Isolate diffractogram of the Fe–Ni alloy after solution heat treatment at 980 °C/2 h/water [45].

After short-term (4–8 h) ageing at 715 °C, the first symptoms of the beginning of the precipitation process were observed in the matrix, namely the occurrence of coherent zones and "clusters" of the γ' phase – Ni_3 (Al, Ti). At grain boundaries, the discontinuous precipitates ("chains") of fine, lenticular lamellae of the G phase – $Ni_{16}Ti_6Si_7$ were formed. Increasing the ageing time (in the range of 16–150 h) has increased the kinetics of the homogeneous process of the precipitation of the γ' phase particles. The γ' phase particles had a spheroidal shape with different diameters. Additionally, regular particles of the G phase – $Ni_{16}Ti_6Si_7$ and lenticular $M_{23}C_6$ carbide particles were observed to occur in the austenite near-boundary zone. The longest ageing time (300–500 h) resulted in further growth of the spheroidal γ' phase particles, maintaining strong coherence with the matrix. Moreover, compounds such as TiC carbide, Ti(C, N) carbonitride, $Ti_4C_2S_2$ carbosulphide and MoB boride were also found in the microstructure (Figure 7.28 and Table 7.15).

Increasing the Fe–Ni alloy temperature to 750 °C intensified the diffusion-controlled processes of the growth, coagulation and overageing of the particles of the intermetallic and carbide phases. At this temperature, the short ageing time in the range of 4–8 h caused the formation of coherent zones and ordered γ' phase particles due to the spinodal decomposition of supersaturated austenite. $M_{23}C_6$ carbides appeared at grain boundaries in the form of discontinuous precipitates – "chains". At the same time, at grain boundaries, G phase particles were released. The consequence of longer ageing time (16–50 h) was precipitation in the matrix of spheroidal and lenticular γ' phase particles showing strong coherence with the matrix. In the microstructure of alloys aged for 100–300 h, the first signs of ageing were observed, namely the formation of colonies of the parallel lamellae of the $\eta(Ni_3Ti)$ phase. The $\gamma' \rightarrow \eta$ phase transformation was accompanied by dissolution of the adjacent fine particles

Table 7.15: Phase composition of Fe–Ni alloys after solution heat treatment and ageing at 715 °C [45].

Alloy condition	Phase constituents
Solution heat treatment: 980 °C/2 h/water	TiC; $TiC_{0.3}N_{0.7}$; $TiN_{0.3}$; $Ti_4C_2S_2$; Ni_2Si; MoB
Ageing: 715 °C/0.5 h	TiC; $TiC_{0.3}N_{0.7}$; $TiN_{0.3}$; $Ti_4C_2S_2$; Ni_2Si; MoB; *G* – $Ni_{16}Ti_6Si_7$
Ageing: 715 °C/2–500 h	TiC; $TiC_{0.3}N_{0.7}$; $TiN_{0.3}$; $Ti_4C_2S_2$; Ni_2Si; MoB; G – $Ni_{16}Ti_6Si_7$; γ′ – $Ni_3(Al,Ti)$

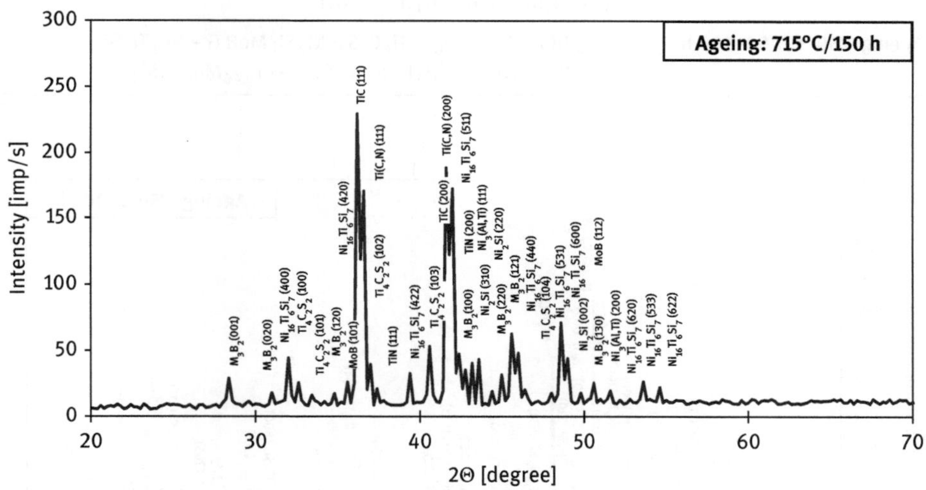

Figure 7.28: Isolate diffractogram of the Fe–Ni alloy after solution heat treatment (980 °C/2 h/water) and ageing at 715 °C/150 h [45].

of the γ′ phase. In border regions, the lamellae of η phase were present in the cellular system. The longest ageing time (400–500 h) progressed in the austenitic matrix the subsequent stages of the growth and coagulation of the γ′ phase particles. At the same time, higher density of dislocations in the matrix grains was noted. Alloy microstructure was reported to contain the following compounds: TiC carbide, $TiC_{0.3}N_{0.7}$ carbonitride, $TiN_{0.3}$ nitride, $Ti_4C_2S_2$ carbosulphide, and particles of Laves Ni_2Si phase, G phase – $Ni_{16}Ti_6Si_7$, β phase – NiTi, σ phase – $Cr_{0.46}$ $Mo_{0.40}Si_{0.14}$ and also MoB boride (Table 7.16 and Figure 7.29).

Analysis of the phase composition of isolates after solution heat treatment and ageing at 780 °C confirmed the presence in the microstructure of TiC carbide, $TiC_{0.3}N_{0.7}$ carbonitride, $TiN_{0.3}$ nitride, $Ti_4C_2S_2$ carbosulphide and particles of Laves Ni_2Si phase, G phase –$Ni_{16}Ti_6Si_7$, *σ* phase – $Cr_{0.46}$ $Mo_{0.40}Si_{0.14}$ and also MoB boride (Figure 7.30 and Table 7.17).

Table 7.16: Phase composition of Fe–Ni alloy isolates after solution heat treatment and ageing at 750 °C [45].

Alloy condition	Phase constituents
Solution HT: 980 °C/2 h/water	TiC; $TiC_{0.3}N_{0.7}$; $TiN_{0.3}$; $Ti_4C_2S_2$; Ni_2Si; MoB
Ageing: 750 °C/2–50 h	TiC; $TiC_{0.3}N_{0.7}$; $TiN_{0.3}$; $Ti_4C_2S_2$; Ni_2Si; MoB; G – $Ni_{16}Ti_6Si_7$; γ' – $Ni_3(Al,Ti)$
Ageing: 750 °C/100–150 h	TiC; $TiC_{0.3}N_{0.7}$; $TiN_{0.3}$; $Ti_4C_2S_2$; Ni_2Si; MoB G – $Ni_{16}Ti_6Si_7$; γ' – $Ni_3(Al,Ti)$; η – Ni_3Ti
Ageing: 750 °C/200–300 h	TiC; $TiC_{0.3}N_{0.7}$; $TiN_{0.3}$; $Ti_4C_2S_2$; Ni_2Si; MoB G – $Ni_{16}Ti_6Si_7$; γ' – $Ni_3(Al,Ti)$; η – Ni_3Ti; β – NiTi
Ageing: 750 °C/400–500 h	TiC; $TiC_{0.3}N_{0.7}$; $TiN_{0.3}$; $Ti_4C_2S_2$; Ni_2Si; MoB G – $Ni_{16}Ti_6Si_7$; γ' – $Ni_3(Al,Ti)$; η – Ni_3Ti; β – NiTi; σ – $Cr_{0.46}Mo_{0.40}Si_{0.14}$

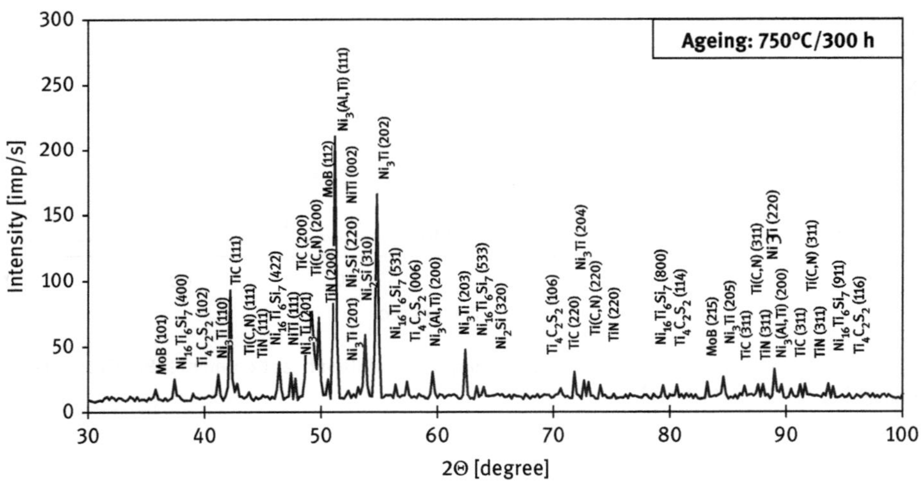

Figure 7.29: Isolate diffractogram of the Fe–Ni alloy after solution heat treatment (980 °C/2 h/water) and ageing at 750 °C/300 h [45].

The results of studies conducted by Ducki [45], presented in this chapter, fulfil all the requirements of speciation. Carefully conducted phase analysis (speciation) revealed all phase constituents present in the alloy with a detailed description of the transformations and their metallurgical and technological conditions. The presence of so many phases in the test material makes their quantification required by the speciation analysis impossible. Only in some cases, and this to a limited extent, it becomes possible to roughly conclude what their quantitative fraction is.

In the comprehensive studies of creep-resistant nickel-based alloys, apart from the presented data on phase extraction and chemical analysis of the composition of

Table 7.17: Phase composition of Fe–Ni alloy isolates after solution heat treatment and ageing at 780 °C [45].

Alloy condition	Phase constituents
Solution HT: 980 °C/2 h/water	TiC; $TiC_{0.3}N_{0.7}$; $TiN_{0.3}$; $Ti_4C_2S_2$; Ni_2Si; MoB
Ageing: 780 °C/2 h	TiC; $TiC_{0.3}N_{0.7}$; $TiN_{0.3}$; $Ti_4C_2S_2$; Ni_2Si; MoB; G – $Ni_{16}Ti_6Si_7$; γ' – $Ni_3(Al,Ti)$
Ageing: 780 °C/4–500 h	TiC; $TiC_{0.3}N_{0.7}$; $TiN_{0.3}$; $Ti_4C_2S_2$; Ni_2Si; MoB; G – $Ni_{16}Ti_6Si_7$; γ' – $Ni_3(Al,Ti)$; σ – $Cr_{0.46}Mo_{0.40}Si_{0.14}$; η – Ni_3Ti

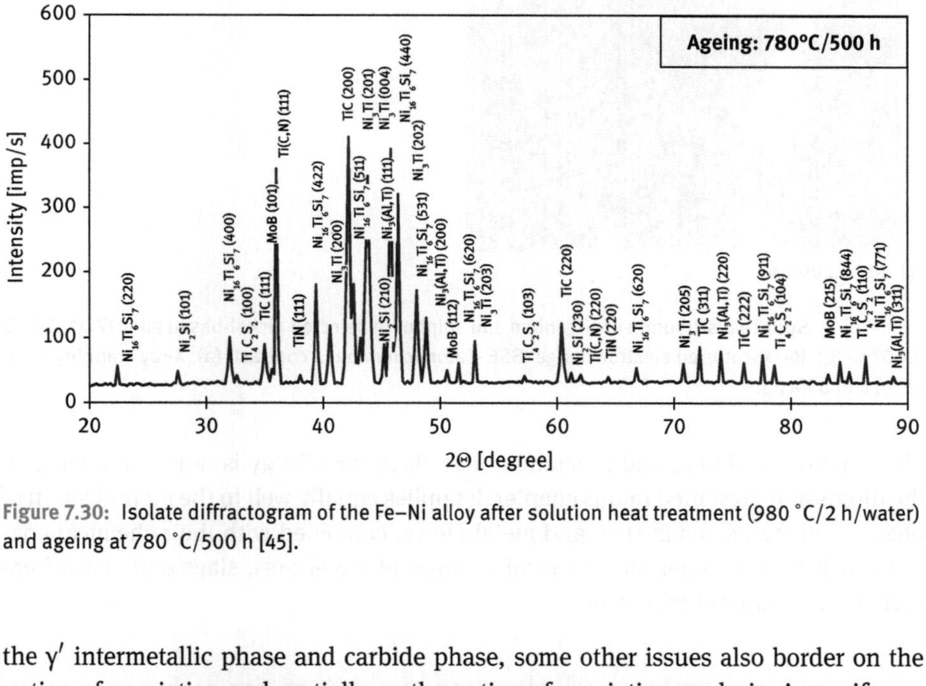

Figure 7.30: Isolate diffractogram of the Fe–Ni alloy after solution heat treatment (980 °C/2 h/water) and ageing at 780 °C/500 h [45].

the γ' intermetallic phase and carbide phase, some other issues also border on the notion of speciation and partially on the notion of speciation analysis. A specific example in this area is the use of zirconium as a microadditive or alloying element, and also as a refining (sulphur binding) agent. Zirconium in these alloys performs precisely this role, forming ZrS_x-type compounds with sulphur (Figure 7.31) [6, 91], but it is not always possible to determine the magnitude of x; sometimes it is only possible to state that x is less than or equal to 2. In a system like this, it would be difficult to talk about speciation analysis in the sense of analytical chemistry, although information about the presence of zirconium compound is very important and valuable to analytical chemists, as it is well known that zirconium compounds with sulphur dissolve in acids, which eliminates the additional and laborious operation of melting, when samples of this type are converted to solution.

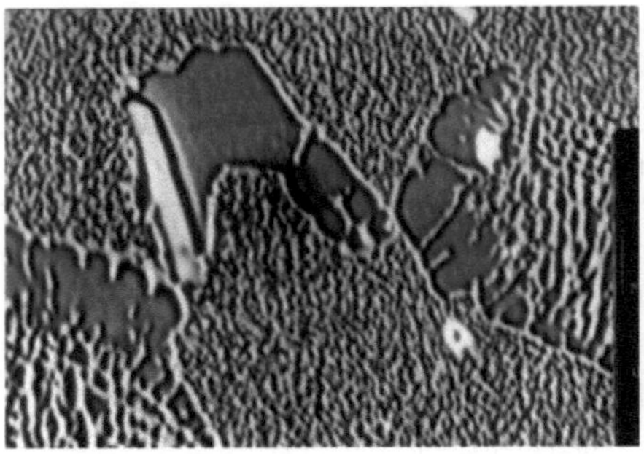

(a) COMPO 2000 x

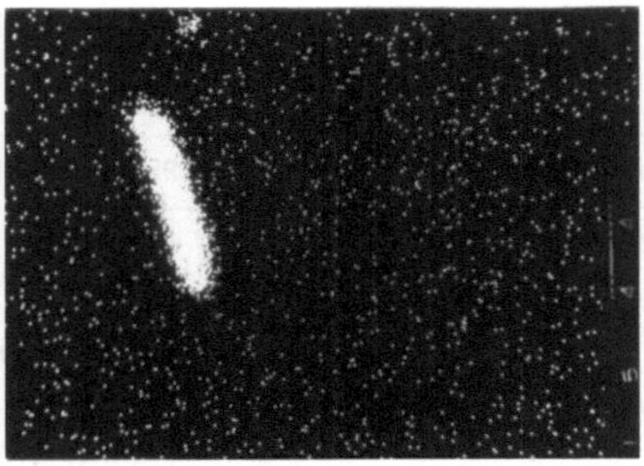

(b) Rtg Zr 2000 x

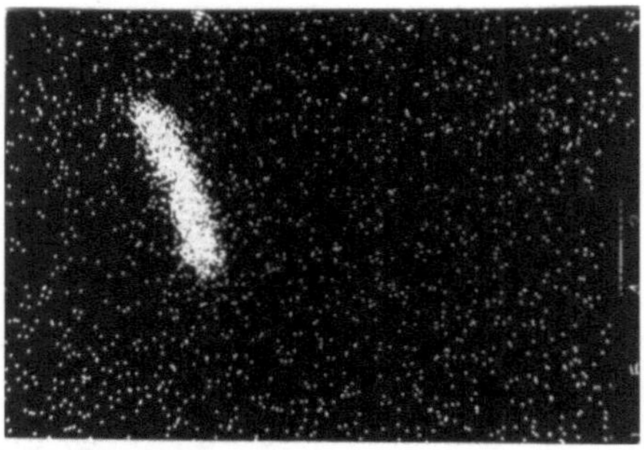

(c) Rtg S 2000 x

Figure 7.31: Surface distribution of zirconium and sulphur in selected nickel-based alloy (Zr 0.53%, S 0.007%) [6]. Backscattered electron image (BSE – atomic number Z contrast) (a), X-ray mapping of Zr La (b) and S Ka (c).

The issue of speciation and speciation analysis in metallurgy is much broader than the discussion presented in this chapter. It applies equally well to the identification of phases and inclusions in steel and metal alloys, combined with their chemical analysis, as it does to other metallurgical samples of metal ores, slags and waste from various metallurgical processes.

7.19 Introduction to speciation and speciation analysis of chromium in welding dust

In metallurgy, the terms *speciation* and *speciation analysis* cover the complete life cycle of metallurgical products, from the manufacture of steel and alloys, through the shaping of products and structures, and in the disposal and recycling of waste and spent materials ending. This chapter briefly describes one selected method, which is meant to serve as an example of processing of steel into products. This is the welding technology discussed in the aspect of the determination of the content of nickel, total chromium and chromium (VI).

As a criterion for the choice of analytical methods to determine the percent content of Ni, Cr and Cr(VI), the following assumptions have been made: the predicted content of individual components in the tested dust ranges from 1% to 6% for nickel, over 5% for total chromium and from 0.1% to about 2% for chromium (VI). Nickel determination was performed by dissolving dust samples in a mixture of acids, followed by nickel precipitation in the form of nickel (II) dimethyl glyoxylate and the determination of its weight. Total chromium was determined by visual titration. This operation consisted in dissolving a sample of dust in acids, chromium oxidation in acid to chromium (VI), and then reduction of chromium (VI) with standard solution of ammonium sulphate (VI) and iron (II) sulphate sexahydrate, followed by titration of excess of this reagent with standard solution of potassium manganate (VII). Chromium (VI) was eluted from the dust sample with an alkaline mixture ($NaOH + Na_2CO_3$) and its content was determined by measurement of atomic absorption while spraying the solution in nitrous oxide–acetylene flame using 357.9 nm wavelength. Detailed analytical methodology is currently being prepared for publication [116].

Welding processes of steel joining conducted in high temperature and with limited possibilities to control metallurgical processes and physical and chemical phenomena belong to the category of processes that have an adverse impact on the work environment. Welding of metals is perceived as a 3D process: dirty, dusty and dangerous. While welding gaseous and dust pollutants containing numerous substances hazardous to workers' health are emitted to work environment, toxicity of those pollutants as a result of undesirable physical and chemical actions after penetration into a human body causes large number of occupational diseases in welders' circles. Welding of metals generates hazardous substances, that is substances that pose a threat to human health and are classified to at least one of the following categories: very toxic, toxic, noxious, irritating, allergenic, carcinogenic or mutagenic. Chemical substances occurring in air in welding processes constitute a polyphase system – an aerosol, and their absorption into the human body depends largely on the form of a particular substance. Welding fume arises from the parent and filler metals, protective coatings of parent metal, shielding gases and ambient air under the influence of high temperature and radiation of welding arc [97–116]. Welding fume (diphase condensation aerosol) is a mixture of fine-dispersed solid particles (welding dusts) and different gases which form a dispersion phase. Welding dusts generated through the action of arc plasma on parent and filler metals [97–116] consist of simple and complex oxides, silicates, fluosilicates, fluorides, chromates, dichromates as well as metal carbonates [97–116]. The size of the average aerodynamic diameter of a particle of welding dusts varies from 0.01 to 1 µm [97–116]. The chemical composition of welding dusts depends on the type of welded metals as well as the method and technological parameters of welding. In case of welding with covered and tubular cored electrodes, the composition of dusts is more complex and has a more complicated structure than in case of dusts generated during gas-shielded welding with solid electrodes. In case of welding of unalloyed steels by means of solid electrodes, the basic components of dusts are iron, manganese and silicon dioxide [97–116], whereas welding of high-alloy steels causes emission of dusts containing also chromium, nickel, molybdenum and niobium compounds [97, 98, 101]. Welding of steel with covered and tubular cored

Table 7.18: Chemical composition of dust during gas-shielded welding of steel [16, 17].

Welding process/filler metal/shielding gas	Parent metal	Chemical composition of dust (% m/m)				
		Fe	Mn	Si	Cr	Ni
MAG/solid electrode/80% Ar + 20% CO_2	Unalloyed steel Low-alloy steel	57.2	11.6	3.57	0.03	0.03
MAG/solid wire/80% Ar + 15% CO_2+ 5% O_2	Unalloyed steel Low-alloy steel	59.7	11.8	3.4	0.03	0.03
MAG/basic cored electrode/CO_2	Unalloyed steel Low-alloy steel	43.6	10.9	3.3	0.03	0.01
MAG/rutile cored electrode/ CO_2	Unalloyed steel Low-alloy steel	34.7	8.3	4.0	0.02	0.01
MAG/metal cored electrode/ 82% Ar + 18% CO_2	Unalloyed steel Low-alloy steel	45.9	14.6	4.4	0.83	1.68
Self-shielded tubular cored electrode	Unalloyed steel Low-alloy steel	50	3.2	0.9	0.02	0.38
MIG/solid electrode/Ar	High-alloy steel	38.0	12.0	5.0	20.0	5.0
MAG/solid electrode/99% Ar + 1% O_2	High-alloy steel	39.0	11.0	5.0	19.0	5.0

Note: MIG, metal inert gas; MAG, metal active gas.

electrodes is additionally accompanied by the emission of sodium, potassium, calcium and magnesium compounds [98, 102, 108]. The source of the aforementioned elements is the coating of an electrode and flux powder composed of various mineral resources (such as silicates, carbonates, simple and complex fluorides, metal oxides, sodium or potassium glass) as well as organic components. The chemical composition of dusts emitted during welding of steel is presented in Table 7.18.

The main sources of emission of gases during welding are as follows [97–116]:
- Decomposition of covering of electrodes, decomposition of fluxes
- Thermal reactions in arc-surrounding atmosphere
- Photochemical reactions in arc-surrounding atmosphere (UV radiation)
- Shielding gas used for protection of an arc

As a result of thermal decomposition of materials protective coatings such as paints, lacquers, plastics and anticorrosive while welding organic compounds, for example, aliphatic alcohols, formaldehyde, phenol, aliphatic ketone, are created [97–116]. Presence of phosgene ($COCl_2$) in air surrounding welding stations is explained by the decomposition under the influence of ultraviolet radiation of some solvents used for cleaning of welded surfaces.

The effect of a long-term exposure of welders to welding fumes are various diseases of a respiratory system [97–116]. The airways are the main route by means of which dusts enter a human body. Pneumoconiosis may affect welders as early as after several years of work and is considerably more common in case of welders working in small or badly ventilated rooms rather than in case of those who work outdoors. Among welders, growing asthma and bronchitis incidence rate can be observed. Apart from bronchus or lung complaints which to a significant extent are the cause of occupational diseases of welders, other illnesses may occur simultaneously, for example, diseases of nervous, cardiovascular or digestive systems [97–102, 106, 108].

International Agency for Research on Cancer has stated that welding fumes belong to the group of pollutants probably carcinogenic to human. It has been confirmed that such components of welding fumes as nickel, chromium VI, beryllium and cadmium have a carcinogenic effect. Silicon dioxide, also present in welding dusts, increases lungs' predisposition to tuberculosis and a number of infectious diseases. Pulmonary tissue's fibrosis is related to the influence of chromium and nickel compounds emitted during welding of high-alloy steels. Inhaling copper, zinc, magnesium or nickel vapours may expose welders to a disease known as zinc-fume fever.

Figure 7.32 presents the general classification of dusts and gases produced during welding of metals as regards their action on a human organism [97–116].

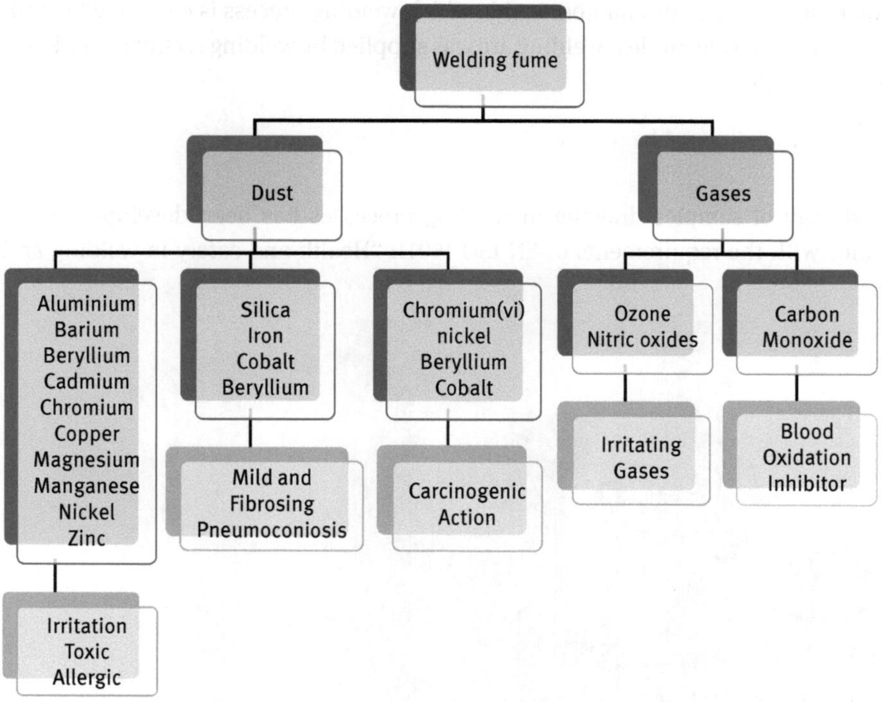

Figure 7.32: Classification of dusts and gases contained in welding fumes with reference to impact on human organism [97–100].

This publication presents laboratory stands for drawing of samples of welding dusts depending on welding methods and consumables used in those processes taking into consideration harmful effect of welding dust on welders' health.

7.19.1 Description and characteristic of laboratory stands

Laboratory experimental stand for drawing samples of dust emitted during MIG/MAG welding of steel consists of experimental hermetic chamber, inside which welding process is conducted. Parent metal is located on a horizontal welding positioner, whereas welding torch is mounted on a stationary basis [111]. Welding process inside the chamber was conducted automatically without anybody's participation from outside. MAGOMIG-401C welding power source was used in the experiments (Figure 7.33).

Stand for samples drawing of dust emitted while TIG welding of steel comprises three subassemblies:
- Experimental chamber
- Exhaust equipment
- TIG welding equipment – KEMPPI PRO 5000 (Figure 7.34)

TIG welding process was conducted manually inside the chamber.

Testing stand for drawing samples of dust emitted during MMA welding consists of experimental hermetic chamber inside which welding process is conducted manually using covered electrodes. Welding arc was supplied by welding rectifier of PSP-630 type (Figure 7.35).

7.19.2 Methodology of total dust emission determination

Methodology of samples drawing in welding processes has been developed in accordance with the requirements of EN ISO 15011: "Health and safety in welding and

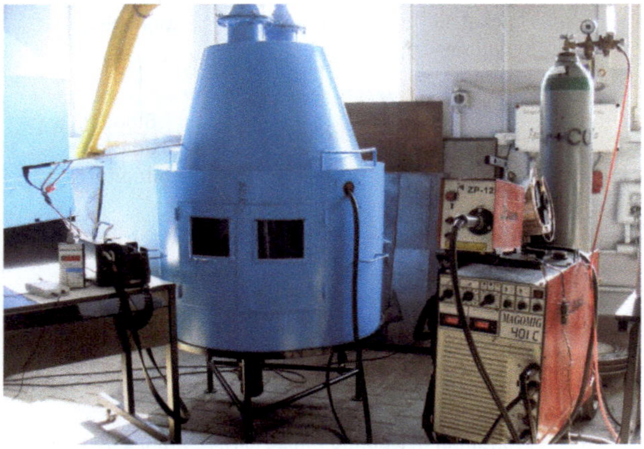

Figure 7.33: Laboratory stand for sampling of dust during MIG/MAG welding of steel [98].

Figure 7.34: Experimental stand for samples drawing of dust emitted while TIG welding of steel [98].

Figure 7.35: Experimental stand for drawing samples of dust generated during MMA welding with covered electrodes [98].

allied processes – Laboratory method for sampling fume and gases generated by arc welding – Part 1: Determination of emission rate and sampling for analysis of particulate fume" [111]. Separate stages of the methodology for total dust estimation during welding are shown in Figure 7.36.

Mass of emitted dust during welding was calculated from formula (7.5):

$$m_p = m_2 - m_1 \tag{7.5}$$

where m_p is the mass of dust (mg) (accuracy of 0.1 mg); m_1 the mass of a clean filter (mg); m_2 the mass of a filter with retained dust ([mg); whereas calculation of generation of dust E_c is conducted according to formula (7.6):

$$E_c = m_p/t \text{ (mg/s)} \text{ (accuracy of 0.01 mg/s)} \tag{7.6}$$

Determination of total dust emission due to MIG/MAG, TIG and MMA welding of stainless steel

Preparation of materials for testing

- Before testing a filler metal (solid electrode, tubular cored electrode, rod, covered electrode), tested plate and shielding gases were prepared
- Type and diameter of filler metal was noted
- Material of tested plate was selected in accordance with testing guidelines
- Samples of diameter of 500 mm for MIG/MAG welding were taken from the plate of 10 mm in thickness
- Samples of dimension of 250 x 100 mm for TIG and MMA welding were taken from the plate of 8 – 10 mm in thickness

Preparation of measuring filters

- Whatman cellulose filters of dimension of 185 mm were prepared
- A set of filters for single measurement (2 separate filters) was weighted on analytical balance with accuracy of readings up to 0.1 mg
- Marked set of filters were placed into an exsiccator
- Before the testing, filters were taken out from exsiccator and weighted again
- Filters closed in a container were carried to testing stand

Testing measures in drawing samples of dust for emission determination

- Process current parameters were selected in accordance with testing guidelines
- Measuring filters were placed on exhaust connectors of the chamber
- Welding process lasted 60 s
- In case of welding with covered electrode the whole electrode was consumed
- Exhaust equipment was turned on for 10 min
- Exhaust equipment was turned off
- Measuring filters after testing were placed in the closed container
- Samples of dust were taken 5 times
- Filters were placed into the exsiccator again
- Set of filters was weighted on analytical balance with accuracy of readings up to 0.1 mg

Figure 7.36: Determination of total dust emission while welding of stainless steel with MIG/MAG, TIG and MMA processes [98, 115].

where m_p represents the mass of retained dust (mg) (accuracy of 0.1 mg); t the time of process duration (s) (accuracy of 1 s).

7.19.3 Methodology of drawing of dust samples for chemical analysis

Testing procedure for dust sampling for chemical analysis comprises the following activities:

- Prepare weight and mark a container for dust
- Prepare Whatman cellulose filters of a diameter of 185 mm
- Place measuring filters on exhaust connectors of the experimental chamber
- Set welding parameters
- Turn on the exhaust system
- Conduct welding for 3 min
- Turn off the exhaust system
- Remove filters with dust
- Transfer dust from filters to a container with the use of small brush
- Weight a container (mass of dust for analysis should amount to at least 2 g, depending on the number of components determined in the dust)
- If it is found that mass of dust is insufficient, then samples drawing should be repeated

Demanded chemical constitution is determined from the mass of dust emitted during welding of steel.

Laboratory stands presented in this chapter make possible to conduct complex testing of dust emitted during welding both in reference to mass and its chemical constitution. It allows for determination of many relations concerning dust emission, in the aspect of applied welding processes, selection of consumables and technological conditions of welding. Quantitative relations determined on this ground and their detailed analysis are the base for modification of welding processes in the aspect of reduction of a content of harmful substances, especially those that are particularly hazardous to welders' health. As a criterion for the choice of analytical methods to determine the percent content of Ni, Cr and Cr(VI), the following assumptions have been made: the predicted content of individual components in the tested dust ranges from 1% to 6% for nickel, over 5% for total chromium, from 0.1% to about 2% for chromium (VI). Nickel determination was performed by dissolving dust samples in a mixture of acids, followed by nickel precipitation in the form of nickel (II) dimethyl glyoxylate and the determination of its weight. Total chromium was determined by visual titration. This operation consisted in dissolving a sample of dust in acids, chromium oxidation in acid to chromium (VI) and then reduction of chromium (VI) with standard solution of ammonium sulphate (VI) and iron (II) sulphate sexahydrate, followed by titration of excess of this reagent with standard solution of potassium manganate (VII). Chromium (VI) was eluted from the dust sample with an alkaline mixture (NaOH + Na_2CO_3), and its content was determined by the measurement of atomic absorption while spraying the solution in nitrous oxide–acetylene flame using 357.9 nm wavelength. Detailed analytical methodology is currently being prepared for publication [116].

7.20 Summary

Apart from the basic data on the chemical composition of metallurgical materials, modern research instruments also provide additional information on the types of phases, chemical compounds and so on. This information, in turn, when transferred to technologists and material engineers allows them to upgrade the properties of the already manufactured products and design new metallurgical materials for specific applications. So, it can serve as a typical example of feedback between "(metallurgical) material analytics development" and "(metallurgical) technology development" (Figure 7.37). This area also includes problems of phase analysis, speciation and speciation analysis in metallurgy [6].

It is worth thinking for a while about the benefits of using in chemical metallurgical laboratories the most modern methods and tools developed by the advanced systems of statistical process control, like Ishikawa diagram, Pareto-Lorenz principle, control charts, detailed analysis of process flowcharts and other universal solutions, among which the most notable is the 5S system of improvement [117–126]. The 5S system is based on Japanese philosophy and is a set of five simple rules that streamline the organization of processes and solutions used in the workplace. The abbreviation "5S" comes from the Japanese words **seiri** (sort, clean, classify), **seiton** (straighten, simplify, set in order, configure), **seiso** (sweep, shine, scrub, clean and check), **seiketsu** (standardize, stabilize) and **shitsuke** (sustain, maintain, practise self-discipline). The framework for the implementation of 5S system in a chemical research laboratory is presented in Ref. [125]. Another issue worth remembering is the role of cross-posting in a properly functioning research laboratory [121]. Cross-posting is a relatively new concept of improvement of the work organization in a research laboratory. It allows the laboratory manager and the laboratory staff to avoid the situation in which only one person can perform a specific task on a specific job. Participants of cross-posting programme have the opportunity to learn about the research process at each workplace. Introduction of cross-posting in chemical research laboratories allows identification of possible errors in analytical cycles and their elimination. In a well-functioning research laboratory, emphasis is also put on ethical issues, including, but not limited to, the confidentiality of research, improper (overestimated or underestimated in bidding processes and tenders) valuation of an executed order,

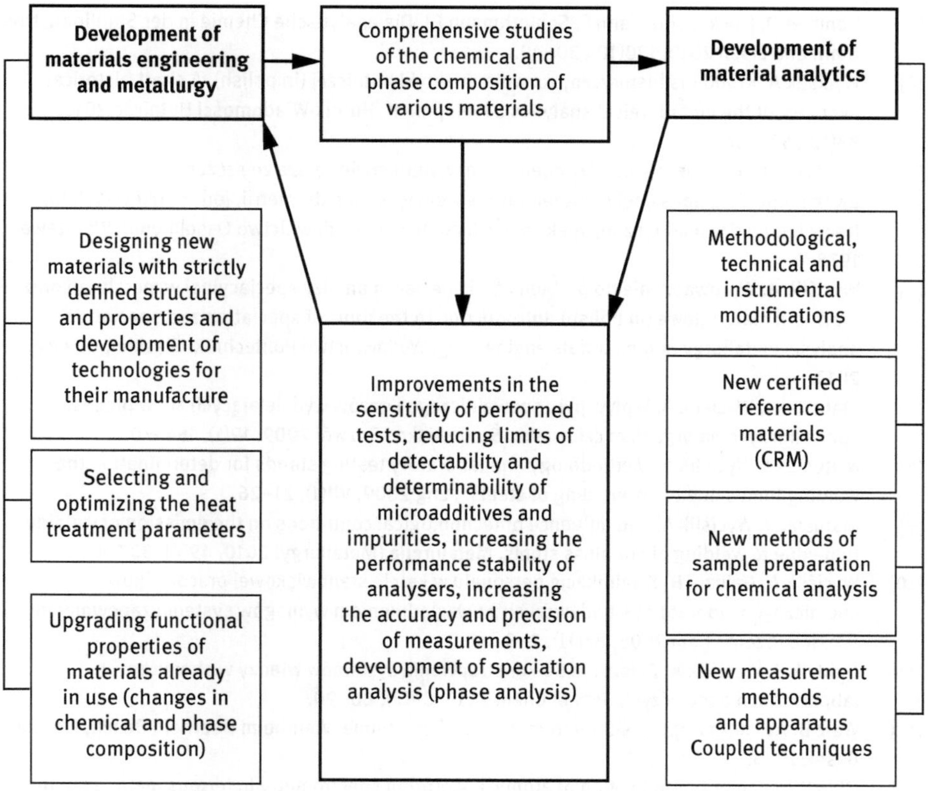

Figure 7.37: Interdependence of the development of material analytics, metallurgy and materials engineering

offering unfair and unreliable terms of execution, for example unrealistic execution times of orders, taking into account the duration of the analytical/measurement cycle, as well as keeping the customer unaware that all or part of the research is subcontracted to another laboratory, using "strong" position in the market segment of relevant researches or services (subcontracting) [123]. In chemical research laboratories or industrial laboratories that operate within companies or R&D centres and are planning to implement a health and safety management system, it is necessary to systematize the basic principles of safe work [126].

References

[1] Sommer D, Mertens R, Schlothmann BJ, Thiemann E, Von Wachtendonk HJ. Aufgaben der Stahlanalytik für die Qualitätssicherung bei der Stahlerzeugung. Stahl und Eisen 1999, 119(6/7), 109–12.

[2] Karl Heinz K. Fortschritte und Zukunft der metallurgischen Analytik. Stahl und Eisen 2002, 122(8), 59–65.

[3] Sommer D, Flock J, Thiemann E, Schlothmann BJ. Die analytische Chemie in der Stahlindustrie. Stahl und Eisen 2010, 130(10), 30–42.

[4] Wycislik A. Krótki rys historyczny rozwoju analityki hutniczej (in polish). A short historical overview of the metallurgical analytics development. Hutnik-Wiadomości Hutnicze 2016, 83(12), 542–544.

[5] Jurczyk J. Automatyzacja analizy chemicznej w hutnictwie żelaza ze szczególnym uwzględnieniem emisyjnej i rentgenofluorescencyjnej spektrometrii. [ed. Fijałkowski J] In: Nowe kierunki w analitycznej spektrometrii atomowej. Wydawnictwo Ossolineum, Warszawa, 1978.

[6] Wyciślik A. Wprowadzenie do problematyki specjacji i analizy specjacyjnej w metalurgii oraz inżynierii materiałowe (In polish). Introduction to the topic of speciation and speciation analysis metallurgy and materials engineering. Wydawnictwo Politechniki Śląskiej, Gliwice, 2013.

[7] Matusiak J, Wyciślik A. Wpływ pyłu spawalniczego i gazów wydzielających się w procesach spawalniczych na organizm człowieka. Środowisko i Rozwój 2009, 19(1), 154–70.

[8] Matusiak J, Wyciślik A. Methodology and laboratory testing stands for determination the welding fumes due to arc welding of steels. J Eng 2009, VII(1), 21–26.

[9] Matusiak J, Wyciślik A. The influence of technological conditions on the emission of welding fume due to welding of stainless steels. Metalurgija (Metallurgy) 2010, 49(4), 307–11.

[10] Wyciślik A, Gajdzik B. Kwalifikacje personelu w karcie stanowiskowej pracowników chemicznych laboratoriów badawczych z uwzględnieniem wymogów systemu zapewnienia jakości. Przem Chem 2008, 87(11), 2–8.

[11] Gajdzik B, Wyciślik A. Założenia do konstruowania systemów wiedzy w chemicznych laboratoriach badawczych. Przem Chem 2010, 89(2), 86–90.

[12] Welz B. Atom-Absorptions-Spektrometrie. Verlag Chemie, Weinheim Deerfield Beach, Florida, Basel, 1983.

[13] Ohls K, Sommer D. Application of atomic absorption spectrometry in ferrous metallurgy. In: Haswell SJ, ed. Atomic Absorption Spectrometry, Theory, Design and Applications. Elsevier, Amsterdam, Oxford, New York, Tokyo, 1991.

[14] Staufenberger O. Die Einsatzmöglichkeiten der Atom-Absorptions-Spektrometrie in einem Laboratorium des Elektromaschinenbaus. Analysentechnische Berichte Z. 41, 1975.

[15] Cantle JE, ed. Atomic Absorption Spectrometry. Elsevier Scientific Publishing Company, Amsterdam, Oxford, New York, 1982.

[16] Van Loon JC. Analytical Atomic Absorption Spectroscopy, Selected Methods, Academic Press, New York, London, Toronto, Sydney, San Francisco, 1980.

[17] Price WJ. Analytical Atomic Absorption Spectrometry. Heyden & Son Ltd, London, New York, 1978.

[18] Koch KH. Fortschritte und Zukunft der metallurgischen Analytik. Stahl und Eisen 2002, 122, 59–65.

[19] Sommer D, Mertens R, Schlothmann B-J, Thieman E, von Wachtendonk H-J. Aufgaben der Stahlanalytik für die Qualitätssicherung bei der Stahlerzeugung. Stahl und Eisen 1999, 119, 109–12.

[20] Wyciślik A. Zastosowanie płomieniowej metody atomowej spektrometrii absorpcyjnej w badaniach materiałoznawczych. Hutnik-Wiadomości Hutnicze 1999, 66, 479–501.

[21] Ohls K, Sommer D. Applications: Metals and Industrial Materials. J. Wiley & Sons, Inc, New York, London, 1987.

[22] Jones PT, Hermans P, Blanpaln B, Wollants P. Optimization of an accurate and precise analysis procedure for metallurgical VOD slags with ICP-OES. At Spectrosc 2000, 21, 86–92.

[23] Hlavackova I, Hlavacek I. Vyuziti atomove emisni spektrometrie s indukcne vazanym plazmatem pri rozboru titanovych slitin a jejich predslitin. Hutnicke Listy 1989, 9, 194–9.

[24] Guy AG. Introduction in Materials Science. McGraw-Hill Book Company, New York, 1981.
[25] Šapovalov ĖT, Baranova LI, Zekcer GO. Elektrochimičeskie metody v metallovedenii i fazovom analize, izd. Metallurgija, Moskva 1988.
[26] Kirchhofer H, Rottmann J, Schubert F, Nickel H. Untersuchungen zum isothermen Ausscheidungsverhalten von Nickelbasislegierungen mit Hilfe der elektrochemischen Phasenisolierung, Institut für Reaktorwerkstoffe, Berichte der Kernforschungsanlage Jülich GmbH, Nr 1903, 1984.
[27] Golubcova RB. Fazovyj analiz nikelevych splavov, izd. Nauka ,Moskva 1969.
[28] Kunath J, Stahlberg R. Methodische Untersuchungen zur metallkundlichen Analyse mittels elektrochemischer Phasenisolation. Arch. Eisenhüttenwesen 1983, 54, 31–38.
[29] Wyciślik A. Zasada elektrochemicznej izolacji faz i metody badań izolatów. Hutnik-Wiadomości Hutnicze 2008, 75(2), 60–66.
[30] Richter J, Wyciślik A. Isolation techniques and methods for isolated phases investigation. Acta Metallurgica Slovaca, 2001, 7, 481–5.
[31] Samsonov GV, Vinickij IM. Tugoplavkije sojedinenija. Metallurgija, Moskva 1976.
[32] Lupis CHP. Chemical Thermodynamics of Materials. North-Holland, New York, Amsterdam, Oxford, 1983.
[33] Headridge JB, Sowerbutts A. Atomic absorption spectrophotometric determination of total aluminium in steel after its dissolution in a pressure bomb. Analyst 1973, 98, 57–60.
[34] Krakovska E, Kuss HM. Rozklady v analytickej chemii, wyd.Vienala, Kosice (Słowacja), 2001.
[35] Del Monte Tamba MG, Dorado Lopez MT, Gomez Coedo A, Microwave digestion system for dissolution of iron, carbon and low-alloy steels, iron ores and slags, In: Nauche R, ed. Progress of Analytical Chemistry in the Iron and Steel Industry, Commission of the European Communities, Directorate-General Science, Research and Development, Brussels Luxembourg, 1992, 293.
[36] Schubert F. Kontrolle der chemischen Zusammensetzung hochwarmfester Nickellegierungen im Hinblick auf ihre Gefügestabilität, Bundesministerium für Forschung und Technologie, Forschungsbericht 79-147, Thyssen Edelstahlwerke AG, Forschungsinstitut Krefeld, 1979.
[37] Sims CT, Hagel WC. The Superalloys. J. Wiley & Sons, New York, 1972.
[38] Richter J, Hetmańczyk M, Cwajna J. Charakterization of carbide phase in nonledeburic high-speed steel containing Ti and Nb. J Mater Process Technol 2010, 53, 341–8.
[39] Kawalla R, Lang C, Lotter U. Beitrag der quantitativen Metallographie zur Optimierung von Gefügen. Stahl und Eisen 1994, 114, 61–70.
[40] Cwajna J. Ilościowy opis struktury stopów narzędziowych i jego zastosowanie. Zeszyty Naukowe Politechniki Śląskiej, Z. 39, Gliwice, 1991.
[41] Richter J. Strukturalne czynniki mikrościerania i dekohezji stopów narzędziowych o zróżnicowanej fazie dyspersyjnej. Zeszyty Naukowe Politechniki Śląskiej, Z. 1674, Gliwice, 2005
[42] Tieking W, Overbosch A, Broersen P, Kooter R, Rijnders M, Alders L. Traces in alloys and their effects on steelmaking and casting. Stahl und Eisen 2013, 133(3), 59–64.
[43] Müller N, Boeckers TME. Das Qualitätssicherungssystem des Stahlwerks in einem ganzheitlichen Management-Konzept. Stahl und Eisen 1999, 119, 115–9.
[44] Humbert J. Steel quality and analytical precision in plate production. Steel Res 1991, 62, 171–5.
[45] Ducki KJ. Mikrostrukturalne aspekty procesów odkształcania, wydzielania i umacniania w austenitycznym nadstopie Fe-Ni, wyd. Politechniki Śląskiej, Gliwice, 2010.
[46] Müller N, Boeckers TME. Das Qualitätssicherungsststem des Stahlwerks in einem ganzheitlichen Managent-Konzept. Stal und Eisen 1999, 119(6/7), 115–9.
[47] Wünnenberg K, Cappel J. Massnahmen zur Verbesserung des oxidischen Reinheitsgrades beim Stranggiessen. Stahl und Eisen 2010, 130, 55–61.

[48] Flossdorf FJ, Wieland HJ. Weiterverarbeitung und Anwendung des Werkstoffes Stahl. Stahl und Eisen 2010, 130(5), 31–42.
[49] Wyciślik A. Some aspects of chemical analysis of isolates by AAS. Mikrochim Acta, 1990, II, 287–294.
[50] Hulanicki A. Współczesna chemia analityczna Wybrane zagadnienia, wyd. PWN, Warszawa, 2001.
[51] Tichonov VN, Jakovlev P. Ja. Opredelenie aljuminija v metallach i splavach, izd. Metallurgija, Moskva, 1978.
[52] Jurczyk J, Jagiełło-Puczka W, Łysakowska M. Badanie izolatów wtrąceń niemetalicznych w stalach metodą spektralną ze wzbudzeniem jarzeniowym. Inżynieria Materiałowa 1993, 14, 88–93.
[53] Wyciślik A, Sosnowski R. Dziedziny zastosowań metody atomowej spektrometrii absorpcyjnej w hutnictwie żelaza. Hutnik-Wiadomości Hutnicze 1999, 66, 204–8.
[54] Wyciślik A. Determination of the chemical composition of intermetallic γ' phase isolates by AAS method using a single specific standard. Chem Anal 1999, 44, 755–61.
[55] Charlamov IP, Eremina GV. Atomno-absorbcjonnyj analiz v cernoj metallurgii. Metallurgija, Moskva, 1982.
[56] Wyciślik A. Zagadnienia metodyczne zastosowania atomowej spektrometrii absorpcyjnej do określania składu chemicznego izolatów. Rudy i Metale Nieżelazne 1995, 40, 194–198.
[57] Sauer KH, Nitsche M. Atomabsorptionsspektrometrische Analyse von oxydischen und karbidischen Isolaten aus Stählen. Arch Eisenhuttenwes 1969, 40, 891–893.
[58] Gomiščék S, Span M, Lavric T, Hudnik V. Beitrag zur Mikroanalyse von Einschlüssen in Stählen mit Hilfe der Atomabsorptionsspektrometrie. Mikrochim Acta 1974, 4, 567–76.
[59] Wyciślik A, Mikuszewski T. Schemat analityczny oznaczania aluminium w stopach na osnowie fazy międzymetalicznej Ni_3Al metodą absorpcji atomowej, Rudy i Metale Nieżelazne 2005, 50, 4, 178–182.
[60] Tuma H, Slama J. Rozbor izolátu po elektrolytické separaci metodou AAS. Hutnicke Listy 1980, 35, 658–661.
[61] Richter J, Wyciślik A, Cwajna J. Investigations of carbide phase and matrix of nonledeburitic high-speed steel with Ti and Nb by atomic absorption spectrometry and EDX methods, Fresenius' J Anal Chem 1994, 349, 153–4.
[62] Wyciślik A. Use of secondary spectral lines in the analysis of alloy-steels and electrochemically isolated structural constituents by AAS method. In: Nauche R, ed. Progress of Analytical Chemistry in the Iron and Steel Industry, Commission of the European Communities, Directorate-General Science, Research and Development EUR 14113 EN, ,Brussels Luxembourg, 1992, 542–6.
[63] Stahlberg R, Steglich F, Zieger M. Zur Genauigkeit sequentieller AAS-Elementbestimmungen in elektro-chemisch isoloierter Gefügebestandteilen differenziert behandelter Stähle. Fresenius Z Anal Chem 1981, 306, 365–71.
[64] Wyciślik A. Some interelement interferences in the chemical analysis of intermetallic γ' phase isolates using AAS method. Mikrochim Acta 1997, 125, 297–300.
[65] Richter J, Wyciślik A, Cwajna J. Analysis of carbide isolates of new high-speed steel by atomic absorption spectrometry. Mikrochim Acta 1998, 130, 117–22.
[66] Wyciślik A. The analysis of carbide and intermetallic phase γ' isolates in nickel alloys by means of injection method of flame atomic absorption spectrometry. Mikrochim. Acta 1987, I 201–9.
[67] Wyciślik A, Jurczyk J, Albrecht H. Prowadnica kapilary do oznaczania zawartości pierwiastków za pomocą techniki iniekcyjnej. Patent PL W-79212, 1987.
[68] Wyciślik A, Sosnowski R. Pokrywka do pionowego wprowadzania kapilary spektrometru absorpcji atomowej podczas analiz małych objętości roztworów próbek, Patent PL W – 107 301.

[69] Wyciślik A. Przykrywka do pionowego wprowadzania kapilary absorpcjometru do teflonowego naczynia kraterowego i jej rola w analizach izolatów węglikowych i międzymetalicznej fazy γ′ metodą AAS. Krzepniecie Metali i Stopów 1998, 36, 301–8.

[70] Sebastiani E, Ohls K, Riemer G. Results of the atomizing method using measured volumes of liquid sample by AAS. Fresenius Z Anal Chem 1973, 264, 105–7.

[71] Manning DC. Aspirating small volume samples in flame atomic absorption spectroscopy. At Absorption Newsl 1975, 14, 99–102.

[72] Berndt H, Jackwerth E. Atomic absorption spectrometric determination of small amounts of substances and analysis of trace element concentrates by the injection method. Spectrochim Acta 1975, 30B, 169–73.

[73] Wyciślik A, Richter J. The role of metallic constituents removal in investigations of the chemical composition of isolated carbide phases. Chem. Anal. 2001, 46, 207–14.

[74] Matusiewicz H. A simple PTFE microsampling device for discrete injection. Int. Lab. 1983, 13, 24–26.

[75] Matusiewicz H. Evaluation of applicability of AAS 1N atomic absorption spectrometer, FLAPHO 4 flame photometer, P 100 automatic sample changer and K 201 recorder of VEB Carl Zeiss Jena for flame discrete nebulization. Acta Chim Hung 1985, 120, 291–301.

[76] Matusiewicz H. Direct determination of normal physiological concentrations of major, minor and trace elements in undiluted microlitre volumes of human body fluids by discrete nebulization and atomic emission spectrometry with a nitrous oxide/acetylene flame. Anal. Chim Acta 1988, 2017, 349–54.

[77] Matusiewicz H. A novel sample introduction system for microwave-induced plasma optical emission spectrometry. Spectrochim Acta, Part B 2002, 57, 485–94.

[78] Rosales AT. The determination of copper from copper flotation products using low sensitivity resonant lines. At Absorption Newsl 1976, 15, 51–52.

[79] Moreiskaya LV, Nemodruk AA, Substoichiometry in atomic absorption spectrophotometry. Determination of nickel in nickel and copper-nickel alloys. Zh Anal Khim 1980, 35, 1310–3.

[80] Manning DC. The determination of aluminum by atomic absorption spectroscopy. At. Absorption Newsl 1964, 3, 84–88.

[81] König P, Schmitz KH, Thiemann E. Determination of aluminum in low and high alloy steels in ores, and slags by atomic absorption. At Absorption Newsl 1970, 9, 103–5.

[82] Bosch H, Büschel E, Lohau KH. Schnellbestimmung des säurelöslichen Aluminiums in Stählen. Analysentechnische Berichte 1970, 20.

[83] Shaw F, Ottaway JM. Determination of trace amounts of aluminium and other elements in iron and steel by atomic absorption spectrometry with carbon furnace atomisation. Analyst 1975, 100, 217–20.

[84] Persson JA, Frech W, Cedergren A. Determination of aluminium in low-alloy and stainless steels by flameless atomic absorption spectrometry. Anal Chim Acta 1977, 89, 119–122.

[85] Persson JA, Frech W, Cedergren A. Investigations of reactions involved in flameless atomic absorption procedures. Part IV A theoretical study of factors influencing the determination of aluminium. Anal Chim Acta 1977, 92, 85–90.

[86] Gregorczyk S, Wyciślik A. Analiza chemiczna wysokostopowych stali manganowo-aluminiowych metodą absorpcji atomowej. Wiadomości Hutnicze 1979, 35, 261–63.

[87] Schinkel H. Störungen der Flammen-ES und -AAS Ursachen Auswirkungen Gegenmassnahmen. Ottweiler Druckerei und Verlag, Berlin, 1997.

[88] Wyciślik A. Oddziaływanie żelaza na pomiary stężeń niobu i tantalu płomieniową metodą absorpcji atomowej. Hutnik-Wiadomości Hutnicze 2006, 73(3), 70–73.

[89] Perina L. Analýza karbidických izolátu metodou "OES-ICP". Hutnicke Listy 1986, 1, 50–52.

[90] Wyciślik A. Kilka praktycznych wskazówek pomocnych w analizie chemicznej stali narzędziowych metodą absorpcji atomowej. Hutnik-Wiadomości Hutnicze 2005, 72(12), 593–8.

[91] Dziedzic K, Wyciślik A. Oddziaływanie mikrododatków cyrkonu na strukturę stopów na osnowie niklu. Inżynieria Materiałowa 1982, 3, 322–7.

[92] Wyciślik A. Dobór kwasów do roztwarzania izolatów faz węglikowych. Hutnik-Wiadomości Hutnicze 2008, 75(9), 556–9.

[93] Wyciślik A. Sposób wzorcowania w płomieniowej metodzie absorpcji atomowej z wykorzystaniem jednego wzorca podstawowego na przykładzie stopu Ti-Al-, Rudy i Metale Nieżelazne 2004, 49, 189–92.

[94] Wyciślik A, Richter J. Dissolution of carbide isolates in acids using microwave technique. Solid State Phenomena 2014, 92, 39–44.

[95] Welcher GG, Kriege OH. The determination of major alloying elements in high-temperature nickel-base alloys by atomic absorption spectrophotometry. At Absorption Newsl 1969, 8, 97–101.

[96] Wyciślik A. The use a single specific standard for the determination of chemical composition of cobalt – base alloys by atomic absorption spectrometry method. J Eng 2007, V, 148–51.

[97] Matusiak J, Rams B. Emission of dust and gases in tubular cored wire welding of steel. Int J Occup Safety Ergon 2003, 9(3), 333–50.

[98] Matusiak J. Wpływ warunków technologicznych spawania łukowego stali nierdzewnych na toksyczność pyłu spawalniczego (The influence of technological conditions of arc welding of stainless steel on welding fume toxicity), Poland, Silesian University of Technology, Katowice, 2007, PhD thesis.

[99] Hewitt P. Occupational health in metal arc welding. Welding World, 1999, 43(5), 12–19, (IIW Doc. VIII-1816-97).

[100] International Agency for Research on Cancer: Monographs on the Evaluation of Carcinogenic Risk to Humans, Vol. 49. Chromium, Nickel and Welding, France, Lyon, 1987–2002.

[101] Jordan DE. Safety and welding of stainless steels, EUROJOIN 4, 4th European Conference on welding and cutting. "Development of welding and allied processes at the beginning of the new millennium", Cavtat-Dubrovnik, 2002, 137–44.

[102] Voitkevich V. Welding Fumes. Formation, Properties and Biological Effects. Abington Publishing, Cambridge, 1995.

[103] Mendez PF, Jenkins NT, Eager TW. Effect of electrode droplet size on evaporation and fume generation in GMA. In: The 21st Century Conference – Proceedings of the gas metal arc welding, Orlando, 2000, 53–9.

[104] Redding C. Fume model for gas metal arc welding. Weld J, Welding Research Supplement 2002, 6, 95–103.

[105] Jenkins NT, Pierce W, Eager TW: Particle size distribution of gas metal and flux cored arc welding fumes. Weld J, Welding Research Supplement 2005, 10, 156–63.

[106] Jenkins NT, Eager TW. Chemical analysis of welding fume particles. Weld J, Welding Research Supplement 2005, 7, 87–93.

[107] Dennis J, Hewitt P, Redding C, Workman D. A model for prediction of fume formation rate in gas metal arc welding (GMAW), globular and spray models, DC electrode positive. Ann Occup Hyg 2001, 45(2), 105–13.

[108] Heile RF, Hill DC. Particulate fume generation in arc welding processes. Weld J, 1975, 54(7), 201–10.

[109] Gray CN, Hewitt PJ, Dare PR. New approach would help control welding fumes at source (MIG, MAG), part 2: MIG fumes. Weld Metal Fab 1982, 11, 393–7.

[110] Gray CN, Hewitt PJ, Dare PR. Control of particulate emissions from electric arc welding by process modification. Ann Occup Hyg 1982, 25, 431–8.

[111] Matusiak J, Wyciślik A. Methodology and laboratory testing stands for determination the welding fumes due to arc welding of steels. J Eng 2009, VII(1), 21–6.

[112] Matusiak J, Wyciślik A. Determination of nickel in welding fume by gravimetric method using dimethylglyoxime. Solid State Phenom 2015, 226, 7–10.

[113] Matusiak J, Wyciślik J, Wyciślik A. Environmental criteria for shielding gas selection turing arc welding of stainless steels. Solid State Phenom 2016, 246, 275–8.

[114] Matusiak J, Wyciślik A. Wpływ pyłu spawalniczego i gazów wydzielających się w procesach spawalniczych na organizm człowieka. Środowisko i Rozwój 2009, 19(1), 154–70.

[115] Matusiak J, Wyciślik A. The influence of technological conditions on the emission of welding fume due to welding of stainless steels. Metalurgija (Metallurgy) 2010, 49(4), 307–11.

[116] Matusiak J, Wyciślik A. Speciation and speciation analysis of chromium in welding dust (in preparation).

[117] Gajdzik B, Wyciślik A. Kompleksowa obsługa klienta w laboratorium chemicznym - implikacje systemowe, proceduralne i narzędziowe. Przem Chem 2011, 90(8), 1467–72.

[118] Gajdzik B, Wyciślik A. Trójsegmentowa analiza kompetencji w chemicznym laboratorium badawczym. Ujęcie personalne, techniczno-metodologiczne i organizacyjne. Przem Chem 2012, 91(6), 1119–23.

[119] Gajdzik B, Wyciślik A. Assessment of environment aspects in a metallurgical enterprises. Metalurgija (Metallurgy) 2012, 51(4), 537–40.

[120] Gajdzik B, Wyciślik A. Planowanie i realizacja inwestycji w chemicznym laboratorium badawczym – wytyczne do kompleksowego zarządzania projektami na przykładzie zakupu aparatury i urządzeń laboratoryjnych. Przem Chem 2012, 91(10), 1002–9.

[121] Gajdzik B, Wyciślik A. Wprowadzenie do problematyki crossingu stanowiskowego w chemicznych laboratoriach badawczych. Przem Chem 2013, 92(7), 1000–3.

[122] Gajdzik B, Wyciślik A. Założenia do konstruowania systemów wiedzy w chemicznych laboratoriach badawczych. Przem Chem 2010, 89(2), 86–90.

[123] Gajdzik B. Kwalifikacje personelu w karcie stanowiskowej pracowników chemicznych laboratoriów badawczych z uwzględnieniem wymogów systemu zapewnienia jakości. Przem Chem 2008, 87(11), 2–8.

[124] Gajdzik B, Wyciślik A. Etyka w funkcjonowaniu chemicznych laboratoriów badawczych. Przem Chem 2015, 94(9), 1001–6.

[125] Gajdzik B, Wyciślik A. Ramowe zasady implementacji metody 5S w laboratorium chemicznym. Przem Chem 2016, 95(2), 1000–3.

[126] Wyciślik A, Gajdzik B. Usystematyzowanie ramowych zasad bezpiecznej pracy w chemicznych laboratoriach badawczych i przemysłowych. Przem Chem 2017, 96(3), 482–6.

[127] Eaton A, Schiemer E. A simple Teflon sampling manifold for use with small injections. At Absorption Newsl 1978, 17, 113–4.

Part III: Inorganic and Bioinorganic Speciation Analysis at Trace Level

Ewa Bulska

8 Quality of Results in Trace Element and Speciation Analysis

8.1 Introduction

Trace element determination as well as speciation analysis requires quality control (QC), which becomes a demanding task for analytical laboratories. In order to achieve the expected quality of analytical results, the harmonization of chemical measurements and technical specification required for validation of analytical procedures have to be established. It is worth to stress that analytical results are used in a number of areas, such as biology, geology, medicine, microbiology, mineralogy, ecology, pharmacy, and toxicology. This is of special importance when considering trace elements and speciation analysis, mainly because of the luck of fully validated procedures and availability of certified reference materials (CRM) for the large variety of matrix. It should be stressed that reference materials (RMs), especially CRMs, are one of the major tools for laboratories to establish the traceability of the results to monitor the performance of their analytical work [1].

Although it is difficult to evaluate accurately the real impact of chemical measurements on all aspects of economic and social activities, it is clear that they are playing an increasingly important role in decision-making at the official, legal, or private level. It has therefore been recognized by those who need analytical data, and in particular, by those interested in environmentally related investigations that the quality of the analytical data should be guaranteed. Clearly, it is important to deliver accurate results and to be able to show that they are correct. The importance of quality assurance (QA) and QC is therefore well established and accepted in analytical chemistry [2, 3].

The need to determine particular chemical forms (species) of elements rather than their total contents has been highlighted decades ago, which was mainly linked with identified toxicity risks to environment and humans. As in any other measurements, speciation analysis requires the availability of suitable RMs for the purpose of verification of accuracy and QA needs. Measurements that were considered as a primary interest in speciation were related to the determination of methylmercury in biological tissues and organotins in environmental matrices. In this respect, the issue of defining the target species in between the variety of the chemical compounds being present in nature and their stability during sampling as well as sample processing caused a special attention of analytical chemists [4].

https://doi.org/10.1515/9783110366730-009

8.2 General aspects of QA and QC

QA is defined as all planned and systematic actions, implemented within a management system and demonstrated as required, that are deemed necessary to engender confidence that a product, process, or service will fulfil the given requirement for quality. In this respect, "quality" is regarded as the totality of features and characteristics of a product or service (in analytical chemistry this means the results delivered by a laboratory) that bear on its ability to satisfy the stated or implied needs of customers. In the particular case of chemical measurements, this could be expressed as follows: QA covers all the actions undertaken for planning the proper execution of the analytical task in order to obtain accurate and precise measurements.

As mentioned earlier, chemical measurements are essential in different fields, for example, environmental protection, geology, medicine, and biology. Important decisions are often based on these measurements, for example, whether environmental compartments are polluted or not, food can be eaten, goods can be sold, a patient should be treated, and also in support of legislation (related to health care and trade), production processes, and social problems. This underscores not only the importance of the chemical measurements themselves but also needs to guarantee their validity to their users by assuring the quality of the results.

To achieve the required quality, the chemist should be involved from the beginning of the process, when the needs of the users of results are defined, until the final report is delivered. In practice, therefore, the analytical chemist has to be consulted at every stage of the process sample selection, sample storage, and transportation procedures, the parameters to be analysed, and the level of accuracy and precision necessary for an adequate response to be given.

It is now internationally recognized that for any laboratory to produce reliable data, appropriate scheme of QA must be implemented. As a minimum, this must ensure that the laboratory is using methods that have been validated as fit for the purpose before their application to a specific task. These methods should be fully documented, staff should be trained, the laboratory infrastructure should be appropriate to the measurements to be made, and the mechanisms ensuring that the procedure is under statistical control should be present. A quality management system in place should include: validation of methods, uncertainty evaluation, the use of primary standards and CRM. Participation in interlaboratory comparison (ILCs) and proficiency testing (PT), all serve to assure and demonstrate the quality of measurements.

QA system describes the overall measures that a laboratory uses to ensure the quality of its operation. Typical items include suitable equipment, trained and skilled staff, documented and validated methods, calibration requirements, standards and RMs, traceability, internal QC, PT, non-conformance management, internal audits, and statistical analysis.

8.3 Validation of analytical procedure

Validation of analytical procedure is regarded as one of the most important issues of QA. Before selecting the measurement procedure (analytical method) for a particular purpose, the laboratory should consider its experience, the technical infrastructure at its disposal, and the expected time frame and financial outlay. Validation of the analytical procedure provides necessary information on its performance characteristics and raises the confidence of users in the results.

According to ISO/IEC 17025, validation is confirmation by examination and provision of objective evidence that the particular requirements for a specific intended use are fulfilled. Performance parameters can be divided into two groups. The first group refers to the properties of the measurement procedure: detection limit and determination limit, working range, linearity, and sensitivity. The second group covers the properties of the results obtained with this particular measurement procedure, that is, traceability and uncertainty (including recovery, robustness, precision, and accuracy) [5–9].

Full validation of an analytical method usually comprises an examination of its characteristics in inter-laboratory method performance studies. However, before a method is subjected to validation by collaborative studies, the method must be validated by a single laboratory, usually by that laboratory that developed or modified this particular measurement procedure. Method validation cam be described as the set of tests used to establish and document the performance characteristics of a method and against which it may be judged, thereby demonstrating that the method is fit for a particular analytical purpose. There are two approaches to single laboratory method validation: the traditional one that identifies and then evaluates the set of analytical parameters, and a more recent one that is based on the evaluation of uncertainty.

Validation should cover the whole analytical procedure – from the preparation of the laboratory sample to the evaluation of the result, that is, the whole range of intended matrices, and should be performed within the expected range of concentrations. The intended use of the analytical results should also be considered. Several techniques can be used for validation, the most highly recommended ones being (i) evaluation of uncertainty (i.e. a systematic assessment of the quantities influencing the result); (ii) performing CRM analysis; (iii) participation in ILCs/PTs; and (iv) comparison of results with other analytical methods. When fully validated methods are available, the analyst can envisage starting a statistical control system, including the follow-up of performance with the aid of control charts.

If the laboratory develops the validation method in-house, there always needs to be some sample to be used for this purpose; a sample that best mimics routine samples is the most suitable. The usual practice is that a routine sample is used for this purpose as knowledge of the true value is not critical issue at this stage. Next, the trueness of a method is usually determined by analysing an appropriate CRM and/or participating in an ILC, one with an externally defined reference value [10].

Table 8.1: Selected examples of reference materials certified for their contents of total content of elements and their particular chemical forms of elements.

CRM identity	Certified parameters and matrices	Producer
SRM 2108	CR(III) and Cr(VI) in solution	NIST
SRM 2109	CR(III) and Cr(VI) in solution	NIST
CRM 544	CR(III) and Cr(VI) in lyophilized solution	BCR
CRM 545	Cr(VI) in welding dust (loaded on filter)	BCR
SRM 1974a	Total mercury and methylmercury in mussel tissue	NIST
SRM 2974	Total mercury and methylmercury in mussel tissue	NIST
SRM 2976	Total mercury and methylmercury in mussel tissue	NIST
DORM-1	Methylmercury in fish muscle (dogfish)	NRCC
CRM 463	Total mercury and methylmercury in fish muscle (tuna)	BCR
CRM 464	Total mercury and methylmercury in fish muscle (tuna)	BCR
LUTS-1	Trace elements and methylmercury in lobster tissue	NRCC
CRM 580	Trace elements and methylmercury in sediment	BCR
IAEA 356	Methylmercury in sediment	IAEA
IAEA 086	Total mercury and methylmercury in human hair	IAEA
PACS-1	Butyltin compounds in marine sediment	NRCC
CRM 462	Butyltin compound in coastal sediment	BCR
CRM 646	Butyl- and phenyltins in freshwater sediment	BCR
CRM 477	Butyltin compounds in mussel tissue	BCR
NIES 11	Total tin and tributyltin compounds in fish tissue	NIES
CRM 627	Organoarsenic compounds in fish muscle	BCR
CRM 605	Trimethyl lead in urban dust	BCR

BCR: Community Bureau of Reference; NIST: National Institute of Standards and Technology; NRCC: National Research Council Canada; IAEA: International Atomic Energy Agency; NIES: National Institute for Environmental Studies.

It should be stressed that in between all recommended methods used for validation, in the case of trace elements and its speciation analysis, RMs play a crucial role at all stages of method validation (development, optimization, and QC). The solution of pure chemical substances could be used as a standard for evaluating performance criteria of the analytical method, such as sensitivity, selectivity, linearity, detection, and quantification limits. The standard must be of the same character as the measured quantity. Extract or matrix-matched solutions may be used for the same purpose, but also to detect possible interferences and check yields of chemical reactions (e.g. derivatization). Finally, matrix RMs will be used for checking performance criteria such as accuracy (trueness, repeatability, and reproducibility) and the application range of the method (Table 8.1).

8.4 Traceability of analytical results

Worldwide acceptance of analytical results requires reliable, traceable, and comparable measurements. A key property of reliable results is their traceability to stated references. Traceability basically means that a laboratory knows what is being measured and how accurately it is measured. It is also an important parameter where comparability of results is concerned and is usually achieved by linking the individual result of chemical measurements to a commonly accepted reference or standard. The result can therefore be compared through its relation to that reference or standard.

Every link in the traceability chain must be based on the comparison of an unknown value with a known value. The stated reference might be an International System of Units (SI) or conventional reference scale such as the pH scale, the delta scale for isotopic measurements, or the octane number scale for petroleum fuel. In order to be able to state the uncertainty of the measurement result, the uncertainty of the value assigned to that standard must be known. Therefore a traceability chain should be designed and then demonstrated using the value of the respective standard with its uncertainty. In chemical measurements, traceability of results to SI units is not always comparability of results between laboratories. During the validation of the analytical procedure, traceability of the results can be demonstrated by comparison against the certified value of a CRM which provides exactly this traceable assigned value with a stated uncertainty [11, 12].

The example of CRM with the certified values for selected trace elements (for their total content) as well as arsenic compounds in rice flour is given in Table 8.2.

8.4.1 Reference materials

8.4.1.1 Pure substances and calibrating materials

RMs consist of pure substances or solutions for use in calibration and/or identification of given parameters, or aimed at testing part or totality of an analytical procedure (e.g. raw or purified extracts, spiked samples, etc.). To name a few of such calibrants for the purpose of QC of speciation, pure organotins, methylmercury chloride and trimethyl lead were used for the certification campaigns. They were not aimed at producing

Table 8.2: The list of analytes certified in white rice flour.

Specification	Result	Analyte
Dimethylarsinic acid (DMAA as As)	(0.0122 ± 0.0009) mg/kg	Arsenic
Inorganic arsenic; arsenate As(V) as As (arsenic acid as As)	(0.0130 ± 0.0009) mg/kg	Arsenic
Inorganic arsenic; arsenite As(III) as As (arsenic trioxide as As)	(0.0711 ± 0.0029) mg/kg	Arsenic
Total element concentration	(0.098 ± 0.007) mg/kg	Arsenic
Total element concentration	(0.194 ± 0.007) mg/kg	Cadmium
Total element concentration	(1.88 ± 0.07) mg/kg	Copper
Total element concentration	(5.42 ± 0.21) mg/kg	Iron
Total element concentration	(9.2 ± 0.4) mg/kg	Manganese
Total element concentration	(20.7 ± 0.9) mg/kg	Zinc

pure substances for commercial purpose, but rather to ensure a firm comparability basis for measurements performed by laboratories participating in corresponding certifications, that is, for verifying that no systematic errors due to calibration were left undetected. Other examples are fish extracts (either raw or spiked with MeHg) and urban dust spiked with trimethyl lead used to test the performance of separation and detection methods prior to certification work. Pure substances have also been prepared to check derivatization reactions in the framework of certification of organotin compounds, that is, secondary organotin calibrants (e.g. ethylated, pentylated forms of butyltin compounds) for verifying derivatization yields. Other materials, of known composition, are used for calibrating certain types of measurement instruments. Matrix-matched RMs are prepared gravimetrically (by specialized laboratories), and examples are known for speciation, for example, synthesis of pure arsenobetaine and preparation/certification of pure calibrating solution of this compound [13, 14].

8.4.1.2 Matrix RMs

Matrix-matched RMs should represent as much as possible the matrix analysed by the laboratory. Due to requirements for homogeneity and stability of an RM, it is almost not possible to exactly match the matrix of the origin sample. It should be however expected that the deviation that is introduced to the matrix during the production of RMs is not too significant. But one has always be aware that even a matrix-matched RM in the best case only comes close to a sample of the respective matrix [15–18].

Several requirements have to be fulfilled for the preparation of matrix RMs, which have to be representative of currently analysed samples, homogeneous and stable over long-term storage, which is of special importance with respect to speciation analysis. In practice, selected preparation procedures were elaborated to each type of samples aiming to be fit for the purpose of the analytical work.

Sound conclusions on the performance of an analytical method can be delivered whenever RMs are used with a composition as close as possible to those of the samples routinely analysed by the laboratory. This means that a matrix RM should, in principle, pose similar difficulties, that is induce the same sources of error, to those encountered when analysing real samples. Requirements for the representativeness of an RM imply in most cases a similarity of matrix composition, concentration range of substances of interest, binding states of the analytes, occurrence of interfering compounds, and physical status of the material. The material should be homogeneous and stable to guarantee that the samples provided to the laboratories are similar, and compromises often have to be made at the stage of preparation to comply with this requirement, which is particularly acute for speciation measurements [19–24].

Several requirements have to be fulfilled for the preparation of matrix RMs, which have to be representative of currently analysed samples, homogeneous and stable over long-term storage (Table 8.3).

Basically, matrix RMs may represent all kinds of natural "matrices" that are currently analysed by laboratories. For speciation studies, focus nowadays is towards the

Table 8.3: Example reference materials with the information on their stability at various conditions.

Matrix	Chemical species	Collection, pre-treatment, and collected amounts
Coastal sediment	DBT and TBT	Stability verified for DBT and TBT from −20 °C to +20 °C over 12 months. Unstable at +40 °C after 3 months and at +20 °C after 24 months
Freshwater sediment	Phenyltins	Stability verified at −20 °C over 12 months. Unstable at +20 °C and +40 °C (also tested over 12 months)
Mussel tissue	Organotins	Stability verified for MBT, DBT and TBT at −20 °C over 44 months. Unstable at +20 °C and +40 °C, and to a lesser extent at +4 °C. For phenyltins, unstable at −20 °C and above
Sediment	Methylmercury	Stability verified from −80 °C to +40 °C over 15 months
Fish tissue	Methylmercury	Stability verified from −20 °C to +40 °C over 12 months
Fish tissue	As species	Stability verified from −20 °C to +40 °C over 9 months
Aqueous solution	Asbetaine	Stability verified from −20 °C to +40 °C over 9 months
Urban dust	Trimethyl lead	Stability verified from +4 °C to +20 °C over 37 months. Instability +40 °C.
Artificial rainwater	Trimethyl lead	Stability verified from +4 °C to +20 °C over 24 months. Unstable at these temperatures after 37 months, and at +40 °C after 1 month
Welding dust on filter	Cr(VI)	Stability verified from +5 °C to +20 °C over 12 months
Lyophilized solution	CR(III) and Cr(VI)	Stability verified from +5 °C to +20 °C over 12 months
Aqueous solution	Se(IV) and Se(VI)	Stability verified at +20 °C over 12 months but instability shown after 24 months
Soil and sediment	Extractable trace element contents	Stability verified from −20 °C to +40 °C. Instability suspected at +40 °C. Extractability changes suspected at −20 °C, stressing the need to carry out stability studies at +4 °C instead of −20 °C

Note: DBT, dibutyltin; TBT, tributyltin.

determination of chemical forms in biological and environmental matrices. Therefore, currently available CRMs concerns matrices such as biological tissues (mussel, fish, lobster, etc.), sediments (freshwater, coastal, and marine) and soils of various compositions, which are certified for a range of chemical forms of elements (e.g. As, Hg, Pb, and Sn) [25, 26]. Details on the available matrix CRMs are given in the following sections.

Among the matrix RMs, a particular category concerns RMs related to operationally defined parameters. In this case, the assigned or certified values are directly linked to a specific method, following a strict analytical protocol. With respect to speciation of trace elements, this type of measurement is referred to as "fractionation". It does not correspond to the determination of chemical forms of elements sensu stricto but rather to the extraction specified procedures, of which results are used for the definition of forms such as mobile, bioavailable, carbonate bound, and so on. Thus, the obtained results refer to the extractable contents of trace metals or element contents determined following a sequential extraction scheme, and these measurement results will be interpreted to correspond forms. Whatever the interpretation given, the only possibility of achieving comparability of measurements based on extraction methods is to use harmonized and adopted by laboratories or standardized methods. CRMs based on harmonized extraction procedures have been produced for soil and sediment analyses and are the only examples to date for ensuring comparability of data in the specific "speciation" field [27–34].

For some chemical forms of elements a definitive methods are available, for example, isotope dilution mass spectrometry (IDMS) [35–39]. The certification of matrix RM using exclusively such a definitive method would not give the user, who does not

Table 8.4: Uncertainties of certified values for chemical forms of elements on selected CRMs.

Matrix/CRM	Chemical species	Certified value ± uncertainty (95% CI)	CV (%)
Sediment/CRM 462	TBT	54 ± 15 µg/kg as TBT	27.8
	DBT	68 ± 12 µg/kg as DBT	17.6
Sediment/BCR-646	TBT	491 ± 65 µg/kg as TBT	13.2
	DBT	770 ± 117 µg/kg as DBT	15.2
	MBT	674 ± 102 µg/kg as MBT	15.1
	TPhT	35 ± 6 µg/kg as TPhT	17.1
	DPhT	38 ± 10 µg/kg as DPhT	26.3
	MPhT	74 ± 22 µg/kg as MPhT	29.7
Mussel tissue/CRM 477	TBT	2.20 ± 0.19 mg/kg as TBT	8.6
	DBT	1.54 ± 0.12 mg/kg as DBT	7.8
	MBT	1.50 ± 0.27 mg/kg as MBT	18.0
Fish tissue/CRM 463 CRM 464	Methylmercury	3.04 ± 0.16 mg/kg as MeHg	5.3
		5.50 ± 0.17 mg/kg as DMeHg	3.1
Sediment/CRM 580	Methylmercury	75.5 ± 3.7 µg/kg as MeHg	4.9
Urban dust/CRM 605	Trimethyl lead	7.9 ± 1.2 µg/kg as TML	15.2
Welding dust/CRM 545	Cr(VI)	40.16 ± 0.56 mg/kg welding dust (as Cr)	1.4
Lyophilized solution/CRM 544	Cr(III)	22.8 ± 1.0 µg/L as Cr	4.4
	Cr(VI)	26.8 ± 1.0 µg/L as Cr	3.7
Fish tissue/CRM 627	Asbetaine	51.5 ± 2.1 mmol/kg as Asbet	4.1
	DMA	2.04 ± 0.27 mmol/kg as DMA	13.2
Aqueous solution/CRM 628	Asbetaine	5.77 ± 0.03 mmol/kg as Asbet	0.5

Note: CRM, certified reference materials; CI, confidence interval; CV, coefficient of variance; TBT, tributyltin; DBT, dibutyltin; MBT, monobutyltin; TPhT, triphenyltin; MPhT, monophenyltin; DPhT, diphenyltin; DMA, dimethylarsinic acid.

Table 8.5: Certified contents of chemical species in various BCR CRMs.

CRM	Compound	Certified content
Tuna fish CRM 463	MeHg	3.04 ± 0.16 mg/kg
Tuna fish CRM 464	MeHg	5.50 ± 0.17 mg/kg
Sediment CRM 580	MeHg	75.5 ± 3.7 µg/kg
Sediment CRM 462	TBT	70 ± 14 µg/kg as TBT
	DBT	128 ± 16 µg/kg as DBT
Mussel CRM 477	TBT	2.20 ± 0.19 mg/kg as TBT
	DBT	1.54 ± 0.12 mg/kg as DBT
	MBT	1.50 ± 0.28 mg/kg as MBT
Urban dust CRM 605	TML	7.9 ± 1.2 µg/kg as TML
Tuna fish CRM 627	DMA	52 ± 3 mmol/kg as DMA
	AsB	2.0 ± 0.3 mmol/kg as AsB
Solution CRM 626	AsB	5.77 ± 0.037 mmol/kg
Lyophilized solution CRM 544	Cr^{III}	26.8 ± 1.0 µg/kg as Cr^{III}
	Cr^{VI}	22.8 ± 1.0 µg/kg as Cr^{VI}
Welding dust CRM 545	Cr^{VI}	40.2 ± 0.6 µg/kg

AsB, arsenobetaine; DBT, dibutyltin; DMA, dimethylarsinic acid; MBT, monobutyltin; MeHg, methylmercury; TBT, tributyltin; TML, trimethyl lead.

apply this technique in routine work, a good estimate of the uncertainty, as it is always much lower than that of more common methods [40–42].

Although definitive methods such as IDMS are used to obtain results of high metrological values, its applicability is limited with respect to the types of matrices and parameters that may be certified. These techniques do not yet exist for certification of organometallic compounds (mainly because isotopically labelled organometallic species are not yet commercially available) [43].

The uncertainty of the certified values is critical for establishing the uncertainty budget in the laboratory as well as for the comparison of the results. This depends not only on the type of matrix but also on the type of chemical form of an element of interest in a given matrix. The example of the variety of the uncertainty values in shown in Table 8.4

The variety of the chemical species of mercury, tin, lead and arsenic together with their certified value accompanying with uncertainty are also listed in Table 8.5. It should be noted that in some cases, more than one species were certified in a given material [44, 45].

8.5 Monitoring of the quality of analytical results

Achieving quality (and being able to demonstrate this) on a daily basis is one of the most difficult tasks for a routine laboratory. This holds true particularly for speciation analysis, due to the problems of low concentration levels, complex matrices, instability of analytes, complicated sample preparation schemes, and risks of contamination. In analytical laboratories that are expected to deliver results with a defined level of accuracy and uncertainty, QC basically involves examining at regular intervals whether the QA system was well designed and executed in such a way as to fulfil the requirements over time. In practice, in accordance with ISO/IEC 17025, the laboratory undertakes QC procedures for monitoring the validity of a test. The resulting data are recorded in such a way that trends are detectable and, where practicable, statistical techniques are applied to the reviewing of the results. Monitoring is planned with respect to the frequency of performing QC measurements and reviewed in order to assure quality over time. Monitoring of the QC process should include the regular use of CRMs or RMs and replicate tests for internal QC, as well as participation in ILC schemes for external quality assessment [46–48].

QC activities mean comparison of results and their uncertainties with quality criteria and/or reference data, and typically are done by

- the use of control charts of the results obtained for RMs, CRMs, in-house control samples, blanks, and so on;
- regular checking of the instrumental performance by calibration or adjustments;
- checking on the purity and stability of reagent and solutions used;
- examination of the repeated results obtained by the same procedure on the same sample in order to examine the influence of any factor on the results.

It is generally recommended that only validated methods are used for speciation analysis [49]. This requirement is, in practice, difficult to comply with, since the influence of the sample matrix on the analytical process is of such over-riding importance that, in principle, the application of a method to a new matrix requires its re-validation. If a complete validation cannot be done, at least the determination of the recovery has to be performed (even if the quantification is done by the standard addition technique). In exceptional cases, it may be acceptable to use external calibration when the results are related to suitable internal standards. These internal standards are spiked in an appropriate way to the matrix (in which they must not be present initially) and undergo the entire sample preparation procedure. For monitoring the stability of reagents of calibrants or for documenting blank values and instrument state, the use of control charts is highly recommended. Control charts plot the result of an analysis as a function of the date or series of analyses. Those used most commonly have a central line (equal to the mean on the analysis results, determined in a pre-period) and, at a distance of 2 standard deviations from this central line the warning limits and $\overline{X} \pm 3$ s, the control limits. An analytical process that is under control will mostly produce results that fall within the interval $\overline{X} \pm 2$ s. If results repeatedly fall in the range between warning and control limits of the control chart, or if they exceed the control limits, there is a reason to believe that the process is out of control (there are a number of exactly defined conditions when situation is considered out of control).

In a case when an out-of-control situation has been encountered, routine analysis should be discontinued until the cause for the deviating results has been identified and eliminated.

Control charts may be used for a number of purposes. The most common use is that of monitoring the analysis result (mean or single value) of a suitable control sample over time. Depending on the nature of the control sample, either the correct operation of the instrument can be demonstrated (e.g. if the control sample is a pure calibrant), or the success of the entire process, if the control sample is a real matrix sample. Control charts may also be used to monitor procedure blanks or recoveries of the entire sample preparation procedure. While in most cases, either simple values or the mean from *n* repetitive measurements are plotted in a mean value chart, and control charts can also be designed to monitor the range of the standard deviation of repetitive measurements. Out-of-control situations can thus not only be detected when there is a significant bias of the result but also when the standard deviation becomes unacceptably large.

For the purpose of monitoring both the instrument and method performance, control samples are used. Suitable control samples may be CRMs, RMs, or laboratory control materials (LCMs) [50]. In order to be suitable, the used materials have to satisfy the following conditions:

- *Stability*: the material must not change its composition over a long period of time under defined storage conditions (e.g. at −20 °C), since the intention is to be able to monitor the method performance for an extended period and to ensure that results that have been measured with long time intervals are still comparable.
- *Availability*: for the same reasons as mentioned earlier, the control material should be available in sufficient quantity.
- *Homogeneity*: the homogeneity, the concentration level, and the minimum sample intake of a control sample are interrelated; the better the homogeneity of the material is, the smaller the recommended sample intake may be, which also depends of course on the concentration level of the analyte, and between-sample variations due to inhomogeneity should be insignificant in comparison with the measurement uncertainty.
- *Assigned/certified value*: a control material may (but need not) have an assigned value for the analyte that is to be measured. If the assigned value has been obtained through the certification procedure of an internationally recognized body (e.g. BCR, NIST, NIES, or NRCC),[98,99] the material is a CRM. When the certification has not been attempted or has not been concluded successfully, the material is distributed as a (not certified) RM. Since the stability and the homogeneity of this material have usually been demonstrated, it may still be of great value to analytical laboratories. The assigned value, however, is not given in the form of a certified value with its associated uncertainty, but instead as an indicative value or concentration range. LCMs may be considered the lowest level in this hierarchy of

control materials. The most important requirements for LCMs are that they should be stable and sufficiently homogenous for the purpose they are used for. They do not need to have an assigned value, at least not through a formal procedure. Since LCMs may, for example, be the remainder of a previously analysed sample batch that is available in sufficient quantity and for which the homogeneity and the stability have been proven, they represent a low-cost alternative to RMs for use in certain cases.

Control materials can be used for different purposes. First, they may serve to monitor the stability of the measurement system or the method. For this purpose, it is not necessary that the assigned value be known or even certified – in principle, the comparison of the results of different series of analyses may already give an indication whether the analytical system is under control. They can also be used to assess whether a new method yields results that are consistent with those obtained using a previous method. The second use of RMs is to assess the accuracy of the procedure. In this case, it is essential that the assigned concentration value for the control material be precisely known with statement of its uncertainty. Only then it is possible to assess recoveries for the method, which may, for example, be applied to the analysis of samples with a similar matrix. Although this approach is, in principle, acceptable, great care has to be exerted. This also holds true when CRMs are used for the calibration of a method whose systematic errors are either not known, or cannot be mastered, but the use of a matrix-matched calibrant of known concentration. Since CRMs are usually not available with analyte concentration at different levels, a one-point calibration is performed. However, particularly in complex matrices, the assumption of a one-point calibration line going through the origin is not always justified. First, recoveries may be different at different concentration levels, which leads to a curvature of the calibration at different concentration levels which in turn to a curvature of the calibration line, and also a one-point calibration is only possible in the absence of any constant, systematic errors.

8.6 Conclusions

The quality of results of trace elements and speciation analysis becomes an important issue in many areas of performing the chemical measurements. The main pillar with the process of QA and QC is validation of an analytical procedure, traceability, as well as uncertainty of the results. In order to execute those requirements, the use of CRMs is recognized as a most important tool in the laboratory practice. This is however not a trivial issue, mainly because of the difficulties in respect of defining the analyte that should be measured, as well as the stability of the particular species. Moreover, the homogeneity of the materials could hamper the usefulness of RM. Nevertheless, the number of well-established CRMs is available for the purpose of trace element determination and speciation analysis, which give solid sounds for performing accurate results of chemical measurements for speciation studies at trace level.

References

[1] Neidhart B, Wegscheider W, eds. Quality in Chemical Measurements. Springer, Berlin, 2001.

[2] Kellner R, Mermet JM, Otto M, Widmer HM, eds. Analytical Chemistry. Wiley-VCH, Weinheim, 1998.

[3] Otto M. Chemometrics: Statistical and Computer Application in Analytical Chemistry. Wiley-VCH, Weinheim, 1999.

[4] Robouch P, Bulska E, Duta S, Lauwaars M, Leito I, Majcen N, Norgaard J, Suchanek M, Vassileva E, Taylor P. TrainMiC-Training in Metrology in Chemistry. European Communities, Luxemburg, 2003. EUR Report 20841 EN.

[5] ISO/IEC 17025. General Requirements for the Competence of Testing and Calibration Laboratories. ISO, Geneva, 2005.

[6] ISO 5725 Part 1-6. Accuracy (Trueness and Precision) of Measurements Method and Results. ISO, Geneva, 1998.

[7] ISO/IEC Guide 99. International Vocabulary of Metrology – Basic and General Concepts and Associated Terms (VIM). ISO, Geneva, 2007.

[8] Eurachem/CITAC Guide. Quality assurance for research and development and non-routine analysis, 1998.

[9] Eurachem/CITAC Guide. A fitness for purpose of analytical measurements: A laboratory guide to method validation and related topics, 1998.

[10] Bulska E, Taylor, P. Do we need education in metrology in chemistry? Anal Bioanal Chem 2003, 377, 588–9.

[11] King B. The practical realization of the traceability of chemical measurements standards. Accred Qual Assur 2000, 5, 429–36.

[12] Prichard E, Barwick, V. Quality Assurance in Analytical Chemistry. Wiley, Chichester, Hoboken, NJ, 2007.

[13] Lobinski R, Chassaigne H, Szpunar J. Analysis for metallothioneins using coupled techniques. Talanta 1998, 46, 271–289.

[14] Quevauviller P, Morabito R. Evaluation of extraction recoveries for organometallic determinations in environmental matrices, Trends in Analytical Chemistry, 2000, 19, 86–96.

[15] Reinecke H, Dunemann L. Does the supplementation of metal ions modify the intrinsic distribution of element species in food extracts? Fresenius J Anal Chem 1990, 338, 630–634.

[16] Caroli S. Speciation of trace elements in human milk by high performance liquid chromatography combined with inductively coupled plasma-mass spectrometry. Trace Elem Electrolyt 1996, 13, 26–32.

[17] Sunaga H, Suzuki KT. J Liquid Chromatography, Methylation of Residual Carboxyl Groups in Gel Permeation Column and Its Effect on Elution and Distribution of Metals and Proteins in Blood Serum, 1988, 11, 701–711.

[18] Raab A, Bratter P. Separation of metalloprotein complexes in serum by size-exclusion chromatography. Optimisation of the separation parameters retention behaviour and recovery employing radiotracers. J Chromatogr Biomed Sci Appl 1998, 707, 17–24.

[19] Micallef S, Couillard Y, Campbell PGC, Tessier A. An evaluation of the HPLC—Gel chromatographic method for analyzing metallothioneins in aquatic organisms. Talanta 1992, 39, 1073–1079.

[20] Lobinski R, Boutron CF, Candelone JP, Hong S, Szpunar-Lobinska J, Adams FC. Speciation Analysis of Organolead Compounds in Greenland Snow at the Femtogram-per-Gram Level by Capillary Gas Chromatography/Atomic Emission Spectrometry. Anal Chem 1993, 65, 2510–2515.

[21] Tao H, Ramaswamy BR, Quetel CR, Nakazato T, Tominaga M, Miyazaki A. Tin Speciation in the Femtogram Range in Open Ocean Seawater by Gas Chromatography/Inductively Coupled Plasma Mass Spectrometry Using a Shield Torch at Normal Plasma Conditions. Anal Chem 1999, 71, 4208–4215.

[22] Makino Y, Nishimura S. High-performance liquid chromatography separation of human apolactoferrin and monofferic and diferric lactoferrins. J Chromatogra B 1992, 117, 246–349.

[23] Leopold I, Fricke B. Inhibition, reactivation, and determination of metal ions in membrane metalloproteases of bacterial origin using high-performance liquid chromatography coupled on-line with inductively coupled plasma mass spectrometry. Anal Biochem 1977, 252, 277–285.

[24] LaFuente JM, Marchante-Gayon JM, Fernandez Sanchez ML, Sanz-Medel A. Urinary selenium speciation by high-performance liquid chromatography – inductively coupled plasma mass spectrometry: advantages of detection with a double-focusing mass analyser with a hydride generation interface. Talanta 1999, 50, 207–217.

[25] Szpunar J, Lobinski R. Speciation in the environmental field – trends in Analytical Chemistry. Fresenius J of Anal Chem 1999, 363, 550–557.

[26] Lobinski R, Potin-Gautier M. Metals and Biomolecules – Bioinorganic Analytical Chemistry. Analysis 1998, 26, 21–24.

[27] McSheehy S, Marcinek M, Chassaigne H, Szpunar J. Identification of dimethylarsinoyl-riboside derivatives in seaweed by pneumatically assisted electrospray tandem mass spectrometry. Anal Chim Acta 2000, 410, 71–84.

[28] Crews HM, Dean JR, Ebdon L, Massey RC. Application of high-performance liquid chromatography–inductively coupled plasma mass spectrometry to the investigation of cadmium speciation in pig kidney following cooking and in vitro gastro-intestinal digestion. Analyst 1989, 114, 895–899.

[29] Szpunar J, Pellerin P, Makarov A, Doco T, Williams P, Medina B, Lobinski R. Speciation analysis for biomolecular complexes of lead in wine by size-exclusion high-performance liquid chromatography-inductively coupled plasma mass spectrometry. J Anal At Spectrom 1998, 13 749–754.

[30] Takatera K, Watanabe T. Individual and synergistic effects of heavy-metal ions on the induction on cyanobacterial metallothionein examined by high-performance liquid chromatography/ inductively coupled plasma mass spectrometry. Anal Sci 1993, 9, 19–23.

[31] High KA, Methven BA, McLaren JW, Siu KWM, Wang J, Klaverkamp JF, Blais JS. "Physico-chemical characterization of metal binding proteins using HPLC-ICP-MS, HPLC-MA-AAS and electrospray-MS". Fresenius J Anal Chem 1995, 351, 393–402.

[32] Szpunar J, Chassaigne H, Donard O, Bettmer J, Lobinski R, Speciation of Protein-Bound Cadmium in Biological Materials by Size Exclusion and Reversed-Phase HPLC with ICP MS Detection, In Holland G, Tanner S (Eds.), Plasma Source Mass Spectrometry, Developments and Applications, Royal Society of Chemistry, Cambridge, 1997.

[33] Snell JP, Stewart II, Sturgeon RE, Frech W. Species specific isotope dilution calibration for determination of mercury species by gas chromatography coupled to inductively coupled plasma- or furnace atomisation plasma ionisation-mass spectrometry. J Anal At Spectrom 2000, 15, 1540–1545.

[34] Arnold CG, Berg M, Muller SR, Dommann U, Schwarzenbach RP. Determination of Organotin Compounds in Water, Sediments, and Sewage Sludge Using Per-deuterated Internal Standards, Accelerated Solvent Extraction, and Large-Volume-Injection GC/MS. Anal Chem 1998, 70, 3094–3101.

[35] Gallus SM, Heumann KG. Development of a gas chromatography inductively coupled plasma isotope dilution mass spectrometry system for accurate determination of volatile element species. Part 1. selenium speciation. J Anal At Spectrom 1996, 11, 887–892.
[36] Hill SJ, Pitts LJ, Fisher AS. High-performance liquid chromatography–isotope dilution inductively coupled plasma mass spectrometry for speciation studies: an overview. Trends Anal Chem 2000, 19, 120–126.
[37] Krupp EM, Pecheyran C, Pinaly H, Motelica-Heino M, Koller D, Young SMM, Brenner IB, Donard OFX. Isotopic precision for a lead species (PbEt4) using capillary gas chromatography coupled to inductively coupled plasma-multicollector mass spectrometry. Spectrochim Acta 2001, 56B, 1233–1240.
[38] Krupp EM, Pecheyran C, Meffan-Main S, Donard OFX. Precise isotope-ratio measurements of lead species by capillary gas chromatography hyphenated to hexapole Multicollector ICP–MS. Fresenius J Anal Chem 2001, 370, 573–580.
[39] Encinar JR, Garcia Alonso JI, Sanz-Medel A. Synthesis and application of isotopically labelled dibutyltin for isotope dilution analysis using gas chromatography-ICP-MS. J Anal At Spectrom 2000, 15, 1233–1239.
[40] Enciar JR, Villar MIM, Santamaria VG, Alonso JIG, Sans-Medel A. Simultaneous Determination of Mono-, Di-, and Tributyltin in Sediments by Isotope Dilution Analysis Using Gas Chromatography–ICPMS. Anal Chem 2001, 73, 3174–3180.
[41] Brown AA, Ebdon L, Hill SJ. Development of a coupled liquid chromatography-isotope dilution inductively coupled plasma mass spectrometry method for lead speciation. Anal Chim Acta 1994, 286, 391–399.
[42] Ebdon L, Hill SJ, Rivas C. Lead speciation in rainwater by isotope dilution-high performance liquid chromatography-inductively coupled plasma-mass spectrometry. Spectrochim Acta 1998, 9, 289–297.
[43] Quevauviller P, Harrison R, Adams F, Ebdson L. An overview of methods for the determination of trimethyllead in rainwater and urban dust reference materials. J Anal Chem 2000, 19, 195–199.
[44] Rivas C, Ebdon L, Evans EH, Hill SJ. An evaluation of reversed-phase and ion-exchange chromatography for use with inductively coupled plasma—mass spectrometry for the determination of organotin compounds. Appl Organomet Chem 1996, 10, 61–68.
[45] Yang L, Mester Z, Sturgeon RE. Species-Specific Isotope Dilution-Based Calibration for Trace Element Speciation and Its Combined Uncertainty Evaluation: Determination of Tributyltin in Sediment by HPLC–ICPMS. Anal Chem 2002, 74, 2968–2976.
[46] Heumann KG, Rottmann L, Vogl J. Elemental speciation with liquid chromatography – inductively coupled plasma isotope dilution mass spectrometry. J Anal At Spectrom 1994, 9, 1351–1355.
[47] Heumann KG, Gallus SM, Raedlinger G, Vogl J. Accurate determination of element species by on-line coupling of chromatographic systems with ICP-MS using isotope dilution technique, Spectrochim Acta 1998, 53B, 273–287.
[48] Kingston HM, Huo D, Lu Y, Chalk S. Accuracy in species analysis: speciated isotope dilution mass spectrometry (SIDMS) exemplified by the evaluation of chromium species. Spectrochim Acta 1998, 53B, 299–309.
[49] Rottmann L, Heumann KG. Development of an on-line isotope dilution technique with HPLC/ICP-MS for the accurate determination of elemental species. Fresenius J Anal Chem 1994, 350, 221–227.
[50] Quevauviller P. CRMs for quality control of determinations of chemical forms of elements in support to EU legislation. Fresenius J Anal Chem 1996, 354, 515–520.

Beata Krasnodębska-Ostręga, Monika Sadowska and Ewa Biaduń

9 Sample Pretreatment for Trace Speciation Analysis

9.1 Introduction

Speciation analysis is already a challenge, and speciation analysis at trace levels is even more difficult [1]. The analytes might be of anthropogenic origin [2] (usually the concentrations are slightly increased) or naturally occurring (geochemical background) [3]. Procedures applied for speciation analysis must ensure low limits of detection as well as high selectivity, especially when the concentration of other sample compounds, potentially interfering, is much higher than the concentration of the analyte. Very often, there is no other choice but to use separation and/or preconcentration techniques. Methods applied for speciation studies should ensure isolation of the analyte from sample matrix without any changes of the original speciation and with highest possible efficiency [4–8]. The lower is the concentration of the analyte, the higher is the uncertainty of the results, because some additional steps have to be included in the analytical procedure. Deep interference in sample composition results in changes of the oxidation states and chemical compounds formed by the element of interest [9]. In case of water, soil or tissues the risk of changes in chemical speciation is created even by the sampling, not to mention chemical modifications or sample storage. Most published studies regarding speciation analysis were focused mainly on methods of separation and detection of the analytes. Determination is usually done using elemental detectors (inductively coupled plasma mass spectrometry [ICP-MS], ICP optical emission spectrometry [OES], graphite furnace atomic absorption spectroscopy [GF-AAS]) or molecular detectors (electrospray ionization [ESI] or time of flight [TOF] coupled with mass spectrometry [MS]) directly after separation on chromatographic column, or indirectly after separation using, for example, solid phase extraction (SPE).

Considering how fragile are the equilibria between speciation forms, the best solution would be to determine speciation directly in the analyzed object, using techniques that can differentiate between oxidation states. In general, only a few techniques allow performing nondestructive speciation analysis of solids or water samples. For water samples, the direct voltammetry technique could be applied [10] but the limit of quantification (LOQ) is too high for trace element speciation. Voltammetric methods can be proposed for (indirect) speciation analysis of As, Tl, and Se. In water samples containing As(III) and As(V) it is possible to perform speciation analysis based on cathodic stripping voltammetry (CSV) measurements. By varying the composition of the supporting electrolyte it is possible to differentiate between As(III) and As(V). Addition of mannitol into the supporting electrolyte leads to detection

https://doi.org/10.1515/9783110366730-010

of only trivalent arsenic, and then As(III) is transferred to As(V) during ultraviolet (UV) oxidation of organic matter and the total As is determined [11]. Thallium speciation can be studied when glassy carbon working electrode is modified with an ion exchange resin, which allows to selectively accumulate trivalent thallium as a chloride complex. This method was applied to determine Tl speciation in water samples modified by an addition of large amounts of chlorides [12]. Another possibility is modification of the supporting electrolyte: addition of diethylenetriaminepentaacetic acid (DTPA) to inactivate Tl(III) (electrochemically) and addition (directly to the voltammetric cell) of anion exchange resin to adsorb the organic matter. In this case, only monovalent thallium is accumulated [13, 14]. Determination of Se(IV) and selenocysteine was done simultaneously in the aqueous phase using HCl as an electrolyte, while determination of dimethyldiselenide was performed in the organic phase after extraction with CH_2Cl_2. Detection was done with differential pulse CSV at a hanging mercury drop electrode [15]. Also chromium speciation in water can be determined indirectly after the addition of an anion exchange resin and a chelating agent to the cell. Then, only Cr(VI) is electroactive in the supporting electrolyte [16, 17].

For solid samples, all methods of speciation analysis, which do not require sample pretreatment, are based on the use of X-rays: Mössbauer spectroscopy, X-ray photoelectron spectroscopy, and X-ray absorption near-edge structure (XANES) spectroscopy [18–21]. Unfortunately, all of them have LOQ too high for the analysis of biotic environmental samples, especially on trace levels. However, XANES was successfully applied for speciation analysis of thallium in polluted soil, and allowed to determine the kind of minerals binding Tl(I) and Tl(III) on different soil layers [20]. In case of plant material, it was found that the dominant form of thallium in leaves of hyperaccumulating *Iberis intermedia* was Tl(I) [19]. An interesting overview on the use of X-ray absorption spectroscopy (XAS) for biological, agricultural, and environmental investigations was published [22]. XAS has been employed for studying of the phytoremediation of heavy metals from polluted soils [23]. The microscopic XANES (μ-XANES) spectroscopy and confocal microscopic X-ray fluorescence analysis were used for the in vivo determination of the distribution of total selenium and for the local speciation of selenium in roots and leaves of onion [24]. The spectroscopic methods, which are nondestructive and offer minor-level sensitivity as well as microscopic lateral resolution [23, 25, 26], can be used in speciation analysis as supporting methods. The results are often an indirect evidence in discussion about speciation of trace elements.

However, even the results obtained with the most reliable methods and procedures mean nothing if the sample does not represent the investigated object or the speciation has changed during the sampling step. Any chemical (pH, oxygenation, UV exposition) or physical changes (drying, fragmentation) has an influence on the speciation. For this reason all steps of analytical procedures (sampling, transport, storage, conservation), especially in case of procedures meant for trace speciation analysis, have to be very carefully studied for potential changes they may cause in sample's properties (Figure 9.1).

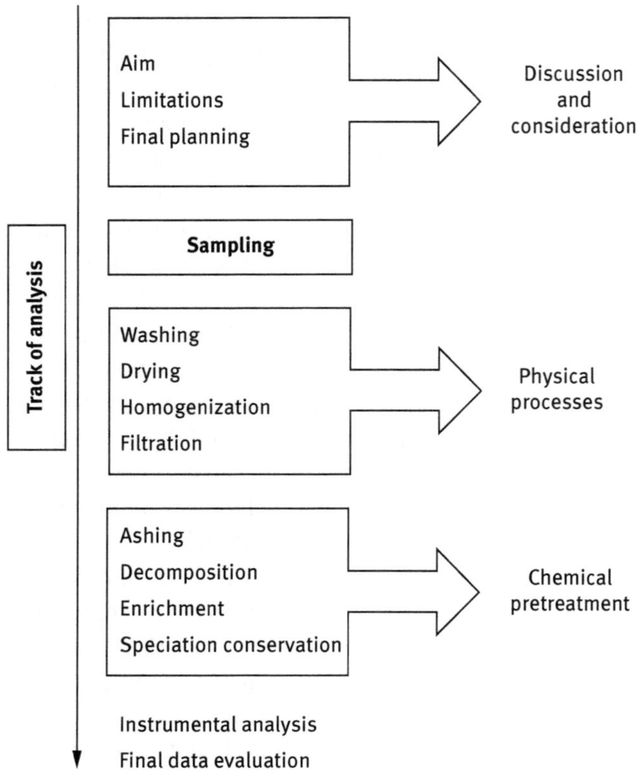

Figure 9.1: Simplified flowchart of environmental sample analysis.

The importance of sample pretreatment can be illustrated on the example of arsenic and its derivatives, where over 50 were identified in biological samples. In regions where significant amounts of organic compounds of As can be expected, groundwater (a deep intake) [2, 27–29] greatly differs in composition from groundwater sampled from soil level [30]. Arsenic comes in a variety of oxidation states and chemical compounds, and the content of organic compounds also differs [2, 3, 31, 32]. It is hard to imagine that there could be one method of sampling and sample conservation that meets all the needs of such challenging analysis.

It seems impossible that the speciation does not change during these critical stages of every trace analysis, but we have to strive for perfection and create procedures that would prevent the changes as much as possible.

9.2 Sampling and sample transport

Sampling of biotic or abiotic elements of the ecosystem is similar for both speciation analysis and the analysis of total content. Standard procedures of sampling are usually appropriate; however, in case of speciation analysis the conservation of the sample should be omitted and the time of transportation limited to a minimum. The unquestionable necessity of the sample to be representative is underlined by all scientists dealing with speciation analysis of traces. For the analysis of chemical speciation, physical speciation, or fractionation studies, it is important to clearly define the aim of the analysis, that is, fraction/phase (e.g, suspension or dissolved fraction in water analysis) and the analyte of interest (e.g., organic or inorganic Hg compounds). This would enable to choose appropriate techniques and methods of sampling and transport, which would guarantee that the physical and chemical properties of the sample are representative for the whole studied object. It is important that there is no contamination or any losses of the analyte and that the fragile equilibrium between various species of the analyte stays intact.

In comparison to the total content analysis, in case of speciation analysis it is even more difficult to assure the immutability of the sample during sampling (e.g., exposition to sunlight of groundwater samples – limitation of photocatalyzed reactions) and transport (e.g., self-reduction of Tl(III) or self-oxidation of Cr(III) – high kinetic effects).

9.2.1 Selection of vessels

All vessels and tools should be made of materials that do not adsorb the studied species. For the speciation analysis of metals recommended are vessels made of polyethylene (PE) or polypropylene (PP). Glass should be avoided because of the risk of adsorption [33, 34]. The United States Environmental Protection Agency (US EPA) allows the use of plastic or glass bottles but plastic is preferred for drinking water sampling (US EPA). In the case of total mercury content or methylmercury compounds (MeHg) determination the collected water samples should be stored in glass bottles [36] or in fluorocarbon polymers and fluoropolymers such as Teflon®, Kynar®, and Tefzel® [37]. In general, mercury is widely considered to be a difficult element to determine, mainly because of its volatility, memory effects, and extraction problems [38, 39]. Also the stainless steel components in the liquid chromatograph system led to adsorption of the mercury compounds, which was more pronounced with $HgCl_2$ than methylmercury [40]. Mercury (II) is quickly lost from all containers except those made of aluminum, which rapidly convert mercury (II) to metallic mercury, which is stable [41]. Similarly to Hg compounds, organic compounds should not be collected nor stored in plastic containers [42], but glass bottles are required (US EPA). In case of both total and speciation analysis, stainless steel devices can cause contamination of the sample with traces of Fe, Ni, Cr, and Mn. Introduction of these elements, when none of them is the analyte, may cause species transformation of other analytes (indirect influence on equilibrium between species of the studied element) [44, 45].

9.2.2 Contamination of the sample with various substances

Uncontrolled introduction of inorganic ions or organic compounds may change not only the total content of the analyte but it can also affect the speciation. Contamination with the analyte can be easily revealed by the analysis of blank samples. However, when the sample gets contaminated with an unexpected substance like weak reducing (e.g., ascorbic acid) or oxidizing agent (e.g., H_2O_2), the oxidation state of many compounds will change (disturbance of the redox equilibria between Fe(III)/Fe(II), Mn(IV)/Mn(II), Tl(III)/Tl(I)) [46]. Not only the changes of the oxidation state of the studied species are problematic but also the increased solubility of compounds binding the analyte, such as MnO_2 and $Fe(OH)_3$ [47, 48], or decrease of Cr solubility [49]. Also the addition of one of the species perturbs the initial speciation. Even trace amounts of reducing agents will completely reduce the trivalent thallium species [13], and the addition of oxidizing agents will change the speciation of Cr and As. Even small amounts of oxidizing agents at pH 6 shift the equilibrium in favor of arsenate (V) [49]. The presence of sulfur compounds (both elemental S and sulfides) at temperature above 22 °C promotes the reduction of Cr(VI) to Cr(III) [50]. Usually, the laboratory vessels are conditioned by washing in acidic solution (pH 1). If the acid is not properly washed out, the sample will be unintentionally acidified, and the pH changes influence directly or indirectly the equilibria between all speciation forms [32, 51]. Sometimes the influence of contaminants is unpredictable. For example, methylation of mercury (II) occurs in the presence of trimethyl lead but the process is inhibited by humic substances [52]. The presence of Fe(III) ions affects thallium speciation – oxidation of Tl(I)(aq) took place when natural water samples were exposed to either sunlight or UV light, notably at low pH [53]. Therefore, any uncontrolled substance introduction or loss should be avoided.

9.2.3 Elimination of UV-Vis irradiation

In practice, limited exposure to sunlight (UV-Vis radiation) is preferred for each environmental sample. UV light (wavelength 100–400 nm) is involved in the degradation of chemical bonds in large organic compounds, and it may cause reduction or oxidation of the analyte [54]. The United States Geological Survey (USGS) suggests in most of the water sampling procedures to keep the sample in dark [37]. Thallium speciation study is a good example of how UV-Vis light affects the speciation. For thallium it was showed that oxidation of Tl(I) in an aqueous solution can take place when the sample is exposed to sunlight irradiation [55]. Ultraviolet irradiation of aqueous solutions containing Tl(III) and being in equilibrium with the atmosphere increases the reduction rate of Tl(III). In systems where photoreduction of Fe(III)(aq) takes place, a quantitative oxidation of Tl(I)(aq) was observed, notably at low pH. The process is reversible, as indicated by the formation of Tl(I) when the irradiated systems were kept in the dark

[53]. Also high instability of Tl(III) in the presence of plant matrix is observed under UV exposition. For example, plant extract with an addition of Tl(III)DTPA standard was exposed to UV for 1 h. It was observed that 95% of Tl(III) added as Tl(III)DTPA was reduced to Tl(I) [14]. The data indicated that Tl(III) extracted as Tl(III)DTPA from plant tissue is not stable under UV irradiation. Therefore, plant extracts should be stored in dark till analysis. Similar study about the stability of Se(IV) extracted from plant tissue showed that it is most likely that Se(IV) is reduced to Se(0) or that an insoluble complex is formed [5]. Most data seem to indicate that darkness is necessary for preservation of mercury species during storage in biological matrices. Methylmercury and inorganic mercury in fish extract solutions were stable for 5 months at 4 °C when stored in the dark [56]. The stability of butyltin species in lyophilized mussel samples was also affected by the light. Significant variations were found in the butyltin content after 3 and 6 months of storage at room temperature in daylight and in the dark, respectively [1].

9.2.4 Oxidation and desorption of carbon dioxide

Deep-sea water samples naturally contain small quantities of oxygen. Contact with air results in rapid dissolution of oxygen in the sample, and the effect is especially intense when the sample is shaken (during transport). The USGS suggests in most of the water sampling procedures to keep the sample without any contact with oxygen, to prevent metal-oxide precipitation [37]. In general, the zone of interest must be isolated, the sample pumped slowly to minimize turbidity, and collected in such conditions that eliminate O_2 and CO_2 exchange with the atmosphere [35, 37]. The redox equilibria are then shifted to oxidizing range, which affects the speciation of free ions within a matrix of a water sample, which is a noticeable effect in samples from reservoirs with low oxygenation of water. Consequently, oxidization of sulfides to sulfates and Fe(II) to Fe(III) will occur, and then precipitation of iron hydroxide with traces of Co, Ni, As, and Pb. Changing of pH and hardness (temporary hardness) of the water sample due to absorption of CO_2 from the air accelerates and increases the precipitation of calcium carbonate minerals [57, 58]. Also the oxidation of Mn(II) by oxygen dissolved in water brings some consequences for chromium speciation. MnO_2 particles, formed as a result of the process of oxidation of Mn(II) to Mn(IV) in dissolved fraction, accelerate the oxidation of Cr(III) to Cr(VI). The product of this reaction – Cr(VI) – is adsorbed on suspended particle matter (SPM) and therefore both physical and chemical speciation is changed in the water sample [59]. But unlike Cr, As speciation was not changed by oxygenating of the sample. After air exposure of wine and beer samples, arsenite, arsenate, monomethylarsonic acid (MMA), and dimethylarsinic acid (DMA) were stable at 4 °C for months, probably due to the acidic pH of the samples [60]. In contrast, storage of the samples for antimony determinations is very difficult, because Sb(III) easily transforms into Sb(V) in the oxidizing environment [61].

9.2.5 Solid sample for fractionation study

A special case of speciation analysis is fractionation, where the fractions are defined according to the physical properties of the solid samples such as soils [62, 63], sediments [64, 65], or solid wastes [66, 67]. Methods of sampling are the same as those used in the analysis of the total content [37, 68–70]. In the case of soil, the top layer (rhizosphere) is discarded, together with large objects such as stones, roots, or zoological specimens. Then, the top layer (0–10 cm) [71–74] or the bottom layer (10–30 cm) [71–73] is taken for the analysis. The samples are transported with natural water content in a cardboard box granting access to the air. The alluvial deposits depending on the size of river's backwaters are collected from various depths, usually 15–20 cm. The oxygenated layer is analyzed separately and it is taken from 1 to 3 cm depth [75]. The samples are then dried in the laboratory. Despite the fact that the process of drying affects the mobility of various metals [76, 77], fractionation studies (distribution evaluation) are rarely carried out on wet samples.

9.2.6 Temperature lowering just after sampling

The lowering of samples' temperature just after sampling and during transport is one of the most widely and longest used methods. The literature data about sample pretreatment indicate that freezing could decrease the efficiency of extraction of the analytes from plant tissues [4, 78–80]. In the case of Tl speciation study in leaves of hyperaccumulating plant, the storage at −18 °C for over one month lead to total reduction of Tl(III) [81]. The content of arsenobetaine in blue mussels was significantly reduced during storage of frozen material. The content of tetramethylarsonium in fishes as well as in seafood was generally lower after deep freezing and some days of storage [82]. Groundwater, which has naturally low temperature, after collection requires transport in cold state, especially if it was not possible to filtrate the samples and separate the SPM from the dissolved fraction. It is not recommended to freeze the water prior to analysis [35]. The increase of sample temperature causes absorption of carbon dioxide, which affects the pH. Also for biological samples lowering of the temperature (4 °C), usually in combination with storage in complete darkness, may well reduce the activity of microorganisms and enzymes. Cooling limits the speed of these processes, but it does not eliminate them completely [83]. Samples of seafood were stored up to 9 months till analysis, and the arsenic organocompounds were not degraded [84]. Sometimes freezing (−17 °C or −20 °C) is recommended, or even rapid cooling in liquid nitrogen followed by lyophilization. However, this procedure is not suggested if the purpose of the analysis is determination of phytochelatins (PCs) in plant tissues, because these compounds are then degraded to peptides [1]. The analysis of PCs or their complexes is carried out using fresh material, as PCs are known to be unstable during sample preparation [85]. If the analysis cannot be done immediately after the extraction of PCs, derivatization has to be done (precolumn derivatization with monobromobimane), which prevents the process of PC degradation [86]. Only samples containing derivatized PCs can be kept refrigerated for a long time [85]. This limits the amount of information obtained, but the procedure is very useful as there is not always a possibility of carrying direct analysis using MS methods [87–91]. Lowering of the temperature is also suggested for physical fractionation of soil, within the SPM and the solid phase of water, but only when the separation of SPM or the specific soil pretreatment (to reduce putrefaction) cannot be performed on the sampling site. In the case of wastewater containing large amount of S(II) compounds, Cr(VI) was reduced to Cr(III) if the sample was not cooled down to 20 °C [50].

A very particular case is transport of the samples at the boiling point of liquid nitrogen (−150 °C). After sampling the sample is inserted into a labeled container and into thermos with liquid nitrogen. Most of the steps of sample preparation should also be carried out at this temperature, for example, cryomilling, lyophilization, and long-term storage. In such conditions, the samples collected by German Environmental Bank (UPB – Umwelt Proben Bank des Bundes) for long-term storage (over 50 years) are transported to laboratories [71, 72, 92–97].

9.3 Sampling with some pretreatment on sampling site

Both solid and liquid samples should be prepared directly after sampling, and this is often done immediately on the sampling site. However, keep in mind that both chemical and physical interventions in a sample have a smaller or greater impact on the equilibrium that we want to study.

The reaction of the analyte with stabilizing agents should be quick [98–100]. Currently, there are proposed some interesting solutions for sample modification (an addition of specific sorbent) which enables indirect speciation study. It can be achieved by an addition of the substance which is permanently binding one form of the analyte. In the case of chromium, the addition of multiwalled carbon nanotubes with coprecipitated MnO_2 selectively removed the trivalent Cr ions from water sample at pH 5 [44]. The addition of chitosan grafted with 2-hydroxyethyltrimethyl ammonium chloride stabilized the speciation, and the adsorption of Cr(VI) is favorable; therefore, only Cr(III) can be detected [101]. To preserve the samples, the researchers often use chelating reagents, such as ethylenediaminetetraacetic acid (EDTA) or DTPA. The authors do not agree on the stability and permanence of arsenic forms in water, especially at different pH and in the presence of other substances [102]. Generally, in river water As(V) is partially converted to As(III), but after 2 days this is followed by gradual oxidation of As(III) into As(V) to reach an equilibrium. Storage at 5 °C delays this oxidation by about 6 days [103]. In the case of thallium, DTPA was used for stabilization of Tl(III). To prevent self-reduction of trivalent forms of thallium in water samples and in plant extracts the addition of DTPA was used [13, 104, 105]. Stabilization with DTPA was also applied in antimony speciation analysis [61]. If the object of study is arsenic speciation, the addition of EDTA must be done immediately after sample filtration on the sampling site [37].

9.3.1 Suspended matter separation – fractionation in water

SPM is an integral part of the hydrosphere. SPM is an important component of the water, and it is responsible for the transport of elements and substances in water currents [106]. SPM is composed of mainly inorganic colloidal particles in the form of oxides, hydroxides, metal carbonates, and organic components. Therefore, a significant effect on the physicochemical properties of SPM has its origin as well as the shape and size of the suspended particles [107]. Due to the fact that processes of coprecipitation, adsorption, and desorption, as well as ion exchange take place between the dissolved fraction (solution) and SPM fraction (suspended solid phase), it is impossible to store water samples even for a short time without changes in the speciation. The border between SPM and the solution phase is conventional, and it is widely accepted that SPM phase is defined as a fraction bigger than 0.45 μm. This idea came to Nürberg et al. in 1988 [108]. In case of physical fractionation of water samples, that is, to determine the content of SPM, the sampling of the water sample does not differ from the sample for other purposes. Usually, a few hundred milliliters is taken but in exceptional cases even some liters (nano-level or radiation study) but no pretreatment is performed. It is forbidden to add commonly used nitric acid, as such addition will disturb the equilibrium between the two phases of water – suspended and dissolved phases. Such analysis requires filtration just after sampling, on the sampling site, before the sample is oxygenated, warmed up, and/or exposed to UV-Vis radiation. The filtration through a filter with a pore size of 0.45 μm is performed [35, 109–114]. The filter retains the suspended phase (SPM).

Often the mass of SPM is quantified as a gram per milliliter of the water, and then the SPM is decomposed in order to determine the total content of the elements. Usually a large volume of the sample is filtered (100–300 mL), and clogging of the pores could be observed during the filtration [35], which is a phenomenon that affects the quality of the obtained results. While the sampling is done in regions with considerable dustiness, the initial separation of "dust fraction" is suggested. The "dust fraction" is defined physically as a fraction of particles bigger than 100–125 μm. Next, the appropriate fraction bigger than 0.45 μm is separated in sequence. The "dust fraction" plays an important role in the distribution of the contaminants during the dry seasons, but not during the rainy seasons [115, 116]. This part, which is associated with large particles, will quickly sediment and thus it is not responsible for the transport of the pollutants. The SPM fraction can be transported and it spreads the pollutants in the environment. Study of the distribution of As, Cd, Co, Cr, Cu, Mn, Ni, Pb, Tl, and Zn indicated that the dissolved fraction contains the highest amount of trace elements [115]. Trivalent chromium in water remains bound to organic matter (SPM component), which decreases its mobility, but the toxic Cr(VI) is highly soluble [117, 118]. In the case of mercury speciation analysis on low level, filtration directly on the sampling field is required to separate SPM, and it is strongly recommended by the USGS [37]. Often, the removal of SPM is a step of preparation of a water sample for the analysis of total content of the elements (it is assumed that the share of SPM in the total content is negligible) [110] or it is the first step of speciation analysis (e.g., determination of chemical speciation of Tl within the dissolved phase [105] – SPM was not analyzed). However, leaving SPM in a nonacidified water sample even for a short time results in binding of the free ions of Cr, Mn, and Cu with the organic compounds suspended in solution, followed by their fast precipitation [119] or to adsorption of the free ions of Pb, As, Ni, and Co on the sediment containing Mn and Fe oxides (coprecipitation), which is formed during storage [94].

9.3.2 Chemical modification of the sample

Acidification with nitric acid is commonly applied in the analysis of trace metals, but in the case of speciation analysis it is not advisable to lower the pH [51]. Sometimes, however, in specific cases such as determination of As(III), acidification is not only permitted, but even required [120, 121].

Storage of water (chemical speciation) and soil extracts (fractionation) without any pretreatment results in coprecipitating of ions of As, Ni, Co, Pb, and Tl with Mn(II) and Fe(III) oxides [47, 72]. Limitation of this phenomenon can be achieved only through a significant acidification of the solution ($pH < 2$) or the addition of complexing ions (e.g., EDTA), but after such a modification, the sample is not suited for broadly defined speciation analysis. Hydrochloric acid (non-oxidizing acid) is sometimes used because of complexing properties of chlorides. Complexation of free ions (+2 and +3) influences the chemical equilibrium, but such modification (addition of complexing ions) is recommended when Mn(II) and Fe(III) are regarded as interferents [45]. The addition of another complexing agent, DTPA, is indispensable to perform speciation analysis of Cr and Tl. The compounds of Tl(III) are unstable and without DTPA they are slowly reduced [13, 122]. For arsenic speciation study the addition of ascorbic acid is recommended [121]. However, it is inadmissible in the case of trace speciation analysis of Cr and Tl [13, 16], because the ions with the highest oxidation state would be totally reduced. Also, the addition of phenol, previously used to prevent microorganisms' activity, leads to slight changes in the selenium speciation [5]. Therefore, the most suitable preservation scheme for determination of As(III), As(V), Cr(III), Cr(VI), Se(IV), and Se(VI) in water samples was found to be refrigeration at 10 °C with no preservative, followed by the analysis as soon as possible (preferably within 24 h) [123]. Also mercury speciation is highly dependent on the presence of ions, such as chlorides, high content of SPM and pH of the solution. Therefore, any modification of water samples leads to a change in chemical equilibrium [124, 125].

To avoid changes of the analyte species it is important to analyze the samples as soon as possible after their collection.

9.4 Short- and long-term storage

The analysis of samples immediately after their collection and transport to the laboratory is not always possible. Usually speciation analysis requires sophisticated methods of separation and detection. A possibility to prolong the time between sampling and chemical analysis would be an important advantage of any analytical procedure. For this reason the methodology of material storage was studied by many authors. Storage of biological samples is not recommended even for such short period as 2 weeks [126]. But in some cases, various species of arsenic in algae were stable up to 12 months in dried material [127]. In the case of Tl the instability of Tl(III) was found, both freezing and drying caused significant changes in thallium speciation [81]. Recent studies of As speciation indicated that plant material containing As species should be kept at a low temperature, while others claim that the freezing/defrosting processes lead to species conversion, and therefore the samples must be kept at room temperature [128, 129].

For long-term storage deep freezing is recommended. Plant samples were stored at −80 °C and the concentration of arsenic-PC complexes remained relatively constant during 21 days [88]. Samples may also be stored at the boiling point of nitrogen (−150 °C). In such conditions soil and sediment samples are stored, as well as other materials collected by German Environmental Bank (UPB) [71, 72, 75, 94]. Long-term storage without any pretreatment exposes the samples to microorganisms' activity. It is important to avoid bacterial growth in the sample as this may cause changes in the speciation. In the case of Hg, for example, some species of mercury (II) may be reduced to volatile elemental mercury [1].

Samples for speciation analysis should not be stored for a long period, independently on the concentration of the analyte. For example, trivalent arsenic compounds should not be stored for longer than 24 h if the sample is stored without modification [130]. For thallium speciation, both freezing and drying cause significant changes in the speciation, so plant samples should not be stored. The quantitative analysis can be performed only on fresh tissues [81].

The time of storage can be slightly extended if the analyte is separated from the sample matrix.

9.4.1 Dehydration of samples

Dehydration is sometimes necessary for solid samples. The so-called drying is intended to remove the solvent from environmental samples. The main aims of this operation are limitation of the microbial activity and achieving a constant mass, which is taken for the analysis (results are usually given as the content in dry mass). The soil should be either stored deep frozen or after dehydration. Losses of the analyte and its mobility are much lower when the soil is kept dry (air dry) [75]. The procedure is well known and accepted, and routinely used in the analysis of soil for total content and speciation [131–135]. Usually, the samples of soil, sediment, or solid waste are brought to the laboratory with natural water content, and then slowly dried at room temperature in open-air conditions for 24 h. If necessary, the samples are oven dried at 40–50 °C or 105 °C for 5–10 h [67, 112, 126, 134–140]. However, in fractionation study the extractability from such prepared samples is different from the original: lower for K, Cr, Mn, and Fe [77], and higher for Pb, Cu, Cd, Zn, and Ni [76]. Soil and sediment samples, according to the BCR-SMT (Bureau of Reference – now the Standards, Measurements and Testing Programme – SMT) scheme (the standard procedure called BCR procedure), are dried at room temperature or in the oven at 30–80 °C. If higher temperature is required for drying, it should be performed in an inert gas atmosphere to limit the oxidation processes [141]. Many authors claim that low-temperature drying in ambient air and freeze-drying are the most "neutral" methods for samples of soil and sediments and that these methods do not considerably influence the distribution (mobility) of metals in soil phases [142].

Storage of animal and plant tissues with natural moisture content does not limit the microbial activity, and the microorganisms are actively involved in changing the speciation. In the case of arsenic, the intrinsic microbial population causes demethylation of organoarsenic compounds [143]. Therefore, drying is considered also in case of biological samples, but usually the temperature does not exceed 95–98°C and not longer than 16 h after sampling [122, 144, 145]. Delicate plants are dried at a temperature not higher than 35 °C [146, 147]. Also some analytes require lowering of the temperature of the drying process. For the analysis of Se, drying should be slow (over 48 h) and run at no more than 45 °C [146]. Drying and storage of plants even at room temperature cause significant changes in the speciation of thallium. In contrast to the extract of fresh material, there was no measurable amount of Tl(III) in any of the stored samples [81]. The situation is opposite in the case of As speciation analysis, where drying of the plants is recommended because plant moisture can cause species conversion [148]. Many authors claim that biological samples should be dehydrated using freeze-drying systems (lyophilization – sublimation of frozen water under conditions of low pressure and temperature). The process consists of freezing of the sample, usually in liquid nitrogen or dry ice in ethanol, and drying at a very low pressure (10 Pa) in a round-bottom flask made of borosilicate glass [126, 149]. Certainly, the material that was frozen after sampling should be dried by lyophilization [42, 75]. The method of sample preparation has an influence on the extraction process. Lyophilization can decrease the extraction efficiency even by 20% [79, 80]. But in some cases (e.g., As speciation study in *Brachiaria brizantha*), the extraction was more efficient from sample aliquots that were lyophilized and ground (extractability: 87–90%) than from those only stored under different temperatures (extractability: 53–66%) [4]. Usually, after drying the environmental samples are stored at room temperature in closed containers made of PE or PP [140, 150, 151] or cooled down to 4 °C [152, 153].

9.5 Extraction as a method of sample pretreatment for speciation analysis

Extraction techniques are commonly used and widely accepted methods for trace analysis. The phenomena occurring during the extraction process in liquid–liquid and solid–liquid systems are well known. This simple technique can be the basis for procedures of selective isolation of a specific chemical form of the analyte. Theoretically, separated species of the analyte can be stored without any pretreatment for a long time. An example of such methodology is leaching of some selenium compounds from soil during solid (sample)–liquid (extractant) extraction. Selective separation of Se(IV) and Se-Cys from $(CH_2)_2Se_2$ was achieved by leaching of selenium species with 0.5 mol/L HCl. Then, the solution was extracted with CH_2Cl_2. In the aqueous phase Se(IV) and Se-Cys were found, while dimethyldiselenide was present in the organic phase [15]. Extraction from plant tissues using water–methanol mixture is proposed for arsenic species (MMA, DMA, As(III), As(V) and arsenosugars) separation from sample matrix [74]. For metal-PC complexes (Me-PCs) determination of a "soft" extraction procedure is required. For example 1% formic acid is applied for As-PCs [85, 154], 3 mL cold (4 °C) aqueous dithiothreitol for Pb-PCs [155] and 0.5% $NaBH_4$ in NaOH solution for Hg-PCs [156]. In the case of Tl speciation in plant tissues the extraction using DTPA in acetic buffer (pH 6.2) is the only way to prevent self-reduction of Tl(III). The chromatograms (liquid chromatography [LC]-ICP-MS) of extracts were recorded just after extraction, and after 4 months of storage (–18 °C). In most cases, the results were consistent (revealed the presence of both Tl forms). In some cases, however, there occurred a significant decrease or increase in the content of Tl(III). Therefore, freezing of the extracts is an option only for semiquantitative analysis [81]. Additionally, high efficiency of the extraction is essential for trace analysis. This parameter can be accelerated using an extraction assisted with microwaves or ultrasounds, especially in the case of soil and sediment samples [157–159]. Acceleration of the extraction is also applied in leaching of xenobiotics from plant tissues [160]. Cell wall is a barrier for the extractant, and grinding of the plant tissues in the presence of liquid nitrogen noticeably increases the extraction efficiency [122]. It is very difficult to reach 100% recovery of the analyte from extractant. A small addition of surfactant such as sodium dodecyl sulfate (SDS) made to the extractant led to leaching of fractions containing not only water-soluble but also water-insoluble protein complexes [104, 161]. The comparison of data obtained for extraction with and without the addition of SDS reveals the fraction of the analyte (e.g., Pt, Tl) bound to insoluble protein complexes. The efficiency of thallium extraction from the above-ground plant organs using SDS was on the same level as in the case of water extracts (50%), which shows that thallium is not bound to hydrophobic proteins [104]. In the case of platinum, there were significant differences between fractions of Pt extracted from plant material with and without SDS. The efficiency of leaching with water was about 55% of total Pt but the leaching with SDS increased it up to 70% [162]. If it is important to distinguish between inorganic and organic arsenic species, the extraction should be done using an organic solvent, for example, methanol [163]. An extraction with hot water was used to leach Se from onion [164], but similar extraction of methyl arsenic acid (III) from tissues of marine animals is not recommended because this compound is not stable at high temperature [165]. In the case of marine organisms (algae and seafood), separation of arsenolipids was achieved by extraction with nonpolar solvents, and the extraction efficiency was about 50% [166]. Preceding of the proper extraction with degreasing using acetone effects in lowering of the extraction efficiency to 30% of the total content of arsenic [167].

An interesting example of simultaneous separation and conservation of the analyte (prevention against the self-reduction) is thallium speciation analysis in SPM separated from wastewater. Trivalent thallium was transferred from its original compounds into Tl(III)DTPA [116]. This complex is very stable, and its stability is practically independent of pH and the presence of other ions. Thus, leaching based on formation of Tl(III)DTPA complex was applied. But even in that case the self-reduction took place, and after 1 week of storage its rate reached 25% [116], while without storage it is usually not higher than 3% [13, 168]. Extraction with DTPA solution is a widely accepted method of isolation of Tl(III) from plant tissues. Usually the material is ground to a fine powder using a mortar and pestle in a liquid nitrogen bath, and extracted using DTPA solution in acetic buffer (pH 6.2) (plant material is shaken with the extractant for 1 h at 37 °C). After leaching the suspensions are filtered through a 0.45 μm filter to separate solid particles [122, 168, 169], and plant extracts are immediately analyzed by chromatographic methods coupled with elemental detectors [104, 122] or by electrochemical methods (after some "soft" modification).

Voltammetric methods are very sensitive and they are characterized by low limits of detection [170]. Stripping voltammetry enables determination of As, Cr, Tl, Se, Sn, Pt, Rh, Pd, and Pb at the level of ppb [13, 17, 171, 172] or even ppt [173]. However, they require special preparation of the extract or mineralization of organic compounds (significant interferent), which are present in significant quantities in water samples and extracts of plant products. Of course, mineralization of the sample is not acceptable in speciation analysis. Conservation of the plant extract or wastewater using DTPA, and an addition of resin (Amberlite XAD-7) just before the step of ion preconcentration on the working electrode allowed to indirectly define the speciation of Tl and Cr (only not complexed forms of the analyte are electroactive) [14, 17, 172, 174]. Such modification does not affect the speciation. The results obtained by voltammetry were confirmed by comparison with the results of LC-ICP-MS study [14].

Relatively new trend is application of extraction as a method of sample pretreatment, especially for the analysis of organic compounds, known as QuEChERS. This technique of separation of the analytes from complicated sample matrix was developed based on Anastassiades et al.'s work [175]. The acronym means **Qu**ick, **E**asy, **Ch**eap, **E**ffective, **R**ugged, and **S**afe. This extraction technique is a combination of extraction of organic compounds, mostly from food matrices, coupled with removing of another organic compounds (interfering substances) and the excess of _water. It is an alternative to traditional liquid–liquid extraction and SPE [176]. The methodology is suggested by AOAC International for the analysis of pesticides [177].

9.5.1 SPE – water analysis

In the analysis of water the major problem is low concentration of the analyte. SPE technique offers an interesting possibility of separation and considerable preconcentration of the analytes [178]. Combination of filtration through a 0.45 µm filter, and next preconcentration of pesticides on SPE column additionally allowed long-term storage of the samples [179, 180]. The sample of seawater collected for speciation analysis of Tl cannot be stored for a long time, as self-reduction of Tl(III) is observed even in the presence of DTPA. But after SPE separation, trivalent thallium as Tl(III)DDTC is selectively retained on the sorbent, and monovalent thallium is in the effluent. Such a way Tl species are collected separately and they can be stored without any problems till analysis [122]. Also Cr(III) was separated from Cr(VI) and preconcentrated using a column containing chelating resin. This method was successfully applied for speciation analysis of chromium in tap water, lake water, spring water, and wastewater samples [51]. Another SPE method was developed for on-site separation of inorganic arsenic from As-rich groundwater and applied for the study of arsenic removal from ferrihydrite in the absence of oxygen [181].

9.5.2 Extraction in fractionation study

A specific example of speciation analysis is fractionation, where the fraction is defined operationally by applying chemical and/or physical modifications to the sample. It is intended to indicate what fraction of the total content of the analyte is bound to particular phases of the soil (extraction under certain conditions). Fractionation is used as a source of information about xenobiotics' mobility and bioavailability, and for potential risk assessment. Several extraction procedures are used [182]. The most popular are the Tessier's procedure [183], and the BCR procedure [184] (Table 9.1). Nowadays, commonly used are single extractions with acetic acid or EDTA solutions, and three-step sequential extraction (in sequence: acetic acid, hydroxylamine hydrochloride, hydrogen peroxide solutions) developed by the former European Community Bureau of Reference (BCR), now going by the name of Standard Measurements and Testing

Table 9.1: Comparison of two sequential extractions applied for fractionation study in solid samples – BCR (BCR-SMT) and Tessier schemes.

	Tessier scheme
1 g of solid sample	
Ion exchange and carbonate fraction	Ion exchange fraction
16 h at 22 ± 5 °C Continuous mixing 40 mL 0.11 M CH_3COOH pH 4	0.5 h at room temperature Continuous mixing 8 mL 0.1 M $MgCl_2$ pH 7(twice)
	Carbonate fraction
	5 h continuous mixing 8 mL 1 M CH_3COONa pH 5
Oxide fraction	Iron and manganese oxide fraction
16 h at 22 ± 5 °C Continuous mixing 40 mL 0.1 M $NH_2OH{\cdot}HCl$ pH 2	5 h at 96 ± 3 °C 20 mL 0.04 M $NH_2OH{\cdot}HCl$ in 25% CH_3COOH pH 2
Organic fraction	Organic fraction
Substep (1) 2 h water bath 85 ± 2 °C 10 mL 8.8 M H_2O_2 (twice) pH 2 or pH 3	Substep (1) 2 h water bath at 85 ± 2 °C 3 mL 0.02 M HNO_3 + 5 mL 30% H_2O_2 pH 2
Substep (2) 16 h 22 ± 5 °C Continuous mixing 50 mL 1 M CH_3COONH_4 pH 2	Substep (2) 3 h water bath at 85 ± 2 °C 3 mL 30% H_2O_2
	Substep (3) after cooling 0.5 h Continuous mixing 5 mL 3.2 M NH_4OAc in 20% HNO_3
Residual fraction	Residual fraction
Mineralization with mixtures of conc. acids (HF, HNO_3, $HClO_4$)	Mineralization with mixtures of conc. acids HCl + HNO_3 (1:3)

Note: M, mol/L.

Programme of the European Committee (SMT or BCR-SMT) [185, 186]. For fractionation study, dried materials are sieved through a 1 mm [67, 112] or 2 mm [135, 139] sieve or milled and sieved. Usually in sediment analysis the fraction smaller than 63 μm is taken [126, 134, 187] (after milling). Some solid samples as dust and flotation tailings are not ground, only the fraction smaller than 125 μm is used [188]. Dried material can be stored in darkness at ambient temperature until analysis for some months.

German Environmental Bank (UPB) stores the samples of soil in liquid nitrogen and then the material is cryogenically ground, sieved, and dried using lyophilization. Only a portion of the material is analyzed for the purpose of long-term monitoring [71, 94].

9.6 Conclusion

There are not many reference materials available, for which sampling and pretreatment of samples for speciation analysis is proposed. Only Cr(III) and Cr(VI) in water and soil samples, as well as volatile Hg and MeHg in water, fish tissues, and soil were studied in the context of routine analysis. In case of other objects it is beneficial to become familiar with guidelines for the analysis of specific groups of analytes for their total content. Recommendations are published by institutions such as US EPA, Joint Research Centre Institute for Reference Materials and Measurements, and USGS[69, 189, 190], together with standard operating procedures for the collection of representative surface water samples from streams, rivers, lakes, ponds, lagoons, and surface impoundments [191]. There are no details regarding speciation analysis, because it is done in routine only for chromium and mercury. In case of other elements one has to choose the procedure of sample pretreatment according to their own best knowledge. Certified materials are not meant for validation of sampling and sample pretreatment before the analysis. Therefore, high experience of the analyst is of crucial importance, together with his awareness of possible mistakes. Adapting methods already published in the literature could prevent us from "reinventing the wheel." However, all of the procedures should be carefully checked, as sampling, sample transport, and conservation of the speciation are really difficult tasks.

Therefore, each sample and each analytic target are always a new challenge.

References

[1] Cornelis R, Caruso J, Crews H, Heumann K. Handbook of Elemental Speciation: Techniques and Methodology. John Wiley & Sons, Ltd, 2003.

[2] Smedley PL, Kinniburgh DG. A review of the source, behaviour and distribution of arsenic in natural waters. Appl Geochem 2002, 17, 517–68.

[3] Morita M, Edmonds JS. Determination of arsenic species in biological and environmental samples (Technical Report). Pure Appl Chem 1992, 64, 575–90.

[4] Amaral CDB, Nóbrega JA, Nogueira ARA. Investigation of arsenic species stability by HPLC-ICP-MS in plants stored under different conditions for 12 months. Microchem J 2014, 117, 122–6.

[5] Cuderman P, Stibilj V. Stability of Se species in plant extracts rich in phenolic substances. Anal Bioanal Chem 2010, 396, 1433–9.

[6] Feldmann J, Lai VW, Cullen WR, Ma M, Lu X, Le XC. Sample preparation and storage can change arsenic speciation in human urine. Clin Chem 1999, 45, 1988–97.

[7] Nischwitz V, Michalke B, Kettrup A. Investigations on extraction procedures for Pt species from spiked road dust samples using HPLC–ICP–MS detection. Anal Chim Acta 2004, 521, 87–98.

[8] Parker JL, Bloom NS. Preservation and storage techniques for low-level aqueous mercury speciation. Sci Total Environ 2005, 337, 253–63.

[9] Jabłońska-Czapla M. Arsenic, antimony, chromium, and thallium speciation in water and sediment samples with the LC-ICP-MS technique, arsenic, antimony, chromium, and thallium speciation in water and sediment samples with the LC-ICP-MS technique. Int J Anal Chem 2015, e171478.

[10] Cai Q, Khoo SB. Determination of trace thallium after accumulation of thallium(III) at a 8-hydroxyquinoline-modified carbon paste electrode. Analyst 1995, 120, 1047–53.

[11] Henze G, Wagner W, Sander S. Speciation of arsenic(V) and arsenic(III) by cathodic stripping voltammetry in fresh water samples. Fresenius J Anal Chem 1997, 358, 741–4.

[12] Zen J-M, Wu J-W. Square-wave voltammetric stripping analysis of thallium(III) at a poly(4-vinylpyridine)/mercury film electrode. Electroanalysis 1997, 9, 302–6.

[13] Krasnodębska-Ostręga B, Pałdyna J, Wawrzyńska M, Stryjewska E. Indirect anodic stripping voltammetric determination of Tl(I) and Tl(III) in the Baltic Seawater samples enriched in Thallium species. Electroanalysis 2011, 23, 605–10.

[14] Pałdyna J, Krasnodębska-Ostręga B, Sadowska M, Gołębiewska J. Indirect speciation analysis of thallium in plant extracts by anodic stripping voltammetry. Electroanalysis 2013, 25, 1926–32.

[15] Ochsenkühn-Petropoulou M, Tsopelas F. Speciation analysis of selenium using voltammetric techniques. Anal Chim Acta 2002, 467, 167–78.

[16] Grabarczyk M. Ultraselective and sensitive determination of Cr(VI) in the presence of a high excess of Cr(III) in natural waters with a complicated matrix. Electroanalysis 2008, 20, 1495–8.

[17] Grabarczyk M. A catalytic adsorptive stripping voltammetric procedure for trace determination of Cr(VI) in natural samples containing high concentrations of humic substances. Anal Bioanal Chem 2007, 390, 979–86.

[18] Peacock CL, Moon EM. Oxidative scavenging of thallium by birnessite: Explanation for thallium enrichment and stable isotope fractionation in marine ferromanganese precipitates. Geochim Cosmochim Acta 2012, 84, 297–313.

[19] Scheckel KG, Lombi E, Rock SA, McLaughlin MJ. In vivo synchrotron study of thallium speciation and compartmentation in Iberis intermedia. Environ Sci Technol 2004, 38, 5095–100.

[20] Voegelin A, Pfenninger N, Petrikis J, Majzlan J, Plötze M, Senn A-C, Mangold S, Steininger R, Göttlicher J. Thallium speciation and extractability in a thallium- and arsenic-rich soil developed from mineralized carbonate rock. Environ Sci Technol 2015, 49, 5390–8.

[21] Wagner B, Bulska E, Stahl B, Heck M, Ortner HM. Analysis of Fe valence states in iron-gall inks from XVIth century manuscripts by 57Fe Mössbauer spectroscopy. Anal Chim Acta 2004, 527, 195–202.

[22] Prange A, Modrow H. X-ray absorption spectroscopy and its application in biological, agricultural and environmental research. Rev Environ Sci Biotechnol 2002, 1, 259–76.

[23] Gardea-Torresdey JL, Peralta-Videa JR, de la Rosa G, Parsons JG. Phytoremediation of heavy metals and study of the metal coordination by X-ray absorption spectroscopy. Coord Chem Rev 2005, 249, 1797–810.

[24] Bulska E, Wysocka IA, Wierzbicka MLGH, Proost K, Janssens K, Falkenberg G. In vivo investigation of the distribution and the local speciation of selenium in Allium cepa L. by means of microscopic X-ray absorption near-edge structure spectroscopy and confocal microscopic X-ray fluorescence analysis. Anal Chem 2006, 78, 7616–24.

[25] Proost K, Janssens K, Wagner B, Bulska E, Schreiner M. Determination of localized Fe2+/Fe3+ ratios in inks of historic documents by means of μ-XANES. Nuclear instruments and methods in physics research section B: Beam interactions with materials and atoms 2004, 213, 723–8.

[26] Proost K, Vincze L, Janssens K, Gao N, Bulska E, Schreiner M, Falkenberg G. Characterization of a polycapillary lens for use in micro-XANES experiments. X-Ray Spectrom. 2003, 32, 215–22.

[27] Mandal BK, Suzuki KT. Arsenic round the world: a review. Talanta 2002, 58, 201–35.

[28] Mattusch J, Wennrich R, Schmidt AC, Reisser W. Determination of arsenic species in water, soils and plants. Fresenius J Anal Chem 2000, 366, 200–3.

[29] Tossell JA. Theoretical studies on arsenic oxide and hydroxide species in minerals and in aqueous solution. Geochim Cosmochim Acta 1997, 61, 1613–23.

[30] Doušová B, Martaus A, Filippi M, Koloušek D. Stability of arsenic species in soils contaminated naturally and in an anthropogenic manner. Water Air Soil Pollut 2007, 187, 233–41.
[31] Caruso JA, Klaue B, Michalke B, Rocke DM. Group assessment: Elemental speciation. Ecotoxicol Environ Saf 2003, 56, 32–44.
[32] Komorowicz I, Barałkiewicz D. Arsenic and its speciation in water samples by high performance liquid chromatography inductively coupled plasma mass spectrometry–last decade review. Talanta 2011, 84, 247–61.
[33] Lavilla I, Capelo JL, Bendicho C. Determination of cadmium and lead in mussels by electrothermal atomic absorption spectrometry using an ultrasound-assisted extraction method optimized by factorial design. Fresenius J Anal Chem 1999, 363, 283–8.
[34] Quevauviller P, Rauret G, Ure A, Rubio R, López-Sánchez J.-F.; Fiedler H, Muntau H. Preparation of candidate certified reference materials for the quality control of EDTA-and acetic acid-extractable trace metal determinations in sewage sludge-amended soil and terra rossa soil. Mikrochim Acta 1995, 120, 289–300.
[35] US EPA O. Ground Water Sampling for Metal Analyses http://www2.epa.gov/remedytech/ground-water-sampling-metal-analyses.
[36] Ahmed R, Stoeppler M. Decomposition and stability studies of methylmercury in water using cold vapour atomic absorption spectrometry. Analyst 1986, 111, 1371–4.
[37] USGS National Field Manual for the Collection of Water-Quality Data. Techniques of Water-Resources Investigations. Book 9 Handbook for Water-Resources Investigations; 2015.
[38] Behlke MK, Uden PC, Schantz MM. Investigations of sulfur interferences in the extraction of methylmercury from marine tissues. Anal Commun 1996, 33, 91–2.
[39] Woller Á, Garraud H, Martin F, Donard OFX, Fodor P. Determination of total mercury in sediments by microwave-assisted digestion-flow injection-inductively coupled plasma mass spectrometry. J Anal At Spectrom 1997, 12, 53–6.
[40] Harrington CF, Catterick T. Problems encountered during the development of a method for the speciation of mercury and methylmercury by high-performance liquid chromatography coupled to inductively coupled plasma mass spectrometry. J Anal At Spectrom 1997, 12, 1053–6.
[41] Bloom NS. Analysis and stability of mercury speciation in petroleum hydrocarbons. Fresenius J Anal Chem 2000, 366, 438–43.
[42] Epstein MS, Diamondstone BI, Gills TE. A new river sediment standard reference material. Talanta 1989, 36, 141–50.
[43] US EPA, R. 08 Quick Guide to Drinking Water Sample Collection http://www2.epa.gov/region8-waterops/quick-guide-drinking-water-sample-collection
[44] Mohammadkhani S, Gholami MR, Aghaie M. Thermodynamic study of Cr+3 ions removal by "MnO2/MWCNT" nanocomposite. Orient J Chem 2015, 31, 1429–36.
[45] Tyson J. High-performance, flow-based, sample pre-treatment and introduction procedures for analytical atomic spectrometry. Journal of Analytical Atomic Spectrometry 1999, 14, 169–78.
[46] Anschutz P, Dedieu K, Desmazes F, Chaillou G. Speciation, oxidation state, and reactivity of particulate manganese in marine sediments. Chem Geol 2005, 218, 265–79.
[47] Chao TT. Selective dissolution of manganese oxides from soils and sediments with acidified hydroxylamine hydrochloride. Soil Sci Soc Am J 1972, 36, 764–8.
[48] Golimowski J, Krasnodębska-Ostręga B. Voltammetric determination of heavy metals leached from ceramics. Fresenius J Anal Chem 1998, 361, 65–8.
[49] Rasmussen RR, Qian Y, Sloth JJ. SPE HG-AAS method for the determination of inorganic arsenic in rice–results from method validation studies and a survey on rice products. Anal Bioanal Chem 2013, 405, 7851–7.

[50] Sahinkaya E, Kilic A, Altun M, Komnitsas K, Lens PNL. Hexavalent chromium reduction in a sulfur reducing packed-bed bioreactor. J Hazard Mater 2012, 219–20, 253–9.

[51] Tokalıoğlu Ş, Arsav S, Delibaş A, Soykan C. Indirect speciation of Cr(III) and Cr(VI) in water samples by selective separation and preconcentration on a newly synthesized chelating resin. Anal Chim Acta 2009, 645, 36–41.

[52] Reinholdsson F, Briche C, Emteborg H, Baxter DC, Frech W. Determination of mercury species in natural waters. In; CANAS '95 Colloquium Analytische Atomspektroskopie, B. Welz (Ed.), Bodenseewerk Perkin-Elmer GmbH, Überlingen 1996, 399–405.

[53] Karlsson U, Karlsson S, Düker A. The effect of light and iron(II)/iron(III) on the distribution of Tl(I)/Tl(III) in fresh water systems. J Environ Monit 2006, 8, 634–40.

[54] Golimowski J, Golimowska K. UV-photooxidation as pretreatment step in inorganic analysis of environmental samples. Anal Chim Acta 1996, 325, 111–33.

[55] Li D, Gao Z, Zhu Y, Yu Y, Wang H. Photochemical reaction of Tl in aqueous solution and its environmental significance. Geochem J 2005, 39, 113–9.

[56] Quevauviller P, Drabaek I, Muntau H, Griepink B. Improvements in methylmercury determination prior to the certification of two tuna fish materials. Appl Organometal Chem 1993, 7, 413–20.

[57] Kebbekus B.B. Chapter 5 , Preparation of samples for metals analysis, p. 227 in S. Mitra (ed.), *Sample preparation techniques in analytical chemistry*, 162 (2003). John Wiley & Sons, Inc., Hoboken 2003.

[58] Zago C, Capodaglio G, Ceradini S, Ciceri G, Abelmoschi L, Soggia F, Cescon P, Scarponi G. Benthic fluxes of cadmium, lead, copper and nitrogen species in the northern Adriatic Sea in front of the River Po outflow, Italy. Sci Total Environ 2000, 246, 121–37.

[59] Eary LE, Rai D. Kinetics of chromium(III) oxidation to chromium(VI) by reaction with manganese dioxide. Environ Sci Technol 1987, 21, 1187–93.

[60] Huang J-H, Hu K-N, Ilgen J, Ilgen G. Occurrence and stability of inorganic and organic arsenic species in wines, rice wines and beers from Central European market. Food Addit Contam 2012, 29, 85–93.

[61] Garboś S, Rzepecka M, Bulska E, Hulanicki A. Microcolumn sorption of antimony(III) chelate for antimony speciation studies. Spectrochim Acta, Part B 1999, 54, 873–81.

[62] Gasparatos D, Mavromati G, Kotsovilis P, Massas I. Fractionation of heavy metals and evaluation of the environmental risk for the alkaline soils of the Thriassio plain: A residential, agricultural, and industrial area in Greece. Environ Earth Sci 2015, 74, 1099–108.

[63] Wang Y, Zeng X, Lu Y, Su S, Bai L, Li L, Wu C. Effect of aging on the bioavailability and fractionation of arsenic in soils derived from five parent materials in a red soil region of Southern China. Environ Pollut 2015, 207, 79–87.

[64] Lou C, Liu X, Nie Y, Emslie SD. Fractionation distribution and preliminary ecological risk assessment of As, Hg and Cd in ornithogenic sediments from the Ross Sea region, East Antarctica. Sci Total Environ 2015, 538, 644–53.

[65] Zan F, Huo S, Zhang J, Zhang L, Xi B, Zhang L. Arsenic fractionation and contamination assessment in sediments of thirteen lakes from the East Plain and Yungui Plateau Ecoregions, China. J Environ Sci 2014, 26, 1977–84.

[66] Krasnodębska-Ostręga B, Pałdyna J, Kowalska J, Jedynak Ł, Golimowski J. Fractionation study in bioleached metallurgy wastes using six-step sequential extraction. J Hazard Mater 2009, 167, 128–35.

[67] Pałdyna J, Krasnodębska-Ostrega B, Kregielewska K, Kowalska J, Jedynak L, Golimowski J, Grobelski T, Farbiszewska-Kiczma J, Farbiszewska T. The assessment of environmental pollution caused by mining and metallurgy wastes from highly polluted post-industrial regions in Southern Poland. Environ Earth Sci 2012, 68, 439–50.

[68] USGS; Smith DB, Cannon WF, Woodruff LG, Solano F, Kilburn JE, Fey DL. Geochemical and mineralogical data for soils of the conterminous United States.
[69] US EPA US Environmental Protection Agency, http://www3.epa.gov/ (accessed October 16, 2015).
[70] ISO – International Organization for Standardization, http://www.iso.org/iso/home.html
[71] Arunachalam J, Emons H, Krasnodebska B, Mohl C. Sequential extraction studies on homogenized forest soil samples. Sci Total Environ 1996, 181, 147–59.
[72] Krasnodębska-Ostrega B, Emons H, Golimowski J. Selective leaching of elements associated with Mn-Fe oxides in forest soil, and comparison of two sequential extraction methods. Fresenius J Anal Chem 2001, 371, 385–90.
[73] Krasnodębska-Ostręga B, Emons H, Golimowski J. Element fractionation in soil from urban-industrialized areas using sequential extraction. J Soils Sediments 2004, 4, 43–8.
[74] Zhang W, Cai Y, Tu C, Ma LQ. Arsenic speciation and distribution in an arsenic hyperaccumulating plant. Sci Total Environ 2002, 300, 167–77.
[75] Rubio R, Ure AM. Approaches to sampling and sample pretreatments for metal speciation in soils and sediments. Int J Environmental Analytical Chemistry 1993, 51, 205–217.
[76] Pickering WF. Metal ion speciation — soils and sediments (a review). Ore Geol Rev 1986, 1, 83–146.
[77] Shuman LM. Separating soil iron- and manganese-oxide fractions for microelement analysis. Soil Sci Soc Am J 1982, 46, 1099–102.
[78] Godlewska-Żyłkiewicz B, Leśniewska BA, Hulanicki A. The study of magnesium speciation in serum by liquid chromatography and graphite furnace atomic absorption techniques. Anal Chim Acta 1998, 358, 185–93.
[79] Emons H. Artefacts and facts about metal(loid)s and their species from analytical procedures in environmental biomonitoring. Trends Anal Chem 2002, 21, 401–11.
[80] Emons H. Challenges from speciation analysis for the development of biological reference materials. Fresenius J Anal Chem 2001, 370, 115–19.
[81] Sadowska M, Biaduń E, Krasnodębska-Ostręga B. Stability of Tl(III) in the context of speciation analysis of thallium in plants. Chemosphere 2016, 144, 1216–23.
[82] Dahl L, Molin M, Amlund H, Meltzer HM, Julshamn K, Alexander J, Sloth JJ. Stability of arsenic compounds in seafood samples during processing and storage by freezing. Food Chem 2010, 123, 720–7.
[83] Godlewska-Żyłkiewicz B, Leśniewska B, Golimowski J. Systematic errors in the determination of trace metals by GFAAS technique, Part I. Microchim Acta 2003, 143, 13–7.
[84] Le SXC, Cullen WR, Reimer KJ. Speciation of arsenic compounds in some marine organisms. Environ Sci Technol 1994, 28, 1598–604.
[85] Kowalska J, Kińska K, Sadowska M, Biesaga M, Krasnodębska-Ostręga B. The role of phytochelatins in Sinapis alba L. response to stress caused by two toxic elements As and Tl. Int J Environ Anal Chem 2015, 95, 1148–56.
[86] Sneller FEC, van Heerwaarden LM, Koevoets PLM, Vooijs R, Schat H, Verkleij JAC. Derivatization of phytochelatins from Silene vulgaris, induced upon exposure to arsenate and cadmium: Comparison of derivatization with Ellman's reagent and monobromobimane. J Agric Food Chem 2000, 48, 4014–19.
[87] Bluemlein K, Raab A, Meharg AA, Charnock JM, Feldmann J. Can we trust mass spectrometry for determination of arsenic peptides in plants: comparison of LC–ICP–MS and LC–ES-MS/ICP–MS with XANES/EXAFS in analysis of Thunbergia alata. Anal Bioanal Chem 2007, 390, 1739–51.
[88] Bluemlein K, Raab A, Feldmann J. Stability of arsenic peptides in plant extracts: off-line versus on-line parallel elemental and molecular mass spectrometric detection for liquid chromatographic separation. Anal Bioanal Chem 2008, 393, 357–66.

[89] Jedynak Ł, Kowalska J, Leporowska A. Arsenic uptake and phytochelatin synthesis by plants from two arsenic-contaminated sites in Poland. Polish J Environ Stud 2012, 21, 1629–33.

[90] Landberg T, Greger M. No phytochelatin (PC2 and PC3) detected in Salix viminalis. Physiol Plant 2004, 121, 481–7.

[91] Namdjoyan S, Namdjoyan S, Kermanian H. Induction of phytochelatin and responses of antioxidants under cadmium stress in safflower (Carthamus tinctorius) seedlings. Turkish J Bot 2012, 36, 495–502.

[92] Backhaus F, Bagschik U, Burow M, Byrne AR, Froning M, Mohl C, Ostapczuk P, Rossbach M, Schladot JD, Stoeppler M, Waidmann E, Zeisler R. Two spruce shoot candidate reference materials from the German environmental specimen bank. Sci Total Environ 1993, 139–40, 447–58.

[93] Klein R, Bartel M, Paulus M, Quack M, Tarricone K, Teubner D, Wagner G, 2010. Guideline for sampling and sample treatment eelpout (Zoarces viviparus) in: Environment Protection Agency (Eds.), Guidelines for sampling, transport, storage and chemical characterisation of environmental- and human organ samples. Environment Protection Agency, Berlin.

[94] Krasnodębska-Ostręga B, Golimowski J. Element fractionation in suspended matter in landfill leachate using single extractions. Microchim Acta 2004, 146, 7–11.

[95] Paulus M, Klein R, Wagner G, Müller P. Biomonitoring and environmental specimen banking. Environ Sci Pollut Res 1996, 3, 169–77.

[96] Rossbach M, Mohl C, Emons H. Analysis of soils and sediments for monitoring purposes within the environment specimen bank of Germany. Heavy Met Environ 1995, 2, 383–6.

[97] Rüdel H, Müller J, Quack M, Klein R. Monitoring of hexabromocyclododecane diastereomers in fish from European freshwaters and estuaries. Environ Sci Pollut Res 2011, 19, 772–83.

[98] Burgess J. Kinetic aspects of chemical speciation. Analyst 1992, 117, 605–11.

[99] Sanz-Medel A. Trace element analytical speciation in biological systems: importance, challenges and trends. Spectrochim Acta, Part B 1998, 53, 197–211.

[100] Templeton D, Ariese F, Cornelis R, Danielsson L, Muntau H, Van Leeuwen H, Lobinski R. Guidelines for terms related to chemical speciation and fractionation of elements. Definitions, structural aspects, and methodological approaches (IUPAC Recommendations 2000). Pure Appl Chem 2000, 72, 1453–70.

[101] Dai J, Ren F, Tao C. Adsorption of Cr(VI) and Speciation of Cr(VI) and Cr(III) in aqueous solutions using chemically modified chitosan. Int J Environ Res Public Health 2012, 9, 1757–70.

[102] Ebdon L, Pitts L, Cornelis R, Crews H, Donard OFX, Quevauviller P. Trace element speciation for environment, food and health. In: Trace Element Speciation for Environment, Food and Health. Royal Society of Chemistry, Cambridge, 2001.

[103] Peterson ML, Carpenter R. Biogeochemical processes affecting total arsenic and arsenic species distributions in an intermittently anoxic fjord. Mar Chem 1983, 12, 295–321.

[104] Krasnodębska-Ostręga B, Asztemborska M, Golimowski J, Strusińska K. Determination of thallium forms in plant extracts by anion exchange chromatography with inductively coupled plasma mass spectrometry detection (IC-ICP-MS). J Anal At Spectrom 2008, 23, 1632–5.

[105] Lin T-S, Nriagu JO. Thallium speciation in river waters with Chelex-100 resin. Anal Chim Acta 1999, 395, 301–7.

[106] Guéguen C, Dominik J. Partitioning of trace metals between particulate, colloidal and truly dissolved fractions in a polluted river: The Upper Vistula River (Poland). Appl Geochem 2003, 18, 457–70.

[107] Moon J, Cho J. Investigation of nano-colloid transport in UF membranes using flow field-flow fractionation (flow FFF) and an irreversible thermodynamic transport model. Desalination 2005, 179, 151–9.

[108] Fresenius W, Quentin KE, Schneider W. Water Analysis. A Practical Guide to Physico-chemical, Chemical and Microbiological Water Examination and Quality Assurance. Springer, Berlin, Heidelberg, 1988.

[109] Banks D, Markland H, Smith PV, Mendez C, Rodriguez J, Huerta A, Sæther OM. The effect of filtration on analyses of surface water samples. A study from the Salars of Coipasa and Uyuni, Bolivian Altiplano. J Geochem Explor 2005, 86, 104–18.

[110] Krasnodębska-Ostrega B, Dmowski K, Stryjewska E, Golimowski J. Determination of thallium and other elements (As, Cd, Cu, Mn, Pb, Se, Sb, and Zn) in water and sediment samples from the vicinity of the zinc-lead smelter in Poland (3 pp). J Soils Sediments 2005, 5, 71–3.

[111] Nakatsuka S, Okamura K, Norisuye K, Sohrin Y. Simultaneous determination of suspended particulate trace metals (Co, Ni, Cu, Zn, Cd and Pb) in seawater with small volume filtration assisted by microwave digestion and flow injection inductively coupled plasma mass spectrometer. Anal Chim Acta 2007, 594, 52–60.

[112] Ospina-Alvarez N, Głaz Ł, Dmowski K, Krasnodębska-Ostręga B. Mobility of toxic elements in carbonate sediments from a mining area in Poland. Environ Chem Lett 2014, 12, 435–41.

[113] Schäfer AI, Fane AG, Waite TD. Fouling effects on rejection in the membrane filtration of natural waters. Desalination 2000, 131, 215–24.

[114] Schäfer AI, Fane AG, Waite TD. Cost factors and chemical pretreatment effects in the membrane filtration of waters containing natural organic matter. Water Res 2001, 35, 1509–17.

[115] Krasnodębska-Ostręga B, Jedynak Ł, Galus M, Pałdyna J, Kowalska J, Golimowski J. Fractionation of selected elements in water samples from the mining and smelting area. Chem Anal 2009, 54(5), 1009–19.

[116] Ospina-Alvarez N, Burakiewicz P, Sadowska M, Krasnodebska-Ostrega B. TlI and TlIII presence in suspended particulate matter: Speciation analysis of thallium in wastewater. Environ Chem 2015, 12, 374–9.

[117] Shanker AK, Cervantes C, Loza-Tavera H, Avudainayagam S. Chromium toxicity in plants. Environ Int 2005, 31, 739–53.

[118] Singh HP, Mahajan P, Kaur S, Batish DR, Kohli RK. Chromium toxicity and tolerance in plants. Environ Chem Lett 2013, 11, 229–54.

[119] Jain CK, Sharma MK. Distribution of trace metals in the Hindon River system, India. J Hydrol 2001, 253, 81–90.

[120] Daus B, Weiss H, Mattusch J, Wennrich R. Preservation of arsenic species in water samples using phosphoric acid – Limitations and long-term stability. Talanta 2006, 69, 430–34.

[121] Kim MJ. Separation of inorganic arsenic species in groundwater using ion exchange method. Bull Environ Contam Toxicol 2001, 67, 46–51.

[122] Krasnodębska-Ostręga B, Sadowska M, Ostrowska S. Thallium speciation in plant tissues—Tl(III) found in Sinapis alba L. grown in soil polluted with tailing sediment containing thallium minerals. Talanta 2012, 93, 326–9.

[123] Wolf RE, Morman SA, Hageman PL, Hoefen TM, Plumlee GS. Simultaneous speciation of arsenic, selenium, and chromium: species stability, sample preservation, and analysis of ash and soil leachates. Anal Bioanal Chem 2011, 401, 2733–45.

[124] Boszke L, Glosinska G, Siepak J. Some aspects of speciation of mercury in water environment. Pol J Environ Stud 2002, 11, 285–98.

[125] Morel FMM, Kraepiel AML, Amyot M. The chemical cycle and bioaccumulation of mercury. Ann Rev Ecol System 1998, 29, 543–66.

[126] Markert B. Sample preparation (cleaning, drying, homogenization) for trace element analysis in plant matrices. Sci Total Environ 1995, 176, 45–61.

[127] García-Salgado S, Quijano MÁ. Stability of toxic arsenic species and arsenosugars found in the dry alga Hijiki and its water extracts. Talanta 2014, 128, 83–91.

[128] Mir KA, Rutter A, Koch I, Smith P, Reimer KJ, Poland JS. Extraction and speciation of arsenic in plants grown on arsenic contaminated soils. Talanta 2007, 72, 1507–18.

[129] Pell A, Márquez A, Rubio R, López-Sánchez JF. Effects of sample processing on arsenic speciation in marine macroalgae. Anal Methods 2013, 5, 2543–50.

[130] Caruso JA, Heitkemper DT, B'Hymer C. An evaluation of extraction techniques for arsenic species from freeze-dried apple samples. Analyst 2001, 126, 136–40.

[131] Berrow ML, Stein WM. Extraction of metals from soils and sewage sludges by refluxing with aqua regia. Analyst 1983, 108, 277–85.

[132] Jansen B, Nierop KGJ, Kotte MC, de Voogt P, Verstraten JM. The applicability of accelerated solvent extraction (ASE) to extract lipid biomarkers from soils. Appl Geochem 2006, 21, 1006–15.

[133] Mossop KF, Davidson CM. Comparison of original and modified BCR sequential extraction procedures for the fractionation of copper, iron, lead, manganese and zinc in soils and sediments. Anal Chim Acta 2003, 478, 111–18.

[134] Quevauviller P, Rauret G, Muntau H, Ure AM, Rubio R, López-Sánchez JF, Fiedler HD, Griepink B. Evaluation of a sequential extraction procedure for the determination of extractable trace metal contents in sediments. Fresenius J Anal Chem 1994, 349, 808–14.

[135] Quevauviller P, Maier EA, Griepink B. Quality Assurance for Environmental Analysis. Volume 17, 1st edition. Elsevier Science 1995.

[136] Das AK, Chakraborty R, Cervera ML, de la Guardia M. Metal speciation in solid matrices. Talanta 1995, 42, 1007–30.

[137] Davidson CM, Delevoye G. Effect of ultrasonic agitation on the release of copper, iron, manganese and zinc from soil and sediment using the BCR three-stage sequential extraction. J Environ Monit 2001, 3, 398–403.

[138] Jaradat QM, Massadeh AM, Zaitoun MA, Maitah BM. Fractionation and sequential extraction of heavy metals in the soil of scrapyard of discarded vehicles. Environ Monit Assess 2006, 112, 197–210.

[139] Kersten M, Förstner U. Effect of sample pretreatment on the reliability of solid speciation data of heavy metals — implications sesfor the study of early diagenetic processes. Mar Chem 1987, 22, 299–312.

[140] Marguí E, Queralt I, Carvalho ML, Hidalgo M. Assessment of metal availability to vegetation (Betula pendula) in Pb-Zn ore concentrate residues with different features. Environ Pollut 2007, 145, 179–84.

[141] Thomas RP, Ure AM, Davidson CM, Littlejohn D, Rauret G, Rubio R, López-Sánchez JF. Three-stage sequential extraction procedure for the determination of metals in river sediments. Anal Chim Acta 1994, 286, 423–29.

[142] Pauwels J, Vandecasteele C. Determination of the minimum sample mass of a solid CRM to be used in chemical analysis. Fresenius J Anal Chem 1993, 345, 121–3.

[143] Lewis J, Stokes P, Brereton N, Baxter M, Macarthur R. Stability of arsenic speciation in fish muscle samples, under different storage and sample preparation conditions. Microchem J 2012, 105, 56–9.

[144] Dybczyński R, Polkowska-Motrenko H, Samczyński Z, Szopa Z. New Polish certified reference materials for multielement inorganic trace analysis. Fresenius J Anal Chem 1993, 345, 99–103.

[145] Marchandise H. Quality and accuracy in analytical chemistry. Fresenius J Anal Chem 1993, 345, 82–6.

[146] MacNaeidhe F. Procedures and precautions used in sampling techniques and analysis of trace elements in plant matrices. Sci Total Environ 1995, 176, 25–31.

[147] Steinnes E, Rühling Å, Lippo H, Mäkinen A. Reference materials for large-scale metal deposition surveys. Accred Qual Assur 1997, 2, 243–9.

[148] Jedynak Ł, Kowalska J, Kossykowska M, Golimowski J. Studies on the uptake of different arsenic forms and the influence of sample pretreatment on arsenic speciation in White mustard (Sinapis alba). Microchem J 2010, 94, 125–9.

[149] Mayeresse Y, de Cupere V, Veillon R, Brende J. Consideration for transferring a bulk freeze-drying process from a glass container to a tray. Pharm Eng 2009, Mar/Apr, 1–8.
[150] Marguí E, Salvadó V, Queralt I, Hidalgo M. Comparison of three-stage sequential extraction and toxicity characteristic leaching tests to evaluate metal mobility in mining wastes. Anal Chim Acta 2004, 524, 151–9.
[151] Pérez-Cid B, Lavilla I, Bendicho C. Use of ultrasonic energy for shortening the sequential extraction of metals from river sediments. Int J Environ Anal Chem 1999, 73, 79–92.
[152] Moturi MCZ, Rawat M, Subramanian V. Distribution and fractionation of heavy metals in solid waste from selected sites in the industrial belt of Delhi, India. Environ Monit Assess 2004, 95, 183–99.
[153] Tokalioglu Ş, Kartal Ş, Elçi L. Determination of heavy metals and their speciation in lake sediments by flame atomic absorption spectrometry after a four-stage sequential extraction procedure. Anal Chim Acta 2000, 413, 33–40.
[154] Schmöger MEV, Oven M, Grill E. Detoxification of arsenic by phytochelatins in plants. Plant Physiol 2000, 122, 793–802.
[155] Andra SS, Datta R, Sarkar D, Saminathan SKM, Mullens CP, Bach SBH. Analysis of phytochelatin complexes in the lead tolerant vetiver grass [Vetiveria zizanioides (L.)] using liquid chromatography and mass spectrometry. Environ Pollut 2009, 157, 2173–83.
[156] Chen L, Yang L, Wang Q. In vivo phytochelatins and Hg–phytochelatin complexes in Hg-stressed Brassica chinensis L. Metallomics 2009, 1, 101–6.
[157] Krasnodębska-Ostręga B, Kowalska J. Use of ultrasound energy for shortening acetic acid extraction of metals from soils. Chem Anal 2003, 48, 967–74.
[158] Krasnodębska-Ostręga B, Kaczorowska M, Golimowski J. Ultrasound-assisted extraction for the evaluation of element mobility in bottom sediment collected at mining and smelting Pb–Zn ores area in Poland. Microchim Acta 2006, 154, 39–43.
[159] Ruiz-Chancho MJ, Sabé R, López-Sánchez JF, Rubio R, Thomas P. New approaches to the extraction of arsenic species from soils. Microchim Acta 2005, 151, 241–8.
[160] Quaghebeur M, Rengel Z, Smirk M. Arsenic speciation in terrestrial plant material using microwave-assisted extraction, ion chromatography and inductively coupled plasma mass spectrometry. J Anal At Spectrom 2003, 18, 128–34.
[161] Kowalska J, Asztemborska M, Bystrzejewska-Piotrowska G. Platinum uptake by mustard (Sinapis alba L.) and maize (Zea mays L.) plants. Nukleonika 2004, 49, 31–4.
[162] Hawieńczyk M, Bystrzejewska-Piotrowska B, Kowalska J, Asztemborska M. Platinum bioaccumulation by mustard plants (Sinapis alba L.). Nukleonika 2005, 59–61.
[163] Byrne AR, Šlejkovec Z, Stijve T, Fay L, Gössler W, Gailer J, Lrgolic KJ. Arsenobetaine and other arsenic species in mushrooms. Appl Organometal Chem 1995, 9, 305–13.
[164] Michalska-Kacymirow M, Kurek E, Smolis A, Wierzbicka M, Bulska E. Biological and chemical investigation of Allium cepa L. response to selenium inorganic compounds. Anal Bioanal Chem 2014, 406, 3717–22.
[165] Biswas D, Banerjee M, Sen G, Das JK, Banerjee A, Sau TJ, Pandit S, Giri AK, Biswas T. Mechanism of erythrocyte death in human population exposed to arsenic through drinking water. Toxicol Appl Pharmacol 2008, 230, 57–66.
[166] Morita M, Shibata Y. Chemical form of arsenic in marine macroalgae. Appl. Organometal Chem 1990, 4, 181–90.
[167] Ybáñez N, Velez D, Tejedor W, Montoro R. Optimization of the extraction, clean-up and determination of arsenobetaine in manufactured seafood products by coupling liquid chromatography with inductively coupled plasma atomic emission spectrometry. J Anal At Spectrom 1995, 10, 459–65.

[168] Nolan A, Schaumlöffel D, Lombi E, Ouerdane L, Łobiski R, McLaughlin M. Determination of Tl-(I) and Tl-(III) by IC-ICP-MS and application to Tl speciation analysis in the Tl hyperaccumulator plant Iberis intermedia. J Anal At Spectrom 2004, 19, 757–61.

[169] Chu Y-L, Wang R-Y, Jiang S-J. Speciation analysis of thallium by reversed-phase liquid chromatography – inductively coupled plasma mass spectrometry. J Chinese Chem Soc 2012, 59, 219–25.

[170] Wang J. Wiley: Analytical Electrochemistry, 3rd Edition – Joseph Wang http://eu.wiley.com/WileyCDA/WileyTitle/productCd-0471678791.html (accessed October 23, 2015).

[171] Bobrowski A, Gawlicki M, Kapturski P, Mirceski V, Spasovski F, Zarębski J. The silver amalgam film electrode in adsorptive stripping voltammetric determination of palladium(II) as its dimethyldioxime complex. Electroanalysis 2009, 21, 36–40.

[172] Grabarczyk M, Korolczuk M. Development of a simple and fast voltammetric procedure for determination of trace quantity of Se(IV) in natural lake and river water samples. J Hazard Mater 2010, 175, 1007–13.

[173] Jakubowska M, Zembrzuski W, Lukaszewski Z. Thallium determination at the single picomole per liter level by flow-injection differential-pulse anodic stripping voltammetry. Electroanalysis 2008, 20, 1073–77.

[174] Grabarczyk M. Stripping voltammetric determination of As(III) in natural water samples containing surface active compounds. Electroanalysis 2010, 22, 2017–23.

[175] Anastassiades M, Lehotay SJ, Stajnbaher D, Schenck FJ. Fast and easy multiresidue method employing acetonitrile extraction/partitioning and "dispersive solid-phase extraction" for the determination of pesticide residues in produce. J AOAC Int 2003, 86, 412–31.

[176] Payá P, Anastassiades M, Mack D, Sigalova I, Tasdelen B, Oliva J, Barba A. Analysis of pesticide residues using the quick easy cheap effective rugged and safe (QuEChERS) pesticide multiresidue method in combination with gas and liquid chromatography and tandem mass spectrometric detection. Anal Bioanal Chem 2007, 389, 1697–14.

[177] AOAC Official Method 2007.01 Pesticide residues in foods by acetonitrile extraction and partitioning with magnesium sulfate. Gas Chromatography/Mass Spectrometry and Liquid Chromatography/Tandem Mass Spectrometry First Action 2007.

[178] Chen M-L, Ma L-Y, Chen X-W. New procedures for arsenic speciation: A review. Talanta 2014, 125, 78–86.

[179] Pérez S, Aga DS. Recent advances in the sample preparation, liquid chromatography tandem mass spectrometric analysis and environmental fate of microcystins in water. Trends Anal Chem 2005, 24, 658–70.

[180] Wolska L, Wiergowski M, Galer K, Górecki T, Namieśnik J. Sample preparation for GC analysis of selected pesticides in surface water. Chemosphere 1999, 39, 1477–86.

[181] Guo H-M, Liu C-H. Separation of inorganic arsenic species from aqueous solution by anion exchange column and its application in study of arsenic removal. Chinese J of Anal Chem 2013, 40, 1092–97.

[182] Gleyzes C, Tellier S, Astruc M. Fractionation studies of trace elements in contaminated soils and sediments: a review of sequential extraction procedures. Trends Anal Chem 2002, 21, 451–67.

[183] Tessier A, Campbell PGC, Bisson M. Sequential extraction procedure for the speciation of particulate trace metals. Anal Chem 1979, 51, 844–51.

[184] Quevauviller P, Ure A, Muntau H, Griepink B. Improvement of analytical measurements within the BCR-programme: Single and sequential extraction procedures applied to soil and sediment analysis. Int J Environ Anal Chem 1993, 51, 129–34.

[185] Rauret G, López-Sánchez JF, Sahuquillo A, Barahona E, Lachica M, Ure AM, Davidson CM, Gomez A, Lück D, Bacon J, Yli-Halla M, Muntau H, Quevauviller P. Application of a modified BCR

sequential extraction (three-step) procedure for the determination of extractable trace metal contents in a sewage sludge amended soil reference material (CRM 483), complemented by a three-year stability study of acetic acid and EDTA extractable metal content. J Environ Monit 2000, 2, 228–33.
[186] Rusnak R, Halasz G, Horvath M, Remeteiova D. Preliminary results on the intensification of the BCR sequential extraction with sonication for sediments, soils, and gravitation dust sediment samples. Toxicol Environ Chem 2010, 92, 443–52.
[187] Wallmann K, Kersten M, Gruber J, Förstner U. Artifacts in the determination of trace metal binding forms in anoxic sediments by sequential extraction. Int J Environ Anal Chem 1993, 51, 187–200.
[188] Torelm I, Croon L-B, Kolar K, Schröder T. Production and certification of "fresh" reference material for macronutrient analysis. Fresenius J Anal Chem 1990, 338, 435–7.
[189] http://ec.europa.eu/ European Commission http://ec.europa.eu/.
[190] http://www.usgs.gov/ Welcome to the USGS – U.S. Geological Survey http://www.usgs.gov/
[191] US EPA R. 08 Standard Operating Procedure: Surface Water Sampling http://www2.epa.gov/region8/standard-operating-procedure-surface-water-sampling.

Pawel Pohl, Anna Szymczycha-Madeja, Maja Welna and Piotr Jamroz

10 Solid-Phase Extraction in Fractionation of Trace Elements

10.1 Background

Elements in environmental and food samples can be present in different physicochemical forms (speciation forms) [1, 2]. These species of elements are distributed over a distinct molecular size and are characterized by the different mobility, availability, absorbability and toxicity to the biota. Unquestionably, possible speciation forms of elements play an important role in assessing their vital functions or toxicity and hazard in the ecosystem. This is because some of the speciation forms of elements are physiologically and nutritionally important for the proper functioning of the flora, fauna and other organisms; other forms can impose a serious hazard to these organisms and poison biological systems. For this particular reason, measurements of specific physicochemical forms of elements appear to be critical to properly find out interactions of elements with leaving organisms.

Unfortunately, the reliable identification and determination of distinct speciation forms of elements is a challenging task [3]. Methodological difficulties related to this subject are a consequence of low concentrations of certain species of elements in analyzed samples and the presence of a wide range of possible species, strongly varying due to a particle size and a chemical nature. Apparently, elements can be present in the form of simple ions, inorganic and organic complexes of different stability, as well as associates with colloidal or particular inorganic and organic matter. In addition, the speciation analysis of elements in mentioned environmental and food samples can be hindered because targeted physicochemical forms of elements typically participate in pH-dependent dynamic equilibrium with other species of elements. In terms of these inconveniences, the qualitative and quantitative determination of individual species of elements, in line with the IUPAC recommendation about the assessment of a complete distribution of elements over separate molecular chemical and physical forms, is very difficult and sometimes impossible [3]. The difficulty of this task is also related to the fact that the determination of separate speciation forms of elements in certain sample matrices requires an individual and competent treatment. The sample collection, the storage and finally the pretreatment (i.e., the sample preparation procedure) prior to the analysis of environmental (e.g., biological tissues, water) and food samples are critical in this case because the distribution and the integrity of species of elements have to be strictly maintained and preserved. Any uncontrolled changes subjected to sampled and analyzed samples can influence the original speciation of studied elements and distort results of the speciation analysis.

https://doi.org/10.1515/9783110366730-011

Therefore, it is recognized that a better approach to identify and determine speciation forms of elements for finding out their status in the environment and related biological systems is to use selective extraction procedures for the separation of classes (fractions) of species of elements that are similar according to specific physical and chemical properties [2, 3]. Such a strategy is termed as the fractionation, that is, the classification of an analyte or a group of analytes from a certain sample according to their given physical and/or chemical characteristics [3]. In this case, the interest in the speciation analysis is limited to binary or ternary differentiations of species of elements, that is, inorganic and organic species, labile and stable complexes, bound and unbound species, particulate and dissolved fraction [4]. This approach can also be used when other properties of species of elements are important, and hence, species and forms of elements can be grouped into two or more distinguishable categories varying due to their functionality, that is, toxic and nontoxic, bound and unbound, inert and bioavailable, inaccessible or bioaccessible, depending on the aim of the fractionation analysis.

Very often, the primary aim of the fractionation is to assess the bioavailability and/or the bioaccessibility of elements as markers of their impact on living organisms [4]. In this case, possible species of elements can be partitioned due to their associations with molecules of a different size (by using the filtration and the ultrafiltration) and/or chemical properties such as the liability, the solubility, the affinity, the charge and the hydrophobicity (using selective extraction and separation procedures like the solid-phase extraction [SPE]). It appears that results of the fractionation analysis have a higher impact and the informative value. They can bring, for example, the important information about the extent of elements that are resistant to pass through cellular membranes or soluble in given conditions, and hence, available and toxic to living organisms. In terms of this, the bioavailability and the accessibility of elements can be readily estimated while a risk of their toxicity due to the exposure in the environment can be conveniently assessed [1, 2].

Using the fractionation analysis by means of different fractionation/classification strategies, the knowledge about possible fractions of elements having different properties and functionalities can be possessed [1, 2]. It should be noted that such fractions and groupings of species of elements are operationally defined because they exhibit a similar reaction to specific procedures or methods of the separation used for the partitioning of existing species of elements. Certainly, the main inconvenience of such fractionation analyses of elements is that it is difficult to compare results obtained by them. This is mostly because there is no harmonization neither standardization rules regarding chemical fractionation procedures utilizing the SPE and other chromatographic and nonchromatographic separation procedures. The best way to verify the reliability of results, that is, operationally defined fractions of elements, would be the use of special certified reference materials (CRMs) with certified or indicative values of fractions of elements in given sample matrices. To date, there is no such kind of CRMs, making it difficult to verify the dependability of results of the fractionation analysis carried out using the SPE and/or other chromatographic and nonchromatographic separation procedures. This also encourages researchers to use different manners to ensure the quality control (QC) and the quality assurance (QA) of outcomes of the fractionation analysis of elements by means of SPE.

10.2 Operationally defined fractionation of elements through SPE

SPE, normally used for the preconcentration of elements prior to the determination of total concentrations of elements by spectrometric detection techniques, is very useful for the purpose of the fractionation and the partitioning of elements. In this case, a great number of commercially available or synthesized selective or specific solid sorbents can be used for the retention and/or the separation of certain groups (classes) of species of elements due to their similar properties. Species of elements can be retained through ion exchange, chelation or adsorption, depending on their chemical and/or physical nature. It is recognized that fractionation protocols based on the SPE can yield the comprehensive knowledge about the solubility, the reachability, the bioavailability and a complexation degree of species of elements. Combining SPE with the filtration and the ultrafiltration or the electrochemical stripping analysis, the fractionation according to the molecular size (dissolved, colloidal, suspended, particulate) or the stability (mobile, labile and electrochemical active) is possible and quite often applied [1]. In reference to the simplicity and a practical use, the fractionation analysis by SPE does not require a sophisticated and expensive equipment, except for properly selected resins and solid sorbents that are commercialized in the form of SPE cartridges and tubes or packed into glass columns.

Chemical fractionation approaches for the partitioning of elements by the SPE can be grouped into two main categories: (1) one-column approaches, in which sample solutions are separately treated with one or two particular SPE sorbents and (2) two- or three-column approaches, in which sample solutions are treated with two or three different SPE sorbents (packed into columns or filled inside syringe tubes or cartridges) connected in a series. In addition, there are mixed protocols that combine the chemical fractionation by the SPE treatment with the physical fractionation by the filtration or the chemical fractionation by the solvent extraction or the precipitation.

10.2.1 Unary and binary fractionations using one-column SPE approaches

Examples of one-column SPE procedures with names of sorbents applied and chemical fractions separated are given in Table 10.1. As can be seen, anion-exchange, cation-exchange and chelating resins are primarily used in these fractionation protocols to retain respective forms of species of elements, correspondingly grouped into fractions of anionic, cationic and labile species along with positively charged complexes. Hydrophobic nonionic polymeric resins or chemically modified silica

Table 10.1: Chemical fractionation analysis with one-column SPE-based procedures.

Element	Sample	SPE sorbents used and fractions determined	Detection technique	References
Nonelution approach, one-column fractionation approach				
Mg	Olives (water extracts)	Dowex 50W×8 (a strong cation exchanger in a column) – noncationic species, cationic species = total – noncationic species Amberlite XAD-2 (a hydrophobic macroreticula adsorbing resin) – unbound species, polyphenol-bound species = total – unbound species	FAAS	[5]
Ca and Mg	Honeys (water solutions), fruit juices, tea (green and black) infusions	Dowex 50W×4 (a strong cation exchanger in a column) – noncationic species, cationic species = total – noncationic species	FAAS	[8]
Al, Ba, Ca, Co, Fe, Mg, Mn, Ni, Rb, Sr and Zn	Tea (black) infusion	Maxi-Clean IC-Na (a strong cation exchanger in a cartridge) – noncationic species, cationic species = total – noncationic species	ICP MS	[9]
Al, Ba, Ca, Cd, Co, Cr, Cu, Fe, Mg, Mn, Ni, Pb, Sr and Zn	Lager beer	Dowex 50W×4 (a strong cation exchanger in a column), digestion of the sorbent – labile species and cationic species	ICP OES	[10]
Cu, Fe and Mn	Lager beer	Isolute SCX (a strong cation exchanger in a cartridge) – noncationic species, cationic species = total – noncationic species Isolute SAX (a strong anion exchanger in a cartridge) – nonanionic species, anionic species = total – nonanionic species	GFAAS	[11]
Al	Yerba mate infusions	Chelex 100 (a chelating resin), a batch sorption procedure – labile and noninert species, inert (organically bound) species = total – labile and noninert species	FAAS	[12]

Elution approach, one-column fractionation				
Al, Cu, Fe, Mn, Ni and Zn	Tea (black) infusions	Amberlite XAD-2 or Amberlite XAD-7 (hydrophobic macroreticular adsorbing resins), a batch sorption and elution procedure, elution with 2 mol L^{-1} HNO_3 – polyphenol-bound species	ICP OES	[6]
Fe	Natural (river) waters	Envi-18 (an octadecyl silica reverse phase in a tube), elution with methanol – humic acid-bound species	ICP MS	[7]
Mn	Tea (black) infusion	LAB II (a strong cation exchanger in a tube), elution with 2 mol L^{-1} HNO_3 – cationic species LAB V (strong anion exchanger in a tube), elution with 2 mol L^{-1} HNO_3 – anionic species	FAAS	[13]
Cd, Co, Cu, Fe, Ni, Pb and Zn	Mineral (nongaseous) waters	Dowex 50W×4 (a strong cation exchanger in a column), elution with 4 mol L^{-1} HCl – labile and cationic species	ICP OES	[14]
Fe and Zn	Honeys (water solutions), fruit juices, tea (black and green) infusions	Dowex 50W×4 (a strong cation exchanger in a column), elution with 4 mol L^{-1} HCl – cationic species Dowex 1 × 4 (a strong anion exchanger in a column), elution with 4 mol L^{-1} HCl – anionic species Residual species = total – (cationic species + anionic species)	FAAS	[15]
Ca, Fe, Mg and Zn	Milks	Dowex 50W×4 (a strong cation exchanger in a column), elution with 2 mol L^{-1} HCl – cationic species Dowex 1 × 4 (a strong anion exchanger in a column), elution with 2 mol L^{-1} HCl – anionic species Residual species = total – (cationic species + anionic species)	FAAS	[16]
Cu, Fe	Fresh (river) waters	Chelite P (a chelating resin in a minicolumn), elution with 2 mol L^{-1} HCl – simple ions and labile complexes	FAAS	[17]

(Continued)

Table 10.1: (Continued)

Element	Sample	SPE sorbents used and fractions determined	Detection technique	References
Cu, Mn, Zn	Seawaters	Serdolit Chelite Che (a chelating resin in a minicolumn), elution with 3 mol L^{-1} HCl – simple ions and labile complexes	FAAS	[18]
Al	Tea (black and green) infusions	Chelex 100 (a chelating resin in a tube), sequential elution with water, 0.02 mol L^{-1} HCl and 2 mol L^{-1} HCl – labile and cationic species Dowex HCR-S (a strong cation-exchanger in a tube), sequential elution with water, 0.02 mol L^{-1} HCl and 2 mol L^{-1} HCl – labile and cationic species AGMP1 (a strong anion-exchanger in a tube), sequential elution with water, 0.02 mol L^{-1} HCl and 2 mol L^{-1} HCl – anionic and neutral species (polyphenol and organically bound)	FAAS	[19]
As, Cd, Cu, Fe, Ni and Pb	Edible oils	Activated carbon (a suspension in a water–MIBK mixture), a batch sorption and elution procedure, drying and elution with 2 mol L^{-1} HNO_3 – organically bound species	FAAS	[20]
Cd and Pb	Medicinal herb (yarrow, chamomile, bearberry leaves, peppermint, oregano, thyme, hibiscus) infusions	Dowex 50W×8 (a strong cation exchanger), a batch sorption and elution procedure, elution with 3 mol L^{-1} HNO_3 – noncationic species (analyzing solutions left after the sorption), cationic species (simple ions with positively charged complexes, labile complexes with organic ligands)	ETAAS, ICP MS	[21]
Pb	Lipsticks	Activated carbon (a suspension in a water–MIBK mixture), a batch sorption and elution procedure, drying and elution with 10% HNO_3 – dye-bound species	ICP MS, ICP OES	[22]

Mn	Spinach leaves	Activated carbon (a suspension in a water–MIBK mixture), a batch sorption and elution procedure, drying and elution with 10% HNO_3 – bound and adsorbable on a surface species (water and organic solvent soluble)	ICP MS	[23]
Cu	Spinach leaves	Activated carbon (a suspension in a water–MIBK mixture), a batch sorption and elution procedure, drying and elution with 10% HNO_3 – bound and adsorbable on a surface species (bioavailable)	ICP MS	[24]
Mn	Hazelnuts	Activated carbon (a suspension in a water–MIBK mixture), a batch sorption and elution procedure, drying and elution with 10% HNO_3 – bound and adsorbable on a surface species (bioavailable)	ICP MS	[25]

Note: FAAS, flame atomic absorption spectrometry; ICP MS, inductively coupled plasma mass spectrometry; ICP OES, inductively coupled plasma optical emission spectrometry; GFAAS, graphite furnace atomic absorption spectrometry; ETAAS, electrothermal atomic absorption spectrometry; MIBK, methyl isobutyl ketone.

C_{18}-based sorbents were used to adsorb hydrophobic species, mostly organically bound species, that is, complexes with phenolic species and humic acids [5–7].

The easiest way to fractionate elements was to pass sample solutions through SPE tubes, cartridges or columns and collect respective effluents, in which fractions of species of elements not retained by given sorbents were determined [8–11]. A batch sorption procedure was also applied [12]. Then, outcomes of such a nonelution approach were used to evaluate the contribution of fractions of species of elements that were retained on given sorbents. This was readily assessed by differences between total concentrations of elements in analyzed samples and their concentrations (not retained or sorbed) determined in effluents achieved after contacting sample solutions with sorbents in column [5, 8, 9, 11] or batch [12] processes. Resin beds could also be removed from columns and wet digested to possess the information about the concentration elements retained by the resin [10]. Normally, species of elements retained by certain sorbents were recovered by using appropriate solvents [13–18]. Resulting eluates were analyzed by different spectrochemical techniques on the content of studied elements, while these results were used for assessing contributions of distinct fractions of species of elements retained by given sorbents. In addition, a sequential elution protocol with solutions of an increasing reactivity (water, diluted and moderately concentrated HCl solutions) was also used to distinguish three separate fractions of species of elements (the case of Al in tea infusions) [19].

When the same samples were separately treated with cation and anion exchangers, an additional fraction of species of elements, attributed to the presence of residual species, was assessed by subtracting the sum of concentrations of species of elements determined in distinguished cationic and anionic fractions from total concentrations of these elements in samples [15, 16]. When the filtration of samples, including water solutions of dissolved samples or water extracts, through 0.45 μm membrane filters was initially applied at the sample preparation stage, it was possible to separate two physical fractions, that is, the fraction of particulate species and the fraction of dissolved species [5, 8, 12, 14, 15, 17, 18]. The filtration was also required to remove suspend particles [7]. As a rule, the dissolved fraction was subsequently analyzed and subjected to fractionation by the SPE.

By combining the SPE on cation and anion exchangers [13], a cation exchanger and an adsorbing resin [5] or activated carbon [20–25] with other separation methods, more extended fractionation procedures were proposed. Accordingly, coprecipitation with $Al(OH)_3$ was used to separate only the fraction of Mn(II) cations [13]. The precipitation with ethanol was used to separate the fraction of polysaccharide-bound species of elements in the dissolved fraction in case of Mg in olives [5] or Mn in spinach leaves [23]. In addition, acetone was used to precipitate proteins and isolate and distinguish the fraction of protein-bound species of elements in case of Cd and Pb in medicinal herb infusions [21] or Mn in spinach leaves [23]. The solvent extraction with different solvents was used to separate a variety of fractions of species of Mg in olives [5], As, Cd, Cu, Fe, Ni and Pb in edible oils [20], Pb in lipsticks [22], Mn [23] and

Cu [24] in spinach leaves or Mn in hazelnuts [25], that is, lipid bound (CCl_4 [5]), free lipid bound (*n*-hexane [22–25]), neutral lipid bound (chloroform [23, 24]), total lipid bound (chloroform with methanol at 2:1 [23–25]), associated with glycolipids (acetone [23, 24]), associated with oils (diethyl ether [23–25]), chlorophyll bound ($CHCl_3$ [5, 23]), polar species, that is, polyphenols, phospholipids (ethanol or methanol [23–25]), water soluble (water [5, 23–25] or HCl and HNO_3 solutions [20]), stomach absorbable and bioavailable (pepsin with HCl [5, 23–25]), intestine absorbable and bioavailable (*n*-octanol [5, 23–25] or bile salts with pancreatin [23–25]), finally, gastrointestinal available (simulated gastrointestinal juices [23–25]).

10.2.2 Binary and tertiary fractionations using two-column SPE approaches

In case of two-column SPE fractionation procedures, two sorbents, placed in cartridges, tubes or columns, were connected in a series. To retain respective species of elements by these assemblages, samples were passed through the first column (1) and then, resulting effluents were further passed through the second column (2). Different combinations of sorbents were reported for this chemical fractionation of elements by the SPE. It included a combination of a strong anion-exchange resin with a chelating resin [26], a weak anion-exchange resin with a strong cation-exchange resin [27], an adsorbing resin with a strong cation-exchange resin [27–38] or an adsorbing resin with a chelating resin [39]. Arrangements of sorbents in reported tandem-column assemblages were arbitrary selected, that is, to avoid labile anionic species to be sorbed on Chelex 100 [26], or, more commonly, organized after examining sorption properties of sorbents toward different classes of species of elements [27–33, 39, 40]. Table 10.2 presents these results.

When the filtration of samples through 0.45 µm filters was initially carried out, fractions of particulate and dissolved species of elements were distinguished [29–31]. In addition, the filtration through membrane filters was applied to degas samples of beer [33]. By applying membrane 0.45 µm filters as well as ultrafiltration filters with molecular weight cut-offs of 100, 50, 30, 10 and 5 kDa, the physical fractionation of elements present in the dissolved fraction (< 0.45 µm) was possible [34–38].

A comprehensive partitioning protocol was reported for the fractionation of Cu, Fe and Zn in red wines [40]. At the beginning, wines were filtered through 0.20 µm membrane filters to separate particle and dissolved fractions of studied elements. Then, the dissolved fraction (filtrates) were treated with a tandem-column system comprising an adsorbing resin Amberlite XAD-8 and a strong cation-exchange resin Dowex 50W×8, a strong anion-exchange resin Dowex 1 × 8 or an adsorbing resin Amberlite XAD-2 (but after a preliminary addition of a 1,10-phenanthroline to Amberlite XAD-8 column effluents). In addition, fractions of species of elements associated with proteins and polysaccharides were precipitated from Amberlite XAD-8 column effluents using acetone and ethanol, respectively.

Table 10.2: Chemical fractionation analysis using two-column SPE-based procedures.

Element	Sample	SPE sorbents used, elution conditions and fractions determined	Detection technique	References
Cu and Mn	Cow milk (a casein-free fraction after precipitation and filtration)	Chelex 100 (1, a chelating resin in a microcolumn) – AG 1 × 8 (2, a strong anion-exchanger in a microcolumn), elution with 2 mol L^{-1} HCl (1 and 2) – cationic species (1), anionic species (2), residual species (analyzing the effluent after passing the sample through 1 and 2)	ICP OES, GFAAS	[26]
Mn and Zn	Tea (black and green) infusions	Amberlite XAD-7 (1, a nonionic macroreticular adsorbing resin in a column) – Dowex 50W×4 (2, a strong cation-exchanger in a column), elution with 2 mol L^{-1} HNO_3 (1) and 4 mol L^{-1} HCl (2) – hydrophobic species (1), cationic species (2), residual species (analyzing effluents after passing samples through 1 and 2)	FAAS	[27]
Mn and Zn	Tea (black and green) infusions	Reilex 402 (1, a weak anion exchanger in a column) – Dowex 50W×4 (2, a strong cation exchanger in a column), nonelution approach (1, analyzing effluents after passing samples through 1), elution with 4 mol L^{-1} HCl – hydrophobic species (1), cationic species (2), residual species (analyzing effluents after passing samples through 1 and 2)	FAAS	[27]
Mn and Zn	Lager beers	Amberlite XAD-7 (1, a nonionic macroreticular adsorbing resin in a column) – Dowex 50W×4 (2, a strong cation exchanger in a column), elution with 2 mol L^{-1} HNO_3 (1) and 4 mol L^{-1} HCl (2) – hydrophobic (polyphenol bound) species (1), cationic species (2), residual species (analyzing effluents after passing samples through 1 and 2)	FAAS	[28]
Fe	Red wines	Amberlite XAD-7 or XAD-16 (1, a nonionic macroreticular adsorbing resin in a column) – Dowex 50W×8 (a strong cation exchanger in a column), nonelution approach (1, analyzing effluents after passing samples through 1), elution with 2 mol L^{-1} HCl (2) – organically (phenolic) bound species (1), cationic (simple ions, stable and labile cationic complexes) species (2), residual (neutral and/or anionic complexes) species (analyzing effluents after passing samples through 1 and 2)	FAAS	[29]

Cu	Lager beers and red wines	Amberlite XAD-16 or XAD-7 (1, a nonionic macroreticular desorbing resin in a column) – Dowex 50W×8 (a strong cation exchanger in a column), nonelution approach (1, analyzing effluents after passing samples through 1), elution with 2 mol L^{-1} HCl (2) – hydrophobic (phenolic-bound) species (1), cationic (simple ions, stable and labile complexes) species (2), residual (nonhydrophobic, ionic) species (analyzing effluents after passing samples through 1 and 2)	FAAS	[30]
Mn	Red wines	Amberlite XAD-16 or XAD-7 (1, a nonionic macroreticular desorbing resin in a column) – Dowex 50W×8 (2, a strong cation exchanger in a column), nonelution approach (1, analyzing effluents after passing samples through 1), elution with 2 mol L^{-1} HCl (2) – phenolic-bound species (1), cationic species (2), residual (nonhydrophobic, ionic) species (analyzing effluents after passing samples through 1 and 2)	FAAS	[31]
Ca and Mg	Honeys (water solutions)	Amberlite XAD-16 (1, a nonionic macroreticular adsorbing resin in a column) – Dowex 50W×8 (2, a strong cation exchanger in a column), nonelution approach (1, analyzing effluents after passing samples through 1), elution with 2 mol L^{-1} HCl (2) – organically (hydrophobic high molecular weight ligands) bound species (1), cationic (simple ions, labile and stable complexes with low molecular weight ligands) species (2), residual species (analyzing effluents after passing samples through 1 and 2)	FAAS	[32]
Cu, Fe and Mn	Lager beers	Amberlite XAD-16 (1, nonionic macroreticular adsorbing resin in a column) – Dowex 50W×8 (a strong cation exchanger in a column), nonelution approach (1, analyzing effluents after passing samples through 1), elution with 2 mol L^{-1} HCl (2) – phenolic bound species (1), cationic species (2), residual species (analyzing effluents after passing samples through 1 and 2)	FAAS	[33]

(*Continued*)

Table 10.2: (Continued)

Element	Sample	SPE sorbents used, elution conditions and fractions determined	Detection technique	References
Ca, Mg Mn and Fe	Honeys (water solutions)	Amberlite XAD-16 (1, nonionic macroreticular adsorbing resin in a column) – Dowex 50W×8 (a strong cation exchanger in a column), nonelution approach (1, analyzing effluents after passing samples through 1), elution with 2 mol L^{-1} HCl (2) – phenolic bound species (1), cationic species (2), residual species (analyzing effluents after passing samples through 1 and 2)	FAAS	[34]
Zn	Honeys (water solutions)	Amberlite XAD-16 (1, nonionic macroreticular adsorbing resin in a column) – Dowex 50W×8 (a strong cation exchanger in a column), nonelution approach (1, analyzing effluents after passing samples through 1), elution with 2 mol L^{-1} HCl (2) – phenolic bound species (1), cationic species (2), residual species (analyzing effluents after passing samples through 1 and 2)	FAAS	[35]
Cu	Honeys (water solutions)	Amberlite XAD-16 (1, nonionic macroreticular adsorbing resin in a column) – Dowex 50W×8 (a strong cation exchanger in a column), nonelution approach (1, analyzing effluents after passing samples through 1), elution with 2 mol L^{-1} HCl (2) – phenolic bound species (1), cationic species (2), residual species (analyzing effluents after passing samples through 1 and 2)	FAAS	[36]
Ca, Fe, Mg, Mn	Soluble coffees (brews)	Discovery DSC-18 (1, an octadecyl silica reverse phase in a tube) – discovery DSC-SCX (a strong cation exchanger in a tube), nonelution approach (1, analyzing effluents after passing samples through 1), elution with 2 mol L^{-1} HCl (2) – hydrophobic species (1), cationic species (2), residual species (analyzing effluents after passing samples through 1 and 2)	FAAS	[37]

Mn	Ground coffees (infusions)	Discovery DSC-18 (1, an octadecyl silica reverse phase in a tube) – discovery DSC-SCX (a strong cation exchanger in a tube), nonelution approach (1, analyzing effluents after passing samples through 1), elution with 2 mol L^{-1} HCl (2) – hydrophobic species (1), cationic species (2), residual species (analyzing effluents after passing samples through 1 and 2)	FAAS	[38]
Al	Tea (black) infusions	Amberlite XAD-7 (1, a nonionic macroreticular resin in a column) – Chelex 100 (2, a chelating resin in a column), elution with 2 mol L^{-1} HNO_3 (1) and 0.08 mol L^{-1} HCl (2) – organically bound species (1), cationic species (2)	FAAS	[39]
Cu, Fe and Zn	Red wines	Amberlite XAD-8 (1, a nonionic macroreticular adsorbing resin in a column) – Dowex 50W×8 (2, a strong cation exchanger in a column), Dowex 1 × 8 (2, a strong anion exchanger in a column) or Amberlite XAD-2 (2, a nonionic macroreticular adsorbing resin in a column), elution with 2 mol L^{-1} HNO_3 (1), 2 mol L^{-1} HCl (2, ion exchangers) or methanol (2, adsorbent) – polysaccharide–phenolic complexes (1), labile species (2, cation exchanger), negatively charged complexes (2, anion exchanger), labile Fe(II) species (2, adsorbent)	FAAS, GFAAS	[40]

ICP OES, inductively coupled plasma optical emission spectrometry; GFAAS, graphite furnace absorption spectrometry; FAAS, flame atomic absorption spectrometry.

10.2.3 Characterization of sorbents

The examination of sorption and desorption properties of sorbents used for the purpose of the chemical fractionation analysis of elements by means of the SPE was an integral part of this kind of studies. Since fractions of species of elements distinguished using SPE were operationally defined, the investigation of sorption properties of sorbents toward different classes of species of elements was helpful in elucidating the distribution of species of elements among different separated fractions.

Accordingly, solutions of simple ions of elements were commonly used to evaluate sorption properties of chelating [12, 17–19, 26, 39] and cation-exchange [8, 10, 14–16, 27–33, 37, 38, 40] resins. These solutions were also used to verify the extent of species of elements that could eventually be retained by other resins, that is, anion-exchange resins [19, 27] or adsorbing resins [27–33, 37, 39, 40], overestimating, respectively, fractions of anionic and hydrophobic species of elements. In a similar way, solutions of elements with added EDTA were used to verify the sorption behavior of anion-exchange resins toward low molecular weight anionic complexes of elements [15, 16, 26]. Tartaric acid was used for the same purpose [40]. In addition, solutions of simple ions with added EDTA were used to check sorption properties of other resins, that is, adsorbents [27, 28], cation exchangers [27, 28] or chelating resins [17, 18], retaining the group of negatively charged low molecular weight spices of elements. For the same purpose, solutions of simple ions with added citric acid [29, 31, 33, 37, 38] and tartaric acid [30, 32] were used. All these experiments assured that given distinguished fractions of species of elements were not overestimated by other species. Tannic [27–33, 39, 40] and gallic [39, 40] acids were used to bind simple ions of elements and form their high molecular weight species. For the same purpose, humic acid [7, 17, 18] or 1,10-phenanthroline [40] was used to complex elements, that is, Al, Cu, Fe, Mn and Zn. Resulting solutions were used to evaluate sorption properties of adsorbing resins. In addition, they were used to verify the extent of an eventual retention of polyphenolic-bound species of elements by other resins, that is, chelating resins [17–19, 39], cation-exchange resins [19, 27–33] or anion-exchange resins [19].

Desorption properties of sorbents used were studied to maintain conditions of the quantitative recovery of species of elements retained by certain sorbents. As can be seen from Tables 10.1 and 10.2, certain species of elements were eluted using moderately concentrated solutions of mineral acids, that is, HCl at concentrations of 0.02 [19], 0.08 [39], 2 [13, 16, 17, 19, 26, 29–38, 40], 3 [18] or 4 mol L^{-1} [14, 15, 27, 28] and HNO_3 at concentrations of 2 [6, 13, 20, 27, 28, 39, 40] and 3 mol L^{-1} [21] or 10% [23–25]. Methanol was used to elute species of elements from beds of hydrophobic sorbents [7, 40]. Finally, the nonelution approach was very popular [5, 8–12, 22, 26–38].

10.3 QA and QC of the chemical fractionation analysis by the SPE

Due to the lack of respective CRMs suitable for the fractionation analysis of elements, the assessment of the reliability of results achieved by using the SPE and other chromatographic and nonchromatographic separation techniques was difficult. The quality of results was confirmed undertaking additional experiments and proposing alternative ways of QA and QC. Accordingly, quite often samples were spiked with simple ions of elements [8, 11, 14, 17, 18, 20–22, 27, 28], complexes of elements with humic acid [7, 17] and other ligands, like EDTA [17], or other metallo-organic compounds [20]. Such spiked samples were subjected to fractionation protocols in order to assess recoveries of added species of elements. In this case, concentrations of added elements were determined in separated fractions and then sum of these concentrations were related to concentrations of added elements in order to verify the reliability of the SPE-based procedure. Sometimes, other sorbents were used for the evaluation of contributions of given fractions of species of elements and results were compared, that is, a weak cation exchanger Diaion WT01S and a strong cation exchanger Dowex 50W×4 [8], a weak anion exchanger Reillex 402 and an adsorbing resin Amberlite XAD-7 [27], an adsorbing resin XAD-7 with an adsorbing resin Amberlite XAD-16 [29–31, 33] or an adsorbing resin Amberlite XAD-2 with an adsorbing resin Amberlite XAD-7 [6]. Finally, in case of dual-column SPE fractionation procedures, resulting effluents were collected and analyzed (in case of one- or two-column fractions) on the content of unretained (residual) species of elements. Then, sums of concentrations of elements in resulting fractions from elution and nonelution approaches were compared to total concentrations of elements in analyzed samples and the information about the reliability of results was possessed [6, 16, 26, 29, 31–38].

10.4 Conclusions

Unique fractionation patterns of studied elements obtained with SPE-based partitioning procedures are a consequence of their speciation in analyzed sample matrices. Being not aimed at unambiguously distinguishing distinct chemical forms of species of elements, operationally defined fractions of species of elements provided by, described in this chapter, SPE-based protocols bring the very important information about properties of groups of species of elements in reference to their functions, the nutritional value, the mobility and the bioavailability. However, a rational and critical examination of results of the fractionation analysis is significant in this case for their interpretation and understanding. The knowledge about sorption and desorption properties of sorbents toward different classes of species of elements, the order of sorbents in two-column SPE fractionation schemes or the general information about the composition of analyzed samples and possible interactions of ligands with ions of elements can substantially help in reasonably assigning separated and distinguished fractions to certain groups of species of elements. Certainly, the information about the fractionation pattern of given elements in environmental and food samples is more appropriate to assess their functionality than the total concentration.

Acknowledgments

The work was financed by a statutory activity subsidy from the Polish Ministry of Science and Higher Education for the Faculty of Chemistry of Wroclaw University of Technology.

References

[1] Pohl P. Application of ion-exchange resins to the fractionation of metals in water. Trends Anal Chem 2006, 25, 31–43.
[2] Pohl P. Fractionation analysis of metals in dietary samples using ion-exchange and adsorbing resins. Trends Anal Chem 2007, 26, 713–26.
[3] Templeton DM, Ariese F, Cornelis R, Danielsson LG, Muntau H, Van Leeuwen H, Lobinski R. Guidelines for terms related to chemical speciation and fractionation of elements: definitions, structural aspects, and methodological approaches. Pure App Chem 200, 72, 1453–70.
[4] Sperling M. Flame and graphite furnace atomic absorption spectrometry in environmental analysis. In: Meyers RA. ed., Encyclopedia of Analytical Chemistry. Applications of Instrumental Methods. John Wiley and Sons, Chichester, UK, 2000, vol. 3, 2441–500.
[5] Yasar SB, Gucer S. Fractionation analysis of magnesium in olive products by atomic absorption spectrometry. Anal Chim Acta 2004, 505, 43–9.
[6] Mucchino C, Musci M. Extraction of Al, Cu, Fe, Mn, Ni and Zn-polyphenol complexes from black tea infusions by Amberlite resins. J Sci Food Agric 2014, 94, 2234–8.
[7] Fang K, Yuan D, Zhang L, Feng L, Chen Y, Wang Y. A solid phase extraction method for analysis of the iron-humic acid complex in natural water. Anal Methods 2014, 6, 5519–26.
[8] Pohl P, Prusisz B. Fractionation of calcium and magnesium in honeys, juices and tea infusions by ion exchange and flame atomic absorption spectrometry. Talanta 2006, 69, 1227–33.
[9] Odegard KE, Lund W. Multi-element speciation of tea infusion using cation-exchange separation and size-exclusion chromatography in combination with inductively coupled plasma mass spectrometry. J Anal At Spectrom 1997, 12, 403–8.
[10] Pohl P, Prusisz B. Pre-concentration of Cd, Co, Cu, Ni and Zn using different off-line ion exchange procedures followed by the inductively coupled plasma atomic emission spectrometric detection. Anal Chim Acta 2004, 502, 83–90.
[11] Svendsen R, Lund W. Speciation of Cu, Fe and Mn in beer using ion exchange separation and size-exclusion chromatography in combination with electrothermal atomic absorption spectrometry. Analyst 2000, 125, 1933–7.
[12] Da Costa AMG, Nogami EM, Visentainer JV, de Souza NE, Garcia EE. Fractionation of aluminum in commercial green and roasted yerba mate samples (Ilex paraguariensis St. Hil.) and their infusions. J Agric Food Chem 2009, 57, 196–200.
[13] Ozdemir Y, Gucer S. Speciation of manganese in tea leaves and tea infusions. Food Chem 1998, 61, 313–7.
[14] Pohl P, Bogdal Z, Prusisz B. Preconcentration and fractionation of Cd, Co, Cu, Ni, Pb and Zn in natural water samples prior to analysis by inductively coupled plasma atomic emission spectrometry. Microchim Acta 2005, 150, 253–9.

[15] Pohl P, Prusisz B. Simple and versatile operational fractionation of Fe and Zn in dietary products by solid phase extraction on ion exchange resins. Talanta 2007, 71, 411–8.
[16] Pohl P, Prusisz B. Determination of Ca, Mg, Fe and Zn partitioning in UHT cow milks by two-column ion exchange and flame atomic absorption spectrometry detection. Talanta 2007, 71, 715–21.
[17] Yebra-Biurrun MC, Castro-Romero JM. Continuous sonoassisted digestion coupled to flow injection minicolumn separation for the determination of total and labile Cu and Fe in fresh waters by flame atomic absorption spectrometry. Intern J Environ Anal Chem 2012, 92, 1239–49.
[18] Yebra-Biurrun, N Carro-Marino. Flow injection flame atomic absorption determination of Cu, Mn and Zn partitioning in seawater by on-line room temperature sonolysis and minicolumn chelating resin methodology. Talanta 2010, 83, 425–30.
[19] Ruszczynska A, Pyrzynska K, Bulska E. On the use of solid-phase extraction for the fractionation of aluminium species. Chem Anal (Warsaw) 2004, 49, 19–28.
[20] Kowalewska Z, Izgi B, Saracoglu S, Gucer S. Application of liquid-liquid extraction and adsorption on activated carbon to the determination of different forms of metals present in edible oils. Chem Anal (Warsaw) 2005, 50, 1007–19.
[21] Arpadjan S, Celik G, Taskesen S, Gucer S. Arsenic, cadmium and lead in medicinal herbs and their infusions. Food Chem Toxicol 2008, 46, 2871–5.
[22] Yilmazcan O, Erdemir US, Izgi B, Ozer ET, Gucer S. Lead fractionation analysis in lipstick samples by inductively coupled plasma-mass spectrometry. Spectrosc Lett 2013, 48, 290–5.
[23] Erdemir US, Gucer S. Fractionation analysis and bioavailability of manganese in spinach (Spinacia oleracea L.) leaves. Chem Spec Bioavailab 2013, 25, 265–72.
[24] Erdemir US, Gucer S. Assessment of copper bioavailability in spinach (Spinacia oleracea L) leaves by chemical fractionation. Food Anal Methods 2014, 7, 994–1001.
[25] Erdemir US, Gucer S. Fractionation analysis of manganese in Turkish hazelnuts (Corylus avellana L.) by inductively coupled plasma-mass spectrometry. J Agric Food Chem 2014, 62, 10792–9.
[26] Abollino O, Aceto M, Bruzzoniti MC, Mentasti E, Sarzanini C. Speciation of copper and manganese in milk by solid-phase extraction inductively-coupled plasma-atomic emission spectrometry. Anal Chim Acta 1998, 375, 299–306.
[27] Pohl P, Prusisz B. Fractionation analysis of manganese and zinc in tea infusions by two-column solid phase extraction and flame atomic absorption spectrometry. Food Chem 2007, 102, 1415–24.
[28] Pohl P, Prusisz B. Fractionation analysis of manganese and zinc in beers by means of two sorbent column system and flame atomic absorption spectrometry. Talanta 2007, 71, 1616–23.
[29] Pohl P, Prusisz B. Application of tandem column solid phase extraction and flame atomic absorption spectrometry for the determination of inorganic and organically bound forms of iron in wine. Talanta 2009, 77, 1732–8.
[30] Pohl P, Sergiel I. Evaluation of the total content and the operationally defined species of copper in beers and wines. J Agric Food Chem 2009, 57, 9378–84.
[31] Pohl P. Suitability of solid phase extraction and flame atomic absorption spectrometry for manganese partitioning in red wines. Food Chem 2009, 114, 996-1001.
[32] Sergiel I, Pohl P. Determination of the total content of calcium and magnesium and their bioavailability in ripened bee honeys. J Agric Food Chem 2010, 58, 7497–501.
[33] Pohl P, Prusisz B. Chemical fractionation of Cu, Fe and Mn in canned Polish beers. J Food Comp Anal 2010, 23, 86–94.
[34] Pohl P, Sergiel I. Direct determination of the total concentrations of magnesium, calcium, manganese, and iron in addition to their chemical and physical fractions in dark honeys. Anal Lett 2011, 44, 2265–79.

[35] Pohl P, Sergiel I, Prusisz B. Direct analysis of honey for the total content of Zn and its fractionation forms by means of flame atomic absorption spectrometry with solid phase extraction and ultrafiltration approaches. Food Chem 2011, 125, 1504–9.
[36] Pohl P, Sergiel I. Direct analysis of dark honeys for total content of copper and its fractionation forms. Food Anal Methods 2012, 5, 113–8.
[37] Pohl P, Stelmach E, Szymczycha-Madeja A. Simplified sample treatment for the determination of total concentrations and chemical fractionation forms of Ca, Fe, Mg and Mn in soluble coffees. Food Chem 2014, 163, 31–6.
[38] Pohl P, Stelmach E, Szymczycha-Madeja A. Determination of total concentrations and chemical and physical fraction forms of manganese in infusions of ground coffees. Food Anal Methods 2014, 7, 676–82.
[39] Erdemoglu SB, Pyrzynska K, Gucer S. Speciation of aluminum in tea infusion by ion-exchange resins and flame AAS detection. Anal Chim Acta 2000, 81, 411–9.
[40] Karadjova I, Izgi B, Gucer S. Fractionation and speciation of Cu, Zn and Fe in wine samples by atomic absorption spectrometry. Spectrochim Acta Part B 2002, 57, 581–90.

Index

https://doi.org/10.1515/9783110366730-012

Bei Fragen zur Produktsicherheit wenden Sie sich bitte an:
If you have any questions regarding product safety,
please contact:

Walter de Gruyter GmbH
Genthiner Straße 13
10785 Berlin
productsafety@degruyterbrill.com